Learning Guide

for Tortora and Grabowski:
Principles of Anatomy and Physiology Seventh Edition

Kathleen Schmidt Prezbindowski

College of Mount St. Joseph and
The University of Cincinnati

Gerard J. Tortora

Bergen Community College

HarperCollinsCollegePublishers

To Laurie and Amy

Acquisitions Editor: Bonnie Roesch
Developmental Editor: Meryl R. G. Muskin
Project Editor: Katharine Glynn
Design Supervisor: Mary Archondes
Cover Photo: © 1989 Howard Sochurek, The Stock Market
Production Manager/Assistant: Kewal K. Sharma/Jeffrey Taub
Compositor: ComCom Division of Haddon Craftsmen, Inc.
Printer and Binder: Courier Companies, Inc.
Cover Printer: The Lehigh Press, Inc.

Learning Guide for Tortora and Grabowski: Principles of Anatomy and Physiology, Seventh Edition

Library of Congress Cataloging-in-Publication Data

Prezbindowski, Kathleen Schmidt.
    Learning guide for Tortora and Grabowski principles of anatomy and
physiology, seventh edition / Kathleen Schmidt Prezbindowski, Gerard
J. Tortora.
        p.   cm.
    Rev. ed. of: Learning guide for Tortora and Anagnostakos
principles of anatomy and physiology, sixth edition. c1990.
    ISBN 0-06-501197-X
    1. Human physiology—Examinations, questions, etc. 2. Human
anatomy—Examinations, questions, etc.   I. Tortora, Gerard J.
II. Tortora, Gerard J. Principles of anatomy and physiology. 7th
ed.   III. Prezbindowski, Kathleen Schmidt. Learning guide for
Tortora and Anagnostakos principles of anatomy and physiology, sixth
edition.   IV. Title.
    [DNLM: 1. Anatomy—problems.   2. Homeostasis—problems.
3. Physiology—problems.   QS 4 T712p 1992 Suppl.]
QP34.5.T67 1992 Suppl.
612′.0076—dc20
DNLM/DLC
for Library of Congress   92-48208
                                                            CIP

93 94 95 96   9 8 7 6 5 4 3 2

Contents

Preface iv
To the Student v
About the Authors vi

UNIT I Organization of the Human Body 1

Chapter 1   An Introduction to the Human Body 3
Chapter 2   The Chemical Level of Organization 19
Chapter 3   The Cellular Level of Organization 41
Chapter 4   The Tissue Level of Organization 65
Chapter 5   The Integumentary System 85

UNIT II Principles of Support and Movement 101

Chapter 6   Bone Tissue 103
Chapter 7   The Skeletal System: The Axial Skeleton 119
Chapter 8   The Skeletal System: The Appendicular Skeleton 139
Chapter 9   Articulations 153
Chapter 10   Muscle Tissue 169
Chapter 11   The Muscular System 191

UNIT III Control Systems of the Human Body 215

Chapter 12   Nervous Tissue 217
Chapter 13   The Spinal Cord and Spinal Nerves 239
Chapter 14   The Brain and Cranial Nerves 257
Chapter 15   Sensory, Motor, and Integrative Systems 283
Chapter 16   The Special Senses 305
Chapter 17   The Autonomic Nervous System 327
Chapter 18   The Endocrine System 345

UNIT IV Maintenance of the Human Body 377

Chapter 19   The Cardiovascular System: The Blood 379
Chapter 20   The Cardiovascular System: The Heart 399
Chapter 21   The Cardiovascular System: Blood Vessels and Hemodynamics 423
Chapter 22   The Lymphatic System, Nonspecific Resistance to Disease, and Immunity 453
Chapter 23   The Respiratory System 481
Chapter 24   The Digestive System 503
Chapter 25   Metabolism 533
Chapter 26   The Urinary System 557
Chapter 27   Fluid, Electrolyte, and Acid-Base Homeostasis 581

UNIT V Continuity 595

Chapter 28   The Reproductive Systems 597
Chapter 29   Development and Inheritance 627

Preface

This *Learning Guide* is intended to be used in conjunction with Tortora and Grabowski's *Principles of Anatomy and Physiology,* Seventh Edition. The emphasis is on active learning, not passive reading, and the student examines each concept through a variety of activities and exercises. By approaching each concept several times from different points of view, the student sees how the ideas of the text apply to real, clinical situations. In this way the student may be helped not simply to study but to *learn*.

The 29 chapters of the *Learning Guide* parallel those of the Tortora and Grabowski text. Each chapter begins with an **Overview** that introduces content succinctly and a **Framework** that visually organizes the interrelationships among key concepts and terms. **Objectives** from the text are included for convenient previewing and reviewing. Objectives are arranged according to major chapter divisions in the **Topic Outline**. **Wordbytes** present word roots, prefixes, and suffixes that facilitate understanding and recollection of key terms. **Checkpoints** offer a diversity of learning activities that follow the exact sequence of topics in the chapter, with cross-references to pages in the text. This format makes the *Learning Guide* "user friendly" or "student friendly" and enhances the effectiveness of the excellent text. **Answers to Selected Checkpoints** provide feedback to students. A 25-question **Mastery Test** includes both objective and subjective questions as a final tool for student self-assessment of learning. **Answers to the Mastery Test** immediately follow the Mastery Test.

This edition of the *Learning Guide* is especially designed to fit the needs of people with different learning styles. Visual learners will identify the *Frameworks* as assets in organizing concepts and key terms. Visual and tactile learners will welcome the increased number of coloring and labeling exercises with color-code ovals. Further variety in *Checkpoints* is offered by short definitions, fill-ins, tables for completion, and multiple-choice, matching, and arrange-in-correct-sequence activities. Learning on multiple levels is made possible by *For Extra Review* sections that provide enrichment or assistance with particularly difficult topics. *Clinical Challenges* introduce applications especially relevant for students in allied health. These are based on my own experiences as an R.N. and on discussions with colleagues. These challenging exercises enable students to apply abstract concepts to actual clinical settings.

The seventh edition of the *Guide* also provides readily available feedback for self-assessment. Answers to most *Checkpoints* are included at the end of the chapter. The placement of *Answers to the Mastery Test* immediately following the *Mastery Test* offers readily accessible feedback.

A number of persons have contributed to the revision of this book. I would particularly like to thank the following instructors for their helpful comments and suggestions while the manuscript was being revised:

George Billman/Ohio State University
Frank Peek/SUNY-Morrisville College
Deborah Pringnitz/University of Maine, Fort Kent

I offer particular thanks to the students and faculty at the College of Mount St. Joseph and the University of Cincinnati for their ongoing sharing and support of teaching and learning. I thank my daughters, Amy and Laurie Prezbindowski, my parents and sister, Norine, Harry, and Maureen Schmidt, and close friends for their constant encouragement and caring. I am thankful for the expertise, enthusiasm, and patience of the production staff at HarperCollins, especially to Meryl Muskin, developmental editor. Special thanks to Katharine Glynn for her skillful editing and creative approach to excellence.

Kathleen Schmidt Prezbindowski

To the Student

This *Learning Guide* is designed to help you do exactly what its title indicates—*learn*. It will serve as a step-by-step aid to help you bridge the gap between goals (objectives) and accomplishment (learning). The 29 chapters of the *Learning Guide* parallel the 29 chapters of Tortora and Grabowski's *Principles of Anatomy and Physiology*, Seventh Edition. Each chapter consists of the following parts: *Overview, Framework, Topic Outline with Objectives, Wordbytes, Checkpoints, Answers to Selected Checkpoints,* and *Mastery Test with Answers.* Take a moment to turn to Chapter 1 and look at the major headings in this *Learning Guide.* Now consider each one.

    Overview. As you begin your study of each chapter, read this brief introduction, which presents the topics of the chapter. The Overview is a look at the "forest" (or the "big picture") before examination of the individual "trees" (or details).

    Framework. The Framework arranges key concepts and terms in an organizational chart that demonstrates interrelationships and lists key terms. Refer back to the Framework frequently to keep sight of interrelationships.

    Topic Outline and Objectives. Objectives that are identical to those in the text organize the chapter into "bite-sized" pieces that will help you identify the principal subtopics that you must understand. Preview objectives as you start the chapter, and refer back to them to review. Objectives are arranged according to major chapter divisions in the **Topic Outline.**

    Wordbytes present word roots, prefixes, and suffixes that will help you master terminology of the chapter and facilitate understanding and remembering of key terms. Study these, then check your knowledge. (Cover meanings and write them in the margins.) Try to think of additional terms in which these important bits of words are used.

    Checkpoints offer a variety of learning activities arranged according to the sequence of topics in the chapter. At the start of each new Checkpoint section, note the related pages in the Tortora and Grabowski text. Read these pages carefully and then complete this group of Checkpoints. Checkpoints are designed to help you to "handle" each new concept, almost as if you were working with clay. Through a variety of exercises and activities, you can achieve a depth of understanding that comes only through active participation, not passive reading alone. You can challenge and verify your knowledge of key concepts by completing the Checkpoints along your path of learning.

    A diversity of learning activities are included in Checkpoints. You will focus on specific facts in the chapter by doing exercises that require you to provide definitions and comparisons, and you will fill in blanks in paragraphs with key terms. You will also color and label figures. Fine color felt-tipped pens or pencils will be helpful since you can easily fill in related color-code ovals. Your understanding will also be tested with matching exercises, multiple-choice questions, and arrange-in-correct-sequence activities.

    Two special types of Checkpoints are *For Extra Review* and *Clinical Challenges. For Extra Review* provides additional help for difficult topics or refers you by page number to specific activities earlier in the *Learning Guide* so that you may better integrate related information. Applications for enrichment and interest in areas particularly relevant to nursing or other health sciences students are found in selected *Clinical Challenge* exercises.

    When you have difficulty with an exercise, refer to the related pages (listed with each Checkpoint of the *Learning Guide*) of *Principles of Anatomy and Physiology* before proceeding further. You will find specific references by page number for many text figures and exhibits.

    Answers to Selected Checkpoints. As you glance through the activities, you will notice that a number of questions are marked with solid boxes (■). Answers to these checkpoints are given at the end of each chapter, in order to provide you with some immediate feedback about your progress. Incorrect answers alert you to review related objectives in the *Learning Guide* and corresponding text pages. Each *Learning Guide* (LG) figure included in *Answers to Selected Checkpoints* is identified with an "A," such as Figure LG 13-1A.

    Activities without answers are also purposely included in this book. The intention is to encourage you to verify some answers independently by consulting your textbook and also to stimulate discussion with students or instructors.

    Mastery Test. This 25-question self-test provides an opportunity for a final review of the chapter as a whole. Its format will assist you in preparing for standardized tests or course exams that are objective in nature, since the first 20 questions are multiple-choice, true-false, and arrangement questions. The final five questions of each test help you to evaluate your learning with fill-in and short answer questions. Answers for each mastery test are placed at the end of each chapter.

    I wish you success and enjoyment in *learning* concepts and relevant applications of your own anatomy and physiology as well as applications of this information in clinical settings.

Kathleen Schmidt Prezbindowski

About the Authors

KATHLEEN SCHMIDT PREZBINDOWSKI, PH.D., R.N., is a professor of biology at the College of Mount St. Joseph, where she was recently named Distinguished Teacher of the Year. Also a visiting professor at the University of Cincinnati, she has taught anatomy and physiology, pathophysiology, and biology of aging for 23 years. Dr. Prezbindowski received a B.S.N. in 1985, and recently completed requirements for the M.S.N. in both Gerontological and Psychiatric/Mental Health Nursing. Through a grant from the U.S. Administration on Aging, she produced the first video series on the human body ever designed especially for older persons. She is also the author of two other learning guides.

GERARD J. TORTORA is a professor of biology and teaches human anatomy and physiology and microbiology at Bergen Community College in Paramus, New Jersey. He has just completed 30 years as a teacher, the past 24 at Bergen CC. He received his B.S. in biology from Fairleigh Dickinson University in 1962 and his M.A. in biology from Montclair State College in 1965. He has also taken graduate courses in education and science at Columbia University and Rutgers University. He belongs to numerous biology organizations, such as the American Association for the Advancement of Science (AAAS), *The Human Anatomy and Physiology Society* (HAPS), the American Association of Microbiology (ASM), and the Metropolitan Association of College and University Biologists (MACUB). Jerry is the author of a number of best-selling anatomy and physiology, anatomy, and microbiology textbooks and several laboratory manuals.

Organization of the Human Body

FRAMEWORK 1

An Introductory Tour of the Human Body

GETTING STARTED

- ANATOMY AND PHYSIOLOGY DEFINED (A)
  - Branches of anatomy and physiology

- LEVELS OF STRUCTURAL ORGANIZATION (B)
  - Chemicals
  - Cells
  - Tissues (4)
  - Organs
  - Systems (11)
  - Organisms

WHERE TO GO, HOW TO GET THERE

- ANATOMICAL AND DIRECTIONAL TERMS (E)
  - Anatomical position
  - Regional names
  - Directional terms
  - Planes, sections

- BODY CAVITIES AND REGIONS (F)
  - Dorsal
  - Ventral
    - thoracic
    - abdominopelvic
      - nine regions
      - four quadrants

HOW IT WORKS: FUNCTIONS

- LIFE PROCESSES (C)
  - Metabolism
  - Responsiveness
  - Movement
  - Growth
  - Differentiation
  - Reproduction

- HOMEOSTASIS (D)
  - ECF
  - ICF
  - Stressor
  - Feedback systems (loops)

WHEN THE BODY DOES NOT WORK

- MEDICAL IMAGING (G)
  - Radiography
  - CT
  - DSA
  - PET
  - MRI
  - US

An Introduction to the Human Body

You are about to embark on a tour, one that you will enjoy for at least several months, and hopefully for many years. Your destination: the far-reaching corners and fascinating treasures of the human body. Imagine the human body as an enormous, vibrant world, abundant in architectural wonders, and bustling with activity. You are the traveler, the visitor, the learner on this tour.

Just as the world is composed of individual buildings, cities, countries, and continents, so the human body is organized into cells, tissues, organs, and systems. In fact, a close-up look at a cell (comparable to a detailed study of a building) will reveal structural intricacies—the array of chemicals that comprise cells. Any tour requires a sense of direction and knowledge of the language: directional and anatomical terms that guide the traveler through the body. A tour is sure to enlighten the visitor about the culture in each region—in the human body, the functions of each body part or system. By the time you leave each area (body part or system), you will have gained a sense of how that area contributes to overall stability and well-being (homeostasis).

In planning this trip, an overview of sites to visit will be helpful. So please turn to the Framework for Chapter 1, study it carefully, and refer back to it as often as you wish. It contains the key terms of the chapter and serves as your map and your itinerary. Bon voyage!

TOPIC OUTLINE AND OBJECTIVES

A. Anatomy and physiology defined

1. Define anatomy with its subdivisions, and physiology.

B. Levels of structural organization

2. Define each of the following levels of structural organization that make up the human body: chemical, cellular, tissue, organ, system, and organismic.
3. Identify the principal systems of the human body, list representative organs of each system, and describe the function of each system.

C. Life processes

4. List and define several important life processes of humans.

D. Homeostasis

5. Define homeostasis and explain the effects of stress on homeostasis.
6. Define a feedback system and explain its role in homeostasis.
7. Contrast the operation and results of a negative feedback system and a positive feedback system.
8. Explain the relationship between homeostasis and disease.

E. Anatomical and directional terms

9. Define the anatomical position and compare common and anatomical terms used to describe various regions of the human body.
10. Define several directional terms and anatomical planes used in association with the human body.

F. Body cavities and regions

11. List by name and location the principal body cavities and the organs contained within them.

G. Medical imaging

12. Describe the principle and importance of selected medical imaging techniques in the diagnosis of disease.

WORDBYTES

Now study the following parts of words that may help you better understand terminology in this chapter.

Wordbyte	Meaning	Example	Wordbyte	Meaning	Example
ana-	up	*ana*tomy	physi(o)-	nature	*physio*logy
ante-	before	*ante*rior	post-	after	*post*erior
homeo-	same	*homeo*stasis	sagitta-	arrow	*sagittal*
inter-	between	*inter*cellular	stasis, stat-	stand, stay	homeo*stasis*
intra-	within	*intra*cellular	-tomo	to cut	ana*tomy*
logo-	study	physio*logy*	viscero-	body organs, function	*viscer*al
meter	measure	*centi*meter			
pariet-	wall	*parie*tal			

CHECKPOINTS

A. Anatomy and physiology defined   (page 5)

A1. Contrast each of the following pairs of terms.

a. Anatomy/physiology

b. Gross anatomy/histology

c. Systemic anatomy/regional anatomy

d. Developmental anatomy/embryology

e. Cytology/cell physiology

f. Neurophysiology/renal physiology

A2. Look at the two items in each of the following pairs. Tell how their structural differences explain their functional differences.

a. Spoon/fork

b. Hand/foot

c. Incisors (front teeth)/molars

B. Levels of structural organization **(pages 5–6)**

■ **B1.** Arrange the terms below in order from highest to lowest in level of organization, using the lines provided.

        cell   chemical   organ   system   tissue

Level of Organization

              Organism

(highest) _____

        system

        organ

        tissue

        cell

(lowest) _chemical_

■ **B2.** List the four basic types of tissues:

epitheleal   _connective_

muscle   _nervous_

■ B3. Complete the following table describing systems of the body. Name two or more organs in each system. Then list one or more functions of each system.

System	Organs	Functions
a. *integumentary*	Skin, hair, nails	*vitamin D, protection*
b. Skeletal	*bones, cartilage, joints*	*stores minerals, support, movement, produce red blood cells*
c. Muscular	*smooth muscles, skeletal*	*movement, heat, maintains posture*
d. *nervous*	*brain, spinal cord, nerves, eye, ear*	Regulates body by nerve impulses
e. *Endocrine*	Glands that produce hormones	*regulate target organs*
f. *cardiovascular*	Blood, heart, blood vessels	*moves blood, distributes oxygen, nutrients*
g. Lymphatic	*lymph glands, thymus, spleen, tonsils*	*water balance, immunity, transports fats from digestive to card*
h. *respiratory*	*lungs*	Supplies oxygen, removes carbon dioxide, regulates acid-base balance
i. *Digestive*	*stomach, intestines, liver, gallbladder*	Breaks down food and eliminates solid wastes
j. *urinary*	Kidney, ureters, urinary bladder, urethra	*cleanse waste*
k. Reproductive	*ovaries, testes*	*reproduction*

C. Life processes (pages 6–9)

C1. Six characteristics distinguish you from nonliving things. List these characteristics below and give a brief definition of each. One is done for you.

a. _____

b. _____

c. _____

d. **Growth: increase in size and complexity** _____

e. _____

f. _____

■ **C2.** Refer to the list of terms in the box, all related to life processes. Demonstrate your understanding of these terms by selecting the term that best fits each description provided below.

A. Anabolism	M. Metabolism
C. Catabolism	Respo. Responsiveness
D. Differentiation	Repro. Reproduction

_C_ a. Chemical processes which involve the breakdown of large, complex molecules into smaller ones with release of energy

_A_ b. Energy-requiring chemical processes which build structural and functional components of the body

_M_ c. Sum of all the chemical processes [both (a) and (b) above] in the body

Repro d. Formation of new cells for growth, repair, or replacement, or for production of a new individual

_D_ e. Changes that cells undergo during development from unspecialized ancestor cells to specialized cells such as muscle or bone cells

Respo f. Ability to detect and respond to changes in the environment, for example, production of insulin in response to elevated blood glucose level

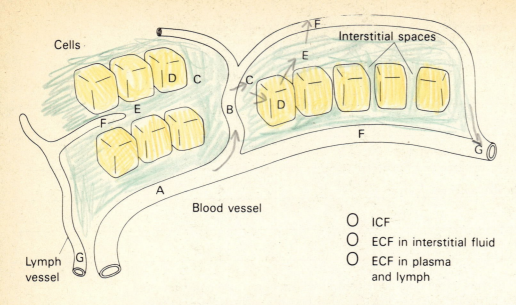

Cells

D C

E

F

F

E

C

B

D

Interstitial spaces

F

G

A

Blood vessel

○ ICF
○ ECF in interstitial fluid
○ ECF in plasma and lymph

Lymph vessel G

Figure LG 1.1 Internal environment of the body: types of fluid. Complete as directed. Areas labeled *A* to *G* refer to Checkpoint D2.

D. Homeostasis: maintaining physiological limits (pages 9–12)

■ **D1.** Describe the types of fluid in the body and their relationships to homeostasis by completing these statements and Figure LG 1.1.

a. Fluid inside of cells is known as _____ICF_____ fluid (ICF). Color areas containing this type of fluid yellow.

b. Fluid in spaces between cells is called _____interstitial_____ . It surrounds and bathes cells and is one form of *(intracellular? extracellular?)* fluid. Color the spaces containing this fluid light green.

c. Another form of extracellular fluid is that located in _____lymph_____

_____ vessels and _____blood vessel_____ vessels. Color these areas dark green.

d. The body's "internal environment" (that is, surrounding cells) is

_____extra_____ -cellular fluid (ECF) (all green areas in your figure). The condition of maintaining ECF in relative constancy is known as

_____homeostasis_____ .

■ **D2.** Refer again to Figure LG 1.1. Show the pattern of circulation of body fluids by drawing arrows connecting letters in the figure in alphabetical order (*A* → *B* → *C*, etc.). Now fill in the blanks below describing this lettered pathway: A (arteries and arterioles) → B (blood capillaries)

→ C (*interstitial*) → D (*intracellular*)

→ E (*interstitial*) → F (*blood or lymph capillaries*)

→ G (*venules & veins or lymph vessels*).

■ **D3.** List six qualities of your ECF that are maintained under optimal conditions when your body is in homeostasis.

contains optimum concentration of gases, nutrients, ions & water, optimal temperature, optimal volume

D4. Consider how your internal environment is affected by your external environment. Which of the six qualities of your ECF listed above would be altered if you were subjected to the following conditions? Compare your answers with those of a study partner.

a. Taking a two-day hike at high altitude without appropriate food or clothing

gases, nutrients, ions, temperature

b. Invasion of your body by an infectious microorganism that causes fever and diarrhea over a prolonged period of time

ions, water, temperature

D5. Suppose you decide to run a quarter mile at your top speed. Tell how your body would respond to this stress created by exercise. List according to system the changes your body would probably make in order to maintain homeostasis.

a. Nervous *detect changes in blood send impulses to brain brain sends impulses to heart*

b. Endocrine

c. Cardiovascular *heart beats faster, sending more oxygen to lungs & gives up carbon dioxide faster*

d. Muscular and skeletal *muscles cells take a lot of oxygen from blood, give off carbon dioxide, Brain sends messages to muscles that control breathing to contract more often*

e. Integumentary *sweat glands cool body*

■ **D6.** Are you seated right now? Stand up quickly. As you do, conceptualize what happens to your blood as you change position: as a response to gravity, it tends to rush to the lower part of your body. If nothing happened to offset that action, your blood pressure (BP) would drop and you would likely faint because of insufficient blood supply to the brain. This usually does not happen because homeostatic mechanisms act to quickly elevate blood pressure back to normal.

Match the answers in the box with factors and functions that helped maintain your blood pressure when you stood up. (*Hint:* Figure 1.4 [page 12 in your text] describes a situation that is just the opposite of this one.)

CCen.	Control center	O.	Output
CCond.	Controlled condition	Recep.	Receptors
E.	Effector	Respo.	Response
I.	Input	S.	Stimulus

C Cond a. Your blood pressure (BP): a factor that must be maintained homeostatically

___S___ b. Standing up; blood flowed by gravity to lower parts of your body

Recep c. Pressure-sensitive nerve cells in large arteries in your neck and chest

___I___ d. Nerve impulses bearing the message "BP is too low" (since less blood was in the upper part of your body as you were standing)

Ccent e. Your brain

___O___ f. Nerve impulses from your brain to your heart, bearing the message "Beat faster!"

___E___ g. Your heart (which beat faster)

respon h. Elevated BP

■ **D7.** *A clinical challenge.* In some people (for example, the elderly, the malnourished, and those with certain neurological disorders), nerve messages necessary for maintaining blood pressure may occur more slowly. As a result, blood pressure may drop as the individual suddenly changes position (from lying to sitting or sitting to standing). This condition, known as postural hypotension, may lead to faintness and falls. What might a person subject to postural hypotension do to minimize the chance of falling?

stand more slowly, when sitting hold onto arms of chair if lying sit up & wait awhile then dangle leg over side of bed then slowly rise

■ **D8.** Consider this example of a feedback system loop. Ten workers on an assembly line are producing handmade shoes. As shoes come off the assembly line, they pile up around the last worker, Worker Ten. When this happens, Worker Ten calls out, "We have hundreds of shoes piled up here." This information (input) is heard by (fed back to) Worker One, who determines when more shoes should be begun. Worker One is therefore the ultimate controller of output. This controller could respond in either of two ways. In a *negative feedback system,* Worker One says, "We have an excess of unsold shoes. Let us slow down or stop production until the excess (stress) is relieved—until these shoes are sold." What might Worker One's response be if this were a *positive feedback system?*

we have excess of shoes make more

D9. Most feedback systems of the body are *(positive? negative?)* Why do you think this might be advantageous and directed toward maintenance of homeostasis?

detects changes that often occur + negates them

■ **D10.** Circle the terms below that are classified as signs, rather than as symptoms.

chest pain   fever   headache   diarrhea   swollen ankles   skin rash   toothache

D11. *A clinical challenge.* Contrast the two terms in each set.

a. Local disease/systemic disease

b. Medical history/diagnosis

c. Absence of disease/wellness

D12. Explain how your own environment and your health practices (behaviors) are related to the health of your:

a. Cardiovascular system

b. Respiratory system

c. Skeletal system

d. Integumentary system

E. Anatomical and directional terms (pages 12–16)

■ **E1.** How would you describe anatomical position? Assume that position yourself.

■ **E2.** Complete the table relating common terms to anatomical terms. (See Figure 1.5, page 13, in your text to check your answers.) For extra practice use common and anatomical terms to identify each region of your own body and that of a study partner.

Common Term	Anatomical Term
a. *armpit*	Axillary
b. Fingers	*digital phalangeal*
c. Arm	*brachial*
d. *hollow behind knee*	Popliteal
e. *Head*	Cephalic
f. Mouth	*oral*
g. *groin*	Inguinal
h. Chest	*Thoracic*
i. *neck*	Cervical
j. *forearm*	Antebrachial
k. Buttock	*gluteal*
l. *heel*	Calcaneal

■ **E3.** Using your own body, a skeleton, a torso, or Figure 1.6 (page 14 in your text), determine relationships among body parts. Write the correct directional term(s) to complete each of these statements.

a. The liver is *inferior* to the diaphragm.

b. Fingers (phalanges) are located *distal* to wrist bones (carpals).

c. The skin on the dorsal surface of your body can also be said to be located on your *posterior* surface.

d. The great (big) toe is *medial* to the little toe.

e. The little toe is ___*lateral*___ to the great toe.

f. The skin on your leg is ___*superficial*___ to muscle tissue in your leg.

g. Muscles of your arm are ___*deep*___ to skin on your arm.

h. When you float face down in a pool, you are lying on your

___*anterior ventral*___ surface.

i. The lungs and heart are located ___*Superior*___ to the abdominal organs.

j. Since the stomach and the spleen are both located on the left side of the abdomen,

they could be described as ___*ipsi*___-lateral.

k. The ___*visceral*___ pleura covers the external surface of the lungs.

✗ ■ **E4.** Match each of the following planes with the phrase telling how the body would be divided by such a plane.

F. Frontal (coronal)	S. Sagittal
M. Midsagittal (median)	T. Transverse

___T___ a. Into superior and inferior portions   ___F___ c. Into anterior and posterior portions

___M___ b. Into equal right and left portions   ___S___ d. Into right and left portions

✗ **F. Body cavities, regions** (pages 16–22)

■ **F1.** After you have studied Figures 1.10 and 1.11 (pages 19–21 in your text), complete this exercise about body cavities. Circle the correct answer in each statement.

a. The *(dorsal? ventral?)* cavity consists of the cranial cavity and the vertebral canal.

b. The viscera, including such structures as the heart, lungs, and intestines, are all located in the *(dorsal? ventral?)* cavity.

c. Of the two body cavities, the *(dorsal? ventral?)* appears to be better protected by bone.

d. Pleural, mediastinal, and pericardial are terms that refer to regions of the *(thorax? abdominopelvis?)*.

e. The *(heart? lungs? esophagus and trachea?)* are located in the pleural cavities.

f. The division between the abdomen and the pelvis is marked by *(the diaphragm? an imaginary line from the symphysis pubis to the superior border of the sacrum?)*.

g. The stomach, pancreas, small intestine, and most of the large intestine are located in the *(abdomen? pelvis?)*.

h. The urinary bladder, rectum, and internal reproductive organs are located in the *(abdominal? pelvic?)* cavity.

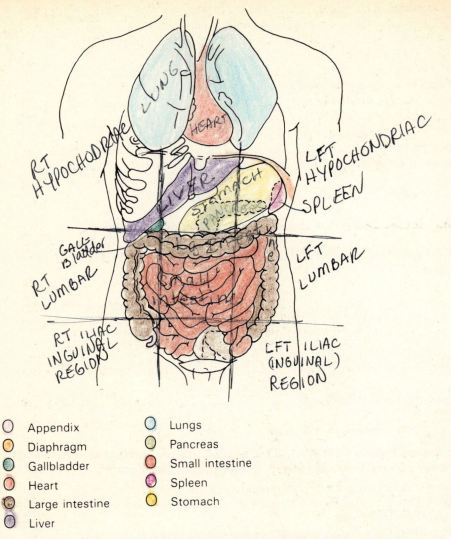

Handwritten labels on figure: RT HYPOCHODRIAC, LUNG, HEART, LFT HYPOCHONDRIAC, LIVER, STOMACH, PANCREAS, SPLEEN, GALL Bladder, RT LUMBAR, LFT LUMBAR, Small intestine, RT ILIAC INGUINAL REGION, LFT ILIAC (INGUINAL) REGION

◯ Appendix		◯ Lungs	
◯ Diaphragm		◯ Pancreas	
◯ Gallbladder		◯ Small intestine	
◯ Heart		◯ Spleen	
◯ Large intestine		◯ Stomach	
◯ Liver			

Figure LG 1.2   Regions of the ventral body cavity. Complete as directed in Checkpoint F2.

■ **F2.** Complete Figure LG 1.2 according to the directions below.

a. Color each of the organs listed on the figure. Select different colors for each organ, and be sure to use the same color for the related color code oval (◯).

b. Next, label each organ on the figure.

c. *For extra review.* Draw lines dividing the abdomen into the nine regions, and note which organs are in each region.

■ **F3.** Using the names of the nine abdominal regions and quadrants, complete these statements. Refer to the figure you just completed and to Figure 1.11 (pages 20–21) in your text.

a. From superior to inferior, the three abdominal regions on the right side are

<u>HYPOCHONDRIAC</u>, <u>LUMBAR</u>, and

<u>ILIAC (INGUINAL)</u>

b. The stomach is located primarily in the two regions named

<u>EPIGASTRIC</u> and <u>LFT HYPOCHONDRIAC</u>.

c. The navel is located in the _UMBILICAL_ region.

d. The region immediately superior to the urinary bladder is named the

HYPOGASTRIC region.

e. The liver and gallbladder are located in the _RT UPPER_ quadrant.

F4. List at least three reasons why an autopsy might be performed.
SUPPORT ACCURACY OF TESTS, EDUCATE HEALTH-CARE STUDENTS, REVEAL CONDITIONS THAT MIGHT EFFECT OFFSPRING OR SIBLINGS, CRIMINAL INVESTIGATION, UNCOVER EXSISTENCE OF DISEASES NOT DETECTED DURING LIFE.

G. Medical imaging (pages 22–25)

■ **G1.** Match the radiographic anatomy techniques listed in the box with the correct descriptions.

CR. Conventional radiograph	MRI. Magnetic resonance imaging
CT. Computed tomography	PET. Positron emission tomography
DSA. Digital subtraction angiography	US. Ultrasound
DSR. Dynamic spatial reconstruction	

_____ a. A recent development in radiographic anatomy, this technique produces moving three-dimensional images of body organs.

_____ b. This method provides a cross-sectional picture of an area of the body by means of an x-ray source moving in an arc. Results are processed by a computer and displayed on a video monitor.

_____ c. This technique produces a two-dimensional image (radiograph) via a single barrage of x-rays. Images of organs overlap, making diagnosis difficult.

_____ d. A method used to study blood vessels by comparing a region before and after injection of a contrast substance (dye).

_____ e. Utilizes injected radioisotopes that assess function as well as structure.

_____ f. Two techniques that are noninvasive and do not utilize ionizing radiation (two answers).

ANSWERS TO SELECTED CHECKPOINTS

B1. From highest to lowest: organism, system, organ, tissue, cell, chemical.

B2. Epithelial tissue, muscle tissue, connective tissue, and nervous tissue.

C2. (a) C. (b) A. (c) M. (d) Repro. (e) D. (f) Respo.

D1. (a) Intracellular. (b) Interstitial (or intercellular or tissue) fluid, extracellular. (c) Blood, lymph. (d) Extra-; homeostasis.

D2. C, interstitial (intercellular) fluid; D, intracellular fluid; E, interstitial (intercellular) fluid again; F, blood or lymph capillaries; G, venules and veins or lymph vessels.

D3. Concentration of gases (such as oxygen or carbon dioxide), nutrients (such as proteins and fats), ions (such as sodium and potassium), and water; blood volume and temperature.

D6. (a) CCond. (b) S. (c) Recep. (d) I. (e) CCen. (f) O. (g) E. (h) Respo.

D7. Change position gradually, for example, by sitting on the side of the bed before getting out of bed. Support body weight on a nightstand or chair for a moment after standing.

D8. "Hurray! We have produced hundreds of shoes; let us go for thousands! Do not worry that the shoes are not selling. Step up production even more."

D10. Fever, diarrhea, swollen ankles, skin rash.

E1. Stand with your arms at your sides and palms of your hands facing forward.

E2. (a) Armpit. (b) Phalanges. (c) Brachial. (d) Hollow area at back of knee. (e) Head. (f) Oral. (g) Groin. (h) Thoracic. (i) Neck. (j) Forearm. (k) Gluteal. (I) Heel.

E3. (a) Inferior (caudad). (b) Distal. (c) Posterior. (d) Medial. (e) Lateral. (f) Superficial. (g) Deep. (h) Anterior (ventral). (i) Superior (cephalad). (j) Ipsi-. (k) Visceral.

E4. (a) T. (b) M. (c) F. (d) S.

F1. (a) Dorsal. (b) Ventral. (c) Dorsal. (d) Thorax. (e) Lungs. (f) An imaginary line from the symphysis pubis to the superior border of the sacrum. (g) Abdomen. (h) Pelvic.

F2.

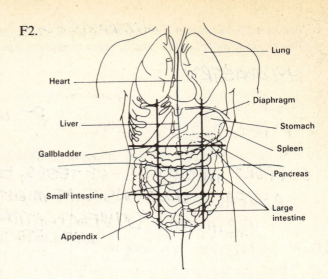

F3. (a) Right hypochondriac, right lumbar, right iliac (inguinal). (b) Left hypochondriac, epigastric. (c) Umbilical. (d) Hypogastric (pubic). (e) Right upper.

G1. (a) DSR. (b) CT. (c) CR. (d) DSA. (e) PET. (f) MRI, US.

MASTERY TEST: Chapter 1

Questions 1–10: Circle the letter preceding the one best answer to each question.

1. In a negative feedback system, when blood pressure decreases slightly, the body will respond by causing a number of changes which tend to:
   A. Lower blood pressure
   B. Raise blood pressure

2. The following structures are all located in the ventral cavity *except*:
   A. Spinal cord
   B. Urinary bladder
   C. Heart
   D. Gallbladder
   E. Esophagus

3. Which pair of common/anatomical terms is mismatched?
   A. Eye/ocular
   B. Skull/cranial
   C. Armpit/brachial
   D. Neck/cervical
   E. Buttock/gluteal

4. Which science studies the "why, when, and where" regarding transmission of diseases?
   A. Endocrinology
   B. Histology
   C. Cytology
   D. Pharmacology
   E. Epidemiology

5. Which is most inferiorly located?
   A. Abdomen
   B. Pelvic cavity
   C. Mediastinum
   D. Diaphragm
   E. Pleural cavity

6. The spleen, tonsils, and thymus are all organs in which system?
   A. Nervous
   B. Lymphatic
   C. Cardiovascular
   D. Digestive
   E. Endocrine

7. Which of the following structures is located totally outside of the upper right quadrant of the abdomen?
   A. Liver
   B. Gallbladder
   C. Transverse colon
   D. Spleen
   E. Pancreas

8. Choose the FALSE statement. (It may help you to mark each statement T for true or F for false as you read it.)
   T A. Stress disturbs homeostasis.
   T B. Homeostasis is a condition in which the body's internal environment remains relatively constant.
   T C. The body's internal environment is best described as extracellular fluid.
   F D. Extracellular fluid consists of plasma and intracellular fluid.

9. The system responsible for providing support, protection, and leverage, for storage of minerals, and for production of most blood cells is:
   A. Urinary
   B. Integumentary
   C. Muscular
   D. Reproductive
   E. Skeletal

10. Which is most proximally located?
    A. Ankle
    B. Hip
    C. Knee
    D. Toe

Questions 11–20: Circle T (true) or F (false). If the statement is false, change the underlined word or phrase so that the statement is correct.

Ⓣ F 11. <u>Anterior and ventral</u> are synonyms.

T Ⓕ 12. In order to assume the anatomical position, you should <u>lie down with arms at your sides and palms turned backwards.</u>

T Ⓕ 13. The appendix is usually located in the <u>left iliac</u> region of the abdomen.

T Ⓕ 14. <u>Anatomy</u> is the study of how structures function.

T Ⓕ 15. In a negative feedback system, the body will respond to an increased blood glucose level by <u>increasing blood glucose to an even higher level.</u>

Ⓣ F 16. A <u>tissue</u> is a group of cells and their intercellular substances that are similar in both origin and function.

T Ⓕ 17. The right kidney is located <u>mostly in the right iliac region of the abdomen.</u>

Ⓣ F 18. The study of the microscopic structure of cells is <u>cytology.</u>

Ⓣ F 19. A sagittal section is <u>parallel to and lateral to</u> a midsagittal section.

T Ⓕ 20. The term homeostasis means that the body exists <u>in a static state.</u>

Questions 21–25: fill-ins. Write the word that best fits the description.

mediastinum 21. Region of thorax located between the lungs. Contains heart, thymus, esophagus.

intracellular 22. Name the fluid located within cells.

differentiation 23. Life process in which unspecialized cells as in early embryo change into specialized cells such as nerve or bone.

ultrasound 24. A painless diagnostic technique in which sound waves are generated and detected by a transducer, resulting in a sonogram.

plasma 25. Name one or more extracellular fluids.

ANSWERS TO MASTERY TEST: ■ Chapter 1

Multiple Choice

1. B	6. B
2. A	7. D
3. C	8. D
4. E	9. E
5. B	10. B

True-False

11. T
12. F. Stand erect facing observer, arms at sides, palms forward
13. F. Right iliac
14. F. Physiology
15. F. Decreasing blood glucose to a lower level
16. T
17. F. At the intersection of four regions of the abdomen: right hypochondriac, right lumbar, epigastric, and umbilical
18. T
19. T
20. F. In a state of dynamic equilibrium

Fill-ins

21. Mediastinum
22. Intracellular
23. Differentiation
24. Ultrasound (US)
25. Plasma, interstitial fluid, lymph, serous fluid, mucus, fluid within eyes or ears, cerebrospinal fluid (CSF)

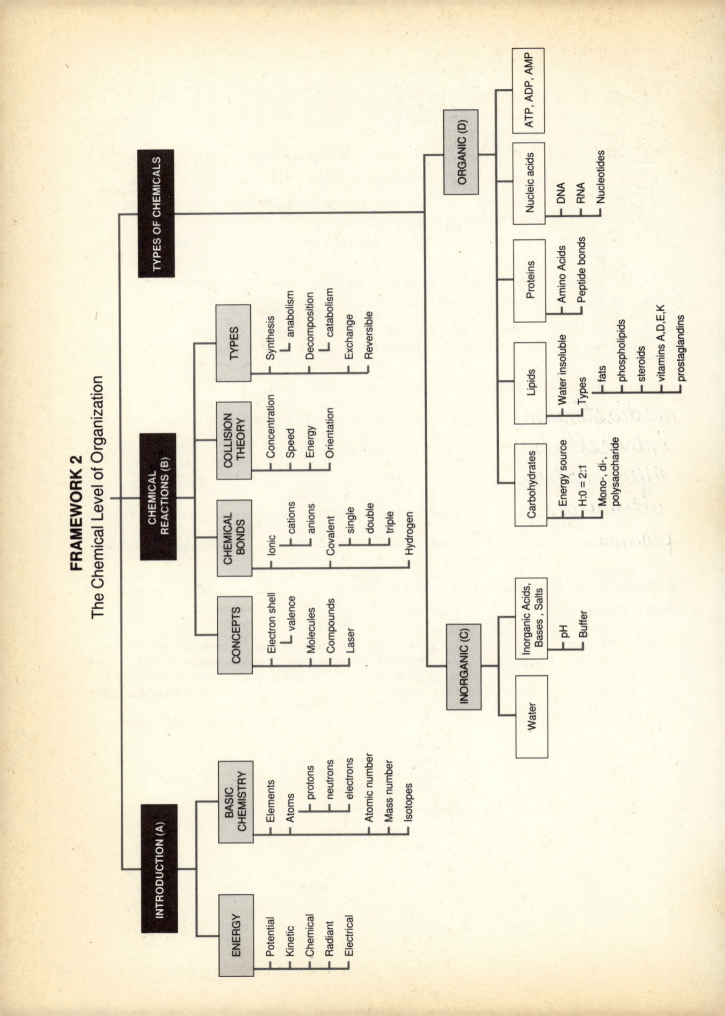

FRAMEWORK 2
The Chemical Level of Organization

The Chemical Level of Organization

Chemicals comprise the ultrastructure of the human body. They are the submicroscopic particles of which cells, tissues, organs, and systems are constructed. Chemicals are the minute entities that participate in all of the reactions that underlie bodily functions. Every complex compound that molds a living organism is composed of relatively simple atoms held together by chemical bonds. Two classes of chemicals contribute to structure and function: inorganic compounds, primarily water, but also inorganic acids, bases, and salts. Organic compounds include carbohydrates, lipids, proteins, nucleic acids, and the energy-storing compound ATP.

As you begin your study of the chemistry of life, carefully examine the Chapter 2 Framework and note the key terms associated with each section.

TOPIC OUTLINE AND OBJECTIVES

A. Introduction to basic chemistry: elements and atoms

1. Distinguish between matter and energy.
2. Describe the structure of an atom.
3. Identify by name and symbol the principal chemical elements of the human body.

B. Chemical reactions

4. Explain how ionic, covalent, and hydrogen bonds form.
5. Define a chemical reaction and explain the basic differences between synthesis, decomposition, exchange, and reversible chemical reactions.

C. Inorganic water, acids, bases, salts; pH, buffers

6. List and compare the properties of water and inorganic acids, bases, and salts.
7. Define pH and explain the role of a buffer system as a homeostatic mechanism that maintains the pH of a body fluid.

D. Organic compounds

8. Compare the structure and functions of carbohydrates, lipids, proteins, deoxyribonucleic acid (DNA), ribonucleic acid (RNA), adenosine triphosphate (ATP), and cyclic AMP.
9. Describe the characteristics and importance of enzymes.

WORDBYTES

Now study the following parts of words that may help you better understand terminology in this chapter.

Wordbyte	Meaning	Example	Wordbyte	Meaning	Example
ase-	enzyme	lip*ase*	pent-	five	*pent*ose
di-	two	*di*saccharide	poly-	many	*poly*unsaturated
hex-	six	*hex*ose	prot-	first	*prot*ein
mono-	one	*mono*saccharide	-saccharide	sugar	poly*saccharide*

CHECKPOINTS

A. Introduction to basic chemistry (pages 29–31)

A1. Why is it important for you to understand the "language of chemistry" in order to communicate about life processes?

■ **A2.** Complete this exercise about matter and energy.

a. Matter exists in three states: solid, ___*liquid*___, or ___*gas*___.

b. When you are on a mountaintop, your *(mass? weight?)* is less than when you are standing on an ocean beach at sea level.

c. ___*Energy*___ is the capacity to do work. Inactive or stored energy is known as *(kinetic? potential?)* energy, whereas the energy of motion is ___*kinetic*___ energy.

d. *(Chemical? Electrical? Radiant?)* energy is the energy released or absorbed as chemicals are either broken apart or formed. An action potential (impulse) in a nerve or muscle cell is an example of ___*electrical*___ energy. Energy that travels in waves, such as microwaves, ultraviolet (UV) rays, or x-rays, is classified as ___*Radiant*___ energy.

A3. Contrast *element* with *atom*.

element is a quantity of matter composed of atoms that are all the same type.

■ **A4.** Refer to Figure LG 2.1, and do the following exercise.

a. Color light blue the four boxes containing the elements that make up about 96 percent of the weight of the body. These four elements are relatively *(simple? complex?)* in atomic structure among the total 109 elements in the periodic table.

b. Color pink the boxes containing calcium and phosphorus. These two elements together contribute about ___*2.5*___ percent of body weight.

Periodic Table

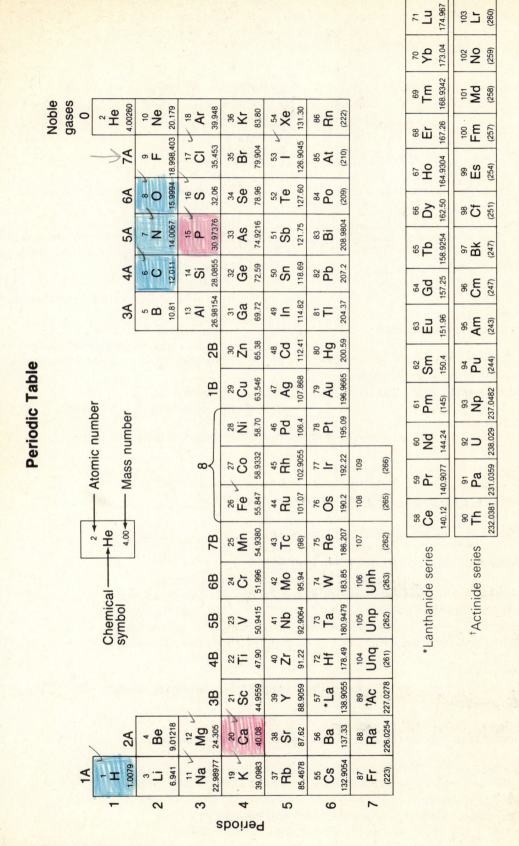

Figure LG 2.1 Periodic table.

■ **A5.** Refer to Exhibit 2.1 (page 30) in your text. Match the elements listed in the box with descriptions of their functions in the body. *For extra review.* Locate each of these elements in the periodic table (Figure 2.1) and write the atomic number of each element after its description. The first one has been done for you.

> ✓C. Carbon   ✓K. Potassium
> ✓Ca. Calcium   ✓Na. Sodium
> ✓Cl. Chlorine   ✓N. Nitrogen
> ✓Fe. Iron   O. Oxygen
> ✓ I. Iodine   P. Phosphorus

___C___ a. Found in every organic molecule

___N___ __6__ b. Component of all protein molecules and DNA and RNA __7__

___I___ c. Vital to normal thyroid gland function __53__

___Fe___ d. Essential component of hemoglobin __26__

___Ca___ e. Constituent of bone and teeth; required for blood clotting and for muscle contraction __20__

___O___ f. Constituent of water; functions in cellular respiration __8__

___K___ g. Most abundant positively charged ion (cation) in intracellular fluid (inside of cells) __19__

___Na___ h. Most abundant cation in extracellular fluid (outside of cells) __11__

___Cl___ i. Most abundant negatively charged ion (anion) in extracellular fluid (outside of cells) __17__

___P___ j. Component of DNA, RNA, ATP; found in bone and teeth __15__

■ **A6.** Complete this exercise about atomic structure.

a. An atom consists of two main parts: __nucleus__ and __electrons__.

b. Within the nucleus are positively charged particles called __protons__ and uncharged (neutral) particles called __neutrons__.

c. Electrons bear *(positive?* ~~negative?~~*)* charges. The number of electrons forming a charged cloud around the nucleus is *(greater than?* (equal to?) *smaller than?)* the number of protons in the nucleus of the atom.

d. Atoms are measured in *(grams?* (daltons) *[atomic mass units = amu]?)*. A proton and a neutron each have the mass of approximately *1.007* dalton. Electrons have a mass of about *.0005* dalton.

■ **A7.** Refer to Figure LG 2.1 and Figure 2.2 in your text (page 32), and complete this exercise about atomic structure.

a. The atomic number of potassium (K) is __19__. Locate this number above the symbol K on Figure LG 2.1. This means that a potassium atom has __19__ protons and __19__ electrons.

b. Identify the number located immediately under the K in Figure LG 2.1. This number is __39.09__, and it represents the mass number of the potassium atom. This indicates the total number of protons and neutrons in the potassium nucleus.

■ **A8.** Complete this exercise about isotopes. *pg 31 2nd column 3rd ¶*

a. If you subtract the atomic number from the mass number, you will identify the number of neutrons in an atom. Most potassium atoms have __20__ *39-19* neutrons. Different isotopes of potassium all have 19 protons, but have varying numbers of *(electrons? neutrons?)*.

b. The isotope ^{14}C differs from ^{12}C in that ^{14}C has __14__ neutrons.

c. Radioisotopes have a nuclear structure that is unstable and can decay, emitting

   __radiation__. Each radioactive isotope has its own distinctive

   __half-life__, which is the time required for the radioactive isotope to emit half of its original amount of radiation.

d. A radioactive form of iodine, ^{131}I, is most often used to indicate size, location, and activity of the *(adrenal? pituitary? thyroid?)* gland. Imaging of the heart and its coronary blood flow most often uses an isotope of *(lead ([Pb]? thallium [Tl]? uranium [U]?)*.

B. Chemical reactions   (pages 32–37)

■ **B1.** The type of chemical bonding that occurs between atoms depends on the number of electrons of the bonding atoms. Do this exercise about electrons and bonding.

a. Electrons are arranged in electron shells. The electron shell closest to the nucleus has a maximum capacity of __2__ electrons, whereas the second shell holds a maximum of __8__ electrons.

b. The combining capacity (or __valence__) refers to the number of

   extra or deficient electrons in the outermost shell (or __valence__ shell). Write the chemical symbols of several atoms in the periodic table (Figure LG 2.1) that have only one electron in the valence shell.

   H, Na, K

c. Now write the chemical symbols of several atoms in the periodic table that have one electron missing from an otherwise complete valence shell.

   F, Cl, Br, I

d. Potassium has one *(extra? missing?)* electron in its valence shell. It is therefore more likely to be an electron *(donor? acceptor?)*. When it gives up an electron (which is a negative entity), it becomes more *(positive? negative?)*; in other words potassium forms K+, which is a(n) *(anion? cation?)*.

e. Chlorine has a valence of *(+1? −1?)*, indicating that it has one *(extra? missing?)* electron in its valence shell. As potassium chloride is formed, chlorine *(accepts? gives up?)* an electron and so becomes the *(anion? cation?)* Cl−.

f. Certain elements have completely filled outer shells and therefore do not tend to gain or lose electrons. These elements, which do not participate in chemical

   reactions, are known as __inert__ elements. Find five such elements in the periodic table.

   He Ne Ar Kr Xe Rn

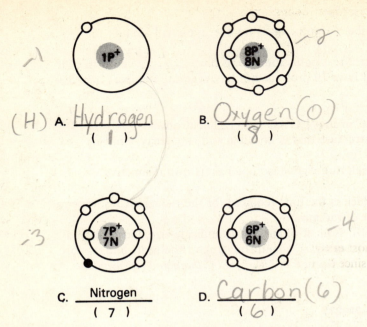

(H) A. Hydrogen
    (1)

B. Oxygen (O)
    (8)

C. Nitrogen
    (7)

D. Carbon (6)
    (6)

Figure LG 2.2 Atomic structure of four common atoms in the human body. Complete as directed in Checkpoint B2.

■ **B2.** Refer to Figure LG 2.2 and complete this exercise.

a. Write atomic numbers in the parentheses ( ) under each atom. (*Hint:* Count the number of electrons or protons = atomic number.) One is done for you.

b. Write the name or symbol for each atom on the line under that atom. (*Hint:* You may identify these by atomic number in Figure LG 2.1.)

c. Based on the number of electrons that each of these atoms is missing from a complete outer electron shell, draw the number of covalent bonds that each of the atoms can be expected to form. Note that this is the valence, or combining capacity, of these atoms. One is done for you.

H   —O—   ＼N／   —C—

d. *For extra review.* Draw the atomic structure of an ammonia molecule (NH_3). Note that this is a covalently bonded compound.

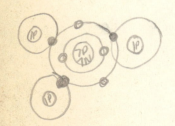

■ **B3.** Match the description with the types of bonds listed in the box.

| C. Covalent | H. Hydrogen | I. Ionic |

_I_ a. Atoms lose electrons if they have just one or two electrons in their outer orbits; they gain electrons if they need just one or two electrons to complete the outer electron shell, as in K^+Cl^- (see Activity B1 above).

_H_ b. This bond is a bridgelike, weak link between a hydrogen atom and another atom such as oxygen or nitrogen.

_C_ c. This type of bond is most easily formed by carbon (C) since its outer orbit is half filled.

_H_ d. These bonds are only about 5 percent as strong as convalent bonds and so are easily formed and broken. They are vital for holding large molecules like protein or DNA in proper configurations.

_C_ e. Double bonds, such as $O = O$ (O_2) or triple bonds, such as $N \equiv N$ (N_2).

_C_ f. These bonds may be polar, such as H_2O (H – OH), or nonpolar, such as $O = O$ (O_2).

■ **B4.** Circle the molecules that are classified as compounds.

H   Glucose

H_2 N_2

H_2O A molecule that contains two or more different kinds of atoms

■ **B5.** Identify the kind of chemical reaction described in each statement below.

| D. Decomposition reaction | R. Reversible reaction |
| E. Exchange reaction | S. Synthesis reaction |

_R_ a. The end product can revert to the original combining molecules.

_S_ b. Two or more reactants combine to form an end product; for example, many glucose molecules bond to form glycogen.

_E_ c. Such a reaction is partly synthesis and partly decomposition.

STOP

_D_ d. All chemical reactions involve making or breaking of bonds. This type of reaction involves only breaking of bonds.

_D_ e. This type of reaction is catabolic, as in digestion of foods, such as starch digestion to glucose.

B6. Write a paragraph describing the collision theory explanation for chemical reactions. Include four factors that determine whether a collision will occur and lead to a chemical reaction.

C. Inorganic compounds, water, acids, bases, salts; pH, buffers (pages 37–40)

■ **C1.** Write *inorganic* after phrases that describe inorganic compounds and *organic* after those that describe organic compounds.

a. Held together almost entirely by covalent bonds: _____

b. Tend to be very large molecules which serve as good building blocks for body

structures: _____

c. The class that includes water, the most abundant compound in the body:

inorganic

d. The class that includes carbohydrates, proteins, fats, and nucleic acids:

organic

✴ ■ **C2.** Answer these questions about the functions of water. *page 37 2nd column*

a. Water normally is a *(solute? solvent?)*.

b. Water requires a *(large? small?)* amount of heat in order to increase its temperature and change it into a gas. As a result, for each small amount of perspiration (sweat) evaporated during heavy exercise, a *(large? small?)* amount of heat can be released. This is fortunate since dehydration is less likely to occur with such heat loss.

c. Name a location where water's function as a lubricant is essential.

chest & abdomen

✴ ■ **C3.** Match the descriptions below with the answers in the box.

A. Acid	B. Base	S. Salt

_B_ a. A substance that dissociates into hydroxide ions (OH^-) and one or more cations. Example: NaOH

_A_ b. A substance that dissociates into hydrogen ions (H^+) and one or more anions. Example: H_2SO_4

_S_ c. A substance that dissolves in water, forming cations and anions neither of which is H^+ or OH^-. Example: $CaCl_2$

✴ **C4.** Draw a diagram showing the pH scale. Label the scale from 0 to 14. Indicate by arrows increasing acidity (H^+ concentration) and increasing alkalinity (OH^- concentration).

alkalinity ↑ *14 OH^-*
13
12
11
10
9
8
7
6
5
acid ↓ *4*
3
2
1
0 H^+

■ **C5.** Choose the correct answers regarding pH.

a. Which pH is most acid?

   (A. 4)   C. 10

   B. 7

b. Which pH has the highest concentration of OH⁻ ions?

   A. 4   (C. 10)

   B. 7

c. Which solution has pH closest to neutral?

   A. Gastric juice (digestive juices of the stomach)

   (B.) Blood

   C. Lemon juice

   D. Milk of magnesia

d. A solution with pH 8 has 10 times *(more? fewer?)* H⁺ ions than a solution with pH 7.

e. A solution with pH 5 has *(10? 20? 100?)* times *(more? fewer?)* H⁺ ions than a solution with pH 7.

C6. Contrast:

a. Strong acid/weak acid

b. Strong base/weak base

C7. Write a sentence or two explaining each of the following:

a. How pH is related to homeostasis

b. How buffers help to maintain homeostasis

D. Organic compounds   (pages 40–53)

D1. Organic compounds all contain carbon (C) and hydrogen (H). State two reasons why carbon is an ideal element to serve as the primary structural component for living systems.

■ **D2.** Carbohydrates carry out important functions in your body.

a. State the principal function of carbohydrates.

b. State two secondary roles of carbohydrates.

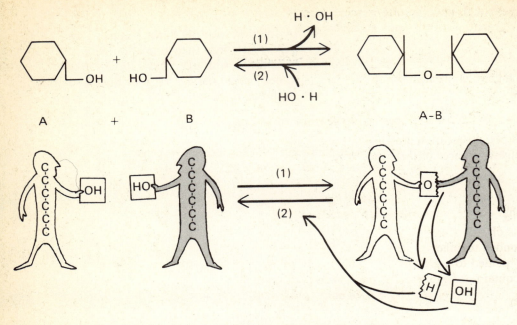

Figure LG 2.3 Dehydration synthesis and hydrolysis reactions. Refer to Checkpoint D3.

■ **D3.** Complete these statements about carbohydrates.

a. The ratio of hydrogen to oxygen in all carbohydrates is _2_ : _1_ . The

   general formula for carbohydrates is _CH_2O_ .

b. The chemical formula for a *hexose* is _____. Two common

   hexoses are _____ and _____.

c. One common pentose sugar found in DNA is _____. Its name
   and its formula ($C_5H_{10}O_4$) indicate that it lacks one oxygen atom from the typical
   pentose formula which is _____.

d. When two glucose molecules ($C_6H_{12}O_6$) combine, a disaccharide is formed. Its for-

   mula is _____, indicating that in a synthesis reaction such as
   this, a molecule of water is *(added? removed?)*. Refer to Figure LG 2.3. Identify
   which reaction *(1? 2?)* on that figure demonstrates synthesis.

e. Continued dehydration synthesis leads to enormous carbohydrates called

   _____. One such carbohydrate is _____.

f. When a disaccharide such as sucrose is broken, water is introduced to split the bond
   linking the two monosaccharides. Such a decomposition reaction is termed

   _____, which literally means "splitting using water." (See Fig-
   ure 2.8, page 42 in your text.) On Figure LG 2.3, hydrolysis is shown by reaction
   (1? 2?).

■ **D4.** Contrast carbohydrates with lipids in this exercise. Write C next to any statement true of carbohydrates, and write L next to any statement true of lipids.

_____ a. These compounds are insoluble in water (hydrophobic). (*Hint:* Water alone is ineffective for washing such compounds out of clothes.)

_____ b. These compounds are organic.

_____ c. These substances have a hydrogen to oxygen ratio of about 2:1.

_____ d. Very few oxygen atoms (compared to the numbers of carbon and hydrogen atoms) are contained in these compounds. (Look carefully at Figure 2.9, page 44 in your text.)

■ **D5.** Circle the answer that correctly completes each statement.
a. A triglyceride consists of:
   A. Three glucoses bonded together
   B. Three fatty acids bonded to a glycerol
b. A typical fatty acid contains (see Figure 2.9, page 44 in your text):
   A. About 3 carbons   B. About 16 carbons
c. A polyunsaturated fat contains _____ than a saturated fat.
   A. More hydrogen atoms   B. Fewer hydrogen atoms
d. Saturated fats are more likely to be derived from:
   A. Plants, for example, corn oil or peanut oil
   B. Animals, as in beef or cheese
e. Oils that are classified as saturated fats include:
   A. Tropical oils such as coconut oil and palm oil
   B. Vegetable oils such as safflower oil, sunflower oil, and corn oil
f. A diet designed to reduce the risk for atherosclerosis is a diet low in:
   A. Saturated fats
   B. Polyunsaturated fats

■ **D6.** Match the types of lipids listed in the box with their functions given below.

B. Bile salts	E. Estrogens
Car. Carotene	Pho. Phospholipids
Cho. Cholesterol	Pro. Prostaglandins

_____ a. Present in egg yolk and carrots; leads to formation of vitamin A, a chemical necessary for good vision

_____ b. The major lipid in cell membranes

_____ c. Present in all cells, but infamous for its relationship to "hardening of the arteries"

_____ d. Substances that break up (emulsify) fats before their digestion

_____ e. Steroid sex hormones produced in large quantities by females

_____ f. Important regulatory compounds with many functions, including modifying responses to hormones

STOP

■ **D7.** Complete this exercise about fat-soluble vitamins.

a. Name four fat-soluble vitamins (consult Exhibit 2.3, page 43 in your text, for help):

_____ _____ _____ _____

b. Since these four vitamins are absorbed into the body along with fats, any problem with fat absorption is likely to lead to:

c. The fat-soluble vitamin that helps with calcium absorption and so is necessary for

normal bone growth is vitamin _____.

d. Vitamin _____ may be administered to a patient before surgery in order to prevent excessive bleeding since this vitamin helps in the formation of clotting factors.

e. Normal vision is associated with an adequate amount of vitamin _____.

f. Vitamin _____ has a variety of possible functions including the promotion of wound healing and prevention of scarring.

■ **D8.** Match the types of proteins listed in the box with their functions given below.

8-11

Cat. Catalytic	R. Regulatory
Con. Contractile	S. Structural
I. Immunological	T. Transport

_____T_____ a. Hemoglobin in red blood cells

_____Con_____ b. Proteins in muscle tissue, such as actin and myosin

_____S_____ c. Keratin in skin and hair; collagen in bone

_____R_____ d. Hormones such as insulin

_____I_____ e. Defensive chemicals such as antibodies and interleukins

_____Cat_____ f. Enzymes such as lipase, which digests lipids

■ **D9.** Contrast proteins with the other organic compounds you have studied so far by completing this exercise.

a. Carbohydrates, lipids, and proteins all contain carbon, hydrogen, and oxygen. A

fourth element, *nitrogen* _____, makes up a substantial portion of proteins.

b. Just as large carbohydrates are composed of repeating units (the

monosaccharides), proteins are composed of building blocks called

amino acids _____.

c. As in the synthesis of carbohydrates or fats, when two amino acids bond together, a water molecule must be *(added? removed?)*. This is another example of *(hydrolysis? dehydration?)* synthesis.

d. The product of such a reaction is called a *Di* _____-peptide. (See Figure 2.13, page 47 in your text.) When many amino acids are linked together in

this way, a *poly* _____-peptide results. One or more polypeptide

chains form a _____.

D10. There are _____ different kinds of amino acids forming the human body, much like a 20-letter alphabet that forms protein "words." It is the specific sequence of amino acids that determines the nature of the protein formed. In order to see the significance of this fact, do the following activity. Arrange the five letters listed below in several ways so that you form different five-letter words. Use all five letters in each word.

e i l m s

smile   limes

miles   slime

Note that although all of your words contain the same letters, the sequence of letters differs in each case. The resulting words have different meanings, or functions. Such is the case with proteins, too, except that protein "words" may be thousands of "letters" (amino acids) long, involving all or most of the 20–amino acid "alphabet." Proteins must be synthesized with accuracy. Think of the drastic consequences of omission or misplacement of even a single amino acid. (How different are *limes, miles, smile, slime!*)

■ **D11.** Complete this exercise about vegetarian diets.

a. A *(lacto-ovo? vegan?)* diet is one that excludes meat, eggs, and dairy products.

b. Of the 20 amino acids needed to form the human body, how many are called "essential," meaning they *must* be included in the human diet?
   A. 2   B. 5   C. 10   D. All 20

c. A "complete protein" is defined as a protein that contains:
   A. All 20 amino acids
   B. All essential amino acids in proportions needed by the human body

d. List three groups of foods that, in combination, are likely to provide an adequate intake of all of the essential amino acids.

e. List two or more vitamins and two or more minerals that are common in animal products and are more difficult to obtain on a vegan diet.

STOP

■ **D12.** Match the description below with the four levels of protein organization in the box.

P. Primary structure	S. Secondary structure
Q. Quaternary structure	T. Tertiary structure

_____ a. Specific sequence of amino acids

_____ b. Twisted and folded arrangement (as in spirals or pleated sheets)

_____ c. Irregular three-dimensional shape

_____ d. Two or more tertiary patterns bonded together

■ **D13.** Describe changes in protein structure by completing this activity.

a. Replacement of one amino acid (valine) with another (glutamate) in the protein

portion of hemoglobin results in a condition known as _____.

b. Cooking an egg white results in destruction (or _____) of the protein albumin. Extreme increases in body temperature, for example, increases

caused by _____, can denature body proteins such as those in skin or blood plasma.

■ **D14.** Match the terms in the box with correct descriptions related to enzymes.

> A. Apoenzyme   Cof. Cofactor
> Coe. Coenzyme

_____ a. Protein portion of an enzyme   _____ c. Organic cofactor

_____ b. Nonprotein portion of an enzyme;
            may be organic molecule or metal ion
            (inorganic) such as Ca^{2+}, Mg^{2+}, or
            Zn^{1+}

D15. Write one or more sentences to explain three factors that affect enzyme function.

a. Specificity

b. Efficiency

c. Control

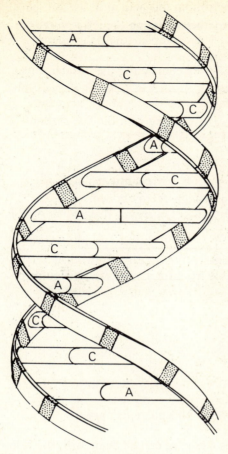

Figure LG 2.4 Structure of DNA. P, phosphate; S, sugar (deoxyribose); A, adenine; T, thymine; G, guanine; C, cytosine. Complete as directed in Checkpoint D16.

16-19

■ **D16.** Complete Figure LG 2.4 as follows:

a. Circle one nucleotide.

b. Label the complementary set of nucleotides on the right side of the figure to produce

the double-stranded structure of DNA. Bases are paired _____ to _____ and

_____ to _____.

c. DNA structure, like that of proteins, depends on sequence. Proteins differ from one

another by their specific sequence of _____, whereas DNA
molecules (for example, DNA determining blood type and DNA controlling eye

color) differ from one another by the sequence of _____. (Note
again, as with protein structure, the dire consequences of an omission or substitution
in this "four-letter alphabet": A C G T.)

d. How does a strand of RNA differ from DNA?

D17. *For extra review.* Cover Figure LG 2.4 and (without glancing at it) extend the length of the DNA polynucleotide by adding three nucleotides on the left side only. Check your drawing and correct where necessary. Then complete the right side.

■ **D18.** Describe the structure and significance of ATP by completing these statements.

a. ATP stands for adenosine triphosphate. Adenosine consists of a base that is a component of DNA, that is, _____, along with the five-carbon sugar named _____.

b. The "TP" of ATP stands for _____. The final *(one? two? three?)* phosphates are bonded to the molecule by high-energy bonds.

c. When the terminal phosphate is broken, a great deal of energy is released as ATP is split into _____ + _____ .

d. ATP, the body's primary energy-storing molecule, is constantly reformed by the reverse of this reaction as energy is made available from foods you eat. Write this reversible reaction.

D19. A compound similar to ATP, but with only one phosphate, is named

_____. A cyclic form of this molecule can be made with the help

of the enzyme _____. Cyclic AMP has important regulatory functions and will be discussed in Chapter 18.

ANSWERS TO SELECTED CHECKPOINTS

A2. (a) Liquid, gas. (b) Weight. (c) Energy; potential; kinetic. (d) Chemical; electrical; radiant.

A4. (a) O (65 percent), C (18.5 percent), H (9.5 percent), N (3.2 percent); simple. (b) 2.5.

A5. (b) N, 7. (c) I, 53. (d) Fe, 26. (e) Ca, 20. (f) O, 8. (g) K, 19. (h) Na, 11. (i) Cl, 17. (j) P, 15.

A6. (a) Nucleus, electrons. (b) Protons (p^+), neutrons (n^0). (c) Negative; equal to. (d) Daltons [atomic mass units = amu]; 1; 1/2000.

A7. (a) 19; 19, 19. (b) 39.0983 (rounded off to 39).

A8. (a) 20 (39 − 19); neutrons. *Note:* varied masses of isotopes account for the uneven mass number, such as 39.0983 for potassium. (b) 8. (c) Radiation; half-life. (d) Thyroid; thallium [Tl].

B1. (a) 2, 8. (b) Valence, valence; examples: H, Na, K. (c) Examples: F, Cl, Br, I. (d) Extra; donor; positive, cation. (e) −1, missing; accepts, anion. (f) Inert; helium (He), neon (Ne), argon (Ar), krypton (Kr), radon (Rn).

B2. (a, b: see Figure LG 2.2A)

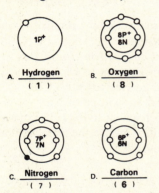

A. Hydrogen (1)   B. Oxygen (8)
C. Nitrogen (7)   D. Carbon (6)

Figure LG 2.2A Atomic structure of four common atoms in the human body.

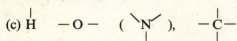

(c) H   — O —   (N),   — C —

(d) Atomic structure of ammonia (NH_3)

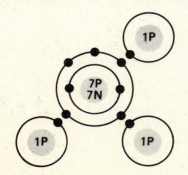

B3. (a) I. (b) H. (c) C. (d) H. (e) C. (f) C.

B4. H_2O, glucose, a molecule that contains two or more different kinds of atoms.

B5. (a) R. (b) S. (c) E. (d) D. (e) D.

C1. (a) Organic. (b) Organic. (c) Inorganic. (d) Organic.

C2. (a) Solvent. (b) Large, large. (c) The mucous lining of digestive or respiratory tract or the serous fluid of pleura, pericardium, or peritoneum.

C3. (a) B. (b) A. (c) S.

C5. (a) A. (b) C. (c) B. (d) Fewer. (e) 100, more.

D2. (a) Provide a readily available source of energy. (b) Converted to other substances; serve as food reserves.

D3. (a) 2:1; $(CH_2O)_n$. (b) $C_6H_{12}O_6$; glucose, fructose, galactose. (c) Deoxyribose; $C_5H_{10}O_5$. (d) $C_{12}H_{22}O_{11}$, removed; 1. (e) Polysaccharides; starch, glycogen, cellulose. (f) Hydrolysis; 2.

D4. (a) L. (b) C, L. (c) C. (d) L.

D5. (a) B. (b) B. (c) B. (d) B. (e) A. (f) A.

D6. (a) Car. (b) Pho. (c) Cho. (d) B. (e) E. (f) Pro.

D7. (a) A, D, E, K. (b) Symptoms of deficiencies of fat-soluble vitamins. (c) D. (d) K. (e) A. (f) E.

D8. (a) T. (b) Con. (c) S. (d) R. (e) I. (f) Cat.

D9. (a) Nitrogen. (b) Monosaccharides, amino acids. (c) Removed; dehydration. (d) Di; poly; protein.

D11. (a) Vegan. (b) 10. (c) B. (d) Legumes, grains, nuts, or seeds. (e) Vitamins: B_{12}, B_2 (riboflavin), and D; minerals: iron (Fe), zinc (Zn), and calcium (Ca).

D12. (a) P. (b) S. (c) T. (d) Q.

D13. (a) Sickle cell anemia. (b) Denaturation; burns or fever (hyperthermia).

D14. (a) A. (b) Cof. (c) Coe.

D16.

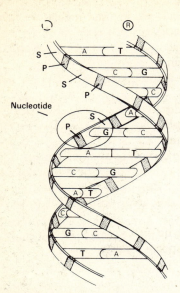

Figure LG 2.4A Structure of DNA. P, phosphate; S, sugar (deoxyribose); A, adenine; T, thymine; G, guanine; C, cytosine.

(a) Phosphate-sugar-base (such as P-S-G or P-S-T in Figure LG 2.4A). (b) A-T, C-G. (c) Amino acids; bases (A, T, C, G). (d) RNA is single-stranded, its sugar is ribose, and it contains the base uracil (U), not thymine (T).

D18. (a) Adenine, ribose. (b) Triphosphate; two. (c) ADP + phosphate (P). (d) ATP → ADP + P + energy.

![bar]

MASTERY TEST: Chapter 2

Questions 1–11: Circle the letter preceding the one best answer to each question.

1. All of the following statements are true *except:*
   A. A reaction in which two amino acids join to form a dipeptide is called a dehydration synthesis reaction.
   B. A reaction in which a disaccharide is digested to form two monosaccharides is known as a hydrolysis reaction.
   C. About 65 to 75 percent of living matter consists of organic matter.
   D. Strong acids ionize more easily than weak acids do.

2. Choose the one *true* statement.
   A. A pH of 7.5 is more acidic than a pH of 6.5.
   B. Anabolism consists of a variety of decomposition reactions.
   C. An atom such as chlorine (Cl), with seven electrons in its outer orbit, is likely to be an electron donor (rather than electron receptor) in ionic bond formation.
   D. Polyunsaturated fats are more likely to reduce cholesterol level than saturated fats are.

3. In the formation of the ionically bonded salt NaCl, Na^+ has:
   A. Gained an electron from Cl^-
   B. Lost an electron to Cl^-
   C. Shared an electron with Cl^-
   D. Formed an isotope of Na^+

4. Over 99 percent of living cells consist of just six elements. Choose the element that is NOT one of these six.
   A. Calcium   E. Nitrogen
   B. Hydrogen   F. Oxygen
   C. Carbon   G. Phosphorus
   D. Iodine

5. Which of the following describes the structure of a nucleotide?
   A. Base-base
   B. Phosphate-sugar-base
   C. Enzyme
   D. Dipeptide
   E. Adenine-ribose

6. Which of the following groups of chemicals includes only polysaccharides?
   A. Glycogen, starch
   B. Glycogen, glucose, galactose
   C. Glucose, fructose
   D. RNA, DNA
   E. Sucrose, polypeptide

7. $C_6H_{12}O_6$ is most likely the chemical formula for:
   A. Amino acid
   B. Fatty acid
   C. Hexose
   D. Polysaccharide
   E. Ribose

8. All of the following answers consist of correctly paired terms and descriptions related to enzymes *except*:
   A. Active site—a place on an enzyme which fits the substrate
   B. Substrate—molecule(s) upon which the enzyme acts
   C. Turnover number—number of substrate molecules converted to product per enzyme molecule per second
   D. Holoenzyme—another name for a cofactor
   E. -ase—ending of most enzyme names

9. Which of the following substances are used mainly for structure and regulatory functions and are not normally used as energy sources?
   A. Lipids   C. Carbohydrates
   B. Proteins

10. Which is a component of DNA, but not of RNA?
    A. Adenine   D. Ribose
    B. Phosphate   E. Thymine
    C. Guanine

11. All of the following answers consist of pairs of proteins and their correct functions *except*:
    A. Contractile—actin and myosin
    B. Immunological—collagen of connective tissue
    C. Catalytic—enzymes
    D. Transport—hemoglobin

Questions 12–20: Circle T (true) or F (false). If the statement is false, change the underlined word or phrase so that the statement is correct.

T F 12. Oxygen can form three bonds since it requires three electrons to fill its outer electron shell.

T F 13. Oxygen, water, NaCl, and glucose are inorganic compounds.

T F 14. There are about four different kinds of amino acids found in human proteins.

T F 15. The number of protons always equals the number of neutrons in an atom.

T F 16. K+ and Cl− are both cations.

T F 17. A strong acid has more of a tendency to contribute H+ to a solution than a weak acid has.

T F 18. Carbohydrates constitute about 18 to 25 percent of the total body weight.

T F 19. ATP contains more energy than ADP.

T F 20. Prostaglandins, cyclic AMP, steroids, and fats are all classified as lipids.

Questions 21–25: fill-ins. Identify each organic compound below. Write names of compounds on lines provided.

21. _____

22. _____

23. _____

24. _____

25. _____

Multiple Choice

1. C   7. C
2. D   8. D
3. B   9. B
4. D   10. E
5. B   11. B
6. A

True-False

12. F. Two bonds since it requires two electrons
13. F. Water and NaCl (Oxygen is not a compound; glucose is organic.)
14. F. 20
15. F. Electrons
16. F. K^+ is a cation.
17. T
18. F. 2–3%
19. T
20. F. Prostaglandins, steroids, and fats

Fill-ins

21. Amino acid
22. Monosaccharide or hexose or glucose
23. Polysaccharide or starch or glycogen
24. Adenosine triphosphate (ATP)
25. Fat or lipid or triglyceride (unsaturated)

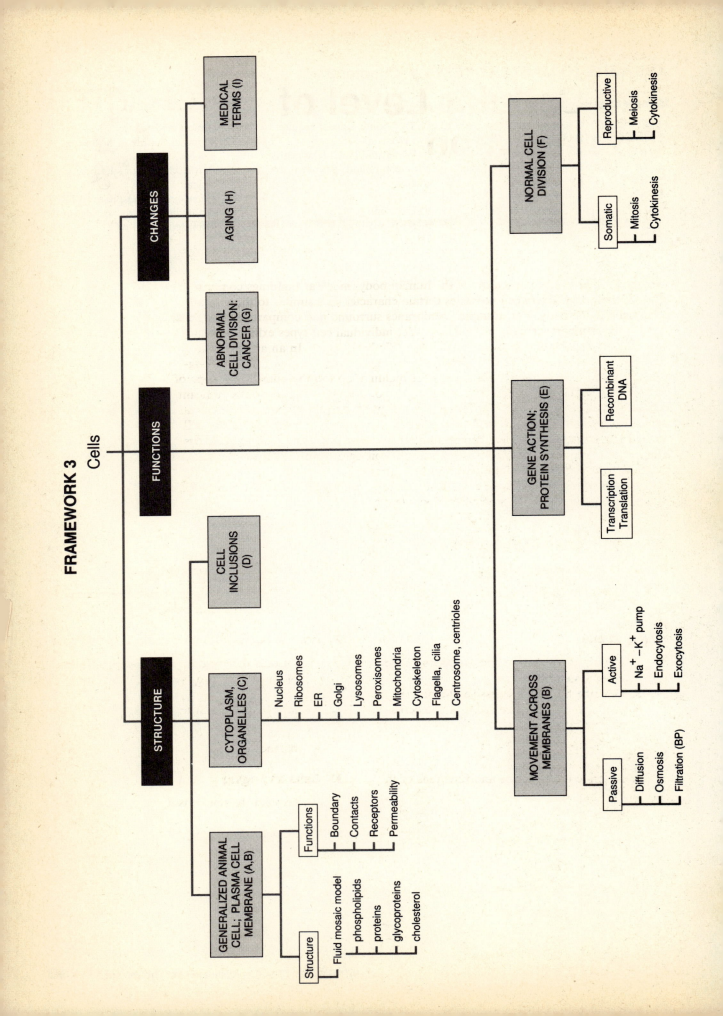

FRAMEWORK 3

Cells

STRUCTURE | FUNCTIONS | CHANGES

STRUCTURE

GENERALIZED ANIMAL CELL; PLASMA CELL MEMBRANE (A,B)

Structure
- Fluid mosaic model
- phospholipids
- proteins
- glycoproteins
- cholesterol

Functions
- Boundary
- Contacts
- Receptors
- Permeability

CYTOPLASM, ORGANELLES (C)
- Nucleus
- Ribosomes
- ER
- Golgi
- Lysosomes
- Peroxisomes
- Mitochondria
- Cytoskeleton
- Flagella, cilia
- Centrosome, centrioles

CELL INCLUSIONS (D)

CHANGES

ABNORMAL CELL DIVISION: CANCER (G)

AGING (H)

MEDICAL TERMS (I)

MOVEMENT ACROSS MEMBRANES (B)

Passive
- Diffusion
- Osmosis
- Filtration (BP)

Active
- Na$^+$–K$^+$ pump
- Endocytosis
- Exocytosis

GENE ACTION; PROTEIN SYNTHESIS (E)

Transcription Translation

Recombinant DNA

NORMAL CELL DIVISION (F)

Somatic
- Mitosis
- Cytokinesis

Reproductive
- Meiosis
- Cytokinesis

The Cellular Level of Organization

Cells are the basic structural units of the human body, much as buildings comprise the cities of the world. Every cell possesses certain characteristics similar to those in all other cells of the body. For example, membranes surround and compartmentalize cells, just as walls support virtually all buildings. Yet individual cell types exhibit variations that make them uniquely designed to meet specific body needs. In an architectural tour, the structural components of cathedral or mansion, lighthouse or fort, can be recognized as specific for activities of each type of building. So too the number and type of organelles differ in muscle or nerve, cartilage or bone cells. Cells carry out significant functions such as movement of substances into, out of, and throughout the cell; synthesis of the chemicals that cells need (like proteins, lipids, and carbohydrates); and cell division for growth, maintenance, repair, and formation of offspring. Just as buildings change with time and destructive forces, cells exhibit aging changes. In fact, sometimes their existence is haphazard and harmful, as in cancer.

As you begin your study of cells, first carefully study the Chapter 3 Framework, noting relationships among concepts and key terms.

TOPIC OUTLINE AND OBJECTIVES

A. Generalized animal cell

    1. Define a cell and list its principal parts.

B. Movement of materials across plasma membrane

    2. Explain the structure and functions of the plasma membrane.

    3. Describe the various passive and active processes by which materials move across plasma membranes.

C. Cytoplasm, organelles

    4. Describe the structure and functions of the following cellular structures: cytosol, nucleus, ribosomes, endoplasmic reticulum, Golgi complex, lysosomes, peroxisomes, mitochondria, cytoskeleton, flagella, cilia, and centrosome.

D. Cell inclusions, extracellular materials

    5. Define a cell inclusion and give several examples.

E. Gene action: protein synthesis

    6. Define a gene and explain the sequence of events involved in protein synthesis.

F. Normal cell division: somatic, reproductive

    7. Discuss the stages, events, and significance of somatic and reproductive cell division.

G. Abnormal cell division: cancer (CA)

    8. Describe cancer (CA) as a homeostatic imbalance of cells.

H. Cells and aging

    9. Explain the relationship of aging to cells.

I. Medical terminology

    10. Define medical terminology associated with cells.

WORDBYTES

Now study the following parts of words that may help you better understand terminology in this chapter.

Wordbyte	Meaning	Example	Wordbyte	Meaning	Example
cyto-	cell	*cyto*logist	-oma	tumor	carcin*oma*
-elle	small	organ*elle*	-philic	loving	hydro*philic*
hydro-	water	*hydro*static	-phobic	fearing	hydro*phobic*
hyper-	above	*hyper*tonic	stasis, -static	stand, stay	meta*stasis*
hypo-	under	*hypo*tonic	-tonic	tension or	*hyper*tonic
iso-	equal	*iso*tonic		pressure	
meta-	beyond	*meta*stasis			

CHECKPOINTS

A. Generalized animal cell (page 57)

■ **A1.** List the four principal parts of a generalized animal cell.

inclusions ✓cell membrane
cytosol ✓organelles

A2. Contrast: *cytoplasm* with *cytosol*.

cytoplasm contains cytosol which is a liquid
cytosol is just the suspenion solution

B. Plasma cell membrane, movement of materials across plasma membranes (pages 57–68)

■ **B1.** Refer to Figure LG 3.1 to complete the following exercise about the fluid mosaic model of the plasma membrane. First color each part of the membrane listed below (a–e) with a different color; then use the same colors to fill in the color code ovals next to the letters below. Next label these parts on the figure. Now use the lines provided next to the name of each membrane part to write one or more functions for each part.

○ a. Phospholipid polar "head" *water soluble + facing the membrane surface, compatable with water environment*

○ b. Phospholipid nonpolar "tail" *forms water insoluble barrier between outside + inside cell nonpolar molecules like lipids O_2 + CO_2 pas*

○ c. Integral protein *Channel for passage of water or ions, carriers pas*

○ d. Glycoprotein *Receptors*

○ e. Cholesterol *strengthen the membrane but decrease its flexibility and permeability*

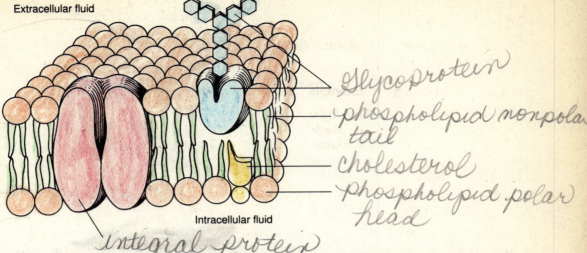

Extracellular fluid

Glycoprotein

phospholipid nonpolar tail

cholesterol

Phospholipid polar head

Intracellular fluid

integral protein

Figure LG 3.1 Plasma membrane, according to the fluid mosaic model. Color as directed in Checkpoint B1.

■ **B2.** *For extra review.* Complete this exercise about the plasma membrane.

a. Plasma membranes are made up of about __50__ percent proteins and __50__ percent lipids by weight. About 25 percent of the lipids are *(cholesterol? glycolipids? phospholipids?).*

b. About 75 percent of the lipids are *(cholesterol? glycolipids? phospholipids?).* Since phospholipids have polar and nonpolar parts, they are said to be

__amphipathic__. The polar part of a phospholipid molecule faces toward the *(interior? surface?)* of the membrane.

c. Glycolipids appear only in the layer that faces *(extracellular? intracellular?)* fluid. This location is logical since these lipids may be important in

__cell to cell recognition & communication and cell adhesion__

d. Which type of membrane molecule is larger? *(Lipid? Protein?).* In different types of membranes, there is more variety in *(lipids? proteins?).* For example, some integral

proteins serve as receptors for ligands such as __hormones__,

__neurotransmitters__, or __nutrients__. Some *(integral? peripheral?)* proteins serve as cytoskeletal anchors. Other membrane proteins known as

__enzymes__ form catalysts for chemical reactions.

integral proteins - extend across the phospholipid bilayer among the fatty acid tails

ligand - a molecule that specifically binds to a receptor (hormones, nutrients, neurotransmitters) by forces other than covalent bonds.

peripheral proteins - do not extend across the phospholipid bilayer

glycoprotein - integral protein

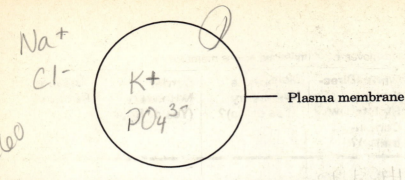

Na+
Cl-

K+
PO₄³⁻

Plasma membrane

page 60

Figure LG 3.2   Outline of a cell. Complete as directed in Checkpoint B3.

B3. Complete this activity about electrochemical gradient.

a. Refer to Figure LG 3.2. Write inside the plasma membrane the symbols for the major ions or electrolytes concentrated in intracellular fluid (ICF). Write outside the plasma membrane the symbols for major ions or electrolytes concentrated in extracellular fluid (ECF). Use these symbols:

$$K^+ \qquad Na^+ \qquad Cl^- \qquad PO_4^{3-}$$

b. This difference between chemicals inside and outside the plasma membrane is known as a(n) *(chemical? electrical?)* gradient. A difference in charge across the membrane accounts for a(n) *(chemical? electrical?)* gradient.

c. The inside of the plasma membrane usually has a more *(positive? negative?)* charge.

Therefore the membrane maintains a voltage called the ~~membrane potential~~,

which is measured in ~~mV (milivolts)~~. This voltage is a measure of *(kinetic? potential?)* energy available to accomplish cell work, such as nerve transmission.

B4. Complete this activity about selective permeability of membranes.

a. Plasma membranes are more permeable to *(nonpolar chemicals such as O_2 and CO_2? most polar molecules and ions?)* since these molecules dissolve readily in the *(protein? phospholipid?)* portion of the membrane.

b. Charged chemicals (such as Na^+, K^+, or Ca^{2+}) and most polar chemicals pass through the *(protein? phospholipid?)* portions of the membrane. Two such proteins

are ~~channels~~ and ~~transporters~~.

c. Most large molecules *(can? cannot?)* pass across the plasma membrane.

B5. Explain why substances must move across your plasma membranes in order for you to survive and maintain homeostasis. (What types of chemicals must be allowed into cells or kept out of cells?)

Certain substances most move into the cell to support needed biochemical reactions while waste materials and harmful substances must be moved out.

X ■ **B6.** Complete parts a–d of Table LG 3.1.

Table LG 3.1 Summary of mechanisms for movement of materials across membranes

Name of Process	What Moves (Particles or Water?)	Where (Direction, e.g., High to Low Concentration)?	Membrane Necessary (Yes or No)?	Carrier Necessary (Yes or No)?	Active or Passive?
a. Simple diffusion	PARTICLES	HIGH to LOW	no	no	passive
b. OSMOSIS	Water	From high to low concentration of water across semipermeable membrane	yes	no	passive
c. Filtration *kidney*	particles or water	From high pressure area to low pressure area	yes	no	passive
d. Facilitated diffusion	particles like glucose	High to low	yes	yes	passive
e. Primary active transport	particles solutes glucose + ions	From low to high concentration of solute	yes	yes pump proteins that split ATP	active
f. Phago-cytosis	particles large	from out through plasma membrane to inside	Yes	no	active
g. pinocytosis	Water	From outside cell to inside cell	yes	no	active
h. Exocytosis	large molecules + particles	from inside to outside	yes	no	Active

■ **B7.** Select from the following list of terms to identify passive transport processes described below.

Fac. Facilitated diffusion	O. Osmosis
Fil. Filtration	SD. Simple diffusion

SD a. Net movement of any substance (such as cocoa powder in hot milk) from region of higher concentration to region of lower concentration; membrane not required.

FAC b. Same as (a) except movement across a semipermeable membrane with help of a carrier; ATP not required.

O c. Net movement of water from region of high water concentration (such as 2 percent NaCl) to region of low water concentration (such as 10 percent NaCl) across semipermeable membrane; important in maintenance of normal cell size and shape.

FIL d. Movement of molecules from high pressure zone to low pressure zone, for example, in response to force of blood pressure.

■ **B8.** Complete the following exercise about osmosis in blood.

a. Human red blood cells (RBCs) contain intracellular fluid that is osmotically similar

to _*B*_ NaCl.
A. 2.0%   C. 0% (pure water)
B. 0.90%

b. A solution that is hypertonic to RBCs contains *(more? fewer?)* solute particles and *(more? fewer?)* water molecules than blood.

c. Which of these solutions is hypertonic to RBCs?
A. 2.0% NaCl   C. pure water
B. 0.90% NaCl

d. If RBCs are surrounded by hypertonic solution, water will tend to move *(into? out of?)* them, so they will *(crenate? hemolyze?)*.

e. A solution that is _*ISO*_-tonic to RBCs will maintain the shape

and size of the RBC. An example of such a solution is _*saline*_.

f. Which solution will cause RBCs to hemolyze?
A. 2.0% NaCl   C. pure water
B. 0.90% NaCl

g. Which solution has the highest osmotic pressure?
A. 2.0% NaCl   C. pure water
B. 0.90% NaCl

■ **B9.** A large part of the human diet is starch. When starch is digested, it is broken down to glucose. Answer these questions related to movement of glucose into body cells.

a. Explain why glucose cannot readily pass through plasma membranes by simple diffusion.

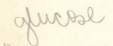

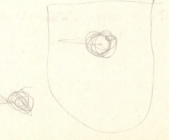

b. By what process does glucose cross plasma membranes as it is absorbed into cells lining the digestive tract?

symports that us Na+

c. List three factors that would speed up this rate of movement of glucose.

1) Difference in concentration of the substance on the 2 sides of the membrane 2) the number of transporters available to transport the substance 3) how quickly the transporter + substance combine

d. Name a hormone that increases the rate of facilitated diffusion of glucose from blood into body cells.

insulin

■ **B10.** Describe active transport in the following exercise.

a. List three categories of substances that must enter or leave cells by active processes.

amino acids   large substances
monosaccharides   substance with wrong charge
ions   substance moving against concentration gradient

b. There are two categories of active processes: ___*active*___ transport

and ___*bulk*___ transport. Active transport may be either primary or secondary. The sodium (or Na+/K+) pump is the most abundant form of (primary? secondary?) transport in cells. The effect of it is to concentrate sodium (Na+) (inside? outside?) of cells. (See Figure LG 3.2.) The Na+/K+ pump requires energy from ATP

directly by splitting ATP with an enzyme known as ___*Na+/K+ ATP*___-ase.

In fact, a typical cell may use up to ___*40*___ percent of its ATP for active transport.

c. Two forms of secondary active transport are known. Symport (cotransport) involves movement of a substance (such as dietary glucose) in (the same? opposite?) direction as sodium (Na+). Antiport (counter-transport) involves movement of a substance (such as Ca²⁺ or H⁺) in (the same? opposite?) direction as sodium (Na+).

d. Bulk transport includes ___*endo*___ -cytosis and

___*exo*___ -cytosis.

■ **B11.** Complete parts e–h of Table LG 3.1.   *liposomes release their drug slowly*

B12. State one advantage of use of liposomes with antibiotic or cancer drug therapy.

since it's a fat it helps the drug enter the cell quicker

C. Cytosol, organelles   (pages 68–77)

■ **C1.** The cell is compartmentalized by the presence of organelles. Of what advantage is this?

many chemical activities can go on at same time with out interference between them

■ **C2.** Color and then label all of the parts of Figure LG 3.3 (on next page) listed with color code ovals.

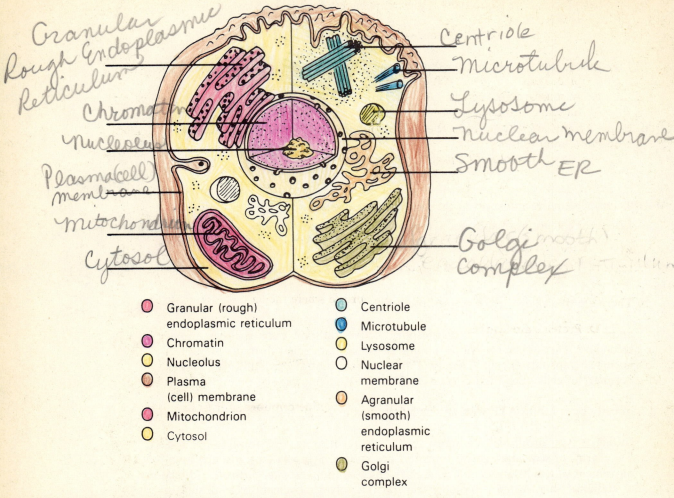

Granular
Rough Endoplasmic
Reticulum

Chromatin
Nucleolus
Plasma(cell)
membrane
mitochondrion
Cytosol

Centriole
Microtubule
Lysosome
nuclear membrane
Smooth ER
Golgi
complex

- ⬤ Granular (rough)
  endoplasmic reticulum
- ⬤ Chromatin
- ⬤ Nucleolus
- ⬤ Plasma
  (cell) membrane
- ⬤ Mitochondrion
- ⬤ Cytosol

- ⬤ Centriole
- ⬤ Microtubule
- ⬤ Lysosome
- ⬤ Nuclear
  membrane
- ⬤ Agranular
  (smooth)
  endoplasmic
  reticulum
- ⬤ Golgi
  complex

Figure LG 3.3 Generalized animal cell. Color and label as directed in Checkpoint C2.

■ **C3.** Describe parts of the nucleus by completing this exercise.

a. The nuclear membrane has pores that are *(larger? smaller?)* than channels in the plasma membrane. Of what significance is this? *Permit passage of large molecules such as RNA + proteins*

b. Ribosomes are assembled at sites known as *nucleoli* where protein, DNA, and RNA are stored.

c. Chromosomes consist of beadlike subunits known as *nucleosomes* which are made of double-stranded *DNA* wrapped around core proteins called *histones*. Nucleosome "beads" are held together, necklace-fashion, by *(histones? linker DNA?)*. Nucleosomes are then held into chromatin fibers and folded into *loops*.

d. Before cell division, DNA duplicates and forms compact, coiled conformations known as *(chromatin? chromatids?)* that may be more easily moved around the cell during the division process.

Read in book

Read in book

■ **C4.** In the blanks below, write *rough ER* or *smooth ER,* as appropriate.

_Sm___ a. This form of ER is also known as agranular ER.

_Sm___ b. Synthesis of fats, such as fatty acids, phospholipids, and steroids, takes place here.

_Sm___ c. In muscle cells, calcium ions (Ca^{2+}) released from this form of ER trigger muscle contraction.

_Sm___ d. Enzymes within these membranes detoxify alcohol and other toxic substances within liver cells.

_Rough__ e. Protein is synthesized and stored in ribosomes located here.

■ **C5.** Describe the sequence of events in synthesis, secretion, and discharge of protein from the cell by numbering these five steps in correct order.

___3___ A. Proteins accumulate and are modified, sorted, and packaged in cisternae of Golgi complex.

___2___ B. Proteins pass within a transport vesicle to *cis* cisternae of Golgi complex.

___5___ C. Secretory granules move toward cell surface where they are discharged.

___1___ D. Proteins are synthesized at ribosomes on rough ER.

___4___ E. Vesicles containing protein pinch off of Golgi complex *trans* cisternae to form secretory vesicles.

■ **C6.** Match the organelles listed in the box with the correct descriptions of their functions. Use each answer once.

Cen. Centrioles	I. Intermediate filaments	Mit. Mitochondria
Cil. Cilia	L. Lysosomes	N. Nucleus
ER. Endoplasmic reticulum	Mf. Microfilaments	P. Peroxisomes
F. Flagella	Mt. Microtubules	R. Ribosomes
G. Golgi complex		

_N___ a. Site of direction of cellular activities by means of DNA located here

_R___ b. Cytoplasmic sites of protein synthesis; may occur attached to ER or scattered freely in cytoplasm

_ER__ c. System of membranous channels providing pathways for transport within the cell and surface areas for chemical reactions; may be granular or agranular

_G___ d. Stacks of cisternae with vesicular ends; involved in packaging and secretion of proteins and lipids and synthesis of glycoproteins

_L___ e. Called "suicide packets" since they may release enzymes which lead to autolysis of the cell

_P___ f. Similar to lysosomes, but smaller; contain the enzyme catalase that uses hydrogen peroxide (H_2O_2) to detoxify harmful chemicals in cells of liver and kidney

_Mit__ g. Cristae-containing structures, called "powerhouses of the cell" since ATP production occurs here

_Mf___ h. Form part of cytoskeleton; involved with cell movement and contraction

_Mt___ i. Part of cytoskeleton, proves support and gives shape to cell; form flagellae, cilia, centrioles, and spindle fibers; made of protein tubulin

_Cen__ j. Help organize mitotic spindle used in cell division

_F___ k. Long, hairlike structures that help move entire cell, for example, sperm cell

_Cil__ l. Short, hairlike structures that move particles over cell surfaces

_I___ m. Contribute to shape of cell; hold organelles (such as nucleus) in place

■ **C7.** *For extra review.* Considering the functions of organelles listed in the above exercise, choose the answers (organelles) that fit the following descriptions. Answers in that list may be used more than once.

_____ *Lys* a. Abundant in white blood cells, which use these organelles to help in phagocytosis of bacteria.

_____ *Mit* b. Present in large numbers in muscle and liver cells, which require much energy.

_____ *Cil* c. Located on surface of cells of the respiratory tract; help to move mucus.

_____ d. Absence of a single enzyme from these organelles causes Tay-Sachs, an inherited disease affecting nerve cells.

_____ e. An acid pH must be present inside these organelles for enzymes to function effectively.

_____ *Mit* f. Sites of cellular respiration requiring oxygen (O_2); contain some DNA used for self-replication.

_____ *ER* g. Function much like railroad tracks to move chemicals and organelles through cytosol; assist in movement of pseudopods of phagocytic cells such as white blood cells.

flagella *cilia* h. Described as having a "nine and two" structure based on arrangement of nine pairs of microtubules around a core of two single microtubules (two answers).

ER *golgi* i. Cystic fibrosis is an inherited disease that involves accumulation of mucus in lungs and pancreatic ducts because of failure of a pump protein synthesized by cells to move out of one of these two types of organelles (two answers).

L + P j. Might be described as a "clean-up service" since enzymes released from these organelles help to remove debris within and around injured cells (two answers). *peroxomes*

D. Cell inclusions (page 77)

D1. Contrast *organelle* with *inclusion.*

■ **D2.** Contrast these three types of cell inclusions in this exercise.

G. Glycogen   L. Lipids   M. Melanin

_____ *L* a. Stored in adipocytes; may be broken down to form ATP

_____ *G* b. A complex carbohydrate stored in cells of the liver, skeletal muscle, uterus, and vagina

_____ *M* c. A pigment that gives brown or black color to skin, hair, or eyes

E. Gene action: protein synthesis (pages 77–81)

E1. State the significance of protein synthesis in the cell.

The proteins determine the physical & chemical characteristics of cells + therefore of organisms

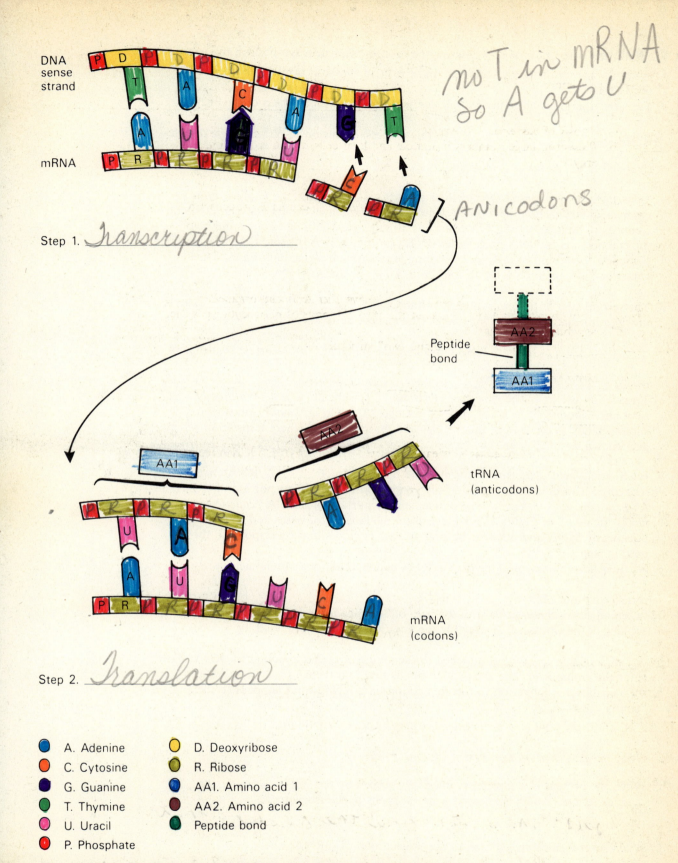

DNA sense strand

mRNA

not in mRNA so A gets U

ANICODONS

Step 1. Transcription

Peptide bond

AA2

AA1

AA2

AA1

tRNA (anticodons)

mRNA (codons)

Step 2. Translation

- ● A. Adenine
- ● C. Cytosine
- ● G. Guanine
- ● T. Thymine
- ● U. Uracil
- ● P. Phosphate
- ● D. Deoxyribose
- ● R. Ribose
- ● AA1. Amino acid 1
- ● AA2. Amino acid 2
- ● Peptide bond

Figure LG 3.4 Protein synthesis. Label and color as directed in Checkpoint E2.

■ **E2.** Refer to Figure LG 3.4 and describe the process of protein synthesis in this exercise.

a. Write the names of the two major steps of protein synthesis on the two lines in Figure LG 3.4.

b. Transcription takes place in the *(nucleus? cytoplasm?)* with the help of the enzyme

   called __RNA polymerase__. In this process, a portion of one side of a double-

   stranded DNA serves as a mold or __template__ and is called the

   (sense? antisense?) strand. Nucleotides of __mRNA__ line up in a
   complementary fashion next to DNA nucleotides (much as occurs between DNA nucleotides in replication of DNA prior to mitosis). Select different colors and color
   the components of DNA nucleotides (P, D, and bases T, A, C, and G).

c. Complete the strand of mRNA by writing letters of complementary bases. Then
   color the parts of mRNA (P, R, and bases U, A, C, and G).

d. The second step of protein synthesis is known as __translation__ since it
   involves translation of one "language" (the nucleotide base code of

   __messenger__ RNA) into another "language" (the correct sequence of

   the 20 __amino acids__ to form a specific protein).

e. Translation occurs in the *(nucleus? cytoplasm?)*, specifically at a *(chromosome?
   lysosome? ribosome?)* to which mRNA attaches. The *(large? small?)* ribosomal
   subunit binds to one end of the mRNA and finds the point where translation will begin, known as *(ATP? the start codon?)*.

f. Each amino acid is transported to this site by a particular __t__ *(transfer)* RNA characterized by a specific three-base unit or *(codon? anticodon?)*. This portion of tRNA,
   for example, U A C, binds to a complementary portion of mRNA *(codon? anticodon?)*. In step 2 of Figure LG 3.4, complete the labeling of bases and color all components of tRNA and mRNA. Be sure you select colors that correspond to those
   used in step 1.

g. As the ribosome moves along each mRNA codon, additional amino acids are

   transferred into place by __t__ RNA. __peptide__ bonds form between adjacent amino acids with the help of enzymes from the *(large? small?)*
   subunit of the ribosome. Energy for the bond formation is provided by

   __ATP__ attached to the amino acid that is bound to tRNA. Now
   color the amino acids and the peptide bonds that join them.

h. As the ribosome moves on to the next codon on mRNA, what happens to the
   "empty" tRNA? __ejected able to pick up
   another molecule of same amino acid__

i. When the protein is complete, synthesis is stopped by the __stop__
   codon. As the protein is released, what happens to the ribosome?
   __splits into its component subunits__

j. When several ribosomes attach to the same strand of mRNA, a

   __polyribosome__ is formed, permitting repeated synthesis of the identical
   protein.

   __DNA Transcription → RNA Translation → Protein__

■ **E3.** Complete this additional activity about protein synthesis.

a. Sections of DNA that do code for parts of proteins are known as *(exons? introns?)*, whereas sections of DNA that do not code for parts of proteins are called

 introns . The mRNA formed initially by transcription contains *(only exons? only introns? both exons and introns?)*.

b. RNA splicing is the process of removal of *(exons? introns?)*, making up over 75 percent of the initial mRNA. This process occurs *(before? after?)* mRNA moves into cytoplasm. In fact, variations of RNA splicing result in changes in function of a single gene to form different proteins as cells undergo the process of

 diferentiation .

c. Define the term *gene*. a group of nucleotides on a DNA molecule that serves as the master mold for manufacturing a specific protein (or small family of similar proteins)

d. The exon portion of a typical gene contains about 300 to 3000 nucleotides

 which code for a protein containing 100 to 1000 amino acids.

E4. List six or more therapeutic substances that have been produced by bacteria containing recombinant DNA.

① human growth hormone (HGH) ⑤ erythropoietin
② insulin ⑥ monoclonal antibodies
③ interferon (IFN) ④ factor VIII

F. Normal cell division: somatic and reproductive   (pages 81–89)

F1. Once you have reached adult size, do your cells continue to divide? *(Yes? No?)* Of what significance is this fact?

 repair old or damaged tissue

■ **F2.** Describe aspects of cell division by completing this exercise.

a. Division of any cell in the body consists of two processes: division of the

 nucleus and division of the cytoplasm

b. In the formation of mature sperm and egg cells, nuclear division is known as

 meosis . In the formation of all other body cells, that is, *somatic*

 cells, nuclear division is called mitosis .

c. Cytoplasmic division in both somatic and reproductive division is known as

 cytokinesis . In *(meiosis? mitosis?)* the two daughter cells have the same hereditary material and genetic potential as the parent cell.

✳ ■ **F3.** Describe interphase in this activity.

  a. Interphase occurs *(during cell division? between cell divisions?)*.

  b. Chromosomes are replicated during phase *(S? G_1? G_2?)* of interphase. In fact "S"

stands for _____ of new DNA. During this time, the DNA helix partially uncoils and the double DNA strand separates at points where

_____ connect so that new complementary nucleotides can at-
tach here. (See Figure LG 2.4, page 33.) In this manner the DNA of a single chromo-

some is doubled to form two identical _____.

  c. Two periods of growth occur during interphase, preparing the cell for cell division.
The *(G_1? G_2?)* phase immediately precedes the S phase, whereas the *(G_1? G_2?)* phase
follows the S portion of interphase. Two types of chemicals formed during G phases

are _____ and _____; organelles formed

then include _____. The "G" of these two phases also refers to

"gaps" since these periods are gaps in _____.

✳ ■ **F4.** Carefully study the phases of the cell cycle shown in Figure 3.24, page 84, in your
text. Then check your understanding of the process by identifying major events in each
phase.

I P M A T

4	A. Anaphase 2 P. Prophase
1	I. Interphase 5 T. Telophase
3	M. Metaphase

_P___ a. This phase immediately follows inter-
       phase.
_P___ b. Chromosomes condense into distinct
       chromosomes (chromatids) each held
       together by a centromere complexed
       with a kinetochore.
_P___ c. Nucleoli and nuclear envelope break
       up; the two centrosomes move to op-
       posite poles of the cell, and formation
       of the mitotic spindle begins.

_m___ d. Chromatids line up with their cen-
       tromeres along the equatorial plate.
_A___ e. Centromeres divide; chromatids (now
       called daughter chromosomes) are
       dragged by microtubules to opposite
       poles of the cell. Cytokinesis begins.
_T___ f. Events of this phase are essentially a
       reversal of prophase; cytokinesis is
       completed.
_P___ g. This phase of cell division is longest.

✳ ■ **F5.** Describe the time intervals involved in cell division by circling the correct answers
below.

  a. Typically, which phase is shortest?
    G_1 S   (G₂)   Mitosis and cytokinesis
  b. Which phase is most variable in duration (from virtually nonexistent to years)?
    G_1 S G_2 Mitosis and cytokinesis

■ **F6.** Do the following exercise about reproductive cell division.

  a. Meiosis is a special type of nuclear division that occurs only in

_____ and is one process in the production of sex cells or

_____. Fusion of gametes during the process of

_____ results in formation of a _____.

b. Cells that begin the process of meiosis are ovarian or testicular cells that have been

undergoing mitosis up to this point. Each cell initially contains _____ chromosomes, which is the *(diploid or 2n? haploid or n?)* number for humans. These chromo-

somes consist of _____ homologous pairs.

c. One pair is known as the *(autosomes? sex chromosomes?)*. In females, this pair consists of *(two X chromosomes? one X and one Y chromosome?)*. The remaining 22

pairs are known as _____.

d. By the time that cells complete meiosis, they will contain *(46? 23?)* chromosomes,
the *(2n? n?)* number.

e. Meiosis differs from mitosis in that meiosis involves *(one? two? three?)* divisions
with replication of DNA only *(once? twice?)*. The first division is known as the

_____ division, whereas the second division is the

_____ division.

f. Replication of DNA occurs *(before? after?)* reduction division, resulting in doubling
of chromosome number in humans from 46 chromosomes to

_____ "chromatids." (Chromatid is a name given to a chromo-
some while the chromosome is attached to its replicated counterpart, that is, in the
doubled state.) Doubled or paired chromatids are held together by means of a *(centromere? centrosome?)*.

g. A unique event that occurs in Prophase I is the lining up of chromosomes in

homologous pairs, a process called _____. Note that this pro-
cess *(does? does not?)* occur during mitosis. The four chromatids of the homologous

pair are known as a _____. The proximity of chromatids per-

mits the exchange or _____ of DNA between maternal and pa-
ternal chromatids. What is the advantage of crossing-over to sexual reproduction?

h. Another factor that increases possibilities for variety among sperm or eggs you can
produce is the random (or chance) assortment of chromosomes as they line up on
either side of the equator in Metaphase *(I? II?)*.

i. Division I reduces the chromosome number to _____ chroma-
tids. Division II results in the separation of doubled chromatids by splitting of

_____ so that resulting cells contain only _____ chromosomes.

j. Meiosis is only one part of the process of forming sperm or eggs. Each original diploid cell in a testis will lead to production of *(1? 2? 4?)* mature sperm, while each
diploid ovary cell produces *(1? 2? 4?)* mature eggs (ova) plus three small cells

known as _____.

F7. Write the full names and a brief description of the roles of the following two chemicals that regulate cell division.

a. MPF

b. cdc2 proteins

G. Abnormal cell division: cancer (CA)   (pages 89–91)

■ **G1.** Complete this exercise about cancer.

a. The term _____ refers to the study of cancer.

_____ and _____ nurses are health care providers who specialize in work with clients who have cancer.

b. A term which means "tumor or abnormal growth" is _____. A cancerous growth is a *(benign? malignant?)* neoplasm, whereas a noncancerous tumor is a *(benign? malignant?)* neoplasm.

c. Which type of growth is more likely to spread and to possibly cause death? *(Benign? Malignant?)* A term which means the "spread" of cancer cells is

_____.

d. In the process of metastasis, cancer cells compete with normal cells for

_____. They also migrate into space previously occupied by

other cells since they do not follow rules of _____. By what routes do cancers reach distant parts of the body?

e. What may cause the pain associated with cancer?

pressure from growing mass

■ **G2.** Match terms in the box with the definitions below.

C. Carcinoma	O. Osteogenic sarcoma
L. Lymphoma	S. Sarcoma

___C___ a. General term for malignant tumor arising from any connective tissue

___O___ b. Malignant tumor derived from bone (a connective tissue)

___S___ c. General term for malignant tumor of epithelial tissue

___L___ d. A malignancy of lymph tissue

■ **G3.** Identify the types of viruses associated with the following types of cancer:

> EBV. Epstein-Barr virus   HPV. Human papilloma virus
> HBV. Hepatitis B virus   HSV. Herpes simplex virus
> HIV. Human immunodeficiency virus   HTLV-1. Human T-cell leukemia-lymphoma virus-1

HTLV-1 _____ a. Cancers of white blood cells (leukemia and lymphoma)

_____ b. Kaposi's sarcoma (KS), a cancer of connective tissues and blood vessels occurring most commonly in AIDS patients, related to impairment of the immune system

_____ c. Hodgkin's disease (a cancer of the lymphatic system)

_____ d. Liver cancer

_____ e. Cervical cancer (two answers)

G4. *A clinical challenge.* Summarize how each of the following aspects of a healthy lifestyle may help to prevent cancer.

a. Healthy diet

b. Elimination or limitation of alcohol

c. Not smoking

d. Safe sex

e. Exercise and weight reduction

f. Stress reduction

g. Reduction of environmental carcinogens

G5. Relate the following terms to cancer:
a. Carcinogens

b. Proto-oncogenes

c. Antioncogene

d. Growth factors

e. Mutation

f. P-glycoprotein

G6. *A clinical challenge.* Are all cancer cells alike within a single tumor? *(Yes? No?)* Of what significance is this in planning cancer treatment?

H. Cells and aging   (pages 91–92)

H1. Describe changes that normally occur in each of the following during the aging process.
a. Number of cells in the body

b. Collagen fibers

c. Elastin

H2. In a sentence or two, summarize the main points of each of these theories of aging.

a. Free radical

b. Immune system

I. Medical terminology (page 82)

■ **I1.** Match the terms in the box describing alterations in cells or tissues with the descriptions below.

D. Dysplasia	HT. Hypertrophy
HP. Hyperplasia	M. Metaplasia

_____ a. Increase in size of tissue or organ by increase in size (not number) of cells, such as growth of your biceps muscle with exercise

_____ b. Increase in size of tissue or organ due to increase in number of cells, such as a callus on your hand or breast tissue during pregnancy

_____ c. Change of one cell type to another normal cell type, such as change from single row of tall (columnar) cells lining airways to multilayers of cells as response to constant irritation of smoking

_____ d. Abnormal change in cells in a tissue as due to irritation or inflammation. May revert to normal if irritant removed, or may progress to neoplasia

■ **I2.** Match the terms in the box with the definitions below.

A. Atrophy	N. Necrosis
B. Biopsy	

_____ a. Death of a group of cells

_____ b. Decrease in size of cells with decrease in size of tissue or organ

_____ c. Removal and examination of tissue from the living body for diagnosis

A1. Plasma (cell) membrane, cytosol, organelles, inclusions.

B1.

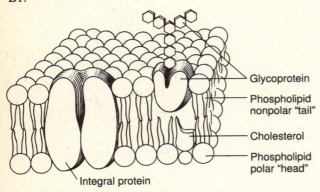

Figure LG 3.1A Plasma membrane.

See Figure LG 3.1A. (a) Water-soluble and facing the membrane surface, so compatible with watery environment surrounding the membrane. (b) Forms a water-insoluble barrier between cell contents and extracellular fluid; nonpolar molecules such as lipids, oxygen, and carbon dioxide pass readily between "tails." (c) Channels for passage of water or ions, carriers. (d) Receptors. (e) Cholesterol molecules strengthen the membrane, but decrease its flexibility and permeability.

B2. (a) 50, 50; cholesterol. (b) Phospholipids; amphipathic; surface. (c) Extracellular; cell adhesion, recognition, and/or communication. (d) Protein; proteins; hormones, neurotransmitters, or nutrients; peripheral; enzymes.

B3. (a)

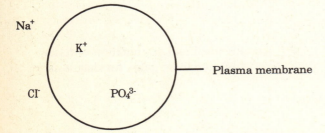

Figure LG 3.2A Major ions in intracellular fluid (ICF) and extracellular fluid (ECF).

(b) Chemical; electrical. (c) Negative; membrane potential, millivolts (mV); potential.

B4. (a) Nonpolar chemicals such as O_2 and CO_2; phospholipid. (b) Protein; channels (pores), transporters (carriers). (c) Cannot.

B6.

Name of Process	What Moves (Particles or Water?)	Where (Direction, e.g., High to Low Concentration)?	Membrane Necessary (Yes or No)?	Carrier Necessary (Yes or No)?	Active or Passive?
a. Simple diffusion	Particles or water	From high to low concentration of solute	No	No	Passive
b. Osmosis	Water	From high to low concentration of water across semipermeable membrane	Yes	No	Passive
c. Filtration	Particles or water	From high pressure area to low pressure area	Yes (or other barrier such as filter paper)	No	Passive
d. Facilitated diffusion	Particle, such as glucose	From high to low concentration of solute	Yes	Yes	Passive

B7. (a) SD. (b) Fac. (c) 0. (d) Fil.

B8. (a) B. (b) More, fewer. (c) A. (d) Out of, crenate. (e) Iso-, 0.90% NaCl (normal saline) or 5.0% glucose. (f) C. (g) A.

B10. (a) Substances that are large (such as proteins and even whole bacteria or red blood cells), have the wrong charge, or must move into an area already highly concentrated with the same substance. (b) Active, bulk; primary; outside; ATP; 40. (c) The same; opposite. (d) Endo-, exo-.

B11.

Name of Process	What Moves (Particles or Water?)	Where (Direction, e.g., High to Low Concentration)?	Membrane Necessary (Yes or No)?	Carrier Necessary (Yes or No)?	Active or Passive?
e. Primary active transport	Solutes, such as glucose or ions (Na^+, K^+)	From low to high concentration of solute	Yes	Yes	Active
f. Phagocytosis	Large particles	From outside of cell to inside of cell	Yes	No	Active
g. Pinocytosis	Water	From outside cell to inside cell	Yes	No	Active
h. Exocytosis	Large molecules or particles	From inside of cell to outside of cell	Yes	No	Active

C2.

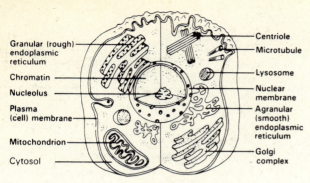

Figure LG 3.3A Generalized animal cell.

C3. (a) Larger; permit passage of large molecules such as RNA and proteins. (b) Nucleoli. (c) Nucleosomes, DNA, histones; linker DNA; loops. (d) Chromatids.
C4. (a–d) Smooth ER. (e) Rough ER.
C5. 3, 2, 5, 1, 4 (or D, B, A, E, C).
C6. (a) N. (b) R. (c) ER. (d) G. (e) L. (f) P. (g) Mit. (h) Mf. (i) Mt. (j) Cen. (k) F. (l) Cil. (m) I.
C7. (a) L. (b) Mit. (c) Cil. (d) L. (e) L. (f) Mit. (g) Mt. (h) Cil, F. (i) ER, G. (j) L, P.
D2. (a) L. (b) G. (c) M.
E1. Proteins serve a number of significant functions. Refer to Chapter 2, page 30 of the Learning Guide, Activity D8.

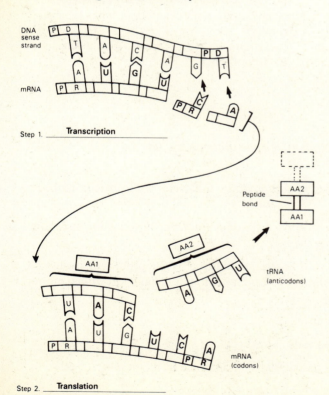

DNA sense strand

mRNA

Step 1. ___Transcription___

Peptide bond

AA2

AA1

AA2

AA1

tRNA (anticodons)

mRNA (codons)

Step 2. ___Translation___

Figure LG 3.4A Protein synthesis.

E2. (a) See Figure LG 3.4A. (b) Nucleus, RNA polymerase; template, sense; messenger RNA (mRNA) (see Figure LG 3.4A). (c) See Figure LG 3.4A. (d) Translation, m (messenger), amino acids. (e) Cytoplasm, ribosome; small, the start codon. (f) t (transfer), anticodon; codon (see Figure LG 3.4A). (g) t (transfer); peptide, large; ATP (see Figure LG 3.4A). (h) It is released and may be recycled to transfer another amino acid. (i) Stop; it splits into its large and small subunits. (j) Polyribosome.
E3. (a) Exons, introns; both exons and introns. (b) Introns; before; differentiation. (c) A group of nucleotides on a DNA molecule that codes for synthesis of a specific protein. (d) 300–3000, 100–1000.
F2. (a) Nucleus, cytoplasm. (b) Meiosis; mitosis, (c) Cytokinesis; mitosis.
F3. (a) Between cell divisions. (b) S; synthesis; nitrogen bases; chromatids. (c) G_1; G_2; RNA and proteins; centrosomes (containing centrioles); DNA synthesis.
F4. (a) P. (b) P. (c) P. (d) M. (e) A. (f) T. (g) P.
F5. (a) Mitosis and cytokinesis (see Figure 3.22, page 82 of text). (b) G_1.
F6. (a) Gonads (ovaries or testes), gametes; fertilization, zygote. (b) 46, diploid or $2n$; 23. (c) Sex chromosomes; two X chromosomes; autosomes. (d) 23, n. (e) Two; once; reduction, equatorial. (f) Before, "46 × 2" (or 46 doubled); centromere. (g) Synapsis; does not; tetrad; crossing-over; permits exchange of DNA between chromatids and greatly increases possibilities for variety among sperm or eggs and thus offspring. (h) I. (i) "23 × 2" (or 23 doubled chromatids); centromeres, 23. (j) 4; 1, polar bodies.
G1. (a) Oncology; oncologists, oncological. (b) Neoplasm; malignant, benign. (c) Malignant; metastasis. (d) Nutrients and space; contact inhibition; blood or lymph, or by detaching from an organ, such as liver, and "seeding" into a cavity (as abdominal cavity) to reach other organs which border on that cavity. (e) Pressure on nerve, obstruction of passageway, loss of function of a vital organ.
G2. (a) S. (b) O. (c) C. (d) L.
G3. (a) HTLV-1. (b) HIV. (c) EBV. (d) HBV. (e) HPV, HSV.
I1. (a) HT. (b) HP. (c) M. (d) D.
I2. (a) N. (b) A. (c) B.

MASTERY TEST: Chapter 3

Questions 1–9: Circle the letter preceding the one best answer to each question.

1. The organelle that carries out the process of autophagy in which old organelles are digested so that their components can be recycled is:
   A. Peroxisome D. Lysosome
   B. Mitochondrion E. Endoplasmic
   C. Centrosome reticulum

2. Which statement about proteins is *false?*
   A. Proteins are synthesized on ribosomes.
   B. Proteins are so large that they normally will not pass out of blood that undergoes filtration in kidneys.
   C. Proteins are so large that they can be expected to create osmotic pressure important in maintaining fluid volume in blood.
   D. Proteins are small enough to pass through pores of typical plasma (cell) membranes.

3. Choose the *false* statement about prophase of mitosis.
   A. It occurs just after metaphase.
   B. It is the longest phase of mitosis.
   C. Nucleoli disappear.
   D. Centrioles move apart and become connected by spindle fibers.

4. Choose the *false* statement about the Golgi complex.
   A. It is usually located near the nucleus.
   B. It consists of stacked membranous sacs known as cisternae.
   C. It is involved with protein and lipid secretion.
   D. It pulls chromosomes toward poles of the cell during mitosis.

5. Choose the *false* statement about genes.
   A. Genes contain DNA.
   B. Genes contain information that controls heredity.
   C. Each cell in the human body has a total of 46 genes.
   D. Genes are transcribed by messenger RNA during the first step of protein synthesis.

6. Choose the *false* statement about protein synthesis.
   A. Translation occurs in the cytoplasm.
   B. Messenger RNA picks up and transports an amino acid during protein synthesis.
   C. Messenger RNA travels to a ribosome in the cytoplasm.
   D. Transfer RNA is attracted to mRNA due to their complementary bases.

7. All of the following structures are part of the nucleus *except:*
   A. Nucleolus C. Chromosomes
   B. Chromatin D. Centrosome

8. The fact that the aroma of cookies baking in the kitchen reaches you in the living room is due to:
   A. Active transport C. Diffusion
   B. Osmosis D. Phagocytosis

9. Bone cancer is an example of what type of cancer?
   A. Sarcoma C. Carcinoma
   B. Lymphoma

Questions 10–20: Circle T (true) or F (false). If the statement is false, change the underlined word or phrase so that the statement is correct.

T F 10. Plasma membranes consist of a double layer of <u>carbohydrate molecules</u> with proteins embedded in the bilayer.

T F 11. Free ribosomes in the cytosol are concerned primarily with synthesis of <u>proteins for insertion in the plasma membrane or for export from the cell.</u>

T F 12. Glucose in foods you eat is absorbed into cells lining the intestine by the process of <u>facilitated transfusion,</u> and then passes into bloodstream by <u>secondary active transport</u> known as <u>antiport (counter-transport).</u>

T F 13. <u>Cristae</u> are folds of membrane in mitochondria.

T F 14. Sperm move by means of lashing their long tails named <u>cilia</u>.

T F 15. Hypertrophy refers to increase in size of a tissue related to increase in the <u>number</u> of cells.

T F 16. <u>Diffusion, osmosis, and filtration</u> are all passive transport processes.

T F 17. A 5 percent glucose solution is <u>hypotonic</u> to a 10 percent glucose solution.

T   F   18. Most healthy cells <u>lack</u> the property of contract inhibition.

T   F   19. The mitotic phase that follows metaphase is <u>anaphase</u>.

T   F   20. <u>Cytokinesis</u> is another name for mitosis.

Questions 21–25: fill-ins. Complete each sentence with the word which best fits.

_____ 21. The first meiotic division is known as the ____ division.

_____ 22. Active processes involved in movement of substances across cell membranes are those that use energy from the splitting of ____.

_____ 23. White blood cells engulf large solid particles by the process of ____.

_____ 24. ____ DNA consists of DNA combined from several sources, such as from human cells and from bacteria.

_____ 25. The sequence of bases of mRNA which would be complementary to DNA bases in the sequence A-T-T-C-A-C would be ____.

ANSWERS TO MASTERY TEST: ■ Chapter 3

Multiple Choice

1 D	6. B
2. D	7. D
3. A	8. E
4. D	9. A
5. C	

True-False

10. F. Phospholipid molecules
11. F. Proteins for use inside of the cell.
12. F. Facilitated transfusion, secondary active transport, symport (co-transport).
13. T
14. F. Flagella
15. F. Size
16. T
17. T
18. F. Do have or possess
19. T
20. F. Nuclear division (Cytokinesis is cytoplasmic division.)

Fill-ins

21. Reduction
22. ATP
23. Phagocytosis
24. Recombinant
25. U-A-A-G-U-G

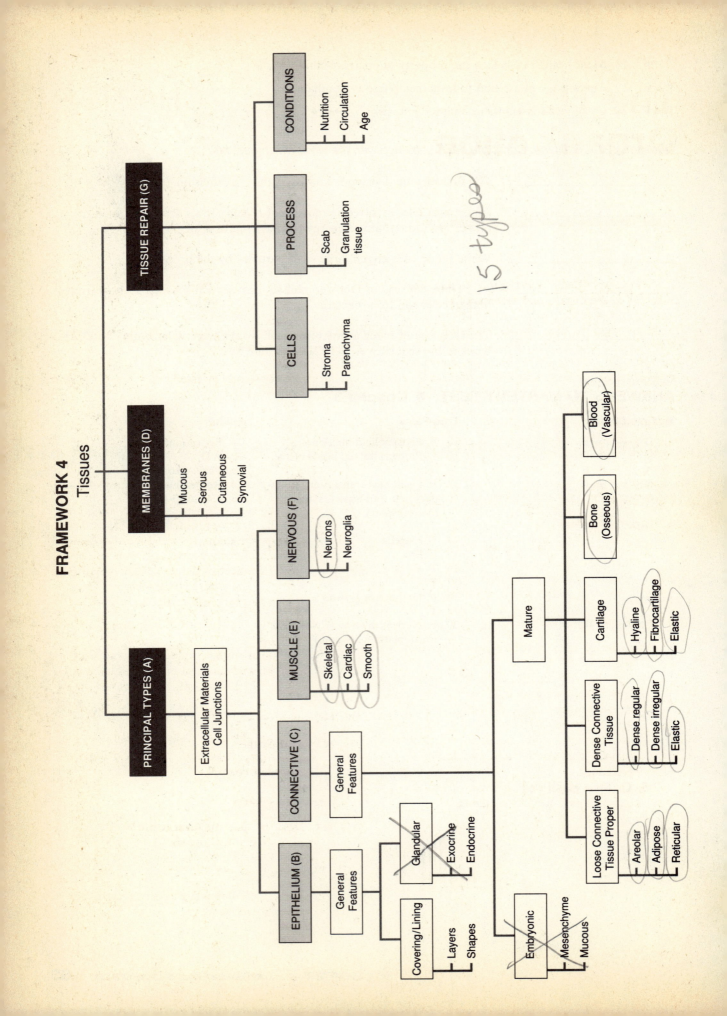

FRAMEWORK 4

Tissues

TISSUE REPAIR (G)

CONDITIONS
- Nutrition
- Circulation
- Age

PROCESS
- Scab
- Granulation tissue

CELLS
- Stroma
- Parenchyma

MEMBRANES (D)
- Mucous
- Serous
- Cutaneous
- Synovial

PRINCIPAL TYPES (A)

Extracellular Materials
Cell Junctions

EPITHELIUM (B)

General Features

Covering/Lining
- Layers
- Shapes

Glandular
- Exocrine
- Endocrine

CONNECTIVE (C)

General Features

Embryonic
- Mesenchyme
- Mucous

Mature

Loose Connective Tissue Proper
- Areolar
- Adipose
- Reticular

Dense Connective Tissue
- Dense regular
- Dense irregular
- Elastic

Cartilage
- Hyaline
- Fibrocartilage
- Elastic

Bone (Osseous)

Blood (Vascular)

MUSCLE (E)
- Skeletal
- Cardiac
- Smooth

NERVOUS (F)
- Neurons
- Neuroglia

The Tissue Level of Organization

CHAPTER 4

Tissues consist of cells and the intercellular (between cells) materials secreted by these cells. All of the organs and systems of the body consist of tissues that may be categorized into four classes or types: epithelium, connective, muscle, and nerve. Each tissue exhibits structural characteristics (anatomy) that determine the function (physiology) of that tissue. For example, epithelial tissues consist of tightly packed cells that form the secretory glands and also provide protective barriers, as in the membranes that cover the surface of the body and line passageways or cavities. Tissues are constantly stressed by daily wear and tear and sometimes by trauma or infection. Tissue repair is therefore an ongoing human maintenance project.

As you begin this chapter, first carefully examine the Chapter 4 Framework, and note relationships among concepts and key terms there.

TOPIC OUTLINE AND OBJECTIVES

A. Types of tissue

1. Define a tissue and classify the tissues of the body into four major types.
2. Describe the structure and functions of the three principal types of cell junctions.

B. Epithelial tissue

3. Describe the general features of epithelial tissue.
4. List the structure, location, and function for the following types of epithelium: simple squamous, simple cuboidal, simple columnar (nonciliated and ciliated), stratified squamous, stratified cuboidal, stratified columnar, transitional, and pseudostratified columnar.
5. Define a gland and distinguish between exocrine and endocrine glands.

C. Connective tissue

6. Describe the general features of connective tissue.
7. Discuss the cells, ground substance, and fibers that compose connective tissue.

8. List the structure, function, and location of mesenchyme; mucous connective tissue; areolar connective tissue; adipose tissue; reticular connective tissue; dense regular and irregular connective tissue; elastic connective tissue; cartilage; bone; and blood.

D. Membranes

9. Define an epithelial membrane and list the location and function of mucous, serous, cutaneous, and synovial membranes.

E. Muscle tissue

10. Contrast the three types of muscle tissue with regard to structure, location, and modes of control.

F. Nervous tissue

11. Describe the structural features and functions of nervous tissue.

G. Tissue repair

12. Describe tissue repair as an attempt to restore homeostasis.

Now study the following parts of words that may help you better understand terminology in this chapter.

Wordbyte	Meaning	Example	Wordbyte	Meaning	Example
acin-	grape	*acin*ar gland	macro-	large	*macro*phage
adip(o)-	fat	*adip*ose	multi-	many	*multi*cellular
areol-	small space	*areol*ar	pseudo-	false	*pseudo*stratified
crine-	to secrete	endo*crine*	squam-	thin plate, scale	*squam*ous
endo-	within	*endo*crine			
exo-	outside	*exo*crine	strat-	layer	*strat*ified
inter-	between	*inter*cellular	uni-	one	*uni*cellular

CHECKPOINTS

A. Introduction to tissues; types of tissues (pages 97–98)

A1. Define the following terms.

a. Tissue

b. Histology

c. Pathologist

d. Biopsy

A2. Complete the table about the four main classes of tissues.

Tissue	General Functions
a. Epithelial tissue	*protection*
b. Connective tissue	*shock absor* *transport of*
c. *muscle*	Movement
d. *nerve*	Initiates and transmits nerve impulses that help coordinate body activities

■ **A3.** Do this matching exercise on the three embryonic germ layers that form all body organs.

Ecto. Ectoderm	Endo. Endoderm	Meso. Mesoderm

_____ a. Epithelial lining of the organs of digestion and of respiratory passageways

_____ b. Brain and nerves

_____ c. Muscles and bones

_____ d. Outer layers of skin

■ **A4.** Complete this exercise about extracellular materials and connections between cells.

a. Two major types of extracellular fluid (ECF) are the _____ located between cells in tissues, and the liquid portion of blood, known as

_plasma_____.

b. The matrix between connective tissues is distinctive for each tissue type. Matrix

consists of different types of _____ embedded in

_____, which may be fluid, gel, or solid.

c. Name one type of cell that normally moves around within extracellular areas.

d. *(Anchoring? Communicating? Tight?)* junctions are common in areas subject to stretching or friction such as the neck of the uterus (which is greatly stretched during childbirth) or the outer layer of skin. A(n) *(desmosome? hemidesmosome? adherens junction?)* is a type of anchoring junction that links cytoskeletons of neighboring cells.

e. *(Anchoring? Communicating? Tight?)* junctions are common in areas where a fluid-tight seal is needed, such as the lining of the stomach or urinary bladder.

f. A gap junction is an example of a(n) *(anchoring? communicating? tight?)* junction.

This junction consists of proteins called _____ that form tunnels between cells. Passage of ions through these tunnels permits spread of action

_potentials_____ within the nervous system or heart muscle. The lack of appropriate communication and coordinated growth of cancer cells is related to the *(lack of? many?)* gap junctions between cancer cells.

B. Epithelial tissue (pages 98–107)

B1. Name the two subtypes of epithelial tissue based on location and function.

ngle layer Absorption + filtration; little wear + tear

high degree wear + tear stratified

B2. Complete the table by describing five or more locations and functions of epithelial tissue. One is done for you.

Location	Function
a. Eyes, nose, outer layer of skin	Sensory reception
b.	
c.	
d.	
e.	

■ **B3.** Describe the structure of epithelium in this exercise.

a. Epithelium consists mostly of *(closely packed cells? intercellular material with few cells?)*.

b. Epithelium is penetrated by *(many? no?)* blood vessels. A term meaning "lacking in blood vessels" is _____. Epithelium *(has? lacks?)* a nerve supply.

c. Epithelium *(does? does not?)* have the capacity to undergo mitosis.

■ **B4.** Complete these sentences describing *basement membrane.*

a. The structure that attaches epithelium to underlying connective tissue is called the

_____. This membrane consists of *(cells? extracellular materials?)*.

b. The portion of this membrane adjacent to epithelium is secreted by *(epithelium? connective tissue?)*. It consists of the protein named *(collagen? reticulin?)* along with proteoglycans, and is known as the *(reticular lamina? basal lamina?)*.

c. The second, underlying layer is secreted by *(epithelium? connective tissue?)*, and is

called the _____ lamina.

B5. Study diagrams of epithelial tissue types (A–F) in Figure LG 4.1 and then write the following information on lines provided on the figure.

■ a. The name of each tissue
b. One or more locations of each type of tissue
c. One or more functions of each type of tissue

Now color the following structures on each diagram:
○ Nucleus   ○ Intercellular material
○ Cytoplasm   ○ Basement membrane

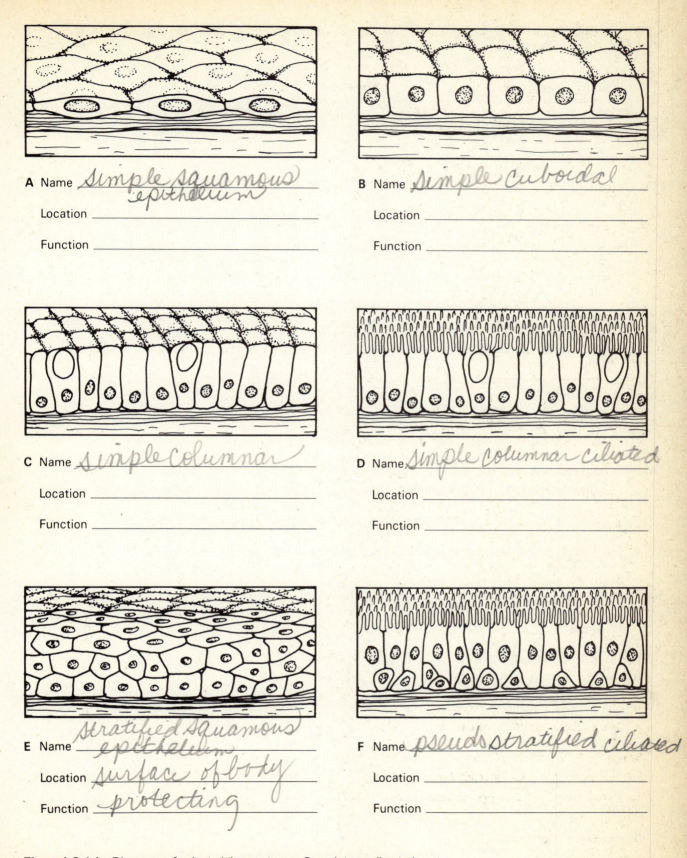

A Name _Simple squamous epithelium_

Location _____

Function _____

B Name _Simple cuboidal_

Location _____

Function _____

C Name _Simple columnar_

Location _____

Function _____

D Name _Simple columnar ciliated_

Location _____

Function _____

E Name _stratified squamous epithelium_

Location _surface of body_

Function _protecting_

F Name _pseudostratified ciliated_

Location _____

Function _____

Figure LG 4.1 Diagrams of selected tissues types. Complete as directed.

G Name _areolar connective_

Location _____

Function _____

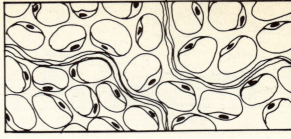

H Name _____

Location _____

Function _storage of lipids for_
energy

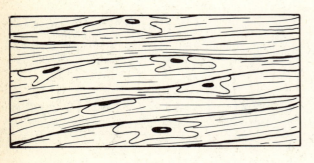

I Name _____

Location _____

Function _____

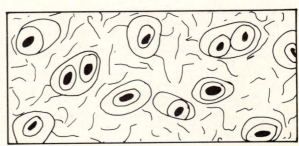

J Name _____

Location _____

Function _____

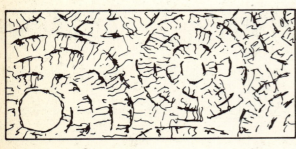

K Name _bone connective_

Location _skeleton_

Function _support movement_
red Blood
ceel

Figure LG 4.1 Continued.

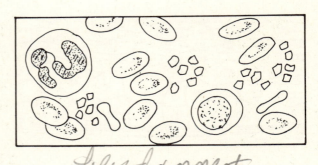

L Name _blood connect_

Location _in_

Function _____

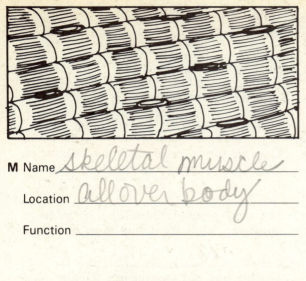

M Name _skeletal muscle_

Location _all over body_

Function _____

N Name _cardiac muscle_

Location _heart_

Function _pump blood_

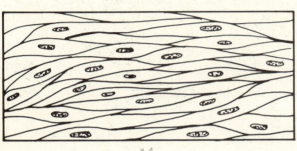

O Name _smooth muscle_

Location _hollow organs_

Function _____

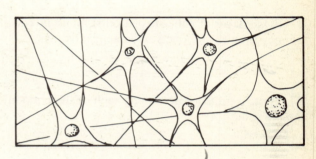

P Name _nervous_

Location _brain + spinal cord_

Function _transmit elec. impulses_

Figure LG 4.1 Continued.

■ **B6.** Check your understanding of the most common types of epithelium listed in the box by writing the name of the type after the phrase which describes it.

> Pseudostratified columnar   Simple squamous
> Simple columnar, ciliated   Stratified squamous
> Simple columnar, unciliated   Stratified transitional
> Simple cuboidal

a. Lines the inner surface of the stomach and

     intestine: _____.

b. Lines urinary tract, as in bladder, permitting

     distension: *transitional* .

c. Lines mouth; present on outer surface of skin:

     _____.

d. Single layer of cube-shaped cells; found in kidney tubules and ducts of some glands:

     Simple cuboidal

e. Lines air sacs of lungs where thin cells are required for diffusion of gases into blood:

     _____.

f. Not a true stratified; all cells on basement membrane, but some do not reach surface of

     tissue: *pseudostratified columnar*

g. Endothelium and mesothelium contain

     _____.

B7. Write a sentence describing each of these modifications of the epithelial lining of the intestine.

a. Microvilli

b. Goblet cells

c. Cilia

■ **B8.** Keratin is a *(carbohydrate? protein?)*. What is its function?

waterproof

List three locations of nonkeratinized stratified squamous epithelium.

B9. What are *glands?* Why are glands studied in this section on epithelium?

■ **B10.** Write EXO before descriptions of *exocrine* glands, and ENDO before descriptions of *endocrine* glands. (Endocrine glands will be studied further in Chapter 18.)

_EX___ a. Their products are secreted into ducts that lead either directly or indirectly to the outside of the body.

_END__ b. Their products are secreted into ECF and then the blood and so stay within the body; they are ductless glands.

_EX___ c. Examples are glands that secrete sweat, oil, mucus, and digestive enzymes.

_END__ d. Examples are glands that secrete hormones.

B11. Complete the table contrasting classes of glands.

Type of Gland	How Secretion Is Released	Example
a. Apocrine		
b.	Cell accumulates secretory product in cytoplasm; cell dies; cell and its contents are discharged as secretion.	
c.		Salivary glands and most other exocrine glands are of this type.

C. Connective tissue   (pages 107–119)

■ **C1.** Write *connective tissue* or *epithelial tissue* following the description below that correctly describes that tissue.

a. Consists of many cells with little intercellular substance (matrix and ground substance): ___epithelial_____

b. Penetrated by blood vessels (vascular): ___connective_____

c. Does not cover body surfaces or line passageways and cavities, but is more internally located; binds, supports, protects: ___connective_____

■ **C2.** List four or more categories of connective tissue matrix. (Note that *inter*cellular means "between cells" or "extracellular.") What is the source of these materials?

■ **C3.** Do this activity about connective tissue cells.

a. Cells in connective tissues derive from *(endo? meso? ecto?)*-derm cells in the embryo.

These cells are called _____ cells.

b. Most matrix is secreted by *(immature? mature?)* connective tissue cells.

c. Names of immature cells end in *(-blast? -cyte?)*, whereas names of mature cells end in *(-blast? -cyte?)*. To check your knowledge further, write correct names for the

following cells: immature bone cell: _____; immature cartilage

cell: _____; mature bone cell: _____.

■ **C4.** Identify characteristics of each connective tissue cell type by writing the name of the correct cell type after its description. (Note that several of these cell types are shown in diagrams G, H, and I of Figure LG 4.1.)

Adipocyte	Mast cell
Fibroblast	Plasma cell
Macrophage	

a. Derived from B lymphocyte, gives rise to antibodies; helpful in defense

_____

b. Phagocytic cell that engulfs bacteria and cleans up debris; important during infection; formed from monocyte:

_____

c. Fat cell: _____

d. Forms collagenous and elastic fibers in

injured tissue: _____

e. Abundant along walls of blood vessels; contains heparin, as well as histamine, which di-

lates blood vessels: _____

■ **C5.** Describe the intercellular matrix of connective tissues in this activity. Match the chemicals and fibers in the box with the descriptions below.

C. Collagen fibers	K. Keratan sulfate
E. Elastic fibers	R. Reticular fibers
H. Hyaluronic acid	

_____ a. Viscous, slippery substance that binds cells together and lubricates joints

_____ b. Chemical component of bone, cartilage, and the cornea of the eye

_____ c. Formed of collagen and glycoproteins; form branching networks that provide stroma for soft organs such as spleen and networks around fat, nerve, and muscle cells

_____ d. Tough fibers in bundles which provide great strength, as needed in bone and tendons; formed of the most abundant protein in the body

_____ e. Can be greatly stretched without breaking, an important quality of connective tissues forming skin, blood vessels, and lungs

■ **C6.** As described earlier, the embryonic connective tissue from which all of the connec-

tive tissues arise is called _____. *(Much? Little?)* of this type of tissue is found in the body after birth. Another embryonic tissue, known as mucous con-

nective tissue or _____ jelly, gives support to the umbilical cord. Tissues such as areolar connective tissue, cartilage, or bone are known as *(embryonic? mature?)* connective tissues.

C7. After you study each mature connective tissue type in the text, refer to diagrams of connective tissues on Figure LG 4.1, G–L. Write the following information on lines provided on the figure.

■ a. The name of each tissue
   b. One or more locations of the tissue
   c. One or more functions of the tissue

Now color the following structures on each diagram, using the same colors as you did for diagrams A–F:

○ Nucleus   ○ Matrix
○ Cytoplasm   ○ Fat globule (a cellular inclusion), diagram H

■ From this activity, you can conclude that connective tissues appear to consist mostly of *(cells? intercellular material* [*matrix*]*?)*.

■ **C8.** Which diagrams in Figure LG 4.1 show *loose connective tissue proper?* _____

C9. Label fibers in Figure LG 4.1, diagrams G, I, and J. (Note that fibers *are* present in bone, but not visible in diagram K.)

■ **C10.** Which two types of tissue form subcutaneous tissue (superficial fascia)?

C11. Explain what accounts for the "signet ring" (like a class ring with a stone) appearance of adipocytes, as in Figure LG 4.1, H.

C12. Write three or more functions of fat in the body.

■ **C13.** Explain how brown fat may be helpful to newborns.

■ **C14.** Match the common types of dense regular connective tissue with the descriptions given.

> A. Aponeurosis   T. Tendon
> L. Ligament

_____ a. Connects muscles to bones

_____ b. Holds bones together at joints

_____ c. Flat band or sheet of tissue connecting muscles to each other or to bones

■ **C15.** Do this activity about cartilage.
  a. In general, cartilage can endure *(more? less?)* stress than other connective tissues studied so far.
  b. Cartilage heals *(more? less?)* rapidly than bone. Explain why this is so.

  c. Growth of cartilage occurs by two mechanisms. In childhood and adolescent years, cartilage increases in size by *(appositional? interstitial?)* growth. This process is also called *(exogenous? endogenous?)* since it is growth by enlargement of existing cells within the cartilage.
  d. Post adolescence, cartilage increases in thickness by *(appositional? interstitial?)*

  growth. This process involves cells in the _____ that covers cartilage. Here cells known as *(chondroblasts? chondrocytes? fibroblasts?)* differentiate

  into _____, which lay down matrix and eventually become mature chondrocytes.

■ **C16.** Match the types of cartilage with the descriptions given.

> E. Elastic   F. Fibrous   H. Hyaline

_____ a. Found where strength and rigidity are needed, as in discs between vertebrae and in symphysis pubis

_____ b. White, glossy cartilage covering ends of bones (articular), covering ends of ribs (costal), and giving strength to nose, larynx, and trachea

_____ c. Provides strength and flexibility, as in external part of ear.

■ **C17.** Complete this exercise about bone and blood tissue.

a. Bone tissue is also known as _____ tissue. Compact bone consists of concentric rings, or *(lamellae? canaliculi?)* with bone cells, called *(chondrocytes? osteocytes?)*, located in tiny spaces called _____. Nutrients in blood reach osteocytes by vessels located in the _____ canal and then via minute, radiating canals called _____ that extend out to lacunae.

b. Blood, or _____ tissue, consists of a fluid called

_____ containing three types of formed elements. Red blood

cells, or _____, transport oxygen and carbon dioxide; white

cells, also known as _____, provide defense for the body;

thrombocytes assist in the function of blood _____.

C18. Now label the following cell types found in diagrams J, K, or I of Figure LG 4.1: *chondrocytes, osteocytes, erythrocytes, leukocytes, thrombocytes.*

D. Membranes (pages 118–119)

D1. Complete the table about the four types of membranes in the body.

Type of Membrane	Location	Example of Specific Location	Function(s)
a.	Lines body cavities leading to exterior		
b. Serous			Allows organs to glide easily over each other
c.		Lines knee and hip joints	
d.	Covers body surfaces	Skin	

■ **D2.** Check your understanding of membrane types by doing this exercise.

a. The serous membrane covering the heart is known as the

_____, whereas that covering the lungs is called the

_____. The serous membrane over abdominal organs is the

_____.

b. The portion of serous membranes that covers organs (viscera) is called the

_____ layer; that portion lining the cavity is named the

_____ layer. See Figure 23.7, page 730 in your textbook.

c. The _____ layer of a mucous membrane binds epithelium to underlying muscle and serves as a route for oxygen and nutrients to the avascular epithelial layer of the membrane.

d. Another name for skin is the _____ membrane.

e. A _____ membrane secretes a lubricating fluid known as synovial fluid. Such a membrane *(does? does not?)* contain epithelium, so it *(is? is not?)* classified as an epithelial membrane.

■ **D3.** *A clinical challenge.* Serous membranes are normally relatively free of microorganisms, whereas mucous and cutaneous membranes have many microorganisms on them. Based on this information, why may antibiotics be administered to patients undergoing stomach or intestinal surgery?

E. Muscle tissue   (pages 119–121)

■ **E1.** These tissues are more *(specific? generalized?)* in their structure and functions than epithelium or connective tissue. (They will be studied in greater detail in Chapters 10 and 12.)

E2. After you study each muscle tissue type in the text, refer to diagrams of muscle tissues in Figure LG 4.1, M–O). Write the following information on lines provided on the figure.

■ a. The name of each muscle tissue
  b. One or more locations of the tissue
  c. One or more functions of the tissue

Now color the following structures on each diagram, using the same colors as you did for diagrams A–L.

O Nucleus
O Cytoplasm
O Intercellular material

■ From this activity, you can conclude that muscle tissues appear to consist mostly of *(cells? intercellular material [matrix]?)*.

■ **E3.** Check your understanding of the three muscle types listed in the box by selecting types that best fit descriptions below.

| C. Cardiac   Sk. Skeletal   Sm. Smooth |

___C___ a. Tissue forming most of the wall of the heart

___Sk___ b. Attached to bones

___Sm___ c. Spindle-shaped cells with ends tapering to points

___C___ d. Contains intercalated discs

___Sm___ e. Found in walls of intestine, urinary bladder, and blood vessels

___Sk___ f. Voluntary muscle

F. Nervous tissue   (page 121)

F1. Refer to diagram P in Figure LG 4.1. Write the following information on lines provided on the figure.

■ a. The name of the tissue
  b. One or more locations of the tissue
  c. One or more functions of the tissue

■ **F2.** *(Axons? Dendrites?)* carry nerve impulses toward the cell body of a neuron,

whereas _____ transmit nerve impulses away from the cell body.

G. Tissue repair   (pages 121–123)

■ **G1.** Circle all of the tissue types that maintain the capacity to undergo mitosis after birth.

(a.) Blood
(b.) Bone
 c. Cardiac muscle
(d.) Epithelium lining the mouth

(e.) Epithelium on the skin surface
 f. Nervous tissue (neurons)
 g. Skeletal muscle

■ **G2.** Choose the correct term to complete each sentence. Write P for *parenchymal* or S for *stromal*.

___P___ a. The part of an organ that consists of functioning cells, such as secreting epithelial cells lining the intestine, is composed of ____ tissue.

___S___ b. The connective tissue cells that support the functional cells of the organ are called ____ cells.

___P___ c. If only the ____ cells are involved in the repair process, the repair will be close to perfect.

___S___ d. If ____ cells are involved in the repair, they will lay down fibrous tissue known as a fibrosis or a scar.

G3. Define each of the following terms related to tissue repair.

a. Regeneration

b. Granulation tissue

c. Adhesions

■ **G4.** You have most likely experienced an injury that led to formation of a *scab*. What causes scabs to form?

blood clot

■ **G5.** List three factors that enhance tissue repair.

nutrition   adequate blood circulation
age

■ **G6.** *A clinical challenge*. Wound healing is likely to be impaired in persons with inadequate nutrition. Identify the vitamins—A, B, C, D, E, or K—that perform the following functions in the process of wound healing:

___D___ a. This vitamin is necessary for healing fractures since it enhances calcium absorption from foods in the intestine.

___C___ b. This is important in repair of connective tissue and walls of blood vessels.

___B___ c. Thiamine, riboflavin, and nicotinic acid are categorized as vitamins of this group. They enhance metabolic processes and the cell division necessary for wound repair.

___A___ d. This vitamin is helpful in replacement of epithelial tissues, for example, in the lining of the respiratory tract.

___E___ e. Believed to promote healing of injured tissues, this vitamin may also help to prevent scarring.

___K___ f. This vitamin aids in blood clotting and so helps to prevent excessive blood loss from wounds.

ANSWERS TO SELECTED CHECKPOINTS

A3. (a) Endo. (b) Ecto. (c) Meso. (d) Ecto.

A4. (a) Interstitial fluid, plasma. (b) Fibers, ground substance. (c) Phagocytes en route to tasty meals. (d) Anchoring; adherens junction. (e) Tight. (f) Communicating; connexons; potentials; lack of.

B3. (a) Closely packed cells. (b) No; avascular; has. (c) Does.

B4. (a) Basement membrane; extracellular material. (b) Epithelium; collagen, basal lamina. (c) Connective tissue, reticular.

B5. (A) Simple squamous epithelium. (B) Simple cuboidal epithelium. (C) Simple columnar epithelium (unciliated). (D) Simple columnar epithelium (ciliated). (E) Stratified squamous epithelium. (F) Pseudostratified columnar epithelium, ciliated.

B6. (a) Simple columnar, unciliated. (b) Stratified transitional. (c) Stratified squamous. (d) Simple cuboidal. (e) Simple squamous. (f) Pseudostratified columnar. (g) Simple squamous.

B8. Protein; waterproofs, resists friction, and repels bacteria. Examples of locations: wet surfaces such as lining of the mouth, tongue, esophagus, and parts of epiglottis and vagina.

B10. (a) EXO. (b) ENDO. (c) EXO. (d) ENDO.

C1. (a) Epithelial tissue. (b) Connective tissue (except cartilage). (c) Connective tissue.

C2. Fluid, semifluid, gelatinous, fibrous, or calcified; secreted by connective tissue (or adjacent) cells.

C3. (a) Meso-; mesenchymal. (b) Immature. (c) -blast, -cyte; osteoblast; chondroblast; osteocyte.

C4. (a) Plasma cell. (b) Macrophage. (c) Adipocyte. (d) Fibroblast. (e) Mast cell.

C5. (a) H. (b) K. (c) R. (d) C. (e) E.

C6. (a) Mesenchyme; little; Wharton's; mature.

C7. (G) Loose (areolar) connective tissue. (H) Adipose tissue. (I) Dense connective tissue. (J) Cartilage. (K) Osseous (bone) tissue. (L) Vascular (blood) tissue. Intercellular material (matrix).

C8. G, H.

C10. Loose (areolar) connective tissue and adipose.

C13. This type of fat has an abundance of mitochondria, and may help to generate heat to maintain proper body temperature in newborns.

C14. (a) T. (b) L. (c) A.

C15. (a) More. (b) Less; cartilage is avascular, so chemicals needed for repair must reach cartilage by diffusion from the perichondrium. (c) Interstitial; endogenous. (d) Appositional; perichondrium; fibroblasts, chondroblasts.

C16. (a) F. (b) H. (c) E.

C17. (a) Osseous; lamellae, osteocytes, lacunae; central (Haversian), canaliculi. (b) Vascular, plasma; erythrocytes; leukocytes; clotting.

D2. (a) Pericardium, pleura; peritoneum. (b) Visceral; parietal. (c) Lamina propria (connective tissue). (d) Cutaneous. (e) Synovial; does not, is not.

D3. Microorganisms from skin or from within the stomach or intestine may be introduced into the peritoneal cavity during surgery; the result may be an inflammation of the serous membrane (peritonitis).

E1. Specific.

E2. (M) Skeletal muscle. (N) Cardiac muscle. (O) Smooth muscle. Cells [known as muscle fibers].

E3. (a) C. (b) Sk. (c) Sm. (d) C. (e) Sm. (f) Sk.

F1. Nervous tissue.

F2. Dendrites, axons (*Hint:* Remember A for Away from cell body).

G1. Blood, bone, epithelium lining the mouth, epithelium on the skin surface.

G2. (a) P. (b) S. (c) P. (d) S.

G4. Fibrin clots form to limit blood loss at wounds. As the clot dries out, the two sides of the wound are drawn together. This process (clot retraction) forms a temporary scab from the dried out fibrin.

G5. Nutrition, adequate blood circulation, and (younger) age.

G6. (a) D. (b) C. (c) B. (d) A. (e) E. (f) K.

MASTERY TEST: Chapter 4

Questions 1–6: Circle the letter preceding the one best answer to each question.

1. A surgeon performing abdominal surgery will pass through the skin, then loose connective tissue (subcutaneous), then fascia covering muscle and muscle tissue itself, to reach the ___ membrane lining the inside wall of the abdomen.
   A. Parietal pleura
   B. Parietal pericardium
   C. Parietal peritoneum
   D. Visceral pleura
   E. Visceral pericardium
   F. Visceral peritoneum

2. The type of tissue that covers body surfaces, lines body cavities, and forms glands is:
   A. Nervous   C. Connective
   B. Muscular   D. Epithelial

3. Neurons are cells that are part of:
   A. Nervous tissue   C. Connective tissue
   B. Muscle tissue   D. Epithelial tissue

4. Which statement about connective tissue is *false?*
   A. Cells are very closely packed together.
   B. Most connective tissues have an abundant blood supply.
   C. Intercellular substance is present in large amounts.
   D. It is the most abundant tissue in the body.

5. Modified columnar cells that are unicellular glands secreting mucus are known as:
   A. Cilia
   B. Microvilli
   C. Goblet cells
   D. Branched tubular glands

6. Which tissues are avascular?
   A. Skeletal muscle and cardiac muscle
   B. Bone and dense connective tissue
   C. Cartilage and epithelium
   D. Areolar and adipose tissues

Questions 7–10: Match tissue types with correct descriptions.

Adipose	Simple columnar epithelium
Cartilage	Simple squamous epithelium
Dense connective tissue	Stratified squamous epithelium
Loose connective tissue	

7. Contains lacunae and chondrocytes:

   Cartilage

8. Forms fascia, tendons, and deeper layers of dermis of skin: _dense connective_

9. Forms thick surface layer of skin on hands and feet, providing extra protection:

   Stratified squamous epithelium

10. Single layer of flat, scalelike cells:

   simple squamous epithelium

Questions 11–20: Circle T (true) or F (false). If the statement is false, change the underlined word or phrase so that the statement is correct.

(T) F 11. Stratified transitional epithelium lines the <u>urinary bladder</u>.

(T) F 12. Epithelial tissue has <u>many</u> cell junctions.

T (F) 13. Hormones are classified as <u>exocrine</u> secretions. *Endocrine*

(T) F 14. Elastic connective tissue, since it provides both stretch and strength, is found in <u>walls of elastic arteries, in the vocal cords, and in some ligaments</u>.

(T) F 15. Epithelial tissue is derived from <u>all three layers: endoderm, mesoderm, and ectoderm</u>.

T (F) 16. A gap junction is an example of <u>an anchoring</u> junction between cells. *communicating*

(T) F 17. The surface attachment between epithelium and connective tissue is called <u>basement membrane</u>.

T (F) 18. <u>Simple</u> squamous epithelium is most likely to line the areas of the body that are subject to wear and tear. *Stratified*

T (F) 19. Marfan syndrome is a <u>muscle tissue</u> disorder that results in persons being <u>short with short extremities</u>. *Connective* *tall long*

(T) F 20. Appositional growth of cartilage <u>starts later in life than interstitial growth and continues throughout life</u>.

Questions 21–25: fill-ins. Complete each sentence with the word that best fits.

biopsy 21. A ____ is a sample of living tissue for microscopic examination, for example, for diagnosis.

Periosteum 22. ____ is a dense, irregular connective tissue covering over bone.

simple Squamous epithelium 23. The kind of tissue that lines alveoli (air sacs) of lungs is ____.

Extra cellular 24. Another term for intercellular is ____.

Epithelium 25. All glands are formed of the tissue type named ____.

ANSWERS TO MASTERY TEST: ■ Chapter 4

Multiple Choice

1. C   4. A
2. D   5 C
3. A   6. C

Matching

7. Cartilage
8. Dense connective tissue
9. Stratified squamous epithelium
10. Simple squamous epithelium

True-False

11. T
12. T
13. F. Endocrine
14. T
15. T
16. F. A communicating
17. T
18. F. Stratified
19. F. Connective tissue; tall with long extremities
20. T

Fill-ins

21. Biopsy
22. Periosteum
23. Simple squamous epithelium
24. Extracellular or interstitial
25. Epithelium

FRAMEWORK 5

Integument

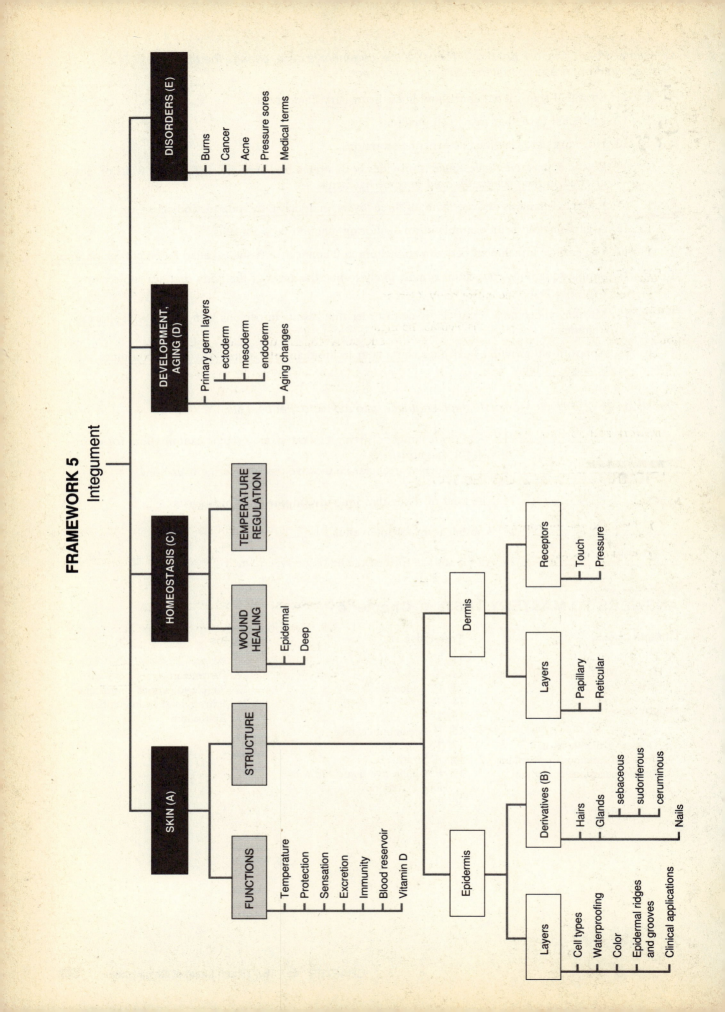

The Integumentary System

CHAPTER
5

In your human body tour, you have now arrived at your first system, the integumentary system, which envelops the entire body. Composed of the skin, hair, nails, and glands, the integument serves as both a barrier and a link with the environment. The many waterproofed layers of cells, as well as the pigmentation of skin, afford protection to the body. Nails, glands, and even hairs offer additional fortification. Sense receptors and blood vessels in skin increase awareness of conditions around the body and facilitate appropriate responses. Even so, this microscopically thin surface cover may be traumatized, invaded, or simply succumb to the passage of time. The integument then demands repair and healing for continued maintenance of homeostasis.

As you begin your study of the integumentary system, first study the Chapter 5 Framework and note key terms associated with each section.

TOPIC OUTLINE AND OBJECTIVES

A. Skin

1. Describe the anatomy and physiology of the skin.
2. Explain the basis for skin color.

B. Epidermal derivatives

3. Compare the anatomy, distribution, and physiology of hair, sebaceous (oil), sudoriferous (sweat), and ceruminous glands.

C. Homeostasis: wound healing, temperature regulation

4. Outline the steps involved in epidermal wound healing and deep wound healing.
5. Explain the role of the skin in helping to maintain the homeostasis of normal body temperature.

D. Aging and development

6. Describe the effects of aging on the integumentary system.
7. Describe the development of the epidermis, its derivatives, and the dermis.

E. Disorders, medical terminology

8. Describe the causes and effects for the following skin disorders: burns, sunburn, skin cancer, acne, and pressure sores.
9. Define medical terminology associated with the integumentary system.

WORDBYTES

Now study the following parts of words that may help you better understand the terminology in this chapter.

Wordbyte	Meaning	Example	Wordbyte	Meaning	Example
cera-	wax	*cera*men	melan-	black	*melan*ocyte
cut-	skin	*cut*aneous	seb(um)-	tallow, fat	*seb*aceous
derm-	skin	*derm*atologist	sub-	under	*sub*cutaneous
epi-	over	*epi*dermis	sudor-	sweat	*sudor*iferous
-fer	carry	sudori*fer*ous			

CHECKPOINTS

A. Skin (pages 127–132)

■ **A1.** Name the structures included in the integumentary system.

Skin, hair, nails + glands nerve endings

A2. *A clinical challenge.* Skin may provide clues about the health of the body; in this case at least, you *can* tell a lot about a book by its cover. Write three examples of how a physical or emotional health state may be detected by inspection of skin.

Sweaty - nervous
dry -
blue - bad circulation

■ **A3.** Answer these questions about the two portions of skin. Label them (bracketed regions) on Figure LG 5.1.

a. The outer layer is named the ___*epidermis*___. It is composed of *(connective tissue? epithelium?)*.

b. The inner portion of skin, called the ___*dermis*___, is made of *(connective tissue? epithelium?)*. The dermis is *(thicker? thinner?)* than the epidermis.

■ **A4.** Most sensory receptors, nerves, blood vessels, and glands are embedded in the *(epidermis? dermis?)*. Fill in all label lines on the right side of Figure LG 5.1. Then color those structures and their related color code ovals.

■ **A5.** The tissue underlying skin is called *subcutaneous,* meaning

___*under skin*___. This layer is also called ___*hypodermis*___.

It consists of two types of tissue, ___*areolar*___ and

___*adipose*___. What functions does subcutaneous tissue serve?

Anchors skin to underlying tissue + organs

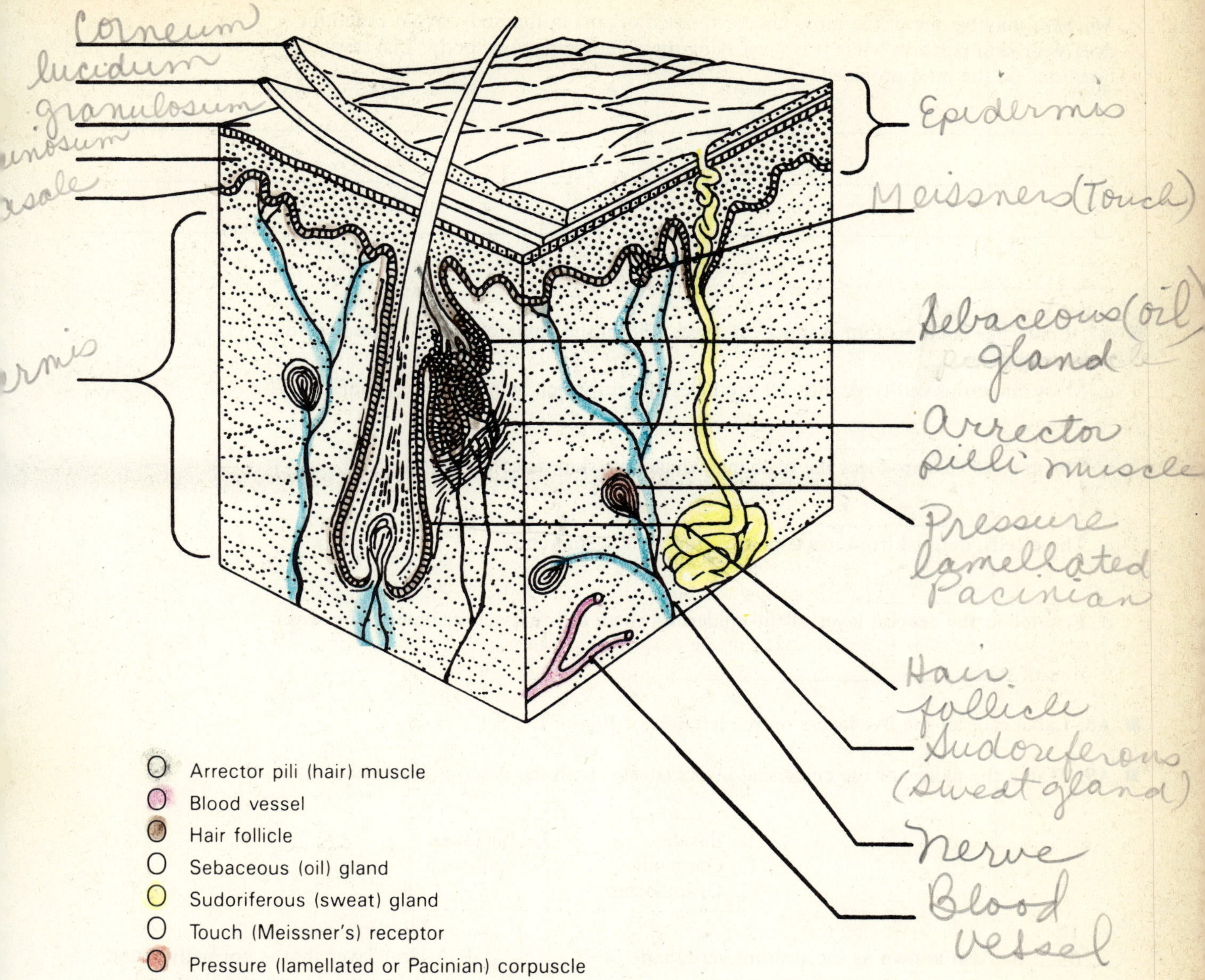
Corneum
lucidum
granulosum
Epidermis
Meissners(Touch)
Sebaceous(oil) gland
Arrector pilli muscle
Pressure lamellated Pacinian
Hair follicle
Sudoriferous (sweat gland)
Nerve
Blood vessel

- Arrector pili (hair) muscle
- Blood vessel
- Hair follicle
- Sebaceous (oil) gland
- Sudoriferous (sweat) gland
- Touch (Meissner's) receptor
- Pressure (lamellated or Pacinian) corpuscle
- Nerve

Figure LG 5.1 Structure of the skin. Label as directed in Checkpoints A3, A4, and A8.

A6. Skin may be one of the most underestimated organs in the body. What functions does your skin perform while it is "just lying there" covering your body? List seven functions on the lines provided.

_____   _____

_____   _____

_____   _____

_____

■ **A7.** Epidermis contains four distinct cell types. Fill in the name of the cell type that fits each description.

a. Most numerous cell type, this cell produces keratin which helps to waterproof skin:

_____.

b. This type of cell produces the pigments which give skin its color:

_____.

c. These cells, derived from macrophages, function in immunity:

_____.

d. Located in the deepest layer of the epidermis, these cells may contribute to the sensa-

tion of touch: _____.

■ **A8.** Label each of the five layers on the left side of Figure LG 5.1.

■ **A9.** Match the names of the epidermal layers (strata) with the descriptions below.

```
B. Basale          L. Lucidum
C. Corneum         S. Spinosum
G. Granulosum
```

_____ a. Also known as the stratum germinati-
         vum, since new cells arise here
_____ b. Stratum immediately superficial to the
         stratum basale; cells look prickly un-
         der microscopic examination
_____ c. Development of keratohyalin, precur-
         sor of keratin, occurs in these cells

_____ d. A clear layer that is normally found
         only in thick layers of palms and
         soles
_____ e. Uppermost layer, consisting of 25 to
         30 rows of flat, dead cells filled with
         keratin.

A10. Briefly describe the relationship of epidermal growth factor (EGF) to skin cancer.

■ **A11.** Describe the dermis in this exercise.

a. The outer one-fifth of the dermis is known as the *(papillary? reticular?)* region. It consists of *(loose? dense?)* connective tissue. Present in fingerlike projections known

as dermal _____ are Meissner's corpuscles, sense receptors sensitive to *(pressure? touch?)*.

b. The remainder of the dermis is known as the _____ region, composed of *(loose? dense?)* connective tissue. Skin is strengthened by *(elastic? collagenous?)* fibers in the reticular layer. Skin is extensible and elastic due to

_____ fibers.

c. Name three areas of the body in which skin is especially thick.

■ **A12.** Explain what accounts for skin color by doing this exercise.

a. Dark skin is due primarily to *(a larger number of melanocytes? greater melanin production per melanocyte?)*. Melanocytes are in greatest abundance in the stratum

_____.

b. Melanin is derived from the amino acid _____. The enzyme

which converts tyrosine to melanin is _____. This enzyme is

activated by _____ light. In the inherited condition known as

_____, melanocytes cannot synthesize tyrosinase, so skin, hair, and eyes lack melanin.

c. What are freckles and liver spots?

Do they tend to become cancerous? _____

d. The yellowish color of skin of persons of Asian origin is due to the combination of

the pigments _____ and _____.

e. What accounts for the pinker color of skin during blushing and as a cooling mechanism during exercise?

This condition is known as *(cyanosis? erythema?)*.

f. Jaundice skin has a more *(blue? red? yellow?)* hue, often due to

_____ problems. Cyanotic skin appears more *(blue? red? yellow?)* due to lack of oxygen and excessive carbon dioxide in the blood vessels of the skin.

■ **A13.** *A clinical challenge.* Identify types of skin grafts that might be used for Tim, a severely burned patient. Choose from these answers; not all will be used.

> Autograft   Homograft
> Autologous skin transplantation   Isograft
> Heterograft

a.   A graft taken from an unburned region of

Tim's skin _____

b.   A section of skin donated by Tim's identical

twin, Tom _____

c.   A portion of Tim's own epidermis grown in the laboratory for several weeks and then transplanted back to the burn site

_____

d.   A section of skin donated by Tim's friend Nick while Tim awaits the graft described in (c); this donated skin will protect the site

from fluid loss _____

B. Epidermal derivatives   (pages 132–135)

■ **B1.** What is the main function of hair? _____

■ **B2.** Complete this exercise about the structure of a hair and its follicle.

a.   Arrange the parts of a hair from superficial to deep: ___ ___ ___
A. Shaft   B. Bulb   C. Root

b.   Arrange the layers of a hair from outermost to innermost: ___ ___ ___
A. Medulla   B. Cuticle   C. Cortex

c.   A hair is composed of *(cells? no cells, but only secretions of cells?)*. What accounts for the fact that hairs are waterproofed?

d.   Surrounding the root of a hair is the hair _____. The follicle consists of two parts. The *(external? internal?)* root sheath is an extension of deeper layers of the *(epidermis? dermis?)*. The internal root sheath consists of cells derived

from the _____.

e.   The _____ is the part of a hair follicle where cells undergo mitosis permitting growth of a new hair. What is the function of the papilla of the hair?

B3. Describe the relationship between the terms in each pair.

a. Arrector pili muscles/"goose bumps"

b. Androgens/hirsutism

c. Sebaceous glands/"blackheads"

B4. In which areas of the body are sebaceous glands largest?

List three functions of these glands.

B5. Describe the composition of perspiration (sweat) and state its functions.

■ **B6.** Check your understanding of skin glands by stating whether the following descriptions refer to *sebaceous, sudoriferous,* or *ceruminous* glands.

a. Sweat glands: _____
b. Glands leading directly to hair follicle; secrete sebum, which keeps hair and skin

from drying out: _____.

c. Line the outer ear canal; secrete earwax: _____.

B7. Look at one of your own nails and Figure 5.4, page 135 in your text. Identify these parts of your nail: *free edge, nail body, lunula,* and *eponychium (cuticle).*

B8. Why does the nail body appear pink, yet the lunula and free edge appear white?

C. Skin and homeostasis: wound healing and temperature regulation (pages 135–138)

■ **C1.** Describe the process of epidermal wound healing in this exercise.

a. State two examples of epidermal wounds.

b. Usually the deepest part of the wound is the *(central? peripheral?)* region.

c. In the process of repair, epidermal cells of the stratum *(corneum? basale?)* break contact from the basement membrane. These are cells at the *(center? periphery?)* of the wound.

d. These basal cells migrate toward the center of the wound, stopping when they meet other similar advancing cells. This cessation of migration is an example of the

phenomenon known as _____. Cancer cells *(do? do not?)* exhibit this characteristic.

e. Both the migrated cells and the remaining epithelial cells at the periphery undergo

_____ to fill in the epithelium up to a normal (or close to normal) level.

C2. The process of deep wound healing involves four phases. List the three or four major events that occur in each phase.

a. Inflammatory

b. Migratory

c. Proliferative

d. Maturation

■ **C3.** *For extra review* of deep wound healing, match the phases in the box with descriptions below.

I. Inflammation	Mig. Migration
Mat. Maturation	P. Proliferation

_____ a. Blood clot temporarily unites edges of wound; blood vessels dilate so neutrophils enter to clean up area.

_____ b. Clot forms a scab; epithelial cells migrate into scab; fibroblasts also migrate to start scar tissue; pink granulation tissue contains delicate new blood vessels.

_____ c. Epithelium and blood vessels grow; fibroblasts lay down many fibers.

_____ d. Scab sloughs off; epidermis grows to normal thickness; collagenous fibers give added strength to healing tissue; blood vessels are more normal.

C4. Humans are *warm-blooded animals*. What is the meaning of that term?

■ **C5.** Complete this exercise about temperature regulation.

a. Human body temperature is maintained at about *(37°C? 38°C? 39°C?)*.

b. The temperature control center of the brain is the *(cerebrum? medulla? hypothalamus?)*.

c. When the body temperature is too hot, as during vigorous exercise, nerve messages from the brain inform sweat glands to *(in? de?)*-crease sweat production. In addition, blood vessels in the skin are directed to *(dilate? constrict?)*.

d. Temperature regulation is an example of a *(negative? positive?)* feedback system since the response (cooling) is *(the same as? opposite to?)* the stimulus (heating).

D. Effects of aging; developmental anatomy (pages 138–139)

■ **D1.** Complete the table relating observable changes in aging of the integument to their causes.

Changes	Causes
a. Wrinkles; skin springs back less when gently pinched.	
b.	Macrophages become less efficient.
c.	Loss of subcutaneous fat
d. Dry, easily broken skin	
e.	Decrease in number and size of melanocytes

■ **D2.** List and briefly describe four or more factors that can help keep your skin healthy throughout your lifetime.

D3. Name the three primary germ layers.

■ **D4.** Describe the development of the integument by completing this learning activity.

a. The epidermis is derived from the *(ecto? meso? endo?)* -derm. All of its layers are formed by the *(second? fourth?)* month of the nine-month human gestation period.

b. The dermis arises from the _____ -derm.

c. Hair, nails, and glands all develop from _____ -derm but grow deeper into the dermis region.

d. Nails begin to develop during the _____ month. The nails reach the end of the digits during the *(sixth? ninth?)* month.

e. When does lanugo form? _____ month. Lanugo *(is? is not?)* usually present on a full-term newborn.

f. Which glands develop from the sides of hair follicles? _____

E. Disorders, medical terminology (pages 139–142)

E1. Contrast systemic effects with local effects of burns.

■ **E2.** Identify characteristics of the three different classes of burns by completing this exercise.

a. In a first-degree burn, only the superficial layers of the *(dermis? epidermis?)* are involved. The tissue appears ___*red*___ in color. Blisters *(do? do not?)* form. Give one example of a first-degree burn.

Typical sunburn

b. Which parts of the skin are injured in a second-degree burn?

epidermis + dermis

Blisters usually *(do? do not?)* form. Epidermal derivatives, such as hair follicles and glands, *(are? are not?)* injured. Healing usually occurs in about three to four *(days? weeks?)*. For most second-degree burns, grafting *(is? is not?)* required.

c. Third-degree burns are called *(partial? full?)*-thickness burns. Such skin appears *(red and blistered? white, brown, or black and dry?)*. Such burned areas are usually *(painful? not painful?)* since nerve endings are destroyed. Grafting *(is? is not?)* required, and scarring *(does? does not?)* result from third-degree burns.

■ **E3.** *A clinical challenge.* Answer these questions about the Lund-Browder method.

a. What does the Lund-Browder method estimate? *(Depth of burn? Amount of body surface area burned?)* This method is based upon differences in body *(size? proportions?)* of age groups.

b. If an adult and a one-year-old each experience burns over the entire anterior surface of both legs, who is more burned? *(Adult? One-year-old?)* If both are burned over the entire anterior surface of the head, who is more burned? *(Adult? One-year-old?)*

■ **E4.** Circle the risk factor for cancer in each pair.

a. Skin type:
A. Fair with red hair   B. Tans well with dark hair

b. Sun exposure:
A. Lives in Canada at sea level
B. Lives in high-altitude mountainous area of South America

c. Age:
A. 30 years old   B. 75 years old

d. Immunological status:
A. Taking immunosuppressive drugs for cancer or following organ transplant
B. Normal immune system

E5. What reasons would you give if you were advising a person against constant over-exposure to sun or the use of tanning salons?

ages skin prematurely
can increase chances of skin cancer

■ **E6.** Match the name of the disorder with the description given.

A. Acne	M. Malignant melanoma
B. Basal cell carcinoma	N. Nevus
D. Decubitus ulcers	P. Pruritus
I. Impetigo	

___D___ a. Pressure sores

___N___ b. Mole

___I___ c. Staphylococcal or streptococcal infection which may become epidemic in nurseries

___A___ d. Inflammation of sebaceous glands especially in chin area; occurs under hormonal influence

___M___ e. Rapidly metastasizing form of cancer

___B___ f. The most common form of skin cancer

___P___ g. Itching

E7. Contrast the following pairs of terms.

a. Topical/intradermal
on top of skin   into skin

b. Corn/wart
build up of skin
virus

ANSWERS TO SELECTED CHECKPOINTS

A1. Skin, its derivatives (hair, nails, and glands), and nerve endings.

A3.

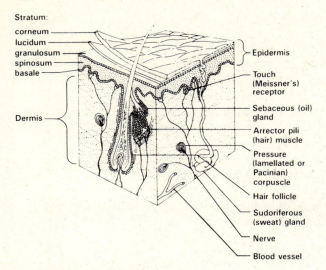

Stratum:
corneum
lucidum
granulosum
spinosum
basale

Dermis

Epidermis

Touch (Meissner's) receptor

Sebaceous (oil) gland

Arrector pili (hair) muscle

Pressure (lamellated or Pacinian) corpuscle

Hair follicle

Sudoriferous (sweat) gland

Nerve

Blood vessel

Figure LG 5.1A Structure of the skin.

(a) Epidermis; epithelium. (b) Dermis, connective tissue; thicker.

A4. Dermis. See Figure LG 5.1A.

A5. Under the skin; superficial fascia or hypodermis; areolar and adipose; anchors skin to underlying tissues and organs.

A7. (a) Keratinocyte. (b) Melanocyte. (c) Langerhans cell. (d) Merkel cell.

A8. See Figure LG 5.1A above.

A9. (a) B. (b) S. (c) G. (d) L. (e) C.

A11. (a) Papillary; loose; papillae, touch. (b) Reticular; dense; collagenous; elastic. (c) Palms, soles, and dorsal surface of the body.

A12. (a) Greater melanin production per melanocyte; basale. (b) Tyrosine; tyrosinase; ultraviolet (UV); albinism. (c) Clusters of melanocytes; no. (d) Carotene and melanin. (e) Widening (dilation) of blood vessels; erythema. (f) Yellow, liver; blue.

A13. (a) Autograft. (b) Isograft. (c) Autologous skin transplantation. (d) Homograft.

B1. Protection.

B2. (a) A C B. (b) B C A. (c) Cells; keratin (especially in cuticle). (d) Follicle; external, epidermis; matrix. (e) Matrix (of bulb of hair follicle); contains blood vessels that nourish the hair.

B6. (a) Sudoriferous. (b) Sebaceous. (c) Ceruminous.

C1. (a) Skinned knee, first- or second-degree burn. (b) Central. (c) Basale; periphery. (d) Contact inhibition; do not. (e) Mitosis.

C3. (a) I. (b) Mig. (c) P. (d) Mat.

C5. (a) 37°C. (b) Hypothalamus. (c) In; dilate. (d) Negative, opposite.

D1.

Changes	Causes
a. Wrinkles; skin springs back less when gently pinched.	**Elastic fibers thicken into clumps and fray**
b. **Increased susceptibility to skin infections and skin breakdown**	Macrophages become less efficient; decrease in number of Langerhans cells.
c. **Loss of body heat; increased likelihood of skin breakdown, as in decubitus ulcers**	Loss of subcutaneous fat
d. Dry, easily broken skin	**Decreased secretion of sebum by sebaceous gland; decreased body fluids**
e. **Gray or white hair; atypical skin pigmentation**	Decrease in number and size of melanocytes

D2. Good nutrition, decreased stress, balance of rest and exercise, not smoking, protection from the sun.

D4. (a) Ecto; fourth. (b) Meso. (c) Ecto. (d) Third; ninth. (e) Fifth or sixth; is not. (f) Sebaceous.

E2. (a) Epidermis; redder; do not; typical sunburn. (b) All of epidermis and upper regions of dermis; do; are not; weeks; is not. (c) Full; white, brown, or black and dry; not painful; is; does.

E3. (a) Amount of body surface burned; proportions. (b) Adult; one-year-old.

E4. (a) A. (b) B. (c) B. (d) A.

E6. (a) D. (b) N. (c) I. (d) A. (e) M. (f) B. (g) P.

MASTERY TEST: Chapter 5

Questions 1–5: Circle the letter preceding the one best answer to each question.

1. "Goose bumps" occur as a result of:
   A. Contraction of arrector pili muscles
   B. Secretion of sebum
   C. Contraction of elastic fibers in the bulb of the hair follicle
   D. Contraction of papillae

2. Select the one *false* statement about the stratum basale.
   A. It is the one layer of cells that can undergo cell division.
   B. It consists of a single layer of squamous epithelial cells.
   C. It is part of the stratum germinativum.
   D. It is the deepest layer of the epidermis.

3. Select the one *false* statement.
   A. Epidermis is composed of epithelium.
   B. Dermis is composed of connective tissue.
   C. Pressure-sensitive Pacinian corpuscles are normally more superficial in location than Meissner's touch receptors.
   D. The amino acid tyrosine is necessary for production of the skin pigment melanin.

4. At about what age do epidermal ridges which cause fingerprints develop?
   A. Months 3–4 in fetal development
   B. Age 3–4 months
   C. Age 3–4 years
   D. Age 30–40 years

5. Which is derived from mesoderm?
   A. Epidermis   B. Dermis

Questions 6–10: Arrange the answers in correct sequence.

_____ _____ _____ 6. From most serious to least serious type of burn:
   A. First-degree
   B. Second-degree
   C. Third-degree

_____ _____ _____ 7. Deep wound healing involves four phases. List in order the phases following the inflammatory phase.
   A. Migration
   B. Maturation
   C. Proliferation

_____ _____ _____ 8. From outside of hair to inside of hair:
   A. Medulla
   B. Cortex
   C. Cuticle

_____ _____ _____ 9. From most superficial to deepest:
   A. Dermis
   B. Epidermis
   C. Superficial fascia

_____ _____ _____ 10. From most superficial to deepest:
   A. Stratum lucidum
   B. Stratum corneum
   C. Stratum germinativum

Questions 11–20: Circle T (true) or F (false). If the statement is false, change the underlined word or phrase so that the statement is correct.

T F 11. Hairs are <u>noncellular structures composed entirely of nonliving substances</u> secreted by follicle cells.

T F 12. The color of skin is due primarily to a pigment named <u>keratin</u>.

T F 13. The <u>outermost layers of epidermis</u> are composed of dead cells.

T F 14. Eccrine sweat glands are <u>more</u> numerous than apocrine sweat glands, and are especially dense on <u>palms and soles</u>.

T F 15. The dermis consists of two regions; the papillary region is more superficial, and the reticular region is deeper.

T F 16. Temperature regulation is a positive feedback system.

T F 17. The internal root sheath is a downward continuation of the epidermis.

T F 18. Both epidermis and dermis contain blood vessels (are vascular).

T F 19. Hair, glands, and nails are all derived from the dermis.

T F 20. Pacinian corpuscles are pressure-sensitive nerve endings most abundant in the subcutaneous tissue, rather than in the epidermis.

Questions 21–25: fill-ins. Complete each sentence with the word or phrase that best fits.

_____ 21. Skin contains a chemical which, under the influence of ultraviolet radiation, leads to formation of vitamin ____.

_____ 22. The cells that are sloughed off as skin cells and undergo keratinization are those of the stratum ____.

_____ 23. Fingerprints are the result of a series of grooves called ____.

_____ 24. The oily glandular secretion which keeps skin and hairs from drying is called ____.

_____ 25. When the temperature of the body increases, nerve messages from brain to skin will decrease body temperature by ____.

ANSWERS TO MASTERY TEST: ■ Chapter 5

Multiple Choice

1. A   4. A
2. B   5. B
3. C

Arrange

6. C B A   9. B A C
7. A C B   10. B A C
8. C B A

True-False

11. F. Composed of different kinds of cells
12. F. Melanin
13. T
14. T
15. T
16. F. Negative
17. F. External root sheath
18. F. Only the dermis
19. F. Epidermis
20. T

Fill-ins

21. D or D_3
22. Corneum
23. Epidermal ridges or grooves
24. Sebum
25. Stimulating sweat glands to secrete and blood vessels to dilate

Principles of Support and Movement

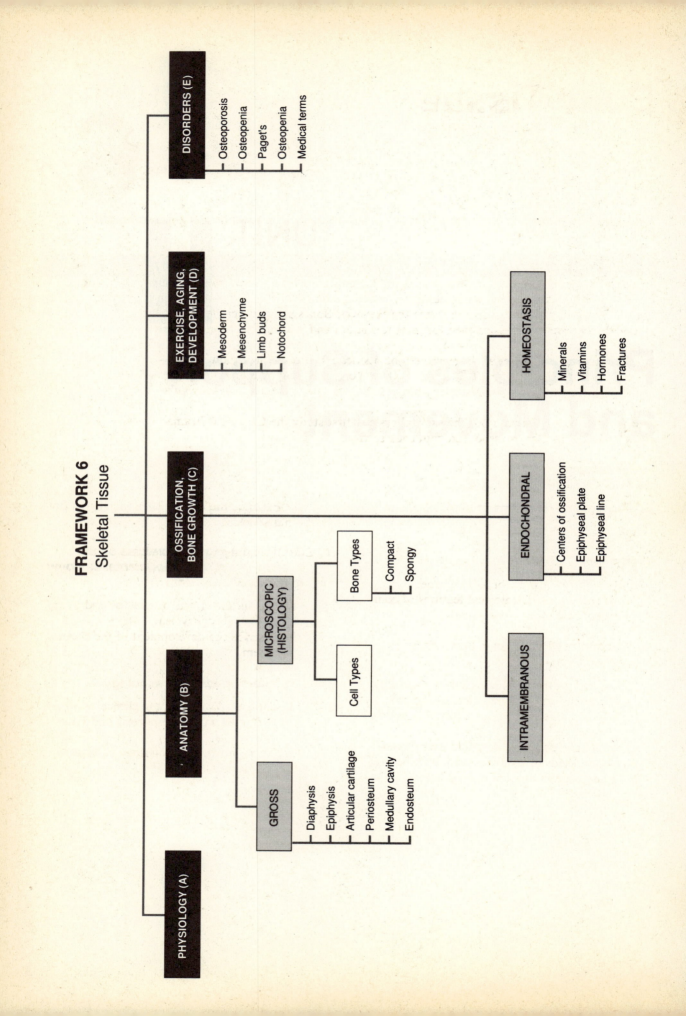

FRAMEWORK 6
Skeletal Tissue

PHYSIOLOGY (A)

ANATOMY (B)

OSSIFICATION, BONE GROWTH (C)

EXERCISE, AGING, DEVELOPMENT (D)
- Mesoderm
- Mesenchyme
- Limb buds
- Notochord

DISORDERS (E)
- Osteoporosis
- Osteopenia
- Paget's
- Osteopenia
- Medical terms

GROSS
- Diaphysis
- Epiphysis
- Articular cartilage
- Periosteum
- Medullary cavity
- Endosteum

MICROSCOPIC (HISTOLOGY)

Cell Types

Bone Types
- Compact
- Spongy

INTRAMEMBRANOUS

ENDOCHONDRAL
- Centers of ossification
- Epiphyseal plate
- Epiphyseal line

HOMEOSTASIS
- Minerals
- Vitamins
- Hormones
- Fractures

Bone Tissue

The skeletal system provides the framework for the entire body, affording strength, support, and firm anchorage for the muscles that move the body. In this initial chapter in Unit II, you will explore tissue that forms the skeleton. Bones provide rigid levers covered with connective tissue designed for joint formation and muscle attachment. Microscopically, bones present a variety of cell types intricately arranged in an osseous sea of calcified intercellular material. Bone growth or ossification occurs throughout life. Essential nutrients, hormones, and exercise regulate the growth and maintenance of the skeleton. Fractures and other disorders may result from a lack of such normal regulation.

As you begin your study of the skeletal system, first study the Chapter 6 Framework and read key terms associated with each section.

TOPIC OUTLINE AND OBJECTIVES

A. Physiology: functions of skeletal tissue

1. Discuss the functions of bone.

B. Anatomy, histology of bone

2. Identify the parts of a long bone.
3. Describe the histological features of compact and spongy bone tissue.

C. Bone formation, bone growth, and homeostasis

4. Contrast the steps involved in intramembranous and endochondral ossification.
5. Describe the processes involved in bone remodeling.
6. Define a fracture, describe several common kinds of fractures, and describe the sequence of events involved in fracture repair.

7. Describe the role of bone in calcium homeostasis.

D. Effects of aging and exercise on the skeletal system; developmental anatomy of the skeletal system

8. Explain the effects of exercise and aging on the skeletal system.
9. Describe the development of the skeletal system.

E. Disorders, medical terminology

10. Contrast the causes and clinical symptoms associated with osteoporosis and Paget's disease.
11. Define medical terminology associated with skeletal tissue.

WORDBYTES

Now study the following parts of words that may help you better understand terminology in this chapter.

Wordbyte	Meaning	Example	Wordbyte	Meaning	Example
-blast	germ, to form	osteo*blast*	os-, osteo-	bone	*os*sification, osteo*cyte*
-clast	to break	osteo*clast*			
chondr-	cartilage	peri*chondr*ium	peri-	around	*peri*osteum

CHECKPOINTS

A. Functions of skeletal tissue   (page 147)

A1. List five or more functions of the skeletal system.

support   blood cell development
protection   mineral storage (calcium, phosphorus)
movement   Storage of energy (yellow marrow)

■ **A2.** What other systems of the body depend on a healthy skeletal system? Explain why in each case.

every system for protection
bones also produce red blood cells
and store calcium

A3. Define *osteology.*

Study of bone structures
and bone disorders

B. Anatomy, histology of skeletal tissue   (pages 147–151)

■ **B1.** The skeletal system consists of four types of connective tissue. Name these.

☆ *Cartilage, bone tissue (osseous) bone marrow*
periosteum

■ **B2.** On Figure LG 6.1, label the *diaphysis, epiphysis, epiphyseal plate, nutrient foramen,* and *medullary (marrow) cavity.* Then color the parts of a long bone listed on the figure.

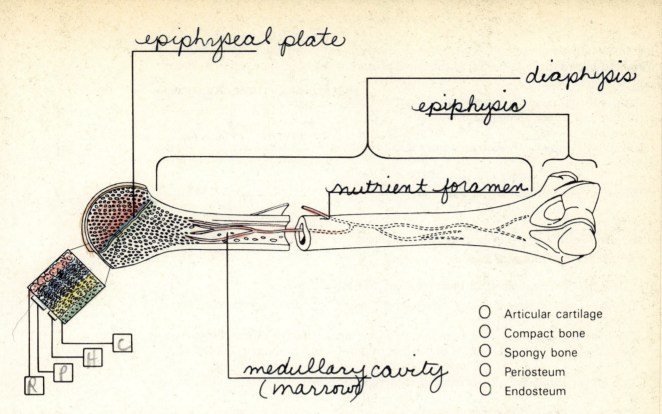

epiphyseal plate

diaphysis

epiphysis

nutrient foramen

medullary cavity
(marrow)

○ Articular cartilage
○ Compact bone
○ Spongy bone
○ Periosteum
○ Endosteum

Figure LG 6.1 Diagram of a long bone that has been partially sectioned lengthwise. Color and label as directed in Checkpoints B2 and C6.

■ **B3.** Match the names of parts of a long bone listed in the box with the description below.

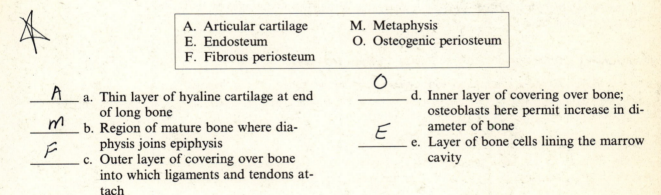

A. Articular cartilage	M. Metaphysis
E. Endosteum	O. Osteogenic periosteum
F. Fibrous periosteum	

_A_ a. Thin layer of hyaline cartilage at end of long bone

_m_ b. Region of mature bone where diaphysis joins epiphysis

_F_ c. Outer layer of covering over bone into which ligaments and tendons attach

_O_ d. Inner layer of covering over bone; osteoblasts here permit increase in diameter of bone

_E_ e. Layer of bone cells lining the marrow cavity

■ **B4.** In the following exercise, bone cells are described in the sequence in which they form and later destroy bone. Write the name of the correct bone type after its description.

a. Derived from mesenchyme, these cells can undergo mitosis and differentiate into osteoblasts: *osteoblasts* osteoprogenitor

b. Form bone initially by secreting mineral salts and fibers: *osteoblasts*

c. Maintain bone tissue; not mitotic: *osteocytes*

d. Resorb (degrade) bone in bone repair, remodeling, and aging: *osteoclasts*

■ **B5.** Describe the components of bone by doing this exercise.

a. Typical of all connective tissues, bone consists mainly of *(cells? intercellular material?)*.

b. The intercellular substance of bone is unique among connective tissues. Protein fibers form about *(25? 50?)* percent of the weight of bone, whereas mineral salts

account for about _50_ percent of the weight of bone.

c. The two main salts present in bone are _Calcium_ and _pottasium_.

d. The *(hardness? tensile strength?)* of bones is related to inorganic chemicals in bone, namely the mineral salts. Resistance to being stretched or torn apart, a property of bone known as *(hardness? tensile strength?)*, is afforded by *(mineral salts? collagen fibers?)*.

e. Bone *(is completely solid? contains some spaces?)*. Write two advantages of that structural feature of bone.

lightness and transfer of nutrients + blood

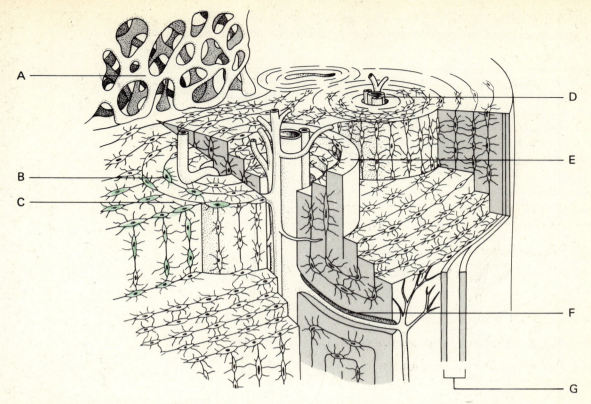

A ————

D ————

B ————

C ————

E ————

F ————

G ————

Figure LG 6.2 Osteons (Haversian systems) of compact bone. Identify lettered structures and color as directed in Checkpoint B6.

■ **B6.** Refer to Figure LG 6.2 and complete this exercise about bone structure.

a. Compact bone is arranged in concentric circle patterns known as

Osteons. Each individual concentric layer of bone is known as a

lamella, labeled with letter _____ in the figure.

b. The osteon pattern of compact bone permits blood vessels and nerves to supply bone cells trapped in hard bone tissue. Blood vessels and nerves penetrate bone from

the periosteum, labeled with letter _G_ in the figure. These structures then pass

through horizontal canals, labeled _E_, and known as perforating (Volkmann's) canals. These vessels and nerves finally pass into microscopic channels, labeled

_____, in the center of each osteon. Color blood vessels red and blue in the figure.

c. Mature bone cells, known as _____, are located relatively far apart in bone tissue. These are present in "little lakes," or

lacunæ, labeled _B_ in the figure. Color ten lacunae green.

d. Minute canals, known as canaliculi, are labeled with letter _____. What is the function of these channels?

■ **B7.** Do this exercise about spongy bone.

a. Spongy bone is arranged in *(osteons? trabeculae?)*, which may be defined as:

_____.

b. Spongy bone makes up most of the *(diaphyses?* ⟨*epiphyses?*⟩*)* of long bones. Spongy bone *(is not?* ⟨*is also?*⟩*)* located within bones that are short, flat, and irregular in shape.

c. *A clinical challenge.* Red blood cell formation (hematopoiesis) normally takes place in *(all? only certain areas of?)* spongy bone tissue. Name four or more bones in which this process takes place.

_____   _____

_____   _____

Of what clinical importance is this information?

C. Ossification, bone growth, and homeostasis   (pages 151–159)

C1. What is mesenchyme? What does it have to do with the skeleton?

■ **C2.** The two main kinds of tissue that compose the "skeleton" of a developing

embryo or fetus are _____ and _____

_____.

■ **C3.** Which bones of the body form by the process of intramembranous ossification?

flat bones of skull

mandible

clavicles

■ **C4.** Refer to Figure 7.6, page 173 in your text. Locate the parietal bone, a typical "flat bone" forming in the fetal skull. Describe intramembranous formation of this bone by using each of the terms in the box once.

~~Compact~~	Mesenchyme	~~Osteoprogenitor~~
~~Fontanel~~	Ossification	Protection
~~Growth~~	~~Osteoblast~~	~~Spongy~~
~~Marrow~~	~~Osteocyte~~	

a. Fibrous membranes surrounding the head of the developing baby contain embryonic *mesenchyme* ~~Osteoprogenitor~~ cells. These cells cluster in areas known as centers of ~~growth~~ *ossification*, and these mark areas where bone tissue will form.

b. Differentiation of these cells leads to *osteoblast* *osteoprogenitor* cells and then into *osteoblasts* and *osteocytes* cells that surround themselves with organic and inorganic matrix forming spongy bone. (For help, look back at Checkpoint B4, page 106.)

c. Eventually, mesenchyme on the outside of the bone develops into a periosteal covering over the skull bone. This lays down some compact bone over spongy bone. The resulting typical flat bone is analogous to a flattened jelly sandwich: two firm (flat) "slices of bread" comparable to *Compact* bone with a "jelly" of *Spongy* bone filled with red *marrow*.

d. By birth, fetal skull bones are still not completed. Some fibrous membranes still remain at points where skull bones meet, known as *fontanel*. Of what advantage are these "soft spots"?

brain growth + baby delivery

■ **C5.** Endochondral ossification refers to bone formation from an initial model made of *(fibrous membranes?* (cartilage?)). Increase your understanding of this process by building your own outline in this exercise.

a. First identify the four major steps in endochondral ossification by filling in the four long lines (I–IV) below with the correct terms selected from those in the box below.

Development	Growth
Diaphysis and epiphyses	Primary ossification center

I. _____ of the cartilage model   ___ ___

II. _____ of the cartilage model   **A** ___ ___ ___

III. Development of the _____   ___ ___

IV. Development of the _____   ___

b. Now fill in details of the process by placing the following statements of events in the correct sequence. Write the letters on the lines to the right of the correct phase (I–IV) above. One is done for you.

A. The cartilage model grows in length as chondrocytes divide and secrete more cartilage matrix; it grows in thickness as new chondroblasts develop within the perichondrium.

B. Cartilage cells in the center of the diaphysis accumulate glycogen, enlarge, and burst, releasing chemicals that alter pH and trigger calcification of cartilage.

C. Cartilage cells die since they are deprived of nutrients; in this way spaces are formed within the cartilage model.

D. Blood vessels penetrate the perichondrium, stimulating perichondrial cells to form osteoblasts. A collar of bone forms and gradually thickens around the diaphysis. The membrane covering the developing bone is now called the periosteum.

E. A hyaline cartilage model of future bone is laid down by differentiation of mesenchyme into chondroblasts.

F. A perichondrium develops around the cartilage model.

G. Secondary ossification centers develop in epiphyses, forming spongy bone there about the time of birth. Hyaline cartilage remains as the epiphyseal plate for as long as the bone grows.

H. Capillaries grow into spaces and osteoblasts deposit bone matrix over disintegrating calcified cartilage. In this way spongy bone is forming within the diaphysis at the primary ossification center.

I. Osteoclasts break down the newly formed spongy bone in the very center of the bone, thereby leaving the medullary (marrow) cavity.

■ **C6.** The epiphyseal plate of a growing bone consists of four zones. First identify each region by placing the letter of the name of that zone next to the correct description. Next place these same four letters in the boxes in the inset on Figure LG 6.1 to indicate the locations of these cartilage zones within the epiphyseal plate of the developing bone.

| C. Calcified | P. Proliferating |
| H. Hypertrophic | R. Resting |

_R_ a. Zone of cartilage which is not involved in bone growth, but anchors the epiphyseal plate (site of bone growth) to the bone of the epiphysis.

_P_ b. New cartilage cells form here by mitosis and are arranged like stacks of coins.

_H_ c. Cartilage cells mature and die here as matrix around them calcifies.

_C_ d. Osteoblasts and capillaries from the diaphysis invade this region to lay down bone upon the calcified cartilage remnants here.

■ **C7.** Complete this summary statement about bone growth at the epiphyseal plate.

Cartilage cells multiply on the *(epiphysis? diaphysis?)* side of the epiphyseal plate, providing temporary new tissue. But cartilage cells then die and are replaced by bone cells on the *(epiphysis? diaphysis?)* side of the epiphyseal plate.

■ **C8.** Complete this activity about maturation of the skeleton.

a. The epiphyseal *(line? plate?)* is the bony region in bones of adults that marks the original cartilaginous epiphyseal *(line? plate?)*.

b. Ossification of most bones is finished by age *(18? 25?)*. Completion of this process usually occurs earlier in *(females? males?)*.

c. Bones must grow in diameter to match growth in length. *(Osteoblasts? Osteoclasts?)* are cells that destroy bone to enlarge the medullary cavity, whereas *(osteoblasts? osteoclasts?)* are cells from the periosteum that increase the thickness of compact bone.

C9. Defend or dispute this statement: "Once a bone, such as your thighbone, is formed, the bone tissue is never replaced unless the bone is broken."

replace every 4 months

■ **C10.** Explain the roles of lysosomes and acids in osteoclastic activity.

Lysosomes release enzymes that may digest collagen

■ **C11.** List factors that appear to be necessary for normal bone growth and remodeling:
a. Four minerals

Calcium   magnesium
phosphorus   manganese

b. Four vitamins

Vit D   Vit A
Vit C   Vit B₁₂

c. Four hormones

hGH   calcitonin
estrogen   parathyroid

C12. Contrast the terms related to types of fractures in each pair.
a. Simple/compound

b. Partial/complete

c. Pott's/Colles'

d. Comminuted/greenstick

C13. *A clinical challenge:* What is the major difference between these two methods of setting a fracture: *closed reduction and open reduction?*

■ **C14.** Describe the four main steps in repair of a fracture by filling in terms from this list in the spaces below. One answer will be used twice.

Cell Type	Process or Stage	Time Period
~~Chondroblasts~~	~~Fracture hematoma~~	Hours
Fibroblasts	~~Inflammation and cell death~~	Months
~~Osteoblasts~~	~~Procallus~~	Weeks
~~Osteoclasts~~	Remodeling	
~~Phagocytes~~		

a. Blood flows from torn blood vessels, forming a clot known as a

fracture hematoma within 6 to 8 _hours_ of the injury.

Because of the trauma and lack of circulation, _inflam + cell death_ occurs in

the area. As a result, cells known as _phagocytes_ and

osteoclasts enter the area and remove debris over a period of

several _weeks_ .

b. Next, granulation tissue known as a _procallus_ forms. Then cells

called _fibroblasts_ produce collagen fibers which connect broken

ends of the bone. Other cells, the _chondroblasts_ , form a soft callus at a
distance from the remaining healthy, vascular bone. This stage lasts about 3

weeks .

c. The hard callus stage occurs next as _osteoblasts_ cells form bone adjacent to the remaining healthy, vascular bone. This stage lasts 3 to 4

months .

d. The callus is reshaped during the final (or _remodeling_) phase.

■ **C15.** Match the hormones listed in the box with their functions in bone growth and re-modeling, and in calcium homeostasis. One answer will be used twice.

> CT. Calcitonin
> hGH. Human growth hormone
> PTH. Parathyroid hormone
>
> S. Sex hormones (testosterone, estrogens)

PTH a. Produced by the parathyroid gland, it increases osteoclast activity, causing bone destruction and increase in blood level of calcium.

CT b. Produced by the thyroid gland, it inhibits osteoclast activity, causing bone formation and decrease in blood level of calcium.

hGH c. A pituitary hormone, it enhances bone growth in general; deficiency leads to dwarfism.

S d. These hormones stimulate osteoblasts to form new bone and so are responsible for growth spurt at puberty; they can also cause premature closure of epiphyseal plate (short stature) if puberty occurs early.

PTH e. Stimulates kidneys to retain calcium (in blood), to eliminate phosphate (in urine), and to activate vitamin D (which helps in absorption of dietary calcium).

C16. Fracture patients may enhance healing of the fractured bone by utilizing coils that create pulsating electromagnetic fields (PEMFs). Explain how this apparatus, which is fastened around a cast for most of the day and night for a number of months, may stimulate bone growth.

D. Effects of exercise and aging on the skeletal system; developmental anatomy of the skeletal system (pages 159–160)

■ **D1.** Circle the correct answers about the effects of exercise upon bones.

a. Exercise (*strengthens? weakens?*) bones. A healing bone that is not exercised is likely to become (*stronger? weaker?*) during the period that it is in a cast.

b. The pull of muscles upon bones, as well as the tension on bones as they support body weight during exercise, causes (*increased? decreased?*) production of the protein collagen.

c. The stress of exercise also stimulates (*osteoblasts? osteoclasts?*) and increases production of the hormone calcitonin, which inhibits (*osteoblasts? osteoclasts?*).

■ **D2.** Complete this exercise about skeletal changes that occur in the normal aging process.

a. The amount of calcium in bones (*de? in?*)-creases with age. As a result, bones of the elderly are likely to be (*stronger? weaker?*) than bones of younger persons. This change occurs at a younger age in (*men? women?*).

b. Another component of bones that decreases with age is ___protein___. What is the significance of this change?

Bones are more britle

■ **D3.** Complete this learning activity about development of the skeleton.

a. Bones and cartilage are formed from (ecto? *meso?* endo?)-derm which later differ-

entiates into the embryonic connective tissue called *mesenchyme*.
Some of these cells become chondroblasts, which eventually form (bone? *cartilage?*),

whereas *osteo*-blasts develop into bone.

b. Limb buds appear during the (*fifth?* seventh? tenth?) week of development. At this
point the skeleton in these buds consists of (bone? *cartilage?*). Bone begins to form

during week *6 (9)*.

c. By week (six? seven? *eight?*) the upper extremity is evident, with defined shoulder,
arm, elbow, forearm, wrist, and hand. The lower extremity is also developing, but at
a slightly (faster? *slower?*) pace.

d. The notochord develops in the region of the (head? *vertebral column?* pelvis?). Most
of the notochord eventually (forms vertebrae? *disappears?*). Parts of it persist in the
intervertebral disk

E. Disorders, medical terminology (pages 160–162)

■ **E1.** Describe osteoporosis in this exercise.

a. Osteoporotic bones exhibit *de*-creased bone mass and *in*-creased susceptibility to
fractures. This disorder is associated with a loss of the hormone

estrogen. This hormone stimulates osteo-

blast activity.

b. Circle the category of persons at higher risk for osteoporosis in each pair:
younger persons/(older persons)
men/(women)
(short, thin persons)/tall, large-build persons
(blacks)/whites

c. List three factors in daily life that will help to prevent osteoporosis. (*Hint:* Focus on
diet and activities.)

exercise weight bar

not smoking

■ **E2.** Check your understanding of bone disorders by matching the terms in the box with
related descriptions.

Osteomyelitis	Osteosarcoma
Osteopenia	

a. Decreased bone mass:
osteopenia

b. Malignant bone tumor:
Osteosarcoma

c. Bone infection, for example, caused by
Staphylococcus: *osteomyelitis*

A2. Essentially all do. For example, muscles need intact bones for movement to occur; bones are site of blood formation; bones provide protection for viscera of nervous, digestive, urinary, reproductive, cardiovascular, respiratory, and endocrine systems; broken bones can injure integument.

B1. Cartilage, bone (osseous) tissue, bone marrow, and periosteum.

B2.

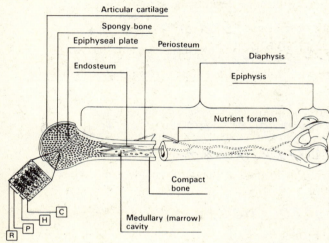

Figure LG 6.1A Diagram of a developing long bone that has been partially sectioned lengthwise.

B3. (a) A. (b) M. (c) F. (d) O. (e) E.

B4. (a) Osteoprogenitor. (b) Osteoblast. (c) Osteocyte. (d) Osteoclast.

B5. (a) Intercellular material. (b) 25, 50. (c) Tricalcium phosphate, calcium carbonate. (d) Hardness; tensile strength, collagen fibers. (e) Contains some spaces; provides channels for blood vessels and makes bones lighter weight.

B6. (a) Osteons (haversian systems); lamella, E. (b) G; F; D. (c) Osteocytes; lacunae, B. (d) C; contain extensions of osteocytes bathed in extracellular fluid (ECF); permit communication between osteocytes at gap junctions; provide routes for nutrients and oxygen to reach osteocytes and for wastes to diffuse away.

B7. (a) Trabeculae, irregular latticework of plates of bone tissue containing osteocytes surrounded by red marrow and blood vessels. (b) Epiphyses; is also. (c) Only certain areas of; hipbones, ribs, sternum, vertebrae, skull bones, and epiphyses of long bones such as femurs. These areas (such as the sternum) may be biopsied to examine for aplas-

tic anemia or for response of bone marrow to antianemia medications. These sites (such as hipbones) may be utilized for marrow transplant (for example, for patients with severe anemia or cancers involving bone marrow).

C2. Hyaline cartilage, fibrous connective tissue membranes.

C3. Flat bones of the skull, mandible (lower jaw bone) and clavicles (collarbones).

C4. (a) Mesenchyme; ossification. (b) Osteoprogenitor, osteoblast, osteocyte. (c) Compact, spongy, marrow. (d) Fontanels (soft spots); permit additional bone growth and safer passage (protection) of the fetal head during birth.

C5. Cartilage. (I) Development: E F. (II) Growth: A B C D. (III) Primary ossification center: H I. (IV) Diaphysis and epiphyses: G.

C6. (a) R. (b) P. (c) H. (d) C. See Figure 6.1A above.

C7. Epiphysis; diaphysis.

C8. (a) Line, plate. (b) 25; females. (c) Osteoclasts, osteoblasts.

C10. Lysosomes release enzymes that may digest the protein collagen, whereas acids may cause minerals of bones to dissolve. Both contribute to bone resorption.

C11. (a) Calcium, phosphorus, magnesium, manganese. (b) D C A B_{12}. (c) hGH, estrogens, progesterone, insulin, thyroid hormone, parathyroid hormone, and calcitonin.

C14. (a) Fracture hematoma, hours; inflammation and cell death, osteoclasts and phagocytes, weeks. (b) Procallus; fibroblasts; chondroblasts; weeks. (c) Osteoblasts; months. (d) Remodeling.

C15. (a) PTH. (b) CT. (c) hGH. (d) S. (e) PTH.

D1. (a) Strengthens; weaker. (b) Increased. (c) Osteoblasts, osteoclasts.

D2. (a) De; weaker; women. (b) Protein; bones are more brittle and vulnerable to fracture.

D3. (a) Meso-, mesenchyme; cartilage, osteo. (b) Fifth; cartilage; six or seven. (c) Eight; slower. (d) Vertebral column; disappears; intervertebral discs.

E1. (a) De, in; estrogen, blast. (b) Older; women; short, thin persons; blacks. (c) Adequate intake of calcium, vitamin D, weight-bearing exercise, and not smoking; estrogen replacement therapy (HRP) may be advised for some women.

E2. (a) Osteopenia. (b) Osteosarcoma. (c) Osteomyelitis.

Questions 1–15: Circle T (true) or F (false). If the statement is false, change the under-lined word or phrase so that the statement is correct.

T (F) 1. Greenstick fractures occur only in <u>adults.</u> *children*

(T) F 2. <u>Appositional</u> growth of cartilage or bone means growth in thickness due to action of cells from the perichondrium or periosteum.

T (F) 3. Calcitriol is an <u>active form of calcium.</u> *Active form of Vit. D*

(T) (F) 4. The epiphyseal plate appears <u>earlier</u> in life than the epiphyseal line.

(T) F 5. Osteons (Haversian systems) are found in <u>compact bone, but not in spongy bone.</u>

T (F) 6. Another name for the epiphysis is the <u>shaft of the bone.</u> *end of the bone*

T (F) 7. A compound fracture is defined as one in <u>which the bone is broken into many pieces.</u> *protrudes through skin*

(T) F 8. Haversian canals run <u>longitudinally (lengthwise)</u> through bone, but perforating (Volkmann's) canals run horizontally across bone.

T (F) 9. Canaliculi are tiny canals containing <u>blood which nourishes bone</u> cells in lacunae. *parts of osteocytes + fluid from blood vessels*

T (F) 10. Compact bone that is of intramembranous origin <u>differs structurally</u> from compact bone devel-oped from cartilage. *only in origin*

(T) F 11. In a long bone the primary ossification center is located in the <u>diaphysis, whereas the secon-dary center of ossification is in the epiphysis.</u>

T (F) 12. <u>Osteoblasts</u> are bone-destroying cells. *Osteoclasts*

(T) F 13. Most bones start out in embryonic life as <u>hyaline cartilage.</u>

(T) F 14. The layer of compact bone is <u>thicker</u> in the diaphysis than in the epiphysis.

T (F) 15. <u>Lamellae</u> are small spaces containing bone cells. *lacunae*

Questions 16–20: Arrange the answers in correct sequence.

__C__ __A__ __B__ 16. Steps in marrow formation during endochondral bone formation, from first to last:
   A. Cartilage cells burst, causing intercellular pH to become more alkaline.
   B. pH changes cause calcification, death of cartilage cells; eventually spaces left by dead cells are penetrated by blood vessels.
   C. Cartilage cells hypertrophy with accumulated glycogen.

__B__ __C__ __A__ 17. From most superficial to deepest:
   A. Endosteum   C. Compact bone
   B. Periosteum

__A__ __C__ __B__ 18. Phases in repair of a fracture, in chronological order:
   A. Fracture hematoma formation   C. Callus formation
   B. Remodeling

__A__ __C__ __B__ 19. Phases in formation of bone in embryonic life, in chronological order:
   A. Mesenchyme cells   C. Osteoblasts
   B. Osteocytes

__A__ __B__ __C__ 20. Portions of the epiphyseal plate, from closest to epiphysis to closest to dia-physis:
   A. Zone of resting cartilage   C. Zone of calcified cartilage
   B. Zone of proliferating cartilage

Questions 21–25: fill-ins. Write the word or phrase that best completes the statement.

diaphysis   21. ____ is a term which refers to the shaft of the bone.

Osteoporosis   22. ____ is a disorder primarily of older women associated with decreased estrogen level; it is characterized by weakened bones.

perichondrium   23. ____ is the connective tissue covering over cartilage in adults and also over embryonic cartilaginous skeleton.

skull clavicles   24. The majority of bones formed by intramembranous ossification are located in the ____.

D   25. Vitamin ____ is a vitamin that is necessary for absorption of calcium from the gastrointestinal tract, and so it is important for bone growth and maintenance.

ANSWERS TO MASTERY TEST: ■ Chapter 6

True-False

1. F. Children
2. T
3. F. Active form of vitamin D
4. T
5. T
6. F. End of the bone (often bulbous)
7. F. Protrudes through the skin
8. T
9. F. Parts of osteocytes and fluid from blood vessels in Haversian canals, but not blood itself
10. F. Only in origin
11. T
12. F. Osteoclasts
13. T
14. T
15. F. Lacunae

Arrange

16. C A B
17. B C A
18. A C B
19. A C B
20. A B C

Fill-ins

21. Diaphysis
22. Osteoporosis
23. Perichondrium
24. Skull; also the clavicles
25. D (or D_3)

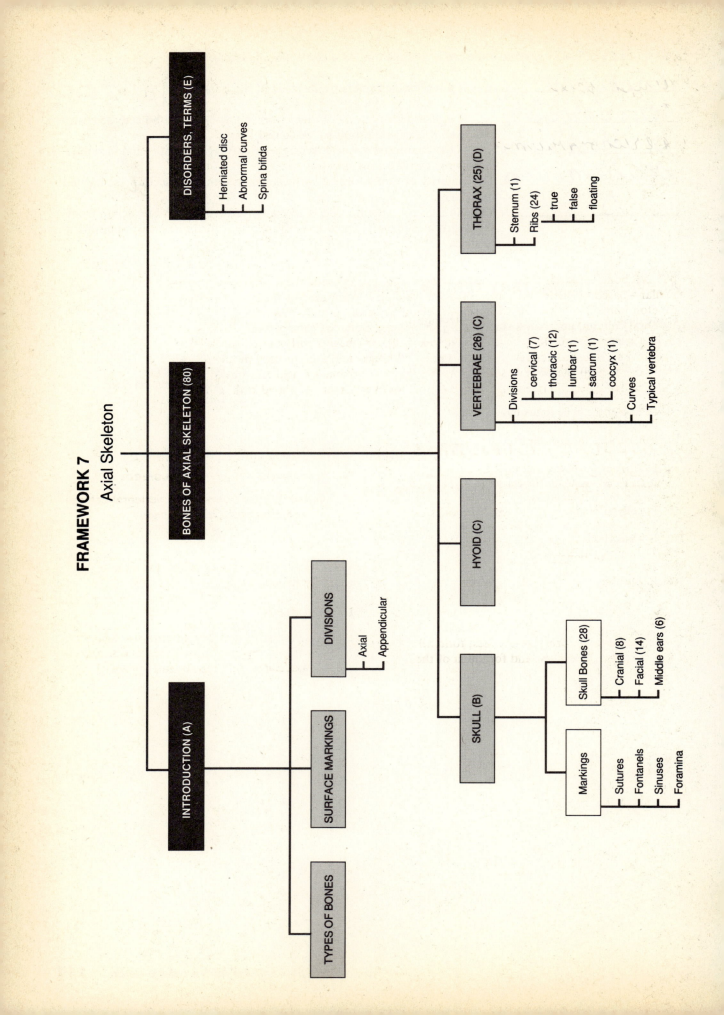

FRAMEWORK 7
Axial Skeleton

DISORDERS, TERMS (E)
— Herniated disc
— Abnormal curves
— Spina bifida

BONES OF AXIAL SKELETON (80)

THORAX (25) (D)
— Sternum (1)
— Ribs (24)
  — true
  — false
  — floating

VERTEBRAE (26) (C)
— Divisions
  — cervical (7)
  — thoracic (12)
  — lumbar (1)
  — sacrum (1)
  — coccyx (1)
— Curves
— Typical vertebra

HYOID (C)

SKULL (B)

Skull Bones (28)
— Cranial (8)
— Facial (14)
— Middle ears (6)

Markings
— Sutures
— Fontanels
— Sinuses
— Foramina

INTRODUCTION (A)

DIVISIONS
— Axial
— Appendicular

SURFACE MARKINGS

TYPES OF BONES

The Skeletal System: The Axial Skeleton

CHAPTER 7

The 206 bones of the human skeleton are classified into two divisions—axial and appendicular—based on their locations. The axial skeleton is composed of the portion of the skeleton immediately surrounding the axis of the skeleton, primarily the skull bones, vertebral column, and bones surrounding the thorax. Bones of the appendages, or extremities, comprise the appendicular skeleton. Chapter 7 begins with an introduction to types and surface characteristics (markings) of all bones and then focuses on the 80 bones of the axial skeleton. Disorders of the vertebral column are also included.

First study the Chapter 7 Framework and note key terms associated with each section.

TOPIC OUTLINE AND OBJECTIVES

A. Introduction: types of bones, surface markings, divisions of the skeletal system

1. Classify the principal types of bones on the basis of shape and location.
2. Describe the various markings on the surfaces of bones.

B. Skull

3. Identify the bones of the skull and the major markings associated with each.
4. Identify the principal sutures, fontanels, paranasal sinuses, and foramina of the skull.

C. Hyoid bone and vertebral column

5. Identify the bones of the vertebral column and their principal markings.

D. Thorax

6. Identify the bones of the thorax and their principal markings.

E. Disorders

7. Contrast herniated (slipped) disc, abnormal curves, and spina bifida as disorders associated with the skeletal system.

Now study the following parts of words that may help you better understand terminology in this chapter.

Wordbyte	Meaning	Example	Wordbyte	Meaning	Example
annulus	ring	*annulus* fibrosus	ethm-	sieve	*ethm*oid bone
costa	rib	*costa*l cartilage	lambd-	L-shaped	*lambd*oidal suture
cribr-	sieve	*cribr*iform plate	pulp-	(soft) flesh	nucleus *pulp*osus
crist-	crest	*crist*a galli			

CHECKPOINTS

A. Introduction: types of bones, surface markings, divisions of the skeletal system (pages 166–169)

A1. Explain how the skeletal system is absolutely necessary for vital life functions of eating, breathing, and movement.

mandible moves to chew   diaphragm pulls lung to breath   bones act as levers to move

A2. Complete the table about the four major and two minor types of bones.

Type of Bone	Structural Features	Examples
a. Long	Slightly curved to absorb stress better	*femur*
b. *short*	*as wide as they are tall highly moveable*	Wrist, ankle bones
c. *flat*	Composed of two thin plates of bone	*large skull bones occipital*
d. Irregular		
e. Sutural (or Wormian)		
f.	Small bones in tendons	

A3. In general, what are the functions of surface markings of bones?

attachment of muscles, allow nutrients in

A4. Contrast the bone markings in each of the following pairs.

a. Tubercle/tuberosity

big bump   little bump

b. Crest/line

c. Fossa/foramen

dent / hole

d. Condyle/epicondyle

bump / bump on bump

■ **A5.** Write the name of the correct answer from the following list of specific bone markings on the line next to its description. The first one has been done for you.

Articular *facet* on vertebra	Maxillary *sinus*
External auditory *meatus*	Optic *foramen*
Greater *trochanter*	Styloid *process*
Head of humerus	Superior orbital *fissure*

a. Air-filled cavity within a bone, connected to nasal cavity: **Maxillary *sinus***

b. Narrow, cleftlike opening between adjacent parts of bone; passageway for blood vessels and nerves: *Superior orbital fissure*

c. Rounded hole, passageway for blood vessels and nerves: *Optic foramen*

d. Tubelike passageway through bone: *External auditory meatus*

e. Large, rounded projection above constricted neck: *head of humerus*

f. Large, blunt projection on the femur: *greater trochanter*

g. Sharp, slender projection: *styloid process*

h. Smooth, flat surface: *articular facet vertabra*

■ **A6.** Describe the bones in the two principal divisions of the skeletal system by completing this exercise. It may help to refer to Figure LG 8.1, page 141.

a. Bones that lie along the axis of the body are included in the ((axial?) appendicular?) skeleton.

b. The axial skeleton includes the following groups of bones. Indicate how many bones are in each category.

22 Skull (cranium, face)

6 Earbones (ossicles)

1 Hyoid

26 Vertebrae

1 Sternum

24 Ribs *12-12*

c. The total number of bones in the axial skeleton is _____.

d. The appendicular skeleton consists of bones in which parts of the body?

e. Write the number of bones in each category. Note that you are counting bones on one side of the body only.

_____ Left shoulder girdle

_____ Left hipbone

_____ Left upper extremity (arm, forearm, wrist, hand)

_____ Left lower extremity (thigh, kneecap, leg, foot)

f. There are _____ bones in the appendicular skeleton.

g. In the entire human body there are _____ bones.

B. Skull (pages 169–183)

■ **B1.** Answer these questions about the cranium.

a. What is the main function of the cranium?

b. Which bones make up the cranium (rather than the face)?

*frontal,   occipital
       parietal*

c. Define *sutures*.

Between which two bones is the sagittal suture located? *parietal*
On Figure LG 7.1a, label the following sutures: *coronal, squamous, lambdoid.*

d. "Soft spots" of a newborn baby's head are known as _____.

What is the location of the largest one? _____

○ Ethmoid bone
○ Frontal bone
○ Lacrimal bone
○ Mandible bone
○ Maxilla bone
○ Nasal bone
○ Occipital bone
○ Parietal bone
○ Sphenoid bone
○ Temporal bone
○ Zygomatic bone

Condylar process

Figure LG 7.1   Skull bones. (a) Skull viewed from right side. Color and label as directed in Checkpoint B6.

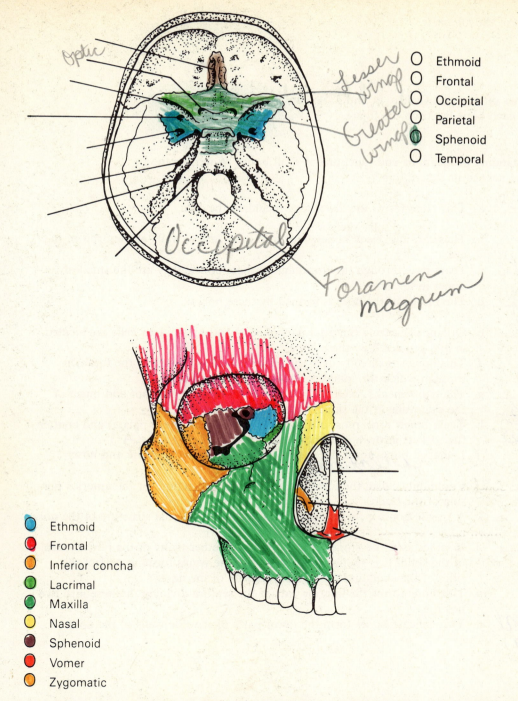

Optic

Lesser wing

Greater wing

O Ethmoid
O Frontal
O Occipital
O Parietal
O Sphenoid
O Temporal

Occipital

Foramen magnum

O Ethmoid
O Frontal
O Inferior concha
O Lacrimal
O Maxilla
O Nasal
O Sphenoid
O Vomer
O Zygomatic

Figure LG 7.1 Skull bones. (b) Floor of the cranial cavity. (c) Anterior view of bones that form the right orbit and the nose. Color and label as directed in Checkpoint B6.

B2. List three functions of the *fontanels*.

■ **B3.** Color the skull bones on Figure LG 7.1a, b, and c. Be sure to color the corresponding color code oval for each bone listed on the figure.

■ **B4.** Check your understanding of locations and functions of skull bones by writing the name of each skull bone next to its description.

✓ *mandible*   a. This bone forms the lower jaw, including the chin.

✓ *zygomatic*   b. These are the cheek bones; they also form lateral walls of the orbit of the eye.

✓ *lacrimal*   c. Tears pass through tiny foramina in these bones; they are the smallest bones in the face.

✓ *nasal*   d. The bridge of the nose is formed by these bones.

✓ *temporal*   e. Organs of hearing (internal part of ears) and mastoid air cells are located in and protected by these bones.

✓ *occipital*   f. This bone sits directly over the spinal column; it contains the foramen through which the spinal cord connects to the brain.

✓ *parietal*   g. The name means "wall." The bones form most of the roof and much of the side walls of the skull.

~~*maxilla*~~ ~~*palatine*~~ ~~*palatine*~~ ~~*lambdoid*~~   h. These bones form most of the roof of the mouth (hard palate) and contain the sockets into which upper teeth are set. *maxilla*

i. L-shaped bones form the posterior parts of the hard palate and nose. *palatine*

✓ *frontal*   j. Commonly called the forehead, it provides protection for the anterior portion of the brain.

✓ *ethmoid*   k. A fragile bone, it forms much of the roof and internal structure of the nose.

✓ *sphenoid*   l. It serves as a "keystone," since it binds together many of the other bones of the skull. It is shaped like a bat, with the wings forming part of the sides of the skull and the legs at the back of the nose.

✓ *vomer*   m. This bone forms the inferior part of the septum dividing the nose into two nostrils.

✓ *inferior nasal concha*   n. Two delicate bones form the lower parts of the side walls of the nose.

■ **B5.** Complete the table describing major markings of the skull.

Marking	Bone	Function
a. Greater wings	Sphenoid	form part of side walls of skull
b. perpendicular plate	ethmoid	Forms superior portion of septum of nose
c. sella turcica	sphenoid	Site of pituitary gland
d. Petrous portion	Temporal	Houses middle ear + inner ear
e. foramen magnum	occipital	Largest hole in skull; passageway for spinal cord
f. alveolar processes	(2) Max Mandible	Bony sockets for teeth
g. External auditory meatus	Temporal	Passageway for sound waves to enter ear
h. Condylar process	Mandible	Articulates w/ temporal bone

■ **B6.** Now label each of the markings listed in the above table on Figure LG 7.1.

■ **B7.** *For extra review.* The 12 pairs of nerves attached to the brain are called cranial nerves (Figure 14.5, page 412 of your text). Holes in the skull permit passage of these nerves to and from the brain. These nerves are numbered according to the order in which they attach to the brain (and leave the cranium) from I (most anterior) to XII (most posterior). To help you visualize their sequence, label the foramina for cranial nerves in order on the left side of Figure LG 7.1b. Complete the table summarizing these foramina. The first one is done for you.

Number and Name of Cranial Nerve	Location of Opening for Nerve
a. I Olfactory	Cribriform plate of ethmoid bone
b. II Optic	

c. III   Oculomotor
   IV   Trochlear
   V   Trigeminal (ophthalmic branch)
   VI Abducens	

d. V Trigeminal (maxillary branch)	
e. V Trigeminal (mandibular branch)	
f. VII Facial VIII Vestibulocochlear	

g. IX   Glossopharyngeal
   X   Vagus
   XI Accessory	

h. XII Hypoglossal	

■ **B8.** Do this exercise about bony structures related to the nose.

a. Name four bones that contain paranasal sinuses.

_____   _____   _____   _____

Practice identifying their locations the next time you have a cold!

b. List three functions of paranasal sinuses.

c. Most internal portions of the nose are formed by a bone that includes these markings: paranasal sinuses, conchae, and part of the nasal septum. Name the bone:

_____ .

d. What functions do nasal conchae perform?

e. On Figure LG 7.1c, label the bone that forms the inferior portion of the nasal septum. The anterior portion of the nasal septum consists of flexible

_____ .

■ **B9.** *A clinical challenge. TMJ syndrome* refers to disorders involving the only movable

joint in the skull, the joint between the _____ and

_____ bones. List three signs or symptoms of this problem.

■ **B10.** A good way to test your ability to visualize locations of important skull bones is to try to identify bones that form the orbit of the eye. On Figure LG 7.lc, locate six bones that form the orbit. (*Note:* The tiny portion of the seventh bone, the superior tip of the palatine bone, is not visible in the figure.)

C. Hyoid bone and vertebral column (pages 183–190)

■ **C1.** Identify the location of the hyoid bone on yourself. Place your hand on your throat and swallow. Feel your larynx move upward? The hyoid sits just *(superior? inferior?)* to the larynx at the level of the mandible.

■ **C2.** In what way is the hyoid unique among all the bones of the body?

it touches no other bone

■ **C3.** The five regions of the vertebral column are grouped (A–E) in Figure LG 7.2. Color vertebrae in each region and be sure to select the same color for the corresponding color code oval. Now write on lines next to A–E the number of vertebrae in each region. One is done for you. Also label the first two cervical vertebrae.

■ **C4.** Note which regions of the vertebral column in Figure 7.2 normally retain an

anteriorly concave curvature in the adult: _____ and

_____. These are considered *(primary? secondary?)* curvatures. This classification is based upon the fact that these curves *(were present originally during fetal life? are more important?)*.

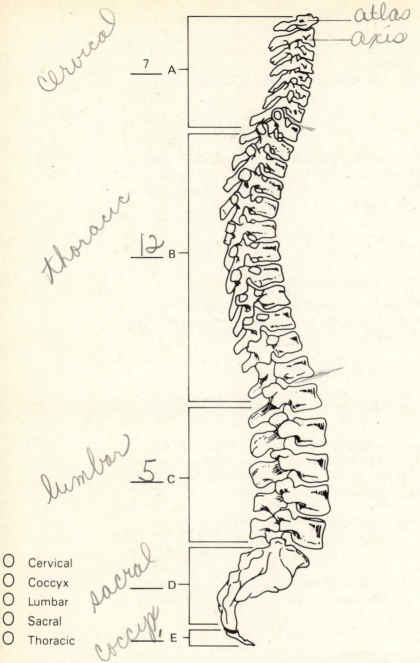

cervical

7 A —

atlas
axis

thoracic

12 B —

lumbar

5 C —

○ Cervical
○ Coccyx
○ Lumbar
○ Sacral
○ Thoracic

sacral

D —

coccyx

E —

Figure LG 7.2 Right lateral view of the vertebral column. Color and label as directed in Checkpoint C3.

C5. Define good posture and list several suggestions for maintaining good posture.

Which sets of muscles are especially important in maintenance of good posture?

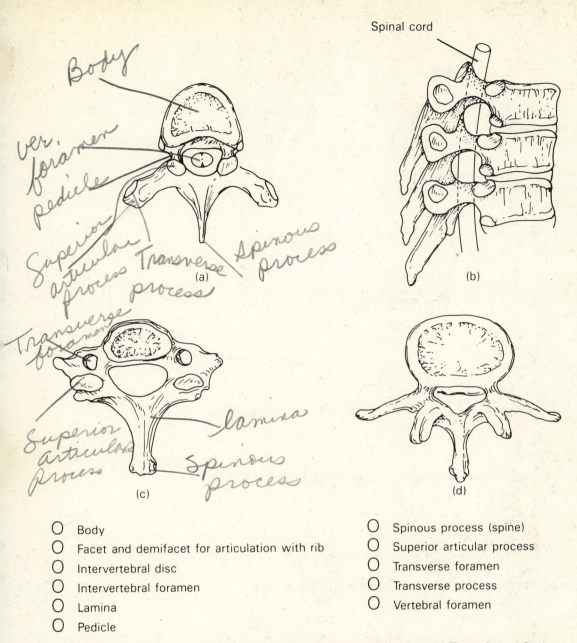

Spinal cord

Body

ver. foramen

pedicle

Superior articular process

Transverse process

Spinous process

Transverse foramen

(a)

(b)

Superior articular process

lamina

Spinous process

(c)

(d)

○ Body
○ Facet and demifacet for articulation with rib
○ Intervertebral disc
○ Intervertebral foramen
○ Lamina
○ Pedicle

○ Spinous process (spine)
○ Superior articular process
○ Transverse foramen
○ Transverse process
○ Vertebral foramen

Figure LG 7.3 Typical vertebrae. (a) Thoracic vertebra, superior view. (b) Thoracic vertebrae, right lateral view. (c) Cervical vertebra, superior view. (d) Lumbar vertebra, superior view. Color and label as directed in Checkpoint C6.

■ **C6.** On Figure LG 7.3, color the parts of vertebrae and corresponding color code ovals. As you do this, notice differences in size and shape among the three vertebral types in the figure.

■ **C7.** Identify distinctive features of vertebrae in each region.

C. Cervical	S. Sacral
L. Lumbar	T. Thoracic

_____ a. Small body, foramina for vertebral blood vessels in transverse processes

_____ b. Only vertebrae that articulate with ribs

_____ c. Massive body, blunt spinous process and articular processes directed medially or laterally

_____ d. Long spinous processes that point inferiorly

_____ e. Articulate with the two hipbones

■ **C8.** *A clinical challenge.* State the clinical significance of these markings.

a. Sacral hiatus

b. Sacral promontory

c. Odontoid process

D. Thorax (pages 190–195)

D1. Name the structures that compose the thoracic cage. (Why is it called a "cage"?)

■ **D2.** On Figure LG 7.4, color the three parts of the sternum.

■ **D3.** *A clinical challenge.* State one reason for performing a sternal puncture.

■ **D4.** Complete this exercise about ribs. Refer again to Figure LG 7.4.

a. There are a total of _____ ribs (_____ pairs) in the human skeleton.
b. Ribs slant in such a way that the anterior portion of the rib is *(superior? inferior?)* to the posterior end of the rib.

c. Posteriorly, all ribs articulate with _____. Ribs also pass *(anterior? posterior?)* to and articulate with transverse processes of vertebrae. What is the functional advantage of such an arrangement?

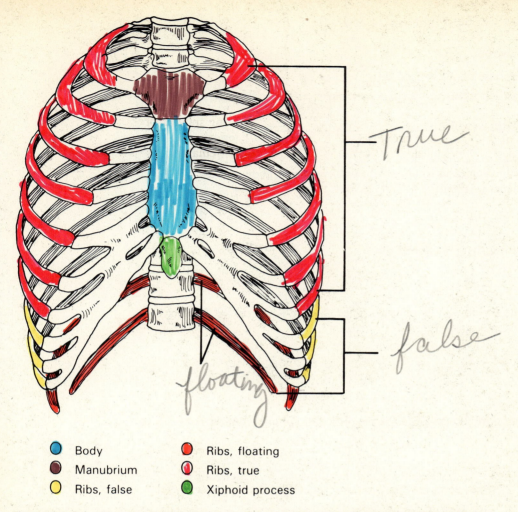

True

false

floating

Body   Ribs, floating
Manubrium   Ribs, true
Ribs, false   Xiphoid process

Figure LG 7.4 Anterior view of the thorax. Color as directed in Checkpoints D2 and D4.

d. Anteriorly, ribs numbered _____ to _____ attach to the sternum directly by

means of strips of hyaline cartilage, called _____ cartilage.
These ribs are called *(true? false?)* ribs. Color these ribs on Figure LG 7.4, leaving
the costal cartilages white.

e. Ribs 8 to 10 are called _____. Color these ribs a different
color, again leaving the costal cartilages white. Do these ribs attach to the sternum?
(Yes? No?) If so, in what manner?

f. Ribs _____ and _____ are called "floating ribs." Use a third color for these
ribs. Why are they so named?

g. What function is served by the costal groove?

h. What occupies intercostal spaces?

■ **D5.** At what point are ribs most commonly fractured?

E. Disorders   (page 195)

■ **E1.** Complete this exercise about slipped discs.

a. The normal intervertebral disc consists of two parts: an outer ring of *(hyaline?*

elastic? fibro-?) cartilage called _____ and a soft, elastic, inner

portion called the _____.

b. Ligaments normally keep discs in alignment with vertebral bodies. What may happen if these ligaments weaken?

c. Why might pain result from a slipped disc?

d. In what part of the vertebral column are slipped discs most common? What symptoms may result from a slipped disc in this region?

■ **E2.** Match types of abnormal curvatures of the vertebral column with descriptions below.

K. Kyphosis	L. Lordosis	S. Scoliosis

_____ a. Exaggerated lumbar curvature; "sway-back"

_____ b. Exaggerated thoracic curvature; "hunchback"

_____ c. S- or C-shaped lateral bending

E3. Imperfect union of the vertebral arches at the midline is the condition known as

_____. Why is it crucial that the vertebral foramen be completely surrounded by bone? What problems may result from incomplete closure?

ANSWERS TO SELECTED CHECKPOINTS

A5. (b) Superior orbital *fissure.* (c) Optic *foramen.* (d) External auditory *meatus.* (e) *Head* of humerus. (f) Greater *trochanter.* (g) Styloid *process.* (h) Articular *facet* on vertebra.

A6. (a) Axial. (b) 22 skull, 6 earbones (studied in Chapter 16), 1 hyoid, 26 vertebrae, 1 sternum, 24 ribs. (c) 80. (d) Shoulder girdles, upper extremities, hipbones, lower extremities. (e) 2 left shoulder girdle, 30 left upper extremity, 1 left hipbone, 30 left lower extremity. (f) 126. (g) 206.

B1. (a) Protects the brain. (b) Frontal, parietals, occipital, temporals, ethmoid, sphenoid. (c) Immovable, fibrous joints between skull bones; between parietal bones; see Figure LG 7.1A. (d) Fontanels; between frontal and parietal bones.

B3. See Figure LG 7.1A.

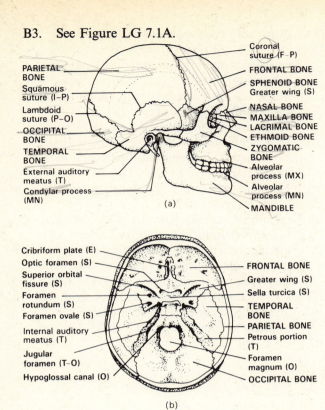

(a)

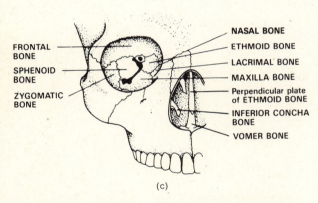

(b)

FRONTAL BONE

SPHENOID BONE

ZYGOMATIC BONE

NASAL BONE

ETHMOID BONE

LACRIMAL BONE

MAXILLA BONE

Perpendicular plate of ETHMOID BONE

INFERIOR CONCHA BONE

VOMER BONE

(c)

Figure LG 7.1A Skull bones. (a) Skull viewed from right side. (b) Floor of the cranial cavity. (c) Anterior view of bones that form the right orbit and the nose. Bone labels are capitalized; markings are in lowercase with related bone initial in parentheses.

B4. (a) Mandible. (b) Zygomatic (malar). (c) Lacrimal. (d) Nasal. (e) Temporal. (f) Occipital. (g) Parietal. (h) Maxilla. (i) Palatine. (j) Frontal. (k) Ethmoid. (l) Sphenoid. (m) Vomer. (n) Inferior concha.

B5.

Marking	Bone	Function
a. Greater wings	Sphenoid	**Form part of side walls of skull**
b. **Perpendicular plate**	**Ethmoid**	Forms superior portion of septum of nose
c. **Sella turcica**	**Sphenoid**	Site of pituitary gland
d. Petrous portion	**Temporal**	**Houses middle ear and inner ear**
e. **Foramen magnum**	**Occipital**	Largest hole in skull; passageway for spinal cord
f. **Alveolar processes**	**(2) Maxillae and mandible**	Bony sockets for teeth
g. **External auditory meatus**	**Temporal**	Passageway for sound waves to enter ear
h. Condylar process	Mandible	**Articulates with temporal bone (in TMJ)**

B6. See Figure LG 7.1A
B7. See Figure LG 7.1A

Number and Name of Cranial Nerve		Location of Opening for Nerve
a. I	Olfactory	Cribriform plate of ethmoid bone
b. II	Optic	**Optic foramen of sphenoid bone**
c. III IV V VI	Oculomotor Trochlear Trigeminal (ophthalmic branch) Abducens	**Superior orbital fissure of sphenoid bone**
d. V	Trigeminal (maxillary branch)	**Foramen rotundum of sphenoid bone**
e. V	Trigeminal (mandibular branch)	**Foramen ovale of sphenoid bone**
f. VII VIII	Facial Vestibulocochlear	**Internal auditory meatus**
g. IX X XI	Glossopharyngeal Vagus Accessory	**Jugular foramen**
h. XII	Hypoglossal	**Hypoglossal canal**

B8. (a) Frontal, ethmoid, sphenoid, maxilla. (b) Warm and humidify air since air sinuses are lined with mucous membrane; serve as resonant chambers for speech and other sounds; make the skull lighter weight. (c) Ethmoid. (d) Like paranasal sinuses, they are covered with mucous membrane, which warms, humidifies, and cleanses air entering the nose. (e) Vomer; cartilage.

B9. Temporal, mandible; pain, clicking noise, or limitation of movement associated with the joint.

B10. See on Figure LG 7.1cA; frontal, zygomatic, maxilla, lacrimal, ethmoid, sphenoid.

C1. Superior.

C2. It articulates (forms a joint) with no other bones.

C3. A, cervical (7); B, thoracic (12); C, lumbar (5); D, sacrum (1); E, coccyx (1); C1, atlas; C2, axis.

C4. B (thoracic) and D (sacral); primary; were present originally during fetal life.

C6.

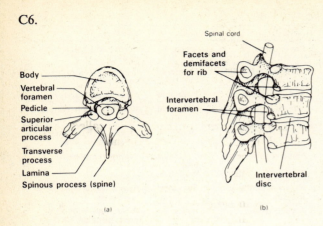

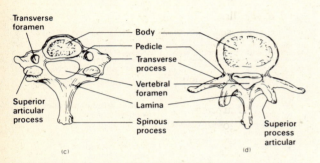

Figure LG 7.3A Typical vertebrae. (a) Thoracic vertebra, superior view. (b) Thoracic vertebrae, right lateral view. (c) Cervical vertebra, superior view. (d) Lumbar vertebra, superior view.

C7. (a) C. (b) T. (c) L. (d) T. (e) S.
C8. (a) Site used for administration of epidural anesthesia. (b) Obstetrical landmark for measurement of size of the pelvis. (c) In whiplash, it may injure the brainstem.
D2.

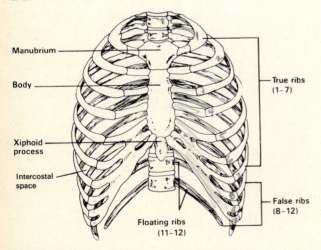

Figure LG 7.4A Anterior view of the thorax.

D3. Biopsy red bone marrow.
D4. (a) 24, 12. (b) Inferior. (c) Bodies of thoracic vertebrae; anterior; prevents ribs from slipping posteriorward. (d) 1, 7, costal; true. (e) False; yes; they attach to sternum indirectly via seventh costal cartilage. (f) 11, 12; they have no anterior attachment to sternum (g) Provides protective channel for intercostal nerve, artery, and vein. (h) Intercostal muscles. Refer also to Figure LG 7.4A.
D5. Just anterior to the costal angle, especially involving ribs 3 to 10.
E1. (a) Fibro-, annulus fibrosus, nucleus pulposus. (b) Annulus fibrosus may rupture, allowing the nucleus pulposus to protrude (herniate), causing a "slipped disc." (c) Disc tissue may press on spinal nerves. (d) L4 to L5 or L5 to sacrum; pain in posterior of thigh and leg(s) due to pressure on sciatic nerve, which attaches to spinal cord at L4 to S3.
E2. (a) L. (b) K. (c) S.

MASTERY TEST: Chapter 7

Questions 1–10: Circle the letter preceding the one best answer to each question.

1. All of these bones contain paranasal sinuses *except:*
   A. Frontal
   B. Maxilla
   C. Nasal
   D. Sphenoid
   E. Ethmoid

2. Choose the one *false* statement.
   A. There are seven vertebrae in the cervical region.
   B. The cervical region normally exhibits a curve that is slightly concave anteriorly.
   C. The lumbar vertebrae are superior to the sacrum.
   D. Intervertebral discs are located between bodies of vertebrae.

3. The hard palate is composed of ____ bones.
   A. Two maxilla and two mandible
   B. Two maxilla and two palatine
   C. Two maxilla
   D. Two palatine
   E. Vomer, ethmoid, and two temporal

4. The lateral wall of the orbit is formed mostly by which two bones?
   A. Zygomatic and maxilla
   B. Zygomatic and sphenoid
   C. Sphenoid and ethmoid
   D. Lacrimal and ethmoid
   E. Zygomatic and ethmoid

5. Choose the one *true* statement.
   A. All of the ribs articulate anteriorly with the sternum.
   B. Ribs 8 to 10 are called true ribs.
   C. There are 23 ribs in the male skeleton and 24 in the female skeleton.
   D. Rib 7 is larger than rib 3.
   E. Cartilage discs between vertebrae are called costal cartilages.

6. Which is the largest fontanel?
   A. Frontal
   B. Occipital
   C. Sphenoid
   D. Mastoid

7. Immovable joints of the skull are called:
   A. Wormian bones
   B. Sutures
   C. Conchae
   D. Sinuses
   E. Fontanels

8. ____ articulate with every bone of the face except the mandible.
   A. Lacrimal bones
   B. Zygomatic bones
   C. Maxillae
   D. Sphenoid bones
   E. Ethmoid bones

9. All of these markings are parts of the sphenoid bone *except:*
   A. Lesser wings
   B. Optic foramen
   C. Crista galli
   D. Sella turcica
   E. Pterygoid processes

10. All of these bones are included in the axial skeleton *except:*
    A. Rib
    B. Sternum
    C. Clavicle
    D. Hyoid
    E. Ethmoid

Questions 11–15: Arrange the answers in correct sequence.

_____ _____ _____ 11. From anterior to posterior:
  A. Ethmoid bone
  B. Sphenoid bone
  C. Occipital bone

_____ _____ _____ 12. Parts of vertebra from anterior to posterior:
  A. Vertebral foramen
  B. Body
  C. Lamina

_____ _____ _____ 13. Vertebral regions, from superior to inferior:
  A. Lumbar
  B. Thoracic
  C. Cervical

_____ _____ _____ 14. From superior to inferior:
  A. Atlas
  B. Axis
  C. Occipital bone

_____ _____ _____ 15. From superior to inferior:
  A. Atlas
  B. Manubrium of sternum
  C. Hyoid

Questions 16–20: Circle T (true) or F (false). If the statement is false, change the underlined word or phrase so that the statement is correct.

T F 16. The space between two ribs is called the <u>costal groove.</u> *intercostal space*

T F 17. The thoracic and sacral curves are called <u>primary</u> curves, meaning that they retain the original curve of the fetal vertebral column.

T F 18. The jugular vein passes through the same foramen as cranial nerves <u>V, VI, and VII.</u>

(T) F 19. In general, <u>foramina, meati, and fissures</u> serve as openings in the skull for nerves and blood vessels.

(T) F 20. The annulus fibrosus portion of an intervertebral disc is a <u>firm ring of fibrocartilage</u> surrounding the nucleus pulposus.

Questions 21–25: fill-ins. Write the word or phrase that best completes the statement.

axis _____ 21. A finger- or toothlike projection called the dens is part of the ____ bone.

mandible 22. The ramus, angle, mental foramen, and alveolar processes are all markings on the ____ bone.

temporal 23. The squamous, petrous, and zygomatic portions are markings on the ____ bone.

ethmoid 24. The perpendicular plate, crista galli, and superior and middle conchae are markings found on the ____ bone.

contains red bone marrow 25. The sternum is often used for a marrow biopsy because ____.

Multiple Choice

1. C   6. A
2. B   7. B
3. B   8. C
4. B   9. C
5. D   10. C

Arrange

11. A B C   14. C A B
12. B A C   15. A C B
13. C B A

True-False

16. F. Intercostal space
17. T
18. F. IX, X, and XI
19. T
20. T

Fill-ins

21. Axis (second cervical vertebra)
22. Mandible
23. Temporal
24. Ethmoid
25. It contains red bone marrow and it is readily accessible.

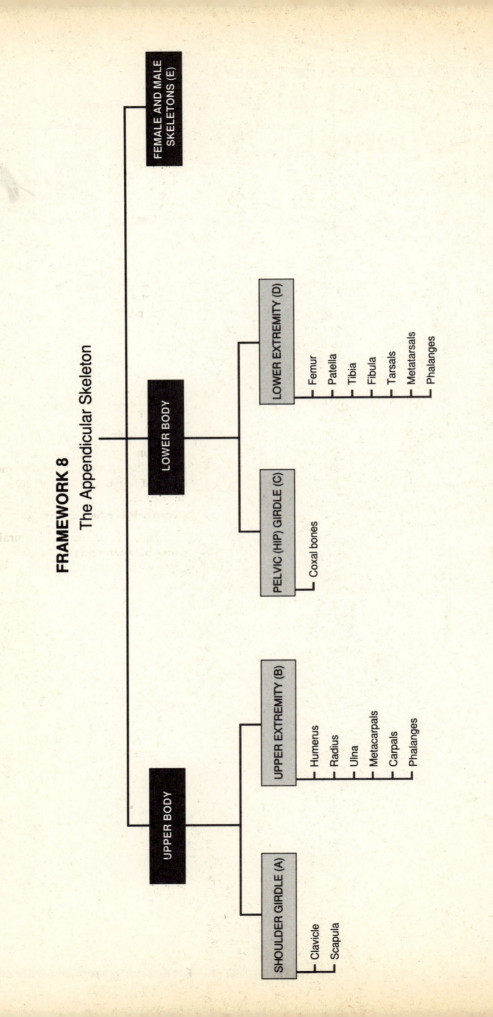

FRAMEWORK 8

The Appendicular Skeleton

FEMALE AND MALE SKELETONS (E)

LOWER BODY

LOWER EXTREMITY (D)
- Femur
- Patella
- Tibia
- Fibula
- Tarsals
- Metatarsals
- Phalanges

PELVIC (HIP) GIRDLE (C)
- Coxal bones

UPPER BODY

UPPER EXTREMITY (B)
- Humerus
- Radius
- Ulna
- Metacarpals
- Carpals
- Phalanges

SHOULDER GIRDLE (A)
- Clavicle
- Scapula

The Skeletal System: The Appendicular Skeleton

8

The appendicular skeleton includes the bones of the limbs, or extremities, as well as the supportive bones of the shoulder and hip girdles. Male and female skeletons exhibit some differences in skeletal structure, noted particularly in bones of the pelvis.

To begin your study of the appendicular skeleton, first refer to the Chapter 8 Framework and study key terms associated with each section.

TOPIC OUTLINE AND OBJECTIVES

A. Pectoral (shoulder) girdles

1. Identify the bones of the pectoral (shoulder) girdle and their major markings.

B. Upper extremities

2. Identify the upper extremity, its component bones, and their markings.

C. Pelvic (hip) girdle

3. Identify the components of the pelvic (hip) girdle and their principal markings.

D. Lower extremities

4. Identify the lower extremity, its component bones, and their markings.
5. Define the structural features and importance of the arches of the foot.

E. Male and female skeletons

6. Compare the principal structural differences between female and male skeletons, especially those that pertain to the pelvis.

Now study the following parts of words that may help you better understand terminology in this chapter.

Wordbyte	Meaning	Example	Wordbyte	Meaning	Example
acro-	tip	*acro*mion process	-physis	to grow	pubic sym*physis*
cap-	head	*cap*itulum	semi-	half	*semi*lunar notch
meta-	beyond	*meta*tarsal	sym-	together	pubic *sym*physis

CHECKPOINTS

A. Pectoral (shoulder) girdles   (pages 199–200)

■ **A1.** Do this exercise about the pectoral girdle.

a. Which bones form the pectoral girdle?

b. Do these bones articulate (form a joint) with vertebrae or ribs? *no*

c. The pectoral girdle is part of the (axial? appendicular?) skeleton. Identify the point (marked by *) on Figure LG 8.1 at which the shoulder girdle articulates with the axial skeleton. Name the two bones forming that joint. Palpate (press and feel) the

bones at this joint on yourself. *sternum*   *clavicle*

■ **A2.** Write the name of the bone that articulates with each of these markings on the scapula.

a. Acromion process: *clavicle*

b. Glenoid cavity: *humerus*

c. Coracoid process: *muscles + ligaments*

■ **A3.** Study a scapula carefully, using a skeleton or Figure 8.3, page 201 in your text. Then match the markings in the box with the descriptions given.

A. Axillary border	M. Medial border
I. Infraspinatus fossa	S. Spine

S a. Sharp ridge on the posterior surface

I b. Depression inferior to the spine; location of infraspinatus muscle

M c. Edge closest to the vertebral column

A d. Thick edge closest to the arm

B. Upper extremities   (pages 200–204)

■ **B1.** List the bone (or groups of bones) in the upper extremity from proximal to distal. Indicate how many of each bone there are. Two are done for you. Refer to Figure LG 8.1 to check your answers.

a. **Humerus** (1)

b. *ulna* (1)

c. *radius* (1)

d. *carpals* ( )

e. **Metacarpals** (5)

f. *phalanges* ( )

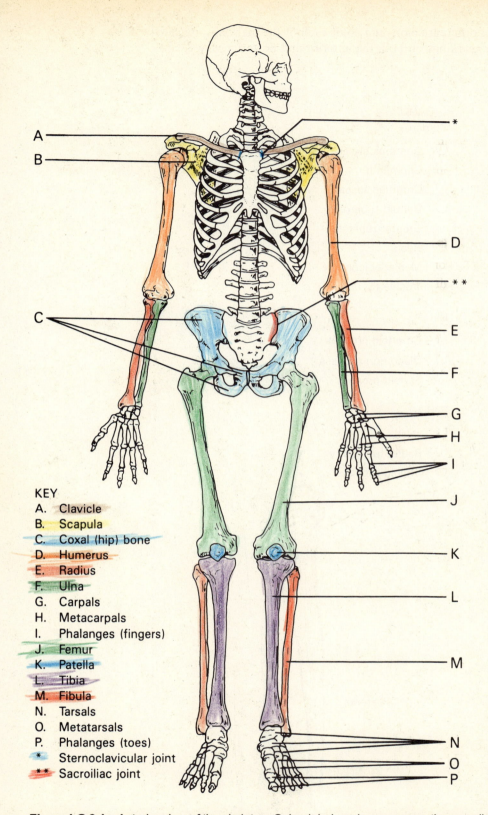

A

B

C

*

D

**

E

F

G

H

I

J

K

L

M

N

O

P

KEY
A. Clavicle
B. Scapula
C. Coxal (hip) bone
D. Humerus
E. Radius
F. Ulna
G. Carpals
H. Metacarpals
I. Phalanges (fingers)
J. Femur
K. Patella
L. Tibia
M. Fibula
N. Tarsals
O. Metatarsals
P. Phalanges (toes)
* Sternoclavicular joint
** Sacroiliac joint

Figure LG 8.1 Anterior view of the skeleton. Color, label, and answer questions as directed in Checkpoints A1, B1, C1, D1, D2, and D7.

■ **B2.** On Figure LG 8.2, select different colors and color each of the markings indicated by O. Where possible, color markings on both the anterior and posterior views.

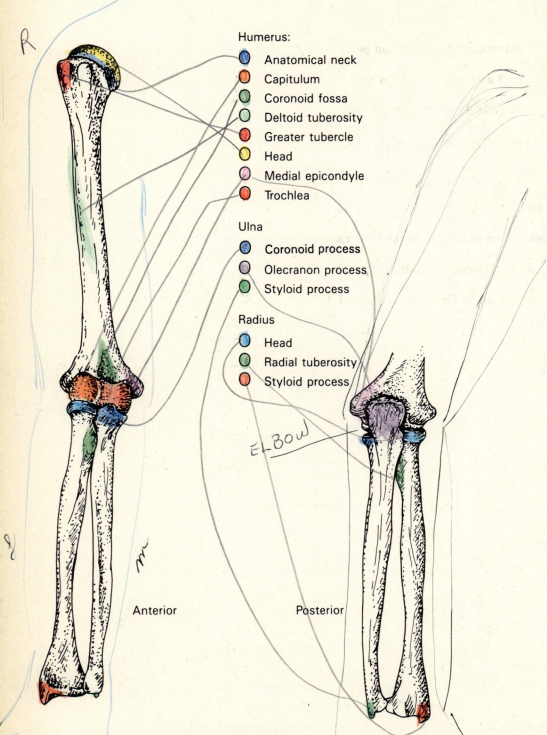

Humerus:

- Anatomical neck
- Capitulum
- Coronoid fossa
- Deltoid tuberosity
- Greater tubercle
- Head
- Medial epicondyle
- Trochlea

Ulna

- Coronoid process
- Olecranon process
- Styloid process

Radius

- Head
- Radial tuberosity
- Styloid process

Anterior

Posterior

ELBOW

Figure LG 8.2 Right upper extremity. (a) Anterior view. (b) Posterior view. Color and label as directed in Checkpoint B2.

■ **B3.** Name the marking that fits each description. Write H, U, or R to indicate whether the marking is part of the humerus (H), ulna (U), or radius (R). One has been done for you.

a. Articulates with glenoid fossa: _____**head**_____ (__**H**__).

b. Rounded head that articulates with radius: *Capitulum* (__H__)

c. Posterior depression that receives the olecranon process: _____ (_____)

d. Half-moon-shaped curved area that articulates with trochlea: _____ (_____)

e. Slight depression in which the head of the radius pivots: _____ (_____)

f. Biceps brachii muscle attaches here: _____ (_____)

■ **B4.** Answer these questions about wrist bones.

a. Wrist bones are called _____*carpals*_____. There are *(5? 7? 8? 14?)* of them in each wrist.

b. The wrist bone most subject to fracture is the bone named

_____, located just distal to the *(radius? ulna?)*.

B5. Trace an outline of your hand. Draw in and label all bones.

C. Pelvic (hip) girdle (pages 204–207)

■ **C1.** Describe the pelvic bones in this exercise.

a. Name the bones that form the pelvic girdle. _____

Which bones form the pelvis? _____
b. Which of these bones is/are part of the axial skeleton?

c. Locate the point (**) on Figure LG 8.1 at which the pelvic girdle portion of the appendicular skeleton articulates with the axial skeleton. Name the bones involved in

that joint. _____ and _____

■ **C2.** Answer the following questions about coxal bones.
a. Each coxal (hip) bone originates as three bones which fuse early in life. These

bones are the _____, _____, and

_____. At what location do the bones fuse?

b. The largest of the three bones is the _____. A ridge along the superior border is called the iliac crest. Locate this on yourself.

c. The iliac crest ends anteriorly as the _____ spine. The crest

ends posteriorly as the _____. This marking causes a dimpling of the skin just lateral to the sacrum, which can be used as a landmark for administering hip injections accurately.

■ **C3.** Complete the table about markings of the coxal bones.

Marking	Location on Coxal Bone	Function
a. Greater sciatic notch		
b. *Ischium*		Supports most of body weight in sitting position
c.		Fibrocartilaginous joint between two coxal bones
d.		Socket for head of femur
e. Obturator foramen	Large foramen surrounded by pubic and ischial rami and acetabulum	

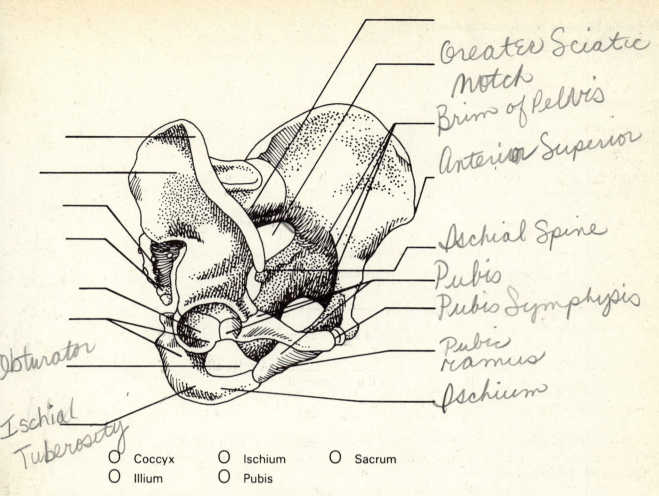

Handwritten labels:
- Greater Sciatic notch
- Brim of Pelvis
- Anterior Superior
- Ischial Spine
- Pubis
- Pubis Symphysis
- Pubic ramus
- Ischium
- Obturator
- Ischial Tuberosity

○ Coccyx   ○ Ischium   ○ Sacrum
○ Illium   ○ Pubis

Figure LG 8.3   Right anterolateral view of the pelvis. Color, label, and answer questions as directed in Checkpoints C4 and C6.

■ **C4.** Refer to Figure LG 8.3 and answer these questions about pelvic markings.

a. Trace with your pencil, and then label, the *brim of the pelvis* on the figure. It is a relatively *(smooth? irregular?)* line that demarcates the *(superior? inferior?)* border of the lesser (true) pelvis. The pelvic brim encircles the pelvic *(inlet? outlet?)*.

b. The bony border of the pelvic outlet is *(smooth? irregular?)*. Why is the pelvic outlet so named?

C5. Contrast the two principal parts of the pelvis. Describe their locations and name the structures which compose them.

a. Greater (false) pelvis

b. Lesser (true) pelvis

■ **C6.** Color the pelvic structures indicated by color code ovals on Figure LG 8.3. Also label the following markings: *acetabulum, anterior superior iliac spine, greater sciatic notch, iliac crest, inferior pubic ramus, ischial tuberosity, obturator foramen, sacral promontory,* and *symphysis pubis.*

D. Lower extremities   (pages 207–213)

■ **D1.** Refer to Figure LG 8.1 and list the bones (or groups of bones) in the lower extremity from proximal to distal. Indicate how many of each bone there are. One is done for you.

a. **Femur** _____ **(1)**   e. _____ **(   )**

b. _____ **(   )**   f. _____ **(   )**

c. _____ **(   )**   g. _____ **(   )**

d. _____ **(   )**

D2. Contrast the size, location, and names of the bones of the upper and lower extremities by coloring the bones on Figure LG 8.1 as follows. Color on one side of the figure only.

Humerus and femur (red)   Carpals and tarsals (blue)
Patella (brown)   Metacarpals and metatarsals (orange)
Ulna and tibia (green)   Phalanges (purple)
Radius and fibula (yellow)

■ **D3.** Circle the term that correctly indicates the location of these parts of the lower extremity.

a. The head is the *(proximal? distal?)* epiphysis of the femur.
b. The greater trochanter is *(lateral? medial?)* to the lesser trochanter.
c. The intercondylar fossa is on the *(anterior? posterior?)* surface of the femur.
d. The tibial condyles are more *(concave? convex?)* than the femoral condyles.
e. The lateral condyle of the femur articulates with the *(fibula? lateral condyle of the tibia?)*.
f. The tibial tuberosity is *(superior? inferior?)* to the patella.
g. The tibia is *(medial? lateral?)* to the fibula.
h. The outer portion of the ankle is the *(lateral? medial?)* malleolus which is part of the *(tibia? fibula?)*.

■ **D4.** The tarsal bone that is most superior in location (and that articulates with the

tibia and fibula) is the _____. The largest and strongest of the

tarsals is the _____.

D5. Answer these questions about the arch of the foot.

a. How is the foot maintained in an arched position?

b. Locate each of these arches on your own foot. Refer to Figure 8.14, page 212 of your text. (*For extra review:* List the bones that form each arch.)
Longitudinal: medial side

Longitudinal: lateral side

Transverse

c. What causes flatfoot?

■ **D6.** Now that you have seen all of the bones of the appendicular skeleton, complete this table relating common and anatomical names of bones.

Common Name	Anatomical Name
a. Shoulder blade	*scapula*
b. *thumb*	Pollex
c. Collarbone	*clavicle*
d. Heel bone	*calcaneus*
e. *elbow*	Olecranon process
f. Kneecap	~~tibia~~ *patella*
g. *Shinbone*	Tibial crest
h. Toes	*phalanges*
i. Palm of hand	*metacarpls*
j. Wrist bones	*carpals*

D7. *For extra review.* Label all bones marked with label lines on Figure LG 8.1.

E. Female and male skeletons (pages 213–214)

E1. State several characteristics of the female pelvis that make it more suitable for childbirth than the male pelvis.

■ **E2.** Identify specific differences in pelvic structure in the two sexes by placing M before characteristics of the male pelvis and F before structural descriptions of the female pelvis.

_____ a. Shallow greater pelvis

_____ b. Heart-shaped inlet

_____ c. Pubic arch greater than 90° angle

_____ d. Acetabulum small

_____ e. Pelvic inlet comparatively small

ANSWERS TO SELECTED CHECKPOINTS

A1. (a) Two clavicles and two scapulas. (b) No. (c) Appendicular; clavicles and manubrium of sternum.

A2. (a) Clavicle. (b) Humerus. (c) None; muscles and ligaments attach here.

A3. (a) S. (b) I. (c) M. (d) A.

B1. (a) Humerus, 1. (b) Ulna, 1. (c) Radius, 1. (d) Carpals, 8. (e) Metacarpals, 5. (f) Phalanges, 14.

B2.

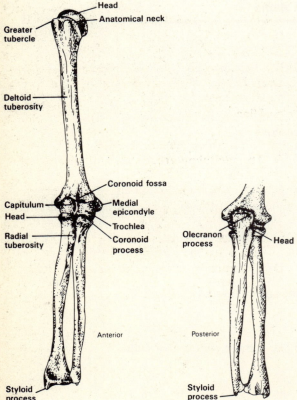

Figure LG 8.2A Right upper extremity. (a) Anterior view. (b) Posterior view.

B3. (b) Capitulum (H). (c) Olecranon fossa (H). (d) Semilunar (trochlear) notch (U). (e) Radial notch (U). (f) Radial tuberosity (R).

B4. (a) Carpals, 8. (b) Scaphoid, radius.

C1. (a) Two coxal (hip) bones; hip bones plus sacrum and coccyx. (b) Sacrum and coccyx. (c) Sacrum and iliac portion of hip bone (sacroiliac).

C2. (a) Ilium, ischium, pubis; acetabulum. (b) Ilium. (c) Anterior superior iliac; posterior superior iliac spine.

C3.

Marking	Location on Coxal Bone	Function
a. Greater sciatic notch	**Inferior to posterior inferior iliac spine**	**Sciatic nerve passes inferior to notch**
b. **Ischial tuberosity**	**Posterior and inferior to obturator foramen**	Supports most of body weight in sitting position
c. **Symphysis pubis**	**Most anterior portion of pelvis**	Fibrocartilaginous joint between two coxal bones
d. **Acetabulum**	**At junction of ilium, ischium, and pubis**	Socket for head of femur
e. Obturator foramen	Large foramen surrounded by pubic and ischial rami and acetabulum	**Blood vessels and nerves pass through**

C4.

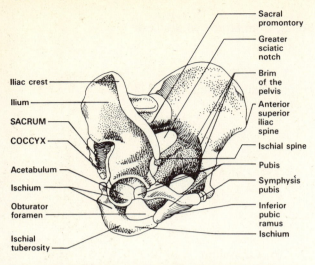

Figure LG 8.3A Right anterolateral view of the pelvis. Bone labels are capitalized; markings are in lowercase.

See Figure LG 8.3A. (a) Smooth, superior; inlet. (b) Irregular; feces, urine, semen, menstrual flow, and baby (at birth) exit via openings in the muscular floor attached to the bony outlet.

C6. See Figure LG 8.3A.

D1. (a) Femur, 1. (b) Patella, 1. (c) Tibia, 1. (d) Fibula, 1. (e) Tarsals, 7. (f) Metatarsals, 5. (g) Phalanges, 14.

D3. (a) Proximal. (b) Lateral. (c) Posterior. (d) Concave. (e) Lateral condyle of the tibia. (f) Inferior. (g) Medial. (h) Lateral, fibula.

D4. Talus; calcaneus.

D6. (a) Scapula. (b) Thumb. (c) Clavicle. (d) Calcaneus. (e) Elbow. (f) Patella. (g) Shinbone. (h) Phalanges. (i) Metacarpals. (i) Carpals.

E2. (a) F. (b) M. (c) F. (d) M. (e) M.

MASTERY TEST: Chapter 8

Questions 1–6: Circle the letter preceding the one best answer to each question.

1. The point at which the upper part of the appendicular skeleton is joined to (articulates with) the axial skeleton is at the joint between:
   A. Sternum and ribs
   B. Humerus and clavicle
   C. Scapula and clavicle
   D. Scapula and humerus
   E. Sternum and clavicle

2. All of the following are markings on the femur *except:*
   A. Acetabulum
   B. Head
   C. Condyles
   D. Greater trochanter
   E. Intercondylar fossa

3. All of the following are bones in the lower extremity *except:*
   A. Talus   D. Ulna
   B. Tibia   E. Fibula
   C. Calcaneus

4. Which structures are on the posterior surface of the upper extremity (in anatomical position)?
   A. Radial tuberosity and lesser tubercle
   B. Trochlea and capitulum
   C. Coronoid process and coronoid fossa
   D. Olecranon process and olecranon fossa

5. The humerus articulates with all of these bones *except:*
   A. Ulna   C. Clavicle
   B. Radius   D. Scapula

6. Choose the *false* statement.
   A. The capitulum articulates with the head of the radius.
   B. The medial and lateral epicondyles are located at the distal ends of the tibia and fibula.
   C. The coronoid fossa articulates with the ulna when the forearm is flexed.
   D. The trochlea articulates with the trochlear notch of the ulna.

_____ _____ _____ 7. According to size of the bones, from largest to smallest:
   A. Femur
   B. Ulna
   C. Humerus

_____ _____ _____ 8. Parts of the humerus, from proximal to distal:
   A. Anatomical neck
   B. Surgical neck
   C. Head

_____ _____ _____ 9. From proximal to distal:
   A. Phalanges
   B. Metacarpals
   C. Carpals

_____ _____ _____ 10. From superior to inferior:
   A. Lesser pelvis
   B. Greater pelvis
   C. Pelvic brim

_____ _____ _____ 11. Markings on coxal (hip) bones in anatomical position, from superior to inferior:
   A. Acetabulum
   B. Ischial tuberosity
   C. Iliac crest

Questions 12–20: Circle T (true) or F (false). If the statement is false, change the underlined word or phrase so that the statement is correct.

T F 12. Another name for the true pelvis is the <u>greater</u> pelvis.

T F 13. The scapulae <u>do</u> articulate with the vertebrae.

T F 14. The olecranon process is a marking on the <u>ulna, and the olecranon fossa is a marking on the humerus.</u>

T F 15. The female pelvis is <u>deeper and more heart-shaped</u> than the male pelvis.

T F 16. The greater tubercle of the humerus is <u>lateral</u> to the lesser tubercle.

T F 17. There are <u>14 phalanges in each hand and also in each foot.</u>

T F 18. The fibula articulates with the <u>femur, tibia, talus, and calcaneus.</u>

T F 19. The organs contained within the right iliac, hypogastric, and left iliac portions (ninths) of the abdomen are located in the <u>true</u> pelvis.

T F 20. The total number of bones in one upper extremity (including arm, forearm, wrist, hand, fingers, but excluding shoulder girdle) is <u>29.</u>

Questions 21–25: fill-ins. Write the word or phrase that best completes the statement.

_____ 21. In about three-fourths of all carpal bone fractures, only the ____ bone is involved.

_____ 22. A fracture of the distal end of the fibula with injury to the tibial articulation is known as a ____ fracture.

_____ 23. The ____ is the thinnest bone in the body compared to its length.

_____ 24. The point of fusion of the three bones forming the coxal bone is the ____.

_____ 25. The kneecap is the common name for the ____.

ANSWERS TO MASTERY TEST: ▪ Chapter 8

Multiple Choice

1. E
2. A
3. D
4. D
5. C
6. B

Arrange

7. A C B
8. C A B
9. C B A
10. B C A
11. C A B

True-False

12. F. Lesser
13. F. Do not
14. T
15. F. Shallower and more oval
16. T
17. T
18. F. Only tibia and talus
19. F. False
20. F. 30

Fill-ins

21. Scaphoid
22. Pott's
23. Fibula
24. Acetabulum
25. Patella

FRAMEWORK 9
Articulations

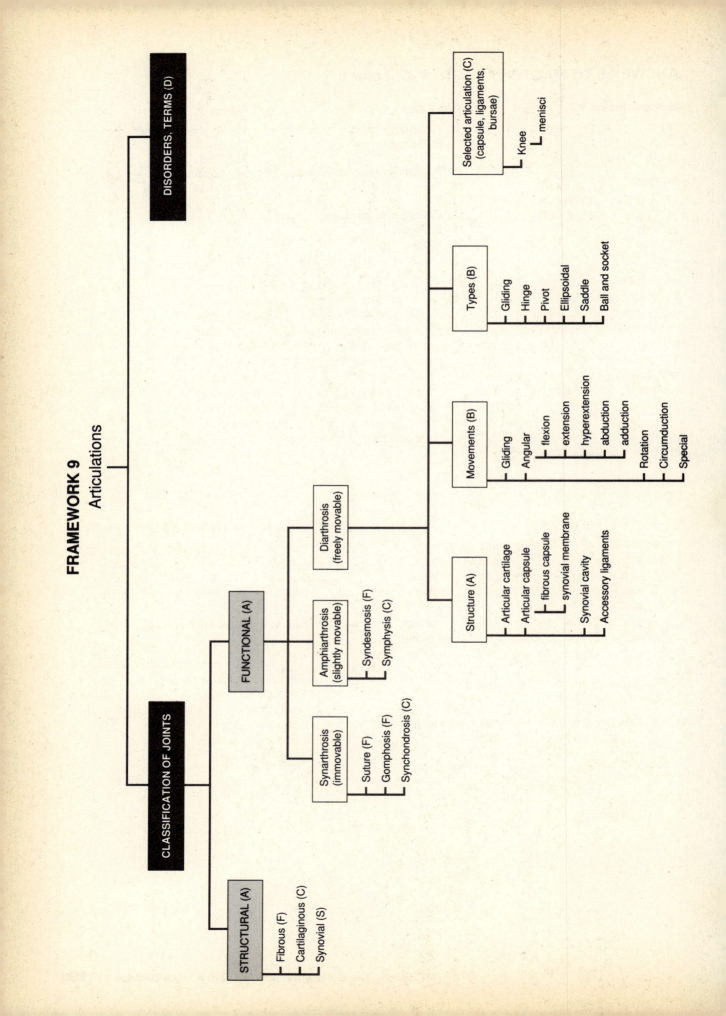

Articulations

CLASSIFICATION OF JOINTS

STRUCTURAL (A)
- Fibrous (F)
- Cartilaginous (C)
- Synovial (S)

FUNCTIONAL (A)

Synarthrosis (immovable)
- Suture (F)
- Gomphosis (F)
- Synchondrosis (C)

Amphiarthrosis (slightly movable)
- Syndesmosis (F)
- Symphysis (C)

Diarthrosis (freely movable)

Structure (A)
- Articular cartilage
- Articular capsule
  - fibrous capsule
  - synovial membrane
- Synovial cavity
- Accessory ligaments

Movements (B)
- Gliding
- Angular
  - flexion
  - extension
  - hyperextension
  - abduction
  - adduction
- Rotation
- Circumduction
- Special

Types (B)
- Gliding
- Hinge
- Pivot
- Ellipsoidal
- Saddle
- Ball and socket

Selected articulation (C) (capsule, ligaments, bursae)
- Knee
  - menisci

DISORDERS, TERMS (D)

Articulations

In the past three chapters you have learned a great deal about the 206 bones in the body. Separated, or disarticulated, these bones would constitute a pile as disorganized as the rubble of a ravaged city deprived of its structural integrity. Fortunately bones are arranged in precise order and held together in specific conformations—articulations or joints—that permit bones to function effectively. Joints may be classified by function (how movable) or by structure (the type of tissue forming the joint). Synovial joints are most common; their structure, movements, and types will be discussed. A detailed examination of one important synovial joint (the knee) is also included. Joint disorders such as arthritis are usually not life-threatening, but plague much of the population, particularly the elderly. An introduction to joint disorders completes this chapter.

Be sure to look over the Chapter 9 Framework with its key terms at the start and finish of your study of the chapter.

TOPIC OUTLINE AND OBJECTIVES

A. Classification of joints

1. Define an articulation (joint) and identify the factors that determine the types and degree (range) of movement at a joint.
2. Classify joints on the basis of structure and function.
3. Contrast the structure, kind of movement, and location of immovable, slightly movable, and freely movable joints.

B. Movements and types of synovial (diarthrotic) joints

4. Describe the structure, types, and movements of freely movable joints.

C. Selected articulations of the body

5. Describe selected articulations of the body with respect to the bones that enter into their formation, structural classification, and anatomical components.

D. Disorders, medical terminology

6. Describe the causes and symptoms of common joint disorders, including rheumatism, rheumatoid arthritis (RA), osteoarthritis (OA), gouty arthritis, Lyme disease, bursitis, ankylosing spondylitis, dislocation, sprain, and strain.
7. Define medical terminology associated with articulations.

WORDBYTES

Now study the following parts of words that may help you better understand terminology in this chapter.

Wordbyte	Meaning	Example	Wordbyte	Meaning	Example
amphi-	both	*amphi*arthrotic	-osis	condition of	syndesm*osis*
arthr-	joint	osteo*arthr*itis	rheum-	water discharge	*rheum*atoid
articulat-	joint	*articulat*ion			arthritis
cruci-	cross	*cruci*ate	syn-	together	*syn*arthrotic
-itis	inflammation	arthr*itis*			

CHECKPOINTS

A. Classification of joints (pages 217–220)

A1. Define the term *articulation (joint)*.

Discuss three structural factors that affect movement at joints.

■ **A2.** Name three classes of joints based on structure.

■ **A3.** Fill in the blanks below to name three classes of joints according to the amount of movement they permit.

a. Synarthrosis: _____

b. _____: slightly movable

c. _____: freely movable

■ **A4.** Describe synarthrotic (immovable) joints by completing this exercise.

a. Fibrous joints *(have? lack?)* a joint cavity. They are held together by

_____ connective tissue.

b. One type of fibrous joint is a _____ found between skull bones.

Such joints are *(freely? slightly? im-?)* movable, or _____-arthrotic.

c. Which of the following is a site of a *gomphosis?*
   A. Joint at distal ends of tibia and fibula
   B. Epiphyseal plate
   C. Attachment of tooth by periodontal ligament to tooth socket in maxilla or mandible

d. Synchondroses involve *(hyaline? fibrous?)* cartilage between regions of bone. An

example is the _____ cartilage between diaphysis and epiphysis of a growing bone. This cartilage *(persists through life? is replaced by bone during adult life?)*. Synchondroses are *(somewhat movable? immovable?)*.

■ **A5.** Describe amphiarthrotic joints by completing this exercise.

a. The distal end of the tibia/fibula joint is a fibrous joint. It is *(more? less?)* mobile

than a suture and is therefore _____-arthrotic. It is called a

_____.

b. Fibrocartilage is present in the type of joint known as a _____.

These joints permit some movement, so are called _____

-arthrotic. Two locations of symphyses are _____ and

_____.

■ **A6.** What structural features of synovial joints make them more freely movable than fibrous or cartilaginous joints?

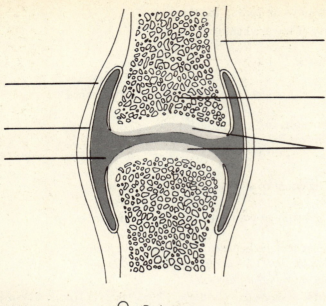

○ Articular cartilage   ○ Periosteum
○ Articulating bone   ○ Synovial (joint) cavity
○ Fibrous capsule   ○ Synovial membrane

Figure LG 9.1 Structure of a generalized synovial joint. Color as directed in Checkpoint A7.

■ **A7.** On Figure LG 9.1, color the indicated structures.

■ **A8.** Select the parts of a synovial joint (listed in the box) that fit the descriptions below.

A. Articular cartilage	SF. Synovial fluid
F. Fibrous capsule	SM. Synovial membrane
L. Ligaments	

_____ a. Hyaline cartilage that covers ends of articulating bones, but does not bind them together.

_____ b. With the consistency of uncooked egg white or oil, it lubricates the joint and nourishes the avascular articular cartilage.

_____ c. Connective tissue membrane that lines synovial cavity and secretes synovial fluid.

_____ d. Parallel fibers in some fibrous capsules; bind bones together.

_____ e. Together these form the articular capsule (two answers).

A9. Explain how joints can produce noises, for example, when you crack your knuckles.

A10. Describe the structure, function, and location of the following structures that may be associated with synovial joints:

a. Articular discs (menisci)

b. Bursae

c. Intracapsular ligaments

B. Movements and types of synovial joints   (pages 220–226)

B1. The design of synovial joints permits free movement of bones. However, if bones moved too freely, they could move right out of their joint cavities (dislocation). Briefly describe four factors that account for limitation of movement at synovial joints.

a. Articulating bones

b. Ligaments

c. Muscles

d. Soft parts

■ **B2.** Relaxin is a(n) *(enzyme? hormone?)*. It is produced by the

_____ and the _____ during pregnancy. Discuss the role of relaxin in preparation for labor and delivery.

■ **B3.** Check your understanding of types of synovial joints by choosing the type of joint that fits the description. (Answers may be used more than once.)

> B. Ball-and-socket   H. Hinge
> E. Ellipsoidal   P. Pivot
> G. Gliding   S. Saddle

_____ a. Monaxial joint; only rotation possible

_____ b. Examples include atlas-axis joint and joint between head of radius and radial notch at proximal end of ulna

_____ c. Triaxial joint, allowing movement in all three planes

_____ d. Hip bone and shoulder joints

_____ e. Spoollike (convex) surface articulated with convex surface, for example, elbow, ankle, and joints between phalanges

_____ f. One type of monoaxial joint in which only flexion and extension are possible

_____ g. Found in joints at sternoclavicular and claviculoscapular joints

_____ h. Thumb joint located between metacarpal of thumb and carpal bone (trapezium)

_____ i. Biaxial joints (two answers)

■ **B4.** From the terms listed in the box, choose the one that fits the type of movement in each case. Not all answers will be used.

> Abd. Abduction   E. Extension   I. Inversion
> Add. Adduction   F. Flexion   P. Plantar flexion
>   C. Circumduction G. Gliding   R. Rotation
>   D. Dorsiflexion

_____ a. Decrease in angle between anterior surfaces of bones (or between posterior surfaces at knee and toe joints)

_____ b. Simplest kind of movement that can occur at a joint; no angular or rotary motion involved; example: ribs moving against vertebrae

_____ c. State of entire body when it is in anatomical position

_____ d. Movement away from the midline of the body

_____ e. Movement of a bone around its own axis

_____ f. Position of foot when heel is on the floor and rest of foot is raised

■ **B5.** Perform the action described. Then write in the name of the type of movement.

a. Describe a cone with your arm, as if you are winding up to pitch a ball. The

movement at your shoulder joint is called _____.

b. Stand in anatomical position (palms forward). Turn your palms backward. This

action is called _____.

c. Move your fingers from "fingers together" to "fingers apart" position. This action is

_____ of fingers.

d. Raise your shoulders, as if to shrug them. This movement is called

_____ of the shoulders.

e. Stand on your toes. This action at the ankle joint is called

_____.

f. Grasp a ball in your hand. Your fingers are performing the type of movement called

_____.

g. Sit with the soles of your feet pressed against each other. In this position, your feet

are performing the action called _____.

h. Thrust your jaw outward (gently!). This action is _____ of the mandible.

■ **B6.** Identify the kinds of movements shown in Figure LG 9.2. Write the name of the movement below each figure. Use the following terms: *abduction, adduction, extension, flexion,* and *hyperextension.*

C. Selected articulations of the body   (pages 226–233)

■ **C1.** Complete this exercise describing the knee (tibiofemoral) joint. Consult Exhibit 9.3 and Figure 9.7 on pages 229–230 in your text.

a. The knee joint actually consists of three joints:
1. Between the femur and the *(patella? fibula?)*
2. Between the lateral condyle of the femur, *(medial? lateral?)* meniscus, and the lateral condyle of the *(tibia? fibula?)*
3. Between the medial condyle of the femur, *(medial? lateral?)* meniscus, and the medial condyle of the *(tibia? fibula?)*

b. The first joint (1) described above is a *(modified hinge? pivot? gliding?)* joint. The

other two joints (2 and 3) are _____ joints.

c. The knee (tibiofemoral) joint *(does? does not?)* include a complete capsule uniting the two bones. This factor contributes to the relative *(strength? weakness?)* of this joint.

d. The medial and lateral patella retinacula and patellar ligament are both tissues that serve as insertions of the *(hamstrings? quadriceps femoris?)* muscles.

e. A number of other ligaments provide strength and support. Identify whether the following ligaments are extracapsular (E) or intracapsular (I).
1. Arcuate popliteal ligament: ____
2. Anterior and posterior cruciate ligaments: ____
3. Tibial (medial) and fibular (lateral) collateral ligaments: ____

f. Which of the ligaments listed in (e) is damaged in most serious knee injuries?

g. Two cartilages, much like two stadiums facing one another, each making an incomplete circle, are located between condyles of the femur and tibia. These articular

discs are called _____. What is their function?

h. There are *(no? three? over a dozen?)* bursae associated with the knee joint.

Inflammation of bursae is called _____.

i. *A clinical challenge.* Which three knee structures beginning with the letter *C* are examined closely when a knee is injured?

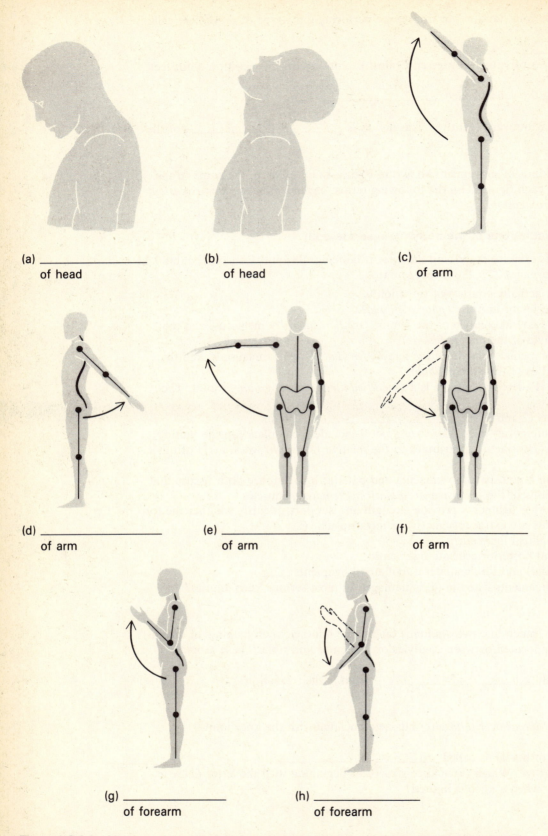

(a) _____
of head

(b) _____
of head

(c) _____
of arm

(d) _____
of arm

(e) _____
of arm

(f) _____
of arm

(g) _____
of forearm

(h) _____
of forearm

Figure LG 9.2   Movements at synovial joints. Answer questions as directed in Checkpoint B6 and in Chapter 11.

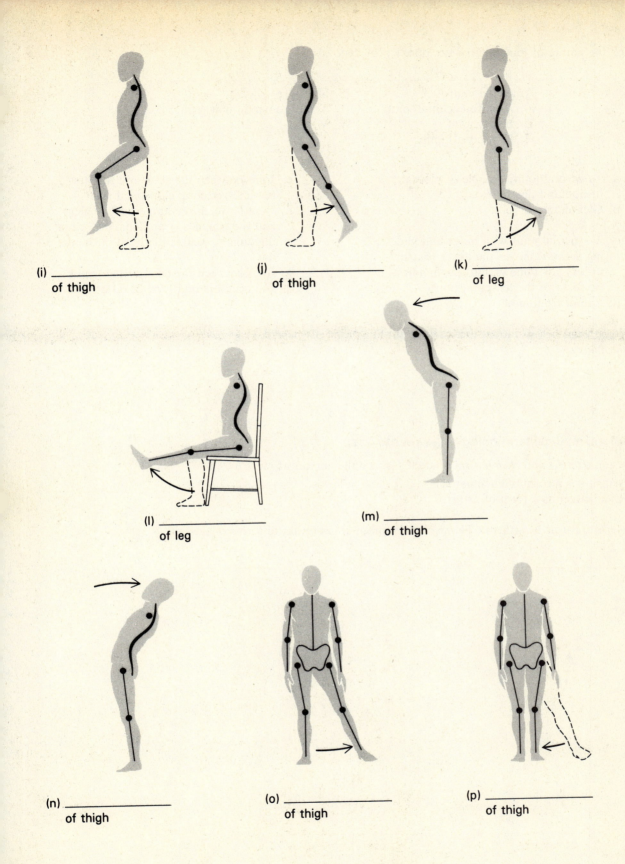

(i) _____
of thigh

(j) _____
of thigh

(k) _____
of leg

(l) _____
of leg

(m) _____
of thigh

(n) _____
of thigh

(o) _____
of thigh

(p) _____
of thigh

■ **C2.** Match the names of joints with descriptions provided.

A. Ankle joint	S. Shoulder
Ao. Atlanto-occipital joint	T. Temporomandibular joint
E. Elbow joint	W. Wrist
H. Hip (coxal) joint	

_____ a. Talocrural joint; capable of plantar flexion and dorsiflexion

_____ b. Glenohumoral joint

_____ c. A synovial joint with hinge and gliding actions; moves the lower (but not upper) jaw bone within the mandibular fossa of the temporal bone

_____ d. Radiocarpal joint

_____ e. Joint between the skull and the first cervical vertebra; ellipsoidal joint

_____ f. Joint between the humerus (trochlea) and the trochlear (semilunar) notch of the ulna as well as the head of the radius

_____ g. Joint between the head of the femur and acetabulum of the coxal bone

C3. *A clinical challenge.* Contrast these two terms as you briefly describe each: *arthroscopy/arthroplasty.*

D. Disorders, medical terminology   (pages 233–235)

■ **D1.** How are *arthritis* and *rheumatism* related? Choose the correct answer.

   A. Arthritis is a form of rheumatism.

   B. Rheumatism is a form of arthritis.

■ **D2.** List several forms of arthritis. Name one symptom common to all forms of this ailment.

■ **D3.** Contrast *rheumatoid arthritis* with *osteoarthritis*. Write *Yes, No, Larger,* or *Smaller* for answers.

	Rheumatoid Arthritis	Osteoarthritis
a. Which is known as "wear-and-tear" arthritis?		
b. Which types of joints are affected?		
c. Is this an inflammatory autoimmune condition?		
d. Is synovial membrane affected?		
e. Does articular cartilage degenerate?		
f. Does fibrous tissue join bone ends?		
g. Is movement limited?		

D4. Suggest several means by which persons with arthritis can manage this common condition and not let arthritis control their lives. Include in your answer the following terms: *pain relief, orthopedic devices, weight, nutrition,* and *range-of-motion (ROM) exercises.*

■ **D5.** Complete this exercise describing disorders involving articulations.

a. Gouty arthritis is a condition due to an excess of _____ in the

blood leading to deposit of _____ in joints. Gout can also

involve damage to joints or to organs such as the _____ be-
cause crystals may be deposited there also. This condition is more common among
(females? males?).

b. An acute chronic inflammation of a bursa is called _____. Inflammation of certain bursae associated with the knee joint is known as

_____ knee.

c. *Luxation,* or _____, is displacement of a bone from its joint
with tearing of ligaments. The most common sites of dislocations are the

_____, _____, and

_____ joints.

d. Pain in a joint is known as _____.

e. _____ is a cluster of conditions named after a town in Connecti-
cut. It is caused by *(bacteria? fungi? viruses?)* known as <u>Borrelia burgdorferi,</u> which
are transmitted by *(fleas? mosquitoes? ticks?)*. Signs and symptoms include

skin _____, then possibly cardiac problems and/or facial paralysis, and then arthritis of *(larger? smaller?)* joints.

f. Ankylosing spondylitis affects primarily *(ankle and wrist? intervertebral and sacroiliac?)* joints.

g. A *(sprain? strain?)* is an overstretching of a muscle. A *(sprain? strain?)* involves
more serious injury to joint structures. The joint most often sprained is the *(ankle?
elbow?)* joint.

ANSWERS TO SELECTED CHECKPOINTS

A2. Fibrous, cartilage, synovial.
A3. (a) Immovable. (b) Amphiarthrosis. (c) Diarthrosis.
A4. (a) Lack; fibrous. (b) Suture; im-, syn. (c) C. (d) Hyaline; epiphyseal; is replaced by bone during adult life; immovable.
A5. (a) More, amphi; syndesmosis. (b) Symphysis; amphi-; discs between vertebrae, symphysis pubis between coxal bones.
A6. The space (synovial cavity) between the articulating bones and the absence of tissue between those bones (which might restrict movement) make the joints more freely movable.
A7.

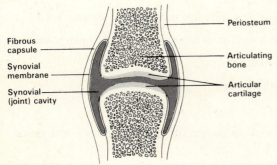

Fibrous capsule

Synovial membrane

Synovial (joint) cavity

Periosteum

Articulating bone

Articular cartilage

Figure LG 9.1A Structure of a generalized synovial joint.

A8. (a) A. (b) SF. (c) SM. (d) L. (e) SM and F.
B2. Hormone; placenta and ovaries; relaxes tissues (pubic symphysis and ligaments) in joints to ease birth of the baby.
B3. (a) P. (b) P. (c) B. (d) B. (e) H. (f) H. (g) G. (h) S. (i) E, S.
B4. (a) F. (b) G. (c) E. (d) Abd. (e) R. (f) D.
B5. (a) Circumduction. (b) Pronation. (c) Abduction. (d) Elevation. (e) Plantar flexion. (f) Flexion. (g) Inversion. (h) Protraction.
B6. (a) Flexion. (b) Hyperextension. (c) Flexion. (d) Hyperextension. (e) Abduction. (f) Adduction. (g) Flexion. (h) Extension. (i) Flex-ion, while leg also slightly flexed. (j) Hyper-extension, while leg extended. (k) Flexion, while thigh extended. (l) Extension, with thigh flexed. (m) Flexion. (n) Hyperexten-sion. (o) Abduction. (p) Adduction.

C1. (a) (1) Patella, (2) Lateral, tibia, (3) Medial, tibia. (Note that neither the femur nor the patella articulates with the fibula.) (b) Glid-ing; modified hinge. (c) Does not; weakness. (d) Quadriceps femoris. (e) (1) E, (2) I, (3) E. (f) Anterior cruciate. (g) Menisci; provide some stability to an otherwise unstable joint. (h) Over a dozen; bursitis. (i) (Tibial) collat-eral ligament, (anterior) cruciate ligament, and (medial meniscus) cartilage.
C2. (a) A. (b) S. (c) T. (d) W. (e) Ao. (f) E. (g) H.

D1. A.
D2. Rheumatoid arthritis (RA), osteoarthritis (OA), and gouty arthritis; pain.
D3.

	Rheumatoid Arthritis	Osteoarthritis
a. Which is known as "wear-and-tear" arthritis?	No	Yes
b. Which types of joints are affected?	Smaller	Larger
c. Is this an inflamma-tory autoimmune con-dition?	Yes	No
d. Is synovial mem-brane affected?	Yes	No
e. Does articular car-tilage degenerate?	Yes	Yes
f. Does fibrous tissue join bone ends?	Yes	No
g. Is movement limited?	Yes	Yes

D5. (a) Uric acid, sodium urate; kidneys; males. (b) Bursitis; housemaid's (carpet layer's) knee. (c) Dislocation; fingers (including thumb) and shoulder. (d) Arthralgia. (e) Lyme disease; bacteria, ticks; rash, larger. (f) Intervertebral and sacroiliac. (g) Strain; sprain; ankle.

Questions 1–14: Circle T (true) or F (false). If the statement is false, change the underlined word or phrase so that the statement is correct.

T F 1. A fibrous joint is one in which there is <u>no joint cavity and bones are held together by fibrous connective tissue.</u>

T F 2. <u>Sutures, syndesmoses, and symphyses</u> are kinds of fibrous joints.

T F 3. Arthritis affects <u>one in 700</u> persons in the United States.

T F 4. <u>Ball-and-socket, gliding, pivot, and ellipsoidal joints</u> are all diarthrotic joints.

T F 5. All fibrous joints <u>are synarthrotic and all cartilaginous joints are amphiarthrotic.</u>

T F 6. In synovial joints synovial membranes <u>cover the surfaces of articular cartilages.</u>

T F 7. Bursae are <u>saclike structures that reduce friction</u> at joints.

T F 8. Synovial fluid becomes <u>more</u> viscous when there is increased movement at a joint.

T F 9. When your arm is in the supine position, your radius and ulna are <u>parallel (not crossed).</u>

T F 10. When you touch your toes, the major action you perform at your hip joint is called <u>hyperextension.</u>

T F 11. The <u>only type of joint that is triaxial</u> is the ball-and-socket.

T F 12. Abduction is movement <u>away from</u> the midline of the body.

T F 13. The elbow, knee, and ankle joints are all <u>hinge</u> joints.

T F 14. Joints that are relatively stable (such as hip joints) tend to have <u>more</u> mobility than joints that are less stable (such as shoulder joints).

Questions 15–16: Arrange the answers in correct sequence.

_____ _____ _____ 15. From most mobile to least mobile:
A. Amphiarthrotic
B. Diarthrotic
C. Synarthrotic

_____ _____ _____ 16. Stages in rheumatoid arthritis, in chronological order:
A. Articular cartilage is destroyed and fibrous tissue joins exposed bone.
B. The synovial membrane produces pannus which adheres to articular cartilage.
C. Synovial membrane becomes inflamed and thickened, and synovial fluid accumulates.

Questions 17–20: Circle the letter preceding the one best answer to each question.

17. Which structure is extracapsular in location?
A. Meniscus
B. Posterior cruciate ligament
C. Synovial membrane
D. Tibial collateral ligament

18. All of these structures are associated with the knee joint *except:*
A. Glenoid labrum
B. Patellar ligament
C. Infrapatellar bursa
D. Medial meniscus
E. Fibular collateral ligament

19. "Housemaid's (carpet layer's) knee" is:
A. Sprained knee
B. Tendinitis
C. Inflammation of a bursa of the knee
D. Torn cartilage

20. A suture is found between:
A. The two pubic bones
B. The two parietal bones
C. Radius and ulna
D. Diaphysis and epiphysis
E. Tibia and fibula (distal ends)

Questions 21–25: fill-ins. Write the word or phrase that best completes the statement.

_____ 21. The term that means total hip replacement is a hip ____.

_____ 22. ____ is the forcible wrenching or twisting of a joint with partial rupture of it, but without dislocation.

_____ 23. The action of pulling the jaw back from a thrust-out position so that it becomes in line with the upper jaw is the movement called ____.

_____ 24. Another name for a freely movable joint is ____.

_____ 25. The type of joint between the atlas and axis and also between proximal ends of the radius and ulna is a ____ joint.

ANSWERS TO MASTERY TEST: ■ Chapter 9

True-False

1. T
2. F. Sutures, syndesmoses, and gomphoses
3. F. One in seven
4. T
5. F. Either synarthrotic or amphiarthrotic, and the same is true of cartilaginous
6. F. Do not cover surfaces of articular cartilages, which may be visualized as floor and ceiling of a room; but synovial membranes do line the rest of the inside of the joint cavity (much like wallpaper covering the four walls of the room).
7. T
8. F. Less
9. T
10. F. Flexion
11. T
12. T
13. T. (Or modified hinge joints)
14. F. Less

Arrange

15. B A C
16. C B A

Multiple Choice

17. D   19. C
18. A   20. B

Fill-ins

21. Arthroplasty
22. Sprain
23. Retraction
24. Diarthrotic
25. Pivot (or synovial or diarthrotic)

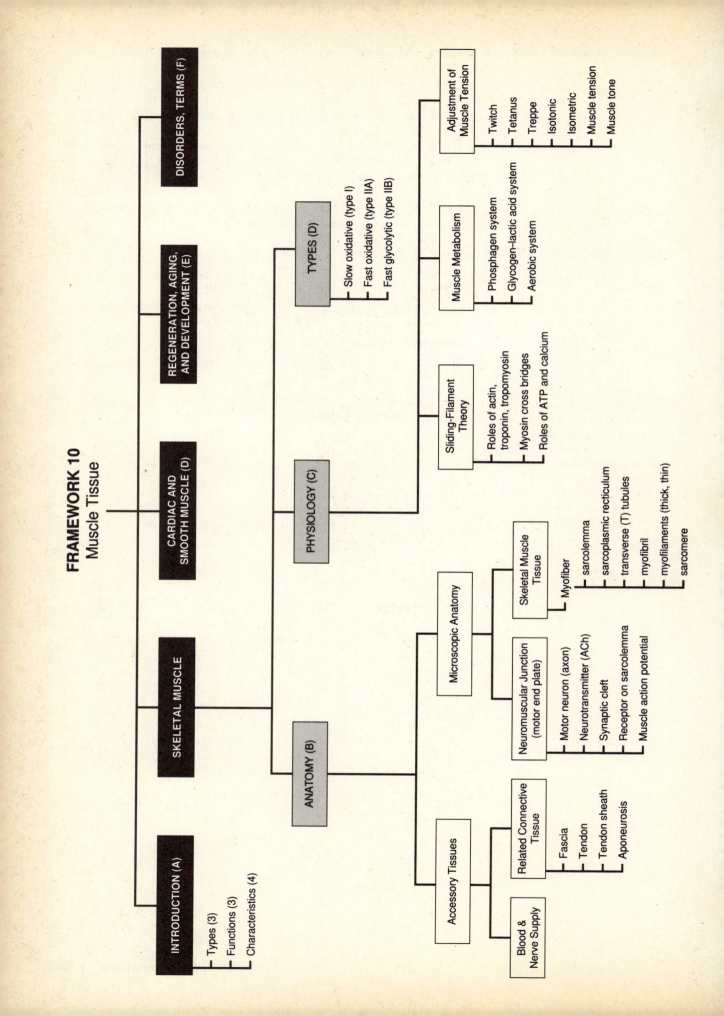

FRAMEWORK 10
Muscle Tissue

INTRODUCTION (A)
- Types (3)
- Functions (3)
- Characteristics (4)

SKELETAL MUSCLE

CARDIAC AND SMOOTH MUSCLE (D)

REGENERATION, AGING, AND DEVELOPMENT (E)

DISORDERS, TERMS (F)

ANATOMY (B)

Accessory Tissues
- Blood & Nerve Supply
- Related Connective Tissue
  - Fascia
  - Tendon
  - Tendon sheath
  - Aponeurosis

Microscopic Anatomy
- Neuromuscular Junction (motor end plate)
  - Motor neuron (axon)
  - Neurotransmitter (ACh)
  - Synaptic cleft
  - Receptor on sarcolemma
  - Muscle action potential
- Skeletal Muscle Tissue
  - Myofiber
    - sarcolemma
    - sarcoplasmic recticulum
    - transverse (T) tubules
    - myofibril
    - myofilaments (thick, thin)
    - sarcomere

PHYSIOLOGY (C)

Sliding-Filament Theory
- Roles of actin, troponin, tropomyosin
- Myosin cross bridges
- Roles of ATP and calcium

Muscle Metabolism
- Phosphagen system
- Glycogen–lactic acid system
- Aerobic system

Adjustment of Muscle Tension
- Twitch
- Tetanus
- Treppe
- Isotonic
- Isometric
- Muscle tension
- Muscle tone

TYPES (D)
- Slow oxidative (type I)
- Fast oxidative (type IIA)
- Fast glycolytic (type IIB)

Muscle Tissue

CHAPTER
10

Muscles are making it possible for you to read this paragraph: facilitating movement of your eyes and head, maintaining your posture, producing heat to keep you comfortable. Muscles do not work in isolation, however. Connective tissue binds muscle cells into bundles and attaches muscle to bones. Blood delivers the oxygen and nutrients for muscle work. And nerves initiate a cascade of events that culminate in muscle contraction.

A tour of the mechanisms of muscle action finds the visitor caught in the web of muscle protein filaments sliding back and forth as the muscle contracts and relaxes. Neurotransmitters, as well as calcium and power-packing ATP, serve as chief regulators. Muscles are observed contracting in a multitude of modes, from twitch to treppe and isotonic to isometric. It may be surprising to hear a tour guide's announcement that most muscle cells present now were actually around at birth. Very few muscle cells can multiply, but fortunately muscle cells can greatly enlarge, as with exercise. With aging, cell numbers decline and so do muscle strength and endurance. Certain disorders decrease muscle function and debilitate even in very early years of life.

To organize your study of muscle tissue, glance over the Chapter 10 Framework now. Be sure to refer to the Framework frequently and note relationships among key terms in each section.

TOPIC OUTLINE AND OBJECTIVES

A. Characteristics, functions, types of muscle tissue

1. List the characteristics and functions of muscle tissue.
2. Compare the location, microscopic appearance, nervous control, functions, and regenerative capacities of the three kinds of muscle tissue.

B. Skeletal muscle tissue and contraction

3. Describe the structure and importance of a neuromuscular junction and a motor unit.
4. Describe the principal events associated with the sliding-filament mechanism of muscle contraction.

5. Identify the sources of energy for muscular contraction.
6. Explain the roles played by muscle in homeostasis of body temperature.

C. Adjusting muscle tension; types of skeletal muscle fibers

7. Explain how muscle tension can be varied.

D. Cardiac muscle tissue, smooth muscle tissue; regeneration

8. Describe the different types of skeletal muscle fibers and compare them to cardiac and smooth muscle fibers.

E. Aging and development of muscles

9. Explain the effects of aging on muscle tissue.
10. Describe the development of the muscular system.

F. Disorders, medical terminology

11. Define such common muscular disorders as fibrosis, fibromyalgia, muscular dystrophies, myasthenia gravis (MG), spasm, tremor, fasciculation, fibrillation, and tic.
12. Define medical terminology associated with the muscular system.

WORDBYTES

Now study the following parts of words that may help you better understand terminology in this chapter.

Wordbyte	Meaning	Example	Wordbyte	Meaning	Example
a-	not	*a*trophy	graph-	to write	electromyo*graphy*
-algia	pain	fibromy*algia*	-lemma	rind, skin	sarco*lemma*
apo-	from	*apo*neurosis	myo-	muscle	*myo*fiber, *myo*sin
dys-	bad	muscular *dys*trophy	mys-	muscle	epi*mys*ium
-gen	to produce	phospha*gen* system	sarco-	flesh, muscle	*sarco*lemma
			troph-	nourishment	muscular dys*troph*y

CHECKPOINTS

A. Types, functions, characteristics of muscle tissue (pages 238–239)

■ **A1.** Match the muscle types listed in the box with descriptions below.

C. Cardiac	Sk. Skeletal	Sm. Smooth

_Sm_ a. Involuntary muscle found in blood vessels and intestine

_C_ b. Involuntary striated muscle

_SK_ c. Striated voluntary muscle attached to bones

_SK_ d. The only type of muscle that is voluntary

■ **A2.** List three functions of muscle tissue that are important for maintenance of homeostasis.

■ **A3.** Match the characteristics of muscles listed in the box with descriptions below.

> ~~Contractility~~   ~~Excitability~~
> Elasticity   Extensibility

a. Ability of muscle to stretch without damaging the tissue *Extensibility*

b. Tendency of stretched or contracted muscle to return to its original shape

    elasticity

c. Ability of muscle cells and nerve cells (neurons) to respond to stimuli by producing action potentials (impulses); irritability

    excitability

d. Ability of muscle tissue to shorten and thicken *Contractility*

B. Skeletal muscle: anatomy (pages 239–246)

■ **B1.** Skeletal muscle tissue *(is)* is not?) vascular tissue. State two or more functions of blood vessels that supply muscles.

oxygen

■ **B2.** Contrast two kinds of fascia by indicating which of the following are characteristics of superficial fascia (S) or deep fascia (D).

S a. Located immediately under the skin (subcutaneous)

D b. Composed of dense connective tissue that extends inward to surround and compartmentalize muscles

S c. A route for nerves and blood vessels to enter muscles; contains much fat, so provides insulation and protection

■ **B3.** Arrange the following terms (connective tissue) in correct sequence according to the amount of muscle surrounded: *endomysium, epimysium, perimysium.*

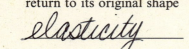

_____ → _____ → _____
(entire muscle)   (bundle of muscle fibers)   (individual muscle fiber)

For extra review. Color these three connective tissues on Figure LG 10.1a.

■ **B4.** Contrast the terms in the following pairs:

a. *Tendon/aponeurosis*

b. *Tendon/tendon sheath*

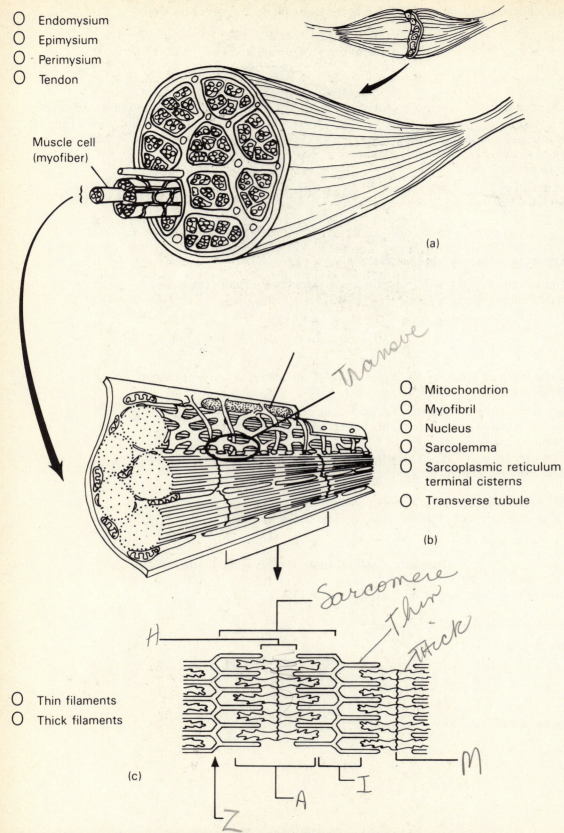

○ Endomysium
○ Epimysium
○ Perimysium
○ Tendon

Muscle cell
(myofiber)

(a)

Transverse

○ Mitochondrion
○ Myofibril
○ Nucleus
○ Sarcolemma
○ Sarcoplasmic reticulum
    terminal cisterns
○ Transverse tubule

(b)

Sarcomere
Thin
Thick

H

○ Thin filaments
○ Thick filaments

(c)

Z

A

I

M

Figure LG 10.1 Diagram of skeletal muscle. (a) Skeletal muscle cut to show cross section and longitudinal section with connective tissues. (b) Section of one muscle cell (myofiber). (c) Detail of sarcomere of muscle cell. Color and label as indicated in Checkpoints B3 and B7.

■ **B5.** Do this activity about the nerve supply of skeletal muscles.

a. In most cases a single nerve cell (neuron) innervates *(one muscle fiber [cell]? an average of 150 muscle fibers [cells]?).* The combination of a neuron plus the muscle fibers

(cells) it innervates is called a <u>motor unit</u>. An example of a motor unit that is likely to consist of just two or three muscle fibers (cells) precisely innervated by one neuron is *(laryngeal muscles controlling speech? calf [gastrocnemius] muscles controlling walking?).*

b. In other words, motor neurons form at least two, and possibly thousands, of

branches called <u>axons</u>, each of which supplies an individual skeletal muscle fiber (cell). When the motor neuron "fires," *(just one muscle fiber supplied by one axon? all muscle fibers within that motor unit?)* will be stimulated to contract.

c. Refer to Figure LG 10.2, in which we zoom in on the portion of a motor unit at which a branch of one neuron stimulates a single muscle fiber (cell). A nerve impulse travels along an axon (at letter **A**) toward one muscle fiber (letter **H**). The axon

is enlarged at its end into a bulb-shaped synaptic <u>terminal</u> (at letter **B**).

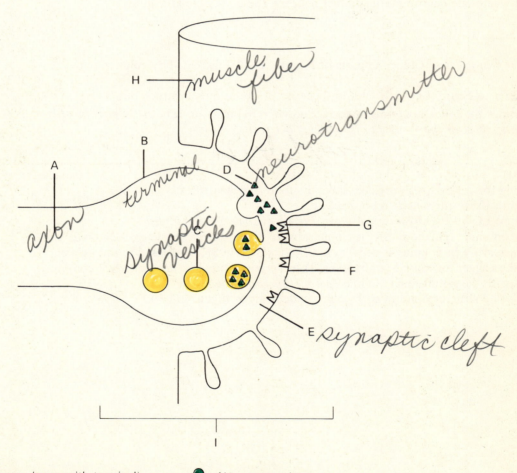

○   Motor neuron (axon with terminal)   ● Neurotransmitter

●   Synaptic vesicle   ○ Skeletal muscle myofiber (cell)

Figure LG 10.2   Diagram of a neuromuscular junction (motor end plate). Refer to Checkpoints B5 and B6.

d. The nerve impulse causes synaptic vesicles (letter _C_) to fuse with the plasma membrane of the axon. Next the vesicles release the neurotransmitter (letter _D_) named _Acetocoline_ into the synaptic cleft (letter _E_).

e. The region of the muscle fiber membrane (sarcolemma) close to the axon terminals is called a ((motor end plate? neuromuscular junction [NMJ]?)). This site contains specific _receptors_ (letter G) that recognize and bind to ACh. A typical motor end plate contains about (30–40? 3000–4000? (30–40 million?)) ACh receptors.

f. The effect of ACh is to cause Na⁺ channels in the sarcolemma to ((open? close?)), so that Na⁺ enters the muscle fiber. As a result, an action potential is initiated, leading to (contraction? relaxation?) of the muscle fiber. The diagnostic technique of recording such electrical activity in muscle cells is called _____.

g. The combination of the axon terminals and the motor end plate is known as a _neuromuscular junction_. At only one site along a muscle fiber (at about its middle) does a nerve approach a muscle cell to innervate it. In other words, there is/are usually (only one? many?) neuromuscular junction(s), labeled _I_ on the figure, for each muscle fiber (cell). (In Chapter 12, we will consider regions similar to NMJ's, but where one neuron meets another neuron. These sites are known as _Synapses_.)

B6. _For extra review._ Color the parts of Figure LG 10.2 indicated by color code ovals.

■ **B7.** Refer to Figure LG 10.1, and do this exercise about muscle structure.

a. Arrange the following terms in correct order from largest to smallest in size: _myofilaments (thick or thin), myofibrils, muscle fiber (cell)._

muscle fiber (cell) → _myofibrils_ → _myofilaments_
(largest)   (smallest) _thick or thin_

b. Using the leader lines provided, label the following structures:
Figure 10.1b: _sarcomere, triad, nucleus_
Figure LG 10.1c: _sarcomere, A band, I band, H zone, M line, Z disc_

c. Sarcoplasmic reticulum (SR) is comparable to ((endoplasmic reticulum? ribosomes? mitochondria?)) in nonmuscle cells. ((Ca²⁺? K⁺?)) stored in SR is released to sarcoplasm as the trigger for muscle contraction. Transverse (T) tubules ((are? are not?)) continuous with the sarcolemma, and lie (parallel? (perpendicular?)) to SR.

d. Color all structures indicated by color code ovals in Figure LG 10.1.

■ **B8.** *For extra review,* Match the correct term from the list in the box with its description below.

A band	Sarcoplasm
I band	Triad
Sarcomere	

a. Cytoplasm of muscle cell

_____

b. A transverse tubule, along with terminal cisterns of sarcoplasmic reticulum on either side

_____

c. Extends from Z disc to Z disc

_____

d. Dark area in striated muscle; contains thick and thin filaments _____

e. Light area on either side of Z disc; location of thin filaments only

_____

■ **B9.** After you study Figure 10.6, page 245 in your text, contrast types of filaments in skeletal muscle fibers.

a. Thin filaments are anchored at *(M lines? Z discs?)* and *(do? do not?)* extend into H zones. These filaments are composed mostly of bean-shaped *(actin? myosin?)* molecules that are twisted into a helix. Thin filaments also contain two other proteins. The protein that covers myosin-binding sites in relaxed muscle is called *(tropomyosin? troponin?)*.

b. Each *(thick? thin?)* filament is composed of about 200 myosin molecules. Each molecule is shaped like two intertwined *(footballs? golf clubs?)*. The ends of the "handles" point toward the *(M lines? Z discs?)*. The rounded head of the "club" is called a

_Cross bridge_____, and it attaches to *(actin? troponin?)* as muscles begin contraction.

c. The third type of filament, called a(n) _____ filament, is

composed of the protein named _____.

C. Skeletal muscle: physiology   (pages 246–258)

■ **C1.** Summarize one theory of muscle contraction in this Checkpoint.

a. In order to effect muscle shortening (or contraction), heads (cross bridges) of

_____ myofilaments act like oars pulling on

_____ molecules of thin filaments.

b. As a result, *(thick? thin?)* myofilaments move toward the center of the sarcomere. Since the thick and thin myofilaments *(shorten? slide?)* to decrease the length of the sarcomere, this theory of muscle contraction is known as the

_____ theory.

■ **C2.** Complete this exercise describing the principal events that occur during muscle contraction and relaxation.

  a. In Checkpoint B5, we discussed stimulation of a muscle fiber at a neuromuscular junction. The action potential (impulse) spreads from the sarcolemma via *(myosin? T tubules?)* to SR.

  b. In a relaxed muscle the concentration of calcium ions (Ca^{2+}) in sarcoplasm is *(high? low?)*. The effect of the nerve impulse and transmitter is to cause Ca^{2+} to pass from

     storage areas in _____ out to the sarcoplasm surrounding myofilaments.

  c. In relaxed muscle myosin cross bridges are not attached to the actin in thin filaments

     and _____ is bound to myosin cross bridges, while the

     tropomyosin- _____ complex blocks binding sites on actin.

  d. The released calcium ions attach to *(myosin? troponin?),* causing a structural change which leads to exposure of binding sites on *(myosin? actin?).*

  e. Breakdown of ATP (on myosin cross bridges) occurs via action of the enzyme

     named _____ (also from the myosin cross bridge). Energy derived from ATP activates myosin cross bridges to bind to and move

     _____. The oarlike action of myosin cross bridges (heads of

     golf clubs) upon actin is called a _____ stroke.

  f. Repeated power strokes slide actin filaments *(toward or even across? away from?)* the H zone and M line, and so shorten the sarcomere (and entire muscle). If this process is compared to running on a treadmill, *(actin? myosin?)* plays the role of the runner staying in one place, whereas the treadmill that *slides* backward (toward H zones) is comparable to the *(thick? thin?)* sliding filament.

  g. Relaxation of a muscle occurs when a synaptic cleft enzyme named

     _____ destroys ACh. This terminates impulse conduction over the muscle. Calcium ions are then sequestered from sarcoplasm back into

     _____ by *(active transport pumps? diffusion?)* and via a

     calcium-binding protein named _____.

  h. With a low level of Ca^{2+} now in the sarcoplasm surrounding myofilaments,

     _____ reforms from ADP on myosin cross bridges, while tropomyosin-troponin complex once again blocks binding sites on

     _____. As a result, thick and thin filaments detach, slip back

     into normal position, and the muscle is said to _____.

  i. The movement of Ca^{2+} back into sarcoplasmic reticulum (C2g) is *(an active? a passive?)* process. *After death, a supply of ATP (is? is not?)* available, so an active transport process cannot occur. Explain why the condition of rigor mortis results.

■ **C3.** *A clinical challenge.* Select terms from the box that are related to the descriptions below.

> Hypertonic   Muscle tone
> Hypotonic

a. Spasticity or rigidity:

    _____

b. Flaccid: _____

c. Sustained, small contractions that cause firm-
    ness in relaxed muscle:

    _____

■ **C4.** Complete this exercise about energy sources for muscle contraction.

a. Breakdown of *(ADP? ATP?)* provides the energy muscles use for contraction. Recall
    from Checkpoint C2e that ATP is attached to *(actin? myosin?)* cross bridges and so
    is available to energize the power stroke. Complete the chemical reaction showing
    ATP breakdown.

    ATP →

b. ATP must be regenerated constantly. One method involves use of ADP and energy
    from food sources. Complete that chemical reaction.

    ADP+

    In essence ADP is serving as a transport vehicle that can pick up and drop off
    energy stored in an extra high energy phosphate bond (~ P).

c. But ATP is used for other cell activities such as _____. To as-
    sure adequate energy for muscle work, muscle cells contain an additional molecule

    for transporting high energy phosphate; this is _____. Com-
    plete the reaction in Figure LG 10.3 showing how the phosphocreatine (PC) and
    ADP transport "vehicles" can meet and transfer the high energy phosphate "trailer"
    so that more ATP is formed for muscle work.

d. How is phosphocreatine (PC) regenerated during time when muscles are at rest?
    Show this on Figure LG 10.3.

e. ATP and PC, together called the _____ system, provide only
    enough energy to power muscle activity for about *(an hour? 10 minutes? 15 sec-
    onds?).* After that, muscles turn first to *(aerobic? anaerobic?)* pathways, and later to
    (aerobic? anaerobic?) pathways. Complete Figure 10.4 to show how muscles get en-
    ergy via these pathways.

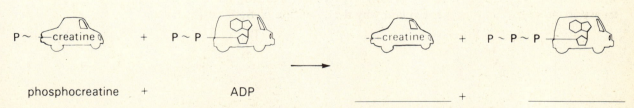

P ~ creatine   +   P ~ P   creatine   +   P ~ P ~ P

phosphocreatine   +   ADP   _____   +   _____

Figure LG 10.3 High energy molecules of the phosphagen system: the (~P) "trailer" tradeoff.
Complete figure as indicated in Checkpoint C4c, d.

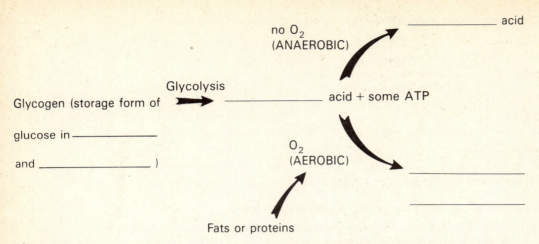

no O₂
(ANAEROBIC)

_____ acid

Glycolysis

Glycogen (storage form of

_____ acid + some ATP

glucose in _____

O₂
(AEROBIC)

and _____)

_____

_____

Fats or proteins

Figure LG 10.4   Anaerobic and aerobic energy sources for muscles. Complete figure as indicated in Checkpoint C4e.

f. During strenuous exercise an adequate supply of oxygen may not be available to

muscles. They must convert pyruvic acid to _____ acid by an *(aerobic? anaerobic?)* process. Excessive amounts of lactic acid in muscle tissue

contribute to some muscle _____ .

g. One substance in muscle that stores oxygen until oxygen is needed by mitochondria

is _____ . This protein is structurally somewhat like the

_____ -globin molecule in blood that also binds to and stores

oxygen. Both of these molecules have a _____ color that accounts for the color of blood and also of red muscle.

h. During exercise lasting more than 10 minutes, more than 90 percent of ATP is provided by the *(anaerobic? aerobic?)* breakdown of pyruvic acid (See Figure LG 10.4.). Athletic training *(de? in?)*-creases the maximal rate at which mitochondria can use oxygen for aerobic catabolism. This rate is known as maximal

_____ _____ .

C5. As you recall an occasion when you exercised vigorously, do this activity.

a. As you exercised, your body temperature *(in? de?)*-creased. What percentage of energy used during muscular contraction is likely to go to fuel contactions? ____% What percentage is used to increase body temperature?`____%

b. After you stopped exercising, your heavy breathing *(stopped immediately? continued for some time?)* List three purposes for which this extra oxygen is needed.

c. Explain what may account for the muscle pain you felt a day or two later.

■ **C6.** When your body is cold, an involuntary increase in muscle tone called

_____ can raise your body temperature. This process is initiated

by the "thermostat" portion of your brain, known as your _____.

■ **C7.** State the all-or-none principle.

This principle applies to *(individual motor units? an entire muscle such as the biceps?)*.

■ **C8.** List three or more factors that can decrease the strength of a muscle contraction.

■ **C9.** Match the terms in the box with definitions below. One answer will be used twice; one answer will not be used.

C. Contraction period	Ref. Refractory period
L. Latent period	Rel. Relaxation period
M. Myogram	T. Twitch

_____ a. Rapid, jerky response to a single stimulus

_____ b. Recording of a muscle contraction

_____ c. Period between application of a stimulus and start of a contraction; Ca^{2+} is being released from the SR during this time.

_____ d. Period when a muscle is not responsive to a stimulus

_____ e. Active transport of Ca^{2+} back into the SR is occurring

_____ f. A short period (about 0.005 second) in skeletal muscle and a long one (about 0.30 second) in cardiac muscle

■ **C10.** Choose the type of contraction that fits each descriptive phrase.

Muscle tone	Treppe	Tetanus

a. Sustained partial contraction of some portion of skeletal muscle; some fibers contracted,

   others not: _____

b. Sustained contraction due to stimulation at a rate of 20–100 stimuli per second:

   _____

c. More forceful contraction of skeletal muscle in response to same strength stimuli after muscle has contracted several times:

   _____

d. Phenomenon that is the principle behind

   athletic warmups: _____

e. Most voluntary contraction of muscles, such

   as biceps: _____

f. Due to a progressive buildup of Ca^{2+} in the sarcoplasm (two answers):

   _____

■ **C11.** Do this activity about changes in the force of muscle contraction.

a. When a muscle is stretched to its optimal length, there is *(much? little or no?)* overlap of myosin cross bridges on thick filaments with actin on thin filaments. In this case, a muscle demonstrates *(minimal? maximal?)* force of contraction. When a skeletal muscle is stretched excessively (such as to 175 percent of its optimal length), then *(many? no?)* myosin cross bridges can bind to actin.

b. As a general rule, the greater a muscle is stretched (within limits), the *(stronger? the weaker?)* the contraction.

C12. Explain how motor unit *recruitment* is related to production of smooth movements.

■ **C13.** Contrast isometric and isotonic contractions by doing this exercise.

a. A contraction in which a muscle shortens while tension (tone) of the muscle remains constant is known as an *(isometric? isotonic?)* contraction.

b. In an isometric contraction the muscle length *(shortens? stays about the same?)*, and tension of the muscle *(increases? stays the same?)*.

c. *A clinical challenge. (Isometric? Isotonic?)* exercise, such as carrying heavy objects, weight lifting, or waterskiing may be dangerous for cardiac patients since blood pressure may increase dramatically during the exercise period.

C14. Contrast terms in each pair below.

a. *Muscular atrophy/muscular hypertrophy*

b. *Disuse atrophy/denervation atrophy*

C15. Describe six or more specific benefits of strength training.

D. Types of skeletal muscle fibers; cardiac and smooth muscle (pages 257–262)

■ **D1.** Refer to the table below and contrast the three types of skeletal muscle by doing this activity.

Type and Rate of Contraction	Color	Myoglobin Concentration	Mitochondria and Blood Vessels	Source of ATP	Fatigue Easily?
I. Slow	Red	High	Many	Aerobic	No
IIA. Fast	Red	Very high	Very high	Aerobic	Moderate
IIB. Fast	White	Low	Few	Anaerobic	Yes

a. Postural muscles (as in neck and back) are used *(constantly? mostly for short bursts of energy?)*. Therefore it is appropriate that they contract at a *(fast? slow?)* rate. Such muscle tissue consists mostly of slow red fibers that *(do? do not?)* fatigue easily. These appear red since they have *(much? little?)* myoglobin and *(many? few?)* blood vessels. They have *(many? few?)* mitochondria and therefore can depend on *(aerobic? anaerobic?)* metabolism for energy. These muscles are classified as *(I? IIA? IIB?)*.

b. Muscles of the arms are used *(constantly? mostly for short bursts of energy?)* as in lifting and throwing. Therefore they must contract at a *(fast? slow?)* rate. The arms consist mainly of fast twitch white fibers. These respond rapidly to nerve impulses and contract *(fast? slowly?)*. They *(do? do not?)* fatigue easily since they are designed for the relatively inefficient processes of *(aerobic? anaerobic?)* metabolism. These muscles are classified as *(I? IIA? IIB?)*.

c. Endurance exercises tend to enhance development of the more efficient (aerobic)

_____ fibers, whereas weight lifting, which requires short bursts

of energy, tends to develop _____ fibers.

d. Most skeletal muscles, such as the biceps brachii, are *(all of one type? mixtures of types I, IIA, and IIB?)*. The muscles of any one motor unit are *(all of one type? mixtures of types I, IIA, and IIB?)*.

D2. State the desired effects as well as disadvantages of uses of anabolic steroids.

■ **D3.** Compare the structure and function of skeletal and cardiac muscle by completing this table. State the significance of characteristics that have asterisks (*).

Characteristic	Skeletal Muscle	Cardiac Muscle
a. Number of nuclei per myofiber	Several (multinucleate)	
b. Number of mitochondria per myofiber (More or Fewer)*		
c. Striated appearance due to alternated actin and myosin myofilaments (Yes or No)	Yes	
d. Arrangement of muscle fibers (Parallel or Branching)		
e. Nerve stimulation required for contraction (Yes or No)*		
f. Length of refractory period (Long or Short)*		

D4. State the significance of intercalated discs in the function of cardiac muscle fibers.

■ **D5.** Contrast smooth muscle with the muscle tissue you have already studied by circling the correct answers in this paragraph.

Smooth muscle fibers are *(cylinder-shaped? tapered at each end?)* with *(several nuclei? one nucleus?)* per cell. They *(do? do not?)* contain actin and myosin. However, due to the irregular arrangement of these filaments, smooth muscle tissue appears *(striated? nonstriated or "smooth"?)*. In general, smooth muscle contracts and relaxes more *(rapidly? slowly?)* than skeletal muscle does, and smooth muscle holds the contraction for a *(shorter? longer?)* period of time than skeletal muscle does. This difference in smooth muscle is due at least partly to the fact that smooth muscle has *(no? many?)* transverse tubules to transmit calcium into the muscle fiber and has a *(large? small?)* amount of SR.

D6. Compare the two types of smooth muscle.

Muscle Type	Structure	Spread of Stimulus	Locations
a. Visceral		Impulse spreads and causes contraction of adjacent fibers.	
b.			Blood vessels, iris of eye

■ **D7.** Smooth muscle *(is? is not?)* normally under voluntary control. List three chemicals released in the body or other factors that can also lead to smooth muscle contraction or relaxation.

D8. *For extra review* of the three muscle types, review Chapter 4, Checkpoints E1–E3, and Figure LG 4.1, M–O, pages 71 and 78–79.

E. Regeneration, aging, and development of muscle tissue (pages 262–264)

■ **E1.** Arrange the three types of your muscle tissue in correct sequence according to ability (from most to least) to regenerate during your lifetime.

_____ _____ _____ A. Heart muscle
                                              B. Biceps muscle
                                              C. Muscle of an artery or intestine

E2. Describe the roles of the following cells in regeneration of muscle tissue:

a. Satellite cells

b. Pericytes

E3. List three changes in skeletal muscle that are likely to occur with normal aging.

■ **E4.** Complete this exercise about the development of muscles.

a. Which of the three types of muscle tissue develop from mesoderm? *(Skeletal? Cardiac? Smooth?)*.

b. Part of the mesoderm forms columns on either side of the developing nervous system. This tissue segments into blocks of tissue called _____.
The first pair of somites forms on day *(10? 20? 30?)* of gestation. By day 30 a total

of _____ pairs of somites are present.

c. Which part of a somite develops into vertebrae? *(Myo-? Derma-? Sclero-?)* derm.
The dermis of the skin and other connective tissues are formed from

_____-tomes, while most skeletal muscles develop from

_____-tomes.

F. Disorders, medical terminology   (pages 264–265)

■ **F1.** Write the correct medical term related to muscles after its description.

a. Muscle or tendon pain and stiffness, such as "charleyhorse":

_____

b. Loss or impairment of motor or muscular function due to nerve or muscle disorder:

_____

c. Inherited, muscle-destroying disease causing atrophy of muscles:

_____

d. A muscle tumor: _____

e. Any disease of muscle: _____

■ **F2.** Describe myasthenia gravis in this exercise.

a. In order for skeletal muscle to contract, a nerve must release the chemical

_____ACH_____ at the myoneural junction. Normally ACh binds to

____receptors____ on the muscle fiber membrane.

b. It is believed that a person with myasthenia gravis produces

____antibodies____ that bind to these receptors, making them unavailable for ACh binding. Therefore, ACh *(can? cannot?)* stimulate the muscle, and it is weakened.

c. One treatment for this condition employs _____ drugs, which enhance muscle contraction by permitting the ACh molecules that *do* bind to act longer (and not be destroyed by AChE).

d. Other treatments include _____-suppressants, which decrease

the patient's antibody production, or _____, which segregates the patient's harmful antibodies.

F3. Define each of these types of abnormal muscle contractions.

a. Spasm

b. Fibrillation

c. Tic

ANSWERS TO SELECTED CHECKPOINTS

A1. (a) Sm. (b) C. (c) Sk. (d) Sk.

A2. 1. Motion or movement. 2. Stabilizing body positions, for example, maintaining posture and regulating organ volume, exemplified by contraction of the heart to pump out blood. 3. Generation of heat to alter body temperature.

A3. (a) Extensibility. (b) Elasticity. (c) Excitability. (d) Contractility.

B1. Is; (1) provide nutrients and oxygen for generation of ATP, (2) remove wastes.

B2. (a) S. (b) D. (c) S.

B3. Epimysium → perimysium → endomysium. See Figure LG 10.1A

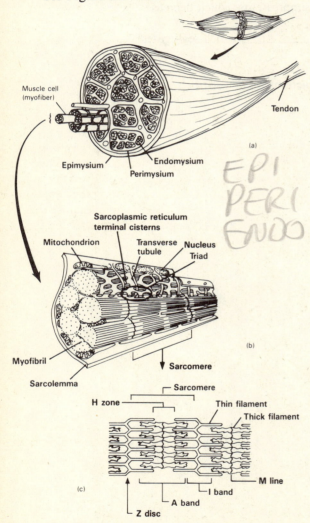

Figure LG 10.1A Diagram of skeletal muscle. (a) Skeletal muscle cut to show cross section and longitudinal section with connective tissues. (b) Section of one muscle cell (myofiber). (c) Detail of sarcomere of muscle cell.

B5. (a) An average of 150 muscle fibers [cells]; motor unit; laryngeal muscles controlling speech. (b) Axons; all muscle fibers within that motor unit. (c) A, H; terminal. (d) C; D, acetylcholine (ACh), E. (e) Motor end plate; receptors; 30–40 million, G. (f) Open; contraction; electromyography (EMG). (g) Neuromuscular junction (NMJ); only one, I; synapses.

B7. (a) Muscle fiber (cell) → myofibrils → myofilaments (thick or thin). (b) See Figure LG 10.1A. (c) Endoplasmic reticulum; Ca^{2+}; are, perpendicular. (d) See Figure LG 10.1A.

B8. (a) Sarcoplasm. (b) Triad. (c) Sarcomere. (d) A band. (e) I band.

B9. (a) Z discs, do not; actin; tropomyosin. (b) Thick; golf clubs; M lines; cross bridge, actin. (c) Elastic, titin (or connectin).

C1. (a) Myosin, actin. (b) Thin; slide, sliding filament.

C2. (a) T tubules. (b) Low; sarcoplasmic reticulum. (c) ATP, troponin. (d) Troponin, actin. (e) ATPase; actin; power. (f) Toward or even across; myosin, thin. (g) Acetylcholinesterase (AChE); SR, active transport pumps, calsequestrin. (h) ATP, actin (or thin filament); relax (or lengthen). (i) An active, is not; myosin cross bridges stay attached to actin and the muscles remain in a state of partial contraction (rigor mortis) for 1–2 days.

C3. (a) Hypertonic. (b) Hypotonic. (c) Muscle tone.

C4. (a) ATP; myosin; ATP → ADP + P + energy. (b) ADP + P + energy (from foods) → ATP. (c) Active transport; creatine (that combines with phosphate to form creatine phosphate or phosphocreatine [PC]); phosphocreatine (PC) + ADP → creatine + ATP. (d) PC + ADP ← C + ATP. (Draw arrow to LEFT in Figure LG 10.3.) (e) Phosphagen, 15 seconds; anaerobic, aerobic. See Figure LG 10.4A. (f) Lactic, anaerobic; fatigue. (g) Myoglobin; hemo-; red. (h) Aerobic; in; oxygen uptake.

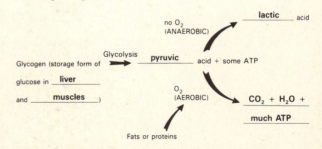

Figure LG 10.4A Anaerobic and aerobic energy sources for muscles.

C6. Shivering; hypothalamus.

C7. A single action potential (nerve impulse) elicits a single contraction in all of the muscle fibers of its motor unit; muscle fibers do not partially contract. Individual motor units.

C8. Decrease in frequency of stimulation by neurons, shorter length muscle fibers just prior to contraction, fewer and smaller motor units recruited, lack of nutrients, lack of oxygen, and muscle fatigue.

C9. (a) T. (b) M. (c) L. (d) Ref. (e) Rel. (f) Ref.

C10. (a) Muscle tone. (b) Tetanus. (c) Treppe. (d) Treppe. (e) Tetanus. (f) Treppe, tetanus.

C11. (a) Much; maximal; no. (b) Stronger.

C13. (a) Isotonic. (b) Stays about the same, increases. (c) Isometric.

D1. (a) Constantly; slow; do not; much, many; many, aerobic; I. (b) Mostly for short bursts of energy; fast; fast; do; anaerobic; IIB. (c) IIA (fast oxidative); IIB (fast glycolytic). (d) Mixtures of types I, IIA, and IIB; all of one type.

D3.

Characteristic	Skeletal Muscle	Cardiac Muscle
a. Number of nuclei per myofiber	Several (multinucleate)	One
b. Number of mitochondria per myofiber (More or Fewer)*	Fewer	More, since heart muscle requires constant generation of energy
c. Striated appearance due to alternated actin and myosin myofilaments (Yes or No)	Yes	Yes
d. Arrangement of muscle fibers (Parallel or Branching)	Parallel	Branching
e. Nerve stimulation required for contraction (Yes or No)*	Yes	No, so heart can contract without nerve stimulation, but nerves can increase or decrease heart rate
f. Length of refractory period (Long or Short)*	Short	Long, to allow heart time to relax between beats

D5. Tapered at each end, one nucleus; do; non-striated or "smooth"; slowly, longer; no, small.

D7. Is not; hormones, pH or temperature changes, O_2 and CO_2 levels, and certain ions.

E1. C B A

E4. (a) All three: skeletal, cardiac, and smooth. (b) Somites; 20; 44. (c) Sclero-; derma-, myo-.

F1. (a) Fibromyalgia. (b) Paralysis. (c) Muscular dystrophy. (d) Myoma. (e) Myopathy.

F2. (a) Acetylcholine (ACh); receptors. (b) Antibodies; cannot. (c) Anticholinesterase (antiAChE). (d) Immuno-, plasmapheresis.

MASTERY TEST: Chapter 10

Questions 1–11: Circle T (true) or F (false). If the statement is false, change the underlined word or phrase so that the statement is correct.

T F 1. Tendons are cords of connective tissue, whereas aponeuroses are broad, flat bands of connective tissue.

T F 2. During contraction of muscle both A bands and I bands get shorter.

T F 3. Fascia is one type of skeletal muscle tissue.

T F 4. Elasticity is the ability of a muscle to be stretched or extended.

T F 5. Myasthenia gravis is caused by an excess of acetylcholine production at the myoneural junction.

T F 6. Muscle fibers remain relaxed if there are few calcium ions in the sarcoplasm.

T F 7. Atrophy means decrease in muscle mass.

T F 8. Muscles that are used mostly for quick bursts of energy (such as those in the arms) contain large numbers of fast glycolytic (type IIB) fibers.

T F 9. Most of the energy released during muscle contraction is used for heat production.

T F 10. In a relaxed muscle fiber thin and thick myofilaments overlap to form the A band.

T F 11. As a result of aging, muscle tissue is largely replaced by fat, and muscle strength decreases.

Questions 12–13: Arrange the answers in correct sequence.

_____ _____ _____ 12. According to the amount of muscle tissue they surround, from most to least:
A. Perimysium
B. Endomysium
C. Epimysium

_____ _____ _____ 13. From largest to smallest:
A. Myofibril
B. Myofilament
C. Muscle fiber (myofiber or cell)

Questions 14–20: Choose the one best answer to each question.

_____ 14. All of the following molecules are parts of thin filaments *except:*
A. Actin   C. Tropomyosin
B. Myosin   D. Troponin

_____ 15. Choose the one statement that is *false:*
A. A band refers to the anisotropic band.
B. The A band is darker than the I band.
C. Thick myofilaments reach the Z line in relaxed muscle.
D. The H zone contains thick myofilaments, but not thin ones.
E. Thick myofilaments are made of myosin.

_____ 16. All of the following terms are correctly matched with descriptions *except:*
A. Denervation atrophy—wasting of a muscle to one-quarter of its size within two years of loss of nerve supply to a muscle
B. Disuse atrophy—decrease of muscle mass in a person who is bedridden or who has a cast on
C. Muscular hypertrophy—increase in size of a muscle by increase in the number of muscle cells
D. Cramp—painful, spasmotic contraction

_____ 17. Which statement about muscle physiology in the relaxed state is *false?*
A. Myosin cross bridges are bound to ATP.
B. Calcium ions are stored in sarcoplasmic reticulum.
C. Myosin cross bridges are bound to actin.
D. Tropomyosin-troponin complex is bound to actin.

_____ 18. The staircase phenomenon refers to ____ contractions.
A. Tetanic   C. Tonic
B. Treppe   D. Isotonic

_____ 19. Choose the *false* statement about cardiac muscle.
A. Cardiac muscle has a long refractory period.
B. Cardiac muscle has one nucleus per fiber.
C. Cardiac muscle cells are called cardiac muscle fibers.
D. Cardiac fibers are separated by intercalated discs.
E. Cardiac fibers are spindle-shaped with no striations.

_____ 20. Most voluntary movements of the body are results of ____ contractions.
A. Isometric   D. Tetanic
B. Fibrillation   E. Spasm
C. Twitch

Questions 21–25: fill-ins. Complete each sentence with the word or phrase that best fits.

$\underline{\text{Smooth}}$ 21. ____ Muscle cells are nonstriated and spindle-shaped with one nucleus per cell.

_____ 22. Cardiac muscle remains contracted longer than skeletal muscle does because ____ is slower in cardiac than in skeletal muscle.

_____ 23. It is during the ____ period of a muscle contraction that calcium ions are released from sarcoplasmic reticulum and myosin cross bridge activity begins to occur.

_____ 24. ____ is the transmitter released from synaptic vesicles of axons supplying skeletal muscle.

_____ 25. ADP + phosphocreatine → ____. (Write the products.)

ANSWERS TO MASTERY TEST: ▪ Chapter 10

True-False

1. T
2. F. A bands but not I bands
3. F. Dense connective tissue (which may surround skeletal muscle)
4. F. Extensibility
5. F. An autoimmune response in which antibodies are produced which bind onto receptors on the sarcolemma
6. T   9. T
7. T   10. T
8. T   11. T

Arrange

12. C A B
13. C A B

Multiple Choice

14. B   18. B
15. C   19. E
16. C   20. D
17. C

Fill-ins

21. Smooth
22. Passage of calcium ions from the extracellular fluid through sarcolemma
23. Latent
24. Acetylcholine
25. ATP + creatine

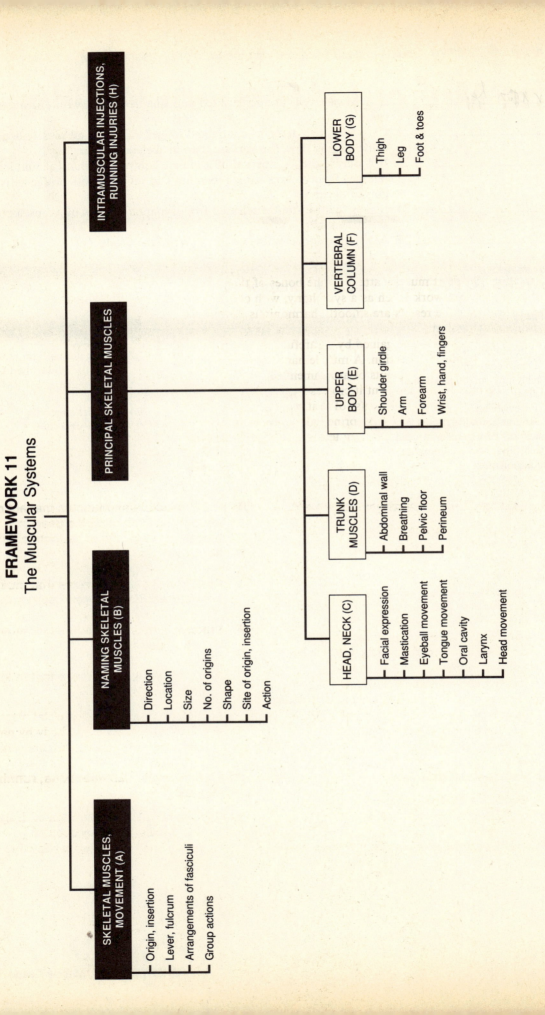

FRAMEWORK 11
The Muscular Systems

SKELETAL MUSCLES, MOVEMENT (A)
- Origin, insertion
- Lever, fulcrum
- Arrangements of fasciculi
- Group actions

NAMING SKELETAL MUSCLES (B)
- Direction
- Location
- Size
- No. of origins
- Shape
- Site of origin, insertion
- Action

PRINCIPAL SKELETAL MUSCLES

INTRAMUSCULAR INJECTIONS, RUNNING INJURIES (H)

HEAD, NECK (C)
- Facial expression
- Mastication
- Eyeball movement
- Tongue movement
- Oral cavity
- Larynx
- Head movement

TRUNK MUSCLES (D)
- Abdominal wall
- Breathing
- Pelvic floor
- Perineum

UPPER BODY (E)
- Shoulder girdle
- Arm
- Forearm
- Wrist, hand, fingers

VERTEBRAL COLUMN (F)

LOWER BODY (G)
- Thigh
- Leg
- Foot & toes

The Muscular System CHAPTER

11

More than 600 different muscles attach to the bones of the skeleton. Muscles are arranged in groups that work much as a symphony, with certain muscles quiet while others are performing. The results are smooth, harmonious movements, rather than erratic, haphazard discord.

Much information can be gained by a careful initial examination of two aspects of each muscle: its name and location. A muscle name may offer such clues as size or shape, direction of fibers, or points of attachment of the muscle. A look at the precise location of the muscle and the joint it crosses—combined with logic—can usually lead to a correct understanding of action(s) of that muscle.

Chapter 11 focuses on all of the principal skeletal muscles of the body. First study the Chapter 11 Framework and the key terms associated with each section.

TOPIC OUTLINE AND OBJECTIVES

A. How skeletal muscles produce movement

1. Describe the relationship between bones and skeletal muscles in producing body movements.
2. Define a lever and fulcrum and compare the three classes of levers on the basis of placement of the fulcrum, effort, and resistance.
3. Identify the various arrangements of muscle fibers in a skeletal muscle and relate the arrangements to the strength of contraction and range of motion.
4. Discuss most body movements as activities of groups of muscles by explaining the roles of the prime mover, antagonist, synergist, and fixator.

B. Naming skeletal muscles

5. Define the criteria employed in naming skeletal muscles.

C. Principal skeletal muscles of the head and neck

D. Principal skeletal muscles that act on the abdominal wall, muscles used in breathing, muscles of the pelvic floor and perineum

E. Principal skeletal muscles that move the shoulder girdle and upper extremity

F. Principal skeletal muscles that move the vertebral column

G. Principal skeletal muscles that move the lower extremity

6. Identify the principal skeletal muscles in different regions of the body by name, origin, insertion, action, and innervation.

H. Intramuscular (IM) injections; running injuries

7. Discuss the administration of drugs by intramuscular (IM) injection.
8. Describe several injuries related to running.

Now study the following parts of words that may help you better understand terminology in this chapter.

Wordbyte	Meaning	Example	Wordbyte	Meaning	Example
bi-	two	*biceps*	grac-	slender	*gracilis*
brachi-	arm	*brachi*alis	-issimus	the most	long*issimus* capitis
brev-	short	flexor digitorum *brevis*	lat-	broad	*lat*issimus dorsi
			maxi-	large	gluteus *maximus*
			mini-	small	gluteus *minimus*
bucc-	mouth, cheek	*bucc*inator	or-	mouth	orbicularis *oris*
cap-, -ceps	head	*capitis*, tri*ceps*	rect-	straight	*rectus* femoris
-cnem-	leg	gastro*cnem*ius	sartor-	tailor	*sartor*ius
delt-	Greek D (Δ)	*delt*oid	serra-	toothed, notched	*serra*tus anterior
gastro-	stomach, belly	*gastro*cnemius			
genio-	chin	*genio*glossus	teres-	round	*teres* major
glossus	tongue	*glossary*, stylo*glossus*	tri-	three	*triceps* femoris
			vast-	large	*vastus* lateralis
glute-	buttock	*glute*us medius			

A. How skeletal muscles produce movement   (pages 270–274)

■ **A1.** What structures constitute the *muscular system?*

skeletal muscles + connective tissues

■ **A2.** Refer to Figure LG 11.1 and consider flexion of your own forearm as you do this learning activity.

a. In flexion your forearm serves as a rigid rod, or *lever*, which moves about a fixed point, called a *fulcrum* (your elbow joint, in this case).

b. Hold a weight in your hand as you flex your forearm. The weight plus your forearm serve as the (*effort? fulcrum? resistance?*) during this movement.

c. The effort to move this resistance is provided by contraction of a *muscle*. Note that if you held a heavy telephone book in your hand while your forearm was flexed, much more *effort* by your arm muscles would be required.

d. In Figure LG 11.1 identify the exact point at which the muscle causing flexion attaches to the forearm. It is the (*proximal?* distal?) end of the (humerus? *radius?* ulna?). Write an E and an I on the two lines next to the arrow at that point in the figure. This indicates that this is the site where the muscle exerts its effort (E) in the lever system, and it is also the insertion (I) end of the muscle. (More about insertions in a minute.)

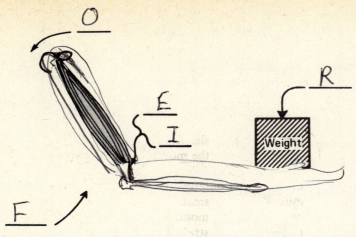

Figure LG 11.1 The lever-fulcrum principle is illustrated by flexion of the forearm. Complete the figure as directed in Checkpoint A2.

e. Each skeletal muscle is attached to at least two bones. As the muscle shortens, one bone stays in place and so is called the *(origin? insertion?)* end of the muscle. What bone in the figure appears to serve as the origin bone?

humerus scapula

Write O on that line at that point in the figure.

f. Now label the remaining arrows in Figure LG 11.1: F at fulcrum and R at resistance. This is an example of a *(first? second? third?)*-class lever.

_____

■ **A3.** Describe the roles of lever systems in the body by completing this exercise.

a. There are *(2? 3? 5? 7?)* classes of lever systems. Levers are categorized according to relative positions of effort (E), fulcrum (F), and resistance (R).

b. Refer to Figure 11.2a, page 271 of your text. Now hyperextend your head as if to look at the sky. The weight of your face and jaw serves as *(E? F? R?)*, while your neck muscles provide *(E? F? R?)*. The fulcrum is the joint between the

_____ and the _____ bones. This is an

example of a _____-class lever.

■ **A4.** After you have read the Chapter 11 *Essays on Wellness,* do this activity about range of motion (ROM).

a. Shorter muscles *(in? de?)*-crease the ROM of a joint. Muscles are more likely to shorten in older persons, causing stiffness. Such stiffness that many older persons experience is due primarily to:
A. The normal aging process
B. Lack of exercise

b. Stretching muscles *(does? does not?)* help to increase flexibility. *Static stretching* is a method of stretching that is done *(slowly and gently? fast?).*

c. PNF is a stretching technique; the letters PNF stand for P_____,

N_____, F_____. The contract-relax with agonist-contraction (CRAC) method involves application of resistance to the muscle group *(that is? opposite to the one?)* being stretched.

d. List two ways that muscles can be "warmed up" to gently stretch the collagenous fibers within fascia surrounding muscles.

■ **A5.** Correlate fascicular arrangement with muscle power and range of motion of muscles.

a. A muscle with *(many? long?)* fibers will tend to have great strength. An example is the *(parallel? pennate?)* arrangement.

b. A muscle with *(many? long?)* fibers will tend to have great range of motion (but less power). An example is the *(parallel? pennate?)* arrangement.

■ **A6.** Refer again to Figure LG 11.1 and do this exercise about how muscles of the body work in groups.

a. The muscle that contracts to cause flexion of the forearm is called a

prime movers. An example of a prime mover in this action would be

the *biceps brachii* muscle.

b. The triceps brachii must relax as the biceps brachii flexes the forearm. The triceps is an extensor. Since its action is opposite to that of the biceps, the triceps is called *(a synergist? an agonist? an antagonist?)* of the biceps.

c. What would happen if the flexors of your forearm were functional, but not the antagonistic extensors? *couldn't lower arm*

d. What action would occur if both the flexors and extensors contracted simultaneously?

staff arm

e. Muscles that assist or cooperate with the prime mover to cause a given action are

known as *synergists*, whereas muscles that stabilize a bone (such as the scapula) so that prime movers and synergists can move another bone (such as

the humerus) are called *fixators*.

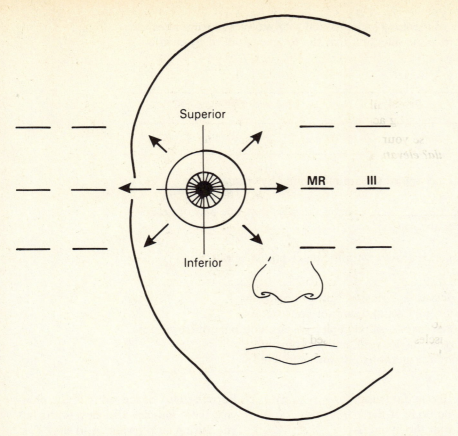

Figure LG 11.3 Right eyeball with arrows indicating directions of eye movements produced by the extrinsic eye muscles. Label as directed in Checkpoint C4.

■ **C5.** *For extra review* of actions of eye muscles, work with a study partner. One person moves the eyes in a particular direction; the partner then names the eye muscles used for that action. *Note:* Will both eyes use muscles of the same name? For example, as you look to your right, will the lateral rectus muscles attached to both eyes contract?

■ **C6.** Actions such as opening your mouth for eating or speaking and also swallowing require integrated action of a number of muscles. You have already studied some that allow you to open your mouth (such as masseter and temporalis). We will now consider muscles of the tongue, floor of the oral cavity, pharynx, and larynx.

Identify locations of three bony structures relative to your tongue. Tell whether each is anterior or posterior and superior or inferior to your tongue. (For help refer to Figure 11.7, page 287 of the text.)

a. Styloid process of temporal bone _____

b. Hyoid bone _____

c. Chin (anterior portion) _____

■ **C7.** Remembering the locations of these bony points and that muscles move a structure (tongue) by pulling on it, determine the direction that the tongue is pulled by each of these muscles.

a. Styloglossus _____

b. Hyoglossus _____

c. Genioglossus _____

■ **C8.** The three muscles listed in Checkpoint C7 are innervated by cranial nerve

_____; the name _____ nerve indicates that this pair of cranial nerves supplies the region "under the tongue."

■ **C9.** Match groups of muscles in each answer with the descriptions below.

> A. Styloglossus, hyoglossus, and genioglossus
> B. Superior, middle, and inferior constrictors
> C. Stylopharyngeus, salpingopharyngeus, and palatopharyngeus
> D. Stylohyoid, mylohyoid, and geniohyoid
> E. Thyrohyoid, sternohyoid, and omohyoid

_D_ a. Located superior to the hyoid (supra-hyoid), these form the floor of the oral cavity. (Recall that the hyoid is directly posterior to the inferior border of the U-shaped mandible.)

_E_ b. These muscles cover the anterior of the larynx and trachea. These permit you to elevate your thyroid cartilage ("Adam's apple") during swallowing to prevent food from entering your larynx. (Try it.)

_C_ c. Forming part of the walls of the phar-ynx, these muscles also elevate the lar-ynx during swallowing. And they close off the nasopharynx (so that food does not back up into the nasal cavity) during swallowing. Finally one of these muscles opens your auditory tube (leading to the middle ear) dur-ing swallowing, an action that helps to "pop" your ears while you are de-scending from a high altitude.

_B_ d. These muscles squeeze the pharynx to propel a bolus of food into the esopha-gus during swallowing.

_A_ e. Muscles permitting tongue movements.

■ **C10.** Using a mirror, find the origin and insertion of your left sternocleidomastoid mus-cle. (See Figure 11.9c, page 292 in your text.) The muscle contracts when you pull your chin down and to the right; this diagonal muscle of your neck will then be readily lo-cated. Note that the left sternocleidomastoid pulls your face toward the *(same? oppo-site?)* side. It also *(flexes? extends?)* the head.

■ **C11.** Alternately flex and extend vertebrae of your neck by looking at your toes and then toward the sky. As you look at the sky, you are *(flexing? extending and then hyper-extending?)* your head and cervical vertebrae. On which surface of the neck would you expect to find extensors of the head and neck? *(Anterior? Posterior?)* Note that these muscles are the most superior muscles of the columns of extensors of the vertebrae. Find them on Figure 11.19 (page 321 in your text). Now name three of these muscles.

D. Principal skeletal muscles that act on the abdominal wall, muscles used in breathing, muscles of the pelvic floor and perineum (pages 294–301)

■ **D1.** Each half of the abdominal wall is composed of *(two? three? four?)* muscles. Describe these in the exercise below.

   a. Just lateral to the midline is the rectus abdominis muscle. Its fibers are *(vertical? horizontal?)*, attached inferiorly to the ___pubis___ and superiorly to the ___sternum___ *ribs 5-7*. Contraction of this muscle permits *(flexion? extension?)* of the vertebral column.

   b. List the remaining abdominal muscles that form the sides of the abdominal wall from most superficial to deepest.

___external oblique___ → ___internal obliq___ ___transverse abdominis___
     most superficial   deepest

   c. Do all three of these muscles have fibers running in the same direction? *(Yes? No?)* Of what advantage is this? *extra support*

■ **D2.** Answer these questions about muscles used for breathing.

   a. The diaphragm is ___dome___-shaped. Its oval origin is located *on the bottom of the rib cage*. Its insertion is not into bone, but rather into dense connective tissue forming the roof of the diaphragm; this tissue is called the ___central tendon___

   b. Contraction of the diaphragm flattens the dome, causing the size of the thorax to *(increase? decrease?)*, as occurs during *(inspiration? expiration?)*.

   c. The name *intercostals* indicates that these muscles are located *between the ribs* ___. Which set is used during expiration? *(Internal? External?)*

D3. *For extra review.* Cover the key to Figure LG 11.2(a) and write labels for muscles F, G, H, I, and X.

D4. Look at the inferior of the human pelvic bones on a skeleton (or refer to Figure 8.8, page 206 in your text). Note that a gaping hole (outlet) is present. Pelvic floor muscles attach to the bony pelvic outlet. Name these muscles and state functions of the pelvic floor.

■ **D5.** Fill in the blanks in this paragraph. *Diaphragm* means literally

_____ *(dia-)* _____ *(-phragm).* You are

familiar with the diaphragm that separates the _____ from the ab-

domen. The *pelvic diaphragm* consists of all of the muscles of the

_____ floor plus their fasciae. It is a "wall" or barrier between the

inside and outside of the body.

E. **Principal skeletal muscles that move the shoulder girdle and upper**
   extremity **(pages 302–318)**

■ **E1.** From the list in the box, select the names of muscles that move the shoulder girdle
as indicated.

LS. Levator scapulae	PM. Pectoralis minor
RMM. Rhomboideus	SA. Serratus anterior
major and minor	T. Trapezius

_____ a. Superiorly (elevation)   _____ c. Towards vertebrae (adduction)

_____ b. Inferiorly (depression)   _____ d. Away from vertebrae (abduction)

■ **E2.** Note on Figure LG 11.2(a) and (b) that the only two muscles listed in Checkpoint

E1 that are superficial are the _____ and a small portion of the

_____. The others are all deep muscles.

■ **E3.** On Figure LG 11.2(a) and (b) identify the pectoralis major, deltoid, and latissimus
dorsi muscles. All three of these muscles are *(superficial? deep?).* They are all directly
involved with movement of the *(shoulder girdle? humerus? radius/ulna?).*

■ **E4.** Write next to each muscle name listed below the letters of *all points of origin and insertion* and *all actions* that apply.

Points of Origin or Insertion	Actions (of Humerus)
C. Clavicle	Ab. Abducts
H. Humerus	Ad. Adducts
I. Ilium	EH. Extension, hyperextension
RC. Ribs or costal cartilages	F. Flexion
Sc. Scapula	
St. Sternum	
VS. Vertebrae and sacrum	

Muscles

a. Pectoralis major _____ _____

b. Deltoid _____ _____

c. Latissimus dorsi _____ _____

■ (E5.) Combine your knowledge of muscle actions with your knowledge of movements at joints from Chapter 9. Return to LG Figure 9.2, page 160 of the guide. Write the name(s) of one or two muscles that produce each of the actions (a–h). Write muscle names next to each figure.

■ (E6.) Complete the table describing three muscles that move the forearm.

Muscle Name	Origin	Insertion	Action on Forearm
a.		Radial tuberosity	
b. Brachialis			
c.			Extension

■ (E7.) Complete this exercise about muscles that move the wrist and fingers.

a. Examine your own forearm, palm, and fingers. There is more muscle mass on the *(anterior? posterior?)* surface. You therefore have more muscles that can *(flex? extend?)* your wrist and fingers.

b. Locate the flexor carpi ulnaris muscle on Figure 11.17, page 313 in your text. What action does it have other than flexion of the wrist? _____
What muscles would you expect to abduct the wrist?

c. What is the difference in location between flexor digitorum superficialis and flexor digitorum profundus?

d. What muscle helps you to point (extend) your index finger?

■ **E8.** *For extra review* of muscles that move the upper extremities, write the name of one or more muscles that fit these descriptions.

a. Covers most of the posterior of the humerus _____

b. Turns your hand from palm down to palm up position _____

c. Originates from upper eight or nine ribs; inserts on scapula; moves scapula laterally

_____

d. Used when a baseball is grasped _____

e. Antagonist to serratus anterior _____
f. Largest muscle of the chest region; used to throw a ball in the air (flex humerus) and

to adduct arm _____
g. Raises or lowers scapula, depending on which portion of the muscle contracts

_____

h. Controls action at the elbow for a movement such as the downstroke in hammering

a nail _____
i. Hyperextends the humerus, as in doing the "crawl" stroke in swimming or exerting a

downward blow; also adducts the humerus _____

F. Principal skeletal muscles that move the vertebral column (pages 318–322)

■ **F1.** Describe the muscles that comprise the sacrospinalis. Locate and label on Figure LG 11.2(b).

a. The sacrospinalis muscle is also called the _____.

b. The muscle consists of three groups: _____ (lateral),

_____ (intermediate), and _____ (medial).

c. In general, these muscles have attachments between _____.
d. They are *(flexors? extensors?)* of the vertebral column, and so are *(synergists? antagonists?)* of the rectus abdominis muscles.

■ **F2.** Explain why it is common for women in their final weeks of pregnancy to experience frequent back pains.

■ **F3.** Choose the correct origin and insertion of the scalene muscles:
A. Ribs—iliac crest
B. Cervical vertebrae—occipital and temporal bones
C. Cervical vertebrae—first two ribs
D. Thoracic vertebrae—sacrum

G. Principal skeletal muscles that move the lower extremity (pages 323–338)

G1. Cover the key in Figure LG 11.2(a) and (b) and identify by size, shape, and location the major muscles that move the lower extremity.

■ G2. Now match muscle names in the box with their descriptions below.

Ad. Adductor group	Ham. Hamstrings
Gas. Gastrocnemius	Il. Iliopsoas
GMax. Gluteus maximus	QF. Quadriceps femoris
GMed. Gluteus medius	Sar. Sartorius

_____ a. Consists of four heads: rectus femoris and three vastus muscles (lateralis, medialis, and intermedius).

_____ b. This muscle mass lies in the posterior (flexor) compartment of the thigh; antagonist to quadriceps femoris.

_____ c. Attached to lumbar vertebrae, anterior of ilium, and lesser trochanter, it crosses anterior to hip joint.

_____ d. Large muscle mass of the buttocks; antagonist to the iliopsoas.

_____ e. The only one of these muscles located in the leg (between knee and ankle), it forms the "calf." Attaches to calcaneus by "Achilles tendon."

_____ f. Forms the medial compartment of the thigh; moves femur medially.

_____ g. Crossing the femur obliquely, it moves lower extremity into "tailor position."

_____ h. Located posterior to upper, outer portion of the ilium, it forms a preferred site for intramuscular (IM) injections.

■ G3. Complete the table by marking an X below each action produced by contraction of the muscles listed. (Some muscles will have two answers.)

Key to actions in table: Ab. Abduct; Ad. Adduct; EH, extend or hyperextend; F, flex

	Movements of Thigh (Hip Joint)				Movement of Leg (Knee)	
	Ab	Ad	EH	F	EH	F
a. Iliopsoas						
b. Gluteus maximus and medius						
c. Adductor mass						
d. Tensor fasciae latae						
e. Quadriceps femoris						
f. Hamstrings						
g. Gracilis						
h. Sartorius						

○ Adductors
○ Hamstrings
○ Quadriceps femoris

Anterior

A
B
C
D
E
Medial

F   N
G
O
H   I   P
J
K   L   M   Q

Lateral

Posterior

KEY

A. Femur
B. Blood vessels and nerves
C. Fascia lata (dense fibrous tissue)
D. Superficial (loose) connective tissue
E. Skin
F. Sartorius
G. Adductor longus
H. Gracilis
I. Adductor brevis

J. Adductor magnus
K. Semitendinosus
L. Semimembranosus
M. Biceps femoris
N. Rectus femoris
O. Vastus medialis
P. Vastus intermedius
Q. Vastus lateralis

Figure LG 11.4 Cross section of the right thigh midway between hip and knee joints. Structures *A* to *E* are nonmuscle; *F* to *Q* are skeletal muscles of the thigh. Color as directed in Checkpoint G4.

■ **G4.** Refer to Figure LG 11.4.

a. Locate the major muscle groups of the right thigh in this cross section. Color the groups as indicated by color code ovals on the figure.

b. Cover the key to Figure LG 11.4 and identify each muscle. Relate position of muscles in this figure to views of muscles in Figure LG 11.2a and b.

■ **G5.** What muscles cause the actions i–p shown in Figure LG 9.2 (pages 160–161)? Write the muscle names next to each diagram.

■ G6. Perform these actions of your foot and toes. Feel which muscles are contracting. Then match names of actions with descriptions.

DF. Dorsiflex	F. Flex toes
Ev. Evert foot	In. Invert foot
Ex. Extend toes	PF. Plantar flex

_____ a. Jump, as if to touch ceiling

_____ b. Walk around on your heels

_____ c. Curl toes down

_____ d. Lift toes upward away from floor

_____ e. Move sole of foot medially

_____ f. Move sole of foot laterally

■ G7. To review details of leg muscles that move the foot, complete this table. Note that muscles within a compartment tend to have similar functions. Muscles with * flex or extend only the great toe (not all toes). Use the same key for foot and toe actions as in the box for Checkpoint G6.

	DF PF	In Ev	F Ex
Posterior compartment: a. Gastrocnemius and soleus b. Tibialis posterior c. Flexor digitorum longus d. Flexor hallucis longus*			
Lateral compartment: e. Peroneus (longus and brevis)			
Anterior compartment: f. Tibialis anterior g. Extensor digitorum longus h. Extensor hallucis longus*			

■ G8. Do this Checkpoint about additional muscles of the foot.

a. Muscles whose main muscle mass lies within the foot (rather than the leg) are known as *(intrinsic? extrinsic?)* muscles of the foot.
b. Most intrinsic foot muscles are on the *(dorsal? plantar?)* surface of the foot. (*Hint:* Feel your own foot.) These muscles are arranged in *(two? four?)* layers. The first layer is most *(superficial? deep?)*.
c. The muscle that moves the great toe medially is known as the

_____

G9. *For extra review* of all muscles, color flexors and extensors using color code ovals on Figure LG 11.2a and b. This activity will allow you to see on which sides of the body most muscles with those actions are located. Omit plantar flexors and dorsiflexors. Note that some muscles are flexors *and* extensors—at different joints. (See Mastery Test question 23 also.)

H. Intramuscular (IM) injections; running injuries   (pages 339–341)

H1. State three reasons why intramuscular (IM) injections may be the method of choice for administration of drugs.

H2. *A clinical challenge.* Why is the gluteus medius considered a safer site for intramuscular injections than the gluteus maximus?

H3. Two other muscles commonly used for intramuscular injections are the

_____ in the thigh and the _____ in the upper extremity.

H4. Answer these questions about running injuries.

a. The most common site of injury for runners is the *(calcaneal tendon? groin? hip? knee?)*.

b. *Patellofemoral stress syndrome* is a technical term for *Runner's Knee*.
   Briefly describe this problem.
   patella glides to the side causing pain

c. *Shinsplint syndrome* refers to soreness along the *(patella? tibia? fibula?)*.

d. Write several suggestions you might make to a beginning runner to help to avoid runners' injuries.

   stretch before running
   wear new supportive shoes
   build up gradually

■ **H5.** Answer these questions about treatment of sports injuries. (See *Essays on Wellness* for Chapter 11.)

a. Initial treatment of sports injuries usually calls for *RICE* therapy. To what does

   RICE therapy refer? R *est*   I *ce*

   C *ompression*   E *levation*.

b. NSAID medications may also be used. NSAID refers to

   N *on*   S *teroidal*

   A *nti*   I *nflammatory*

   D *rugs*.

ANSWERS TO SELECTED CHECKPOINTS

A1. Skeletal muscle tissues and connective tissues.

A2. (a) Lever, fulcrum. (b) Resistance. (c) Muscle, effort. (d) Proximal, radius. (e) Origin, scapula. (f) See Figure LG 11.1A; third.

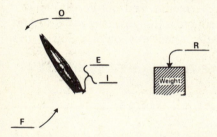

Figure LG 11.1A   The lever-fulcrum principle is illustrated by flexion of the forearm.

A3. (a) 3. (b) *R; E;* atlas, occipital; first.

A4. (a) De; B. (b) Does; slowly and gently. (c) Proprioceptive neuromuscular facilitation; opposite to the one. (d) Heat and aerobic exercise.

A5. (a) Many, pennate. (b) Long, parallel.

A6. (a) Prime mover (agonist); biceps brachii. (b) An antagonist. (c) Your forearm would stay in the flexed position. (d) None: each opposing muscle would negate the action of the other. (e) Synergists, fixators.

B1. (a) D. (b) E. (c) A. (d) C. (e) B. (f) F.

B2. (b) A, P, L. (c) N, L. (d) P. (e) A, S.

C1. (a) Frontalis portion of epicranius. (b) Orbicularis oris. (c) Zygomaticus major. (d) Orbicularis oculi.

C2. Facial, VII.

C3. (a) Elevating the mandible; temporalis, masseter. (b) Close; lateral pterygoid. (c) V, trigeminal. (d) Chewing, facial expression.

C4. (a) They are outside of the eyeballs, not intrinsic like the iris. (b) and (c) See Figure LG 11.3A. (d) III; lateral.

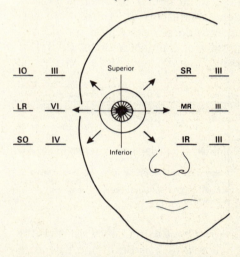

Figure LG 11.3A   Right eyeball with arrows indicating directions of eye movements produced by the extrinsic eye muscles.

C5. No. The left eye contracts its medial rectus, while the right eye uses its lateral rectus and exerts some tension upon both oblique muscles.

C6. (a) Posterior and superior. (b) Inferior. (c) Anterior and inferior.

C7. Same answers as for C6.

C8. XII; hypoglossal.

C9. (a) D. (b) E. (c) C. (d) B. (e) A.

C10. Opposite; flexes.

C11. Extending and then hyperextending; posterior; longissimus capitis, semispinalis capitis and splenius capitis.

D1. Four. (a) Vertical; pubic crest and symphysis pubis; ribs 5 to 7; flexion. (b) External oblique → internal oblique → transversus abdominis. (c) No; strength is provided by the three different directions.

D2. (a) Dome; around the bottom of the rib cage and on lumbar vertebrae; central tendon. (b) Increase, inspiration. (c) Between ribs (costa); internal.

D5. Across, wall; thorax; pelvic.

E1. (a) LS, RMM, upper fibers of T. (b) PM, lower fibers of T. (c) RMM, T. (d) SA.

E2. Trapezius, serratus anterior.

E3. Superficial; humerus.

E4. (a) C, St, RC, H; Ad, F. (b) C, H, Sc; Ab, EH (posterior fibers), F (anterior fibers). (c) I, RC, VS, H; Ad, EH.

E5. (a) Sternocleidomastoid. (b) Three capitis muscles. (c) Pectoralis major, coracobrachialis, deltoid (anterior fibers), biceps brachii. (d) Latissimus dorsi, teres major, deltoid (posterior fibers), triceps brachii. (e) Deltoid and supraspinatus. (f) Pectoralis major, latissimus dorsi, teres major. (g) Biceps brachii, brachialis, brachioradialis. (h) Triceps brachii.

E6.

Muscle Name	Origin	Insertion	Action on Forearm
a. **Biceps brachii**	**Scapula (2 sites)**	Radial tuberosity (anterior)	**Flexion, supination** (*Note:* also flexes humerus)
b. Brachialis	**Anterior humerus**	Ulna (coronoid process)	**Flexion**
c. **Triceps brachii**	**Scapula and 2 sites on posterior of humerus**	**Posterior of ulna (olecranon)**	Extension (**also extension of humerus**)

E7. (a) Anterior; flex. (b) Adducts wrist; those lying over radius, such as flexor and extensor carpi radialis. (c) Superficialis is more superficial, and profundus lies deep. (d) Extensor indicis.

E8. (a) Triceps brachii. (b) Supinator and biceps brachii. (c) Serratus anterior. (d) Flexor digitorum superficialis and profundus. (e) Trapezius. (f) Pectoralis major. (g) Trapezius. (h) Triceps brachii. (i) Latissimus dorsi.

F1. (a) Erector spinae. (b) Iliocostalis, longissimus, spinalis. (c) Hipbone (ilium), ribs and vertebrae. (d) Extensors, antagonists.

F2. Extra weight of the abdomen demands extra support (contraction) by sacrospinalis muscles.

F3. C.

G2. (a) QF. (b) Ham. (c) Il. (d) GMax. (e) Gas. (f) Ad. (g) Sar. (h) Gmed.

G3.

	Movements of Thigh (Hip Joint)				Movement of Leg (Knee)	
	Ab	Ad	EH	F	E	F
a. Iliopsoas				X		
b. Gluteus maximus and medius	X	X				
c. Adductor mass		X		X		
d. Tensor fasciae latae	X			X		
e. Quadriceps femoris				X	X	
f. Hamstrings			X			X
g. Gracilis		X				X
h. Sartorius				X		X

G4. (a) Adductors: G H I J; Hamstrings: KLM; Quadriceps: N O P Q. (b) See key to Figure LG 11.4.

G5. (i) Iliacus + psoas (iliopsoas), rectus femoris, adductors, sartorius; (j) Gluteus maximus, hamstrings, adductor magnus (posterior portion); (k) Hamstrings, gracilis, sartorius, gastrocnemius; (l) Quadriceps femoris; (m) Same as i, but bilateral; (n) Same as j but bilateral; (o) Tensor fasciae latae, gluteus (medius and minimus), piriformis, superior and inferior gemellus, and obturator internus; (p) Adductors (longus, magnus, and brevis), pectineus, quadratus femoris, and gracilis.

G6. (a) PF. (b) DF. (c) F. (d) Ex. (e) In. (f) Ev.

G7.

	DF	PF	In	Ev	F	Ex
Posterior compartment:						
a. Gastrocnemius and soleus		X				
b. Tibialis posterior		X	X			
c. Flexor digitorum longus		X	X		X	
d. Flexor hallucis longus*		X	X		X*	
Lateral compartment:						
e. Peroneus (longus and brevis)		X		X		
Anterior compartment:						
f. Tibialis anterior	X		X			
g. Extensor digitorum longus	X			X		X
h. Extensor hallucis longus*	X		X			X*

G8. (a) Intrinsic. (b) Plantar; 4; superficial. (c) Adductor hallucis.

G9. Figure LG 11.2a: flexors: G, J, K, S, T (anterior portion), U, V, W, AA, BB, EE; extensors: BB, CC, DD. Figure LG 11.2b: flexors: J, K, S; extensors: F, G, H, I, J, K, L, M (posterior fibers), N, O, R.

H2. The gluteus medius is superior and lateral to the sciatic nerve which runs deep to the gluteus maximus muscle.

H3. Vastus lateralis, deltoid.

H4. (a) Knee. (b) "Runner's knee"; the patella tracks (glides) laterally, causing pain. (c) Tibia. (c) Replace worn shoes with new, supportive ones; do stretching and strengthening exercises; build up gradually; get proper rest; consider other exercise if prone to leg injuries since most (70 percent) runners do experience some injuries.

H5. (a) Rest, ice, compression (such as by elastic bandage), and elevation (of the injured part). (b) Nonsteroidal antiinflammatory drugs.

MASTERY TEST: Chapter 11

Questions 1–2: Arrange the answers in correct sequence.

___B___ ___C___ ___A___   1. Abdominal wall muscles, from superficial to deep:
   A. Transversus abdominis
   B. External oblique
   C. Internal oblique

___A___ ___C___ ___B___   2. From superior to inferior in location:
   A. Sternocleidomastoid
   B. Pelvic diaphragm
   C. Diaphragm and intercostal muscles

Questions 3–12: Circle T (true) or F (false). If the statement is false, change the underlined word or phrase so that the statement is correct.

T F   3. In extension of the thigh the hip joint serves as the fulcrum (F), while the hamstrings and gluteus maximus serve as the effort (E).

T F   4. The wheelbarrow and gastrocnemius are both examples of the action of second-class levers.

T F   5. The hamstrings are antagonists to the quadriceps femoris.

T F   6. The name deltoid is based on the action of that muscle.

T F   7. The most important muscle used for normal breathing is the diaphragm.

T F   8. The insertion end of a muscle is the attachment to the bone that does move.

T F   9. In general, adductors (of the arm and thigh) are located more on the medial than on the lateral surface of the body.

T F   10. Both the pectoralis major and latissimus dorsi muscles extend the humerus.

T F   11. The capitis muscles (such as splenius capitis) are extensors of the head and neck.

T F   12. The biceps brachii and biceps femoris are both muscles with two heads of origin located on the arm.

13. All of these muscles are located in the lower extremity *except:*
    A. Hamstrings   D. Deltoid
    B. Gracilis   E. Peroneus longus
    C. Tensor fasciae latae
14. All of these muscles are located on the anterior of the body *except:*
    A. Tibialis anterior   D. Pectoralis major
    B. Rectus femoris   E. Rectus abdominis
    C. Sacrospinalis
15. All of the following muscles are intrinsic muscles of the foot *except:*
    A. Abductor digiti minimi
    B. Extensor digitorum brevis
    C. Extensor digitorum longus
    D. Dorsal interossei
16. All of these muscles are attached to ribs *except:*
    A. Serratus anterior   D. Iliocostalis
    B. Intercostals   E. External oblique
    C. Trapezius
17. All of these muscles are directly involved with movement of the scapulae *except:*
    A. Levator scapulae
    B. Pectoralis major
    C. Pectoralis minor
    D. Rhomboideus major
    E. Serratus anterior

18. All of these muscles have attachments to the coxal bones *except:*
    A. Adductor muscles (longus, magnus, brevis)
    B. Biceps femoris
    C. Rectus femoris
    D. Vastus medialis
    E. Latissimus dorsi
19. The masseter and temporalis muscles are used for:
    A. Chewing
    B. Pouting
    C. Frowning
    D. Depressing tongue
    E. Elevating tongue
20. All of these muscles are used for facial expression *except:*
    A. Zygomaticus major   D. Rectus abdominis
    B. Orbicularis oculi   E. Mentalis
    C. Platysma

Questions 21–25: fill-ins. Refer to Figure LG 11.2(a) and (b). Write the word or phrase or key letters of muscles that best complete the statement or answer the questions.

_____ 21. Muscles G, S, U, V, and W on Figure LG 11.2(a) all have in common the fact that they carry out the action of ____.

_____ 22. Muscles G, N, O, and R on Figure LG 11.2(b) all have in common the fact that they carry out the action ____.

_____ 23. If you colored all flexors red and all extensors blue on these two figures, the view of the ____ surface of the body would appear more blue.

_____ 24. Choose the letters of all of the muscles listed below which would contract as you raise your left arm straight in front of you, as if to point toward a distant mountain: Figure LG 11.2(a): D T U V; Figure LG 11.2(b): N O P.

_____ 25. Choose the letters of all of the muscles listed below that would contract as you raise your knee and extend your leg straight out in front of you, as if you are starting to march off to the distant mountain: Figure LG 11.2(a): AA BB CC DD; Figure LG 11.2(b); H I J K R.

Arrange

1. B C A
2. A C B

True-False

3. T
4. T
5. T
6. F. Shape
7. T
8. T
9. T
10. F. The latissimus dorsi extends, but the pectoralis major flexes.
11. T
12. F. Two heads of origin; but origins of biceps brachii are on the scapula, and origins of biceps femoris are on ischium and femur.

Multiple Choice

13. D
14. C
15. C
16. C
17. B
18. D
19. A
20. D

Fill-ins

21. Flexion
22. Extension
23. Posterior
24. Figure LG 11.2(a): *T* (anterior fibers) *U V;* Figure LG 11.2(b): none.
25. Figure LG 11.2(a): *AA BB CC DD;* Figure LG 11.2(b): *HI.*

UNIT

Control Systems of the Human Body

FRAMEWORK 12
Nervous Tissue

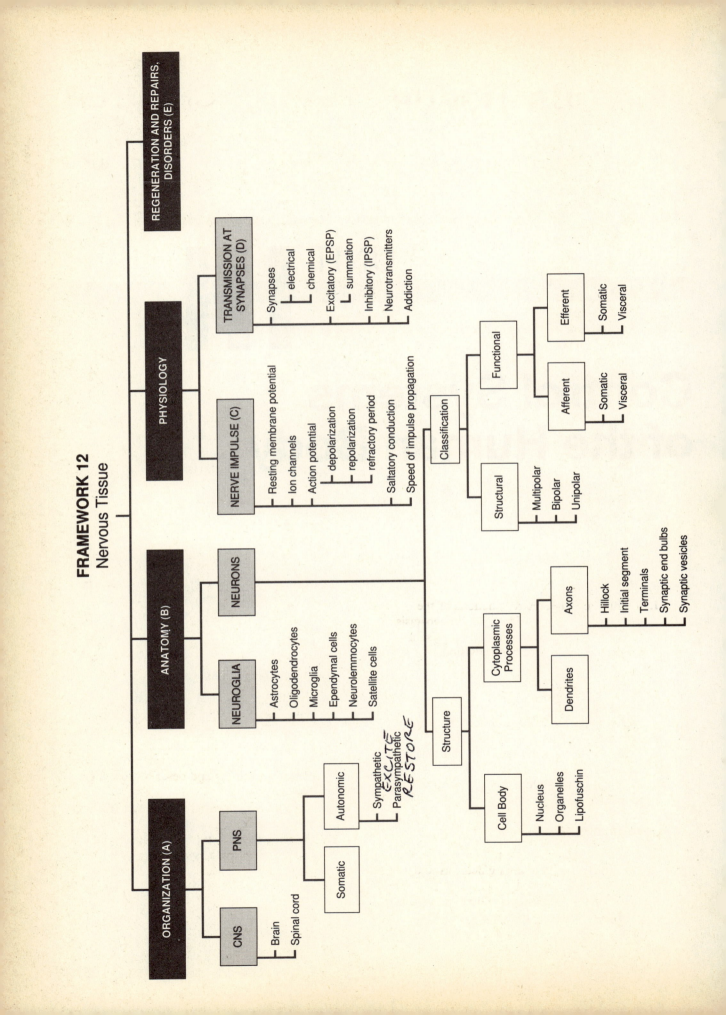

REGENERATION AND REPAIRS, DISORDERS (E)

PHYSIOLOGY

- **TRANSMISSION AT SYNAPSES (D)**
  - Synapses
    - electrical
    - chemical
  - Excitatory (EPSP)
  - summation
  - Inhibitory (IPSP)
  - Neurotransmitters
  - Addiction

- **NERVE IMPULSE (C)**
  - Resting membrane potential
  - Ion channels
  - Action potential
    - depolarization
    - repolarization
    - refractory period
  - Saltatory conduction
  - Speed of impulse propagation

ANATOMY (B)

- **NEURONS**
  - Classification
    - Structural
      - Multipolar
      - Bipolar
      - Unipolar
    - Functional
      - Afferent
        - Somatic
        - Visceral
      - Efferent
        - Somatic
        - Visceral
  - Structure
    - Cell Body
      - Nucleus
      - Organelles
      - Lipofuschin
    - Cytoplasmic Processes
      - Dendrites
      - Axons
        - Hillock
        - Initial segment
        - Terminals
        - Synaptic end bulbs
        - Synaptic vesicles

- **NEUROGLIA**
  - Astrocytes
  - Oligodendrocytes
  - Microglia
  - Ependymal cells
  - Neurolemmocytes
  - Satellite cells

ORGANIZATION (A)

- **PNS**
  - Autonomic
    - Sympathetic *EXCITE*
    - Parasympathetic *RESTORE*
  - Somatic

- **CNS**
  - Brain
  - Spinal cord

Nervous Tissue

CHAPTER **12**

Two systems—nervous and endocrine—are responsible for regulating our diverse body functions. Each system exerts its control with the help of specific chemicals, namely neurotransmitters (nervous system) and hormones (endocrine system). In Unit III both systems of regulation will be considered, starting with the tissue of the nervous system.

Nervous tissue consists of two types of cells: neurons and neuroglia. The name neuroglia (*glia* = glue) offers a clue as to function of these cells: they bind, support, and protect neurons. Neurons perform the work of transmitting nerve impulses. These long and microscopically slender cells sometimes convey information several feet along a single neuron. Their function relies on an intricate balance between ions (Na^+ and K^+) found in and around nerve cells. Neurons release neurotransmitters that bridge the gaps between adjacent neurons and at nerve-muscle or nerve-gland junctions. Analogous to a complex global telephone system, the nervous tissue of the human body boasts a design and organization that permits accurate communication, coordination, and integration of virtually all thoughts, sensations, and movements.

First introduce to your own nervous tissue the Chapter 12 Framework and note the organization of key terms in each section.

TOPIC OUTLINE AND OBJECTIVES

A. Organization

1. Identify the three basic functions of the nervous system in maintaining homeostasis.
2. Classify the organs of the nervous system into central and peripheral divisions.

B. Anatomy

3. Contrast the histological characteristics and functions of neuroglia and neurons.
4. Describe the functions of neuroglia.
5. Describe the structure and functions of neurons.
6. Define gray and white matter and give examples of each.

C. Nerve impulses

7. Describe the cellular properties that permit communication among neurons and muscle fibers.
8. Describe the factors that contribute to generation of a resting membrane potential.

9. Compare the basic types of ion channels and explain how they relate to action potentials and graded potentials.
10. List the sequence of events involved in generation of a nerve impulse.

D. Transmission at synapses

11. Explain the events of synaptic transmission.
12. Distinguish between spatial and temporal summation.
13. Give examples of excitatory and inhibitory neurotransmitters and describe how they may act.
14. List four ways that synaptic transmission may be enhanced or blocked.
15. Describe the various types of neuronal circuits in the nervous system.

E. Regeneration and repair of neurons; disorders

16. List the necessary conditions for the regeneration of nervous tissue.

17. Describe events of damage and repair of peripheral neurons.
18. Describe the symptoms and causes of epilepsy.

WORDBYTES

Now study the following parts of words that may help you better understand terminology in this chapter.

Wordbyte	Meaning	Example	Wordbyte	Meaning	Example
astro-	star	*astrocytes*	neuro-	nerve	*neuro*n
dendr-	tree	*dendr*ite	olig-	few	*olig*odendroglia
-glia	glue	neuro*glia*	syn-	together	*syn*apse
lemm-	sheath	neuri*lemma*			

CHECKPOINTS

A. Organization (page 347)

A1. List three principal functions of the nervous system.

■ **A2.** Check your understanding of organization of the nervous system by selecting answers that best fit descriptions below.

Aff. Afferent	Eff. Efferent
ANS. Autonomic nervous system	PNS. Peripheral nervous system
CNS. Central nervous system	SNS. Somatic nervous system

CNS a. Brain and spinal cord

AFF b. Sensory nerves

SNS EFF somatic efferen c. Carry information from CNS to skeletal muscles

ANS d. Consists of sympathetic and parasympathetic divisions

B. Anatomy (pages 347–356)

■ **B1.** Write *neurons* or *neuroglia* after descriptions of these cells.

a. Conduct impulses from one part of the nervous system to another:

neuron

b. Provide support and protection for the nervous system: _neuroglia_

c. Bind nervous tissue to blood vessels, form myelin, and serve phagocytic functions:

neuroglia

d. Smaller in size, but more abundant in number: _neuroglia_

B2. Complete the table describing neuroglia cells.

Type	Description	Functions
a. *astrocytes*	Star-shaped cells	
b. Oligoden-drocytes *most common*	*smaller than astro. fewer processes*	*produce myelin*
c. *microglia*		Phagocytic
d. *ependymal*		Line ventricles of brain and central canal of spinal cord

■ **B3.** Describe how myelin is laid down on nerve fibers in the:

a. PNS

b. CNS *OLIGO deposit a myelin sheath w/o forming a neurolemma*

■ **B4.** On Figure LG 12.1, label all structures with leader lines. Next draw arrows beside the figure to indicate direction of nerve impulses. Then color structures with color code ovals.

■ **B5.** Match the parts of a neuron listed in the box with the descriptions below.

A. Axon	D. Dendrite
CB. Cell body (soma)	L. Lipofuscin
CS. Chromatophilic substance (Nissl bodies)	M. Mitochondria
	NF. Neurofibrils

_*CB*_ a. Contains nucleus; cannot regenerate since lacks mitotic apparatus

_*L*_ b. Yellowish brown pigment that increases with age; appears to be byproduct of lysosomes

_*M*_ c. Provide energy for neurons

_*NF*_ d. Long, thin filaments that provide support and shape for the cell

_*CS*_ e. Orderly arrangement of rough ER; site of protein synthesis

_*D*_ f. Conducts impulses toward cell body

_*A*_ g. Conducts impulses away from cell body; has synaptic end bulbs that secrete neurotransmitter

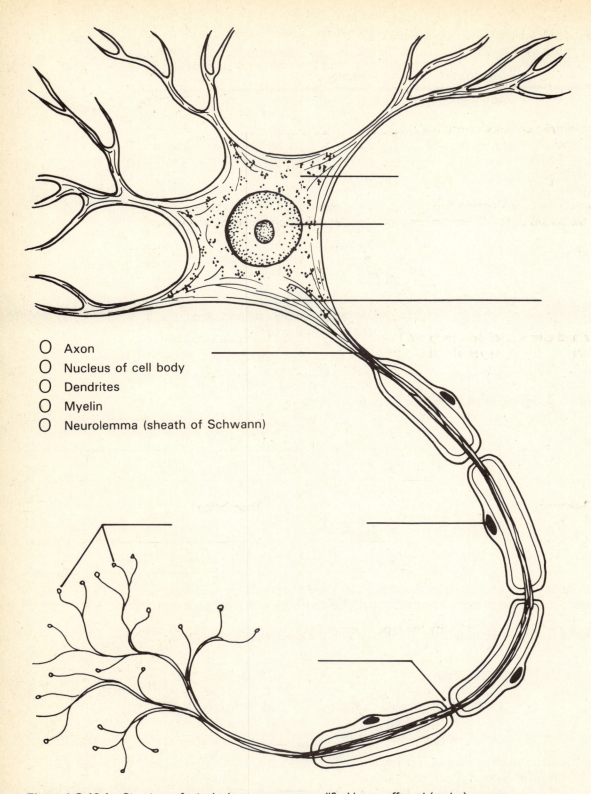

O Axon
O Nucleus of cell body
O Dendrites
O Myelin
O Neurolemma (sheath of Schwann)

Figure LG 12.1 Structure of a typical neuron as exemplified by an efferent (motor) neuron. Complete the figure as directed in Checkpoint B4.

■ **B6.** Identify phrases that describe a *neuron* and those that describe a *nerve*.

a. A single nerve cell; consists of a cell body with axon and dendrites:

neuron

b. Bundle of axons and dendrites of many neurons, both afferent and efferent, somatic

and autonomic; contains no cell bodies: *nerve*

c. Located entirely outside of the CNS and inside the PNS; macroscopic in diameter:

nerve

B7. *A clinical challenge.* Contrast two types of *axonal transport* and tell which type is involved in the spread of the different microorganisms that cause herpes, rabies, and tetanus.

herpes   *rabies*   *tetanus*

B8. Name and give a brief description of three types of neurons based on structural characteristics and three types of neurons classified according to functional differences.

Structural	Description
multi-polar	_____
uni-polar	_____
bi-polar	_____

Functional	Description
a-fferent	*Sensory → to CNS*
e-fferent	*motor away from CNS*
_____	

■ **B9.** Using the list in the box, identify the type of nerve fiber that transmits each of the kinds of nerve impulses listed.

GSA. General somatic afferent	GVA. General visceral afferent
GSE. General somatic efferent	GVE. General visceral efferent

_____ a. All ANS fibers are of this type.

_____ b. Pain from a thorn in your skin is sensed via fibers of this type.

_____ c. Pain from a spasm of smooth muscle in the gallbladder is sensed by means of fibers of this type.

_____ d. With your eyes closed, you can tell the position of your skeletal muscles and joints due to nerve impulses that pass along this type of fiber.

_____ e. In order to increase your heart rate, impulses pass from your brain to your heart via this type of nerve fiber.

_____ f. These nerve fibers carry impulses from the CNS to muscles in your fingers used in writing.

■ **B10.** Briefly explain how myelination relates to:

a. The color of white matter

myelinated fibers myelin is white

b. The color of gray matter

non myelinated and cell bodies neuroglia

c. Speed of nerve impulse transmission

myelination speeds nerve transmission

■ **B11.** Complete the table about structures in the nervous system.

Structure	Color (Gray or White)	Composition (Cell Bodies or Nerve Fibers)	Location (CNS or PNS)
a. Nerve	White		
b. Tract			CNS
c. Nucleus	Gray		

C. Nerve impulses   (pages 356–364)

■ **C1.** Refer to Figure LG 12.2a and complete this Checkpoint.

a. Note that one ion is 14 times more concentrated outside the nerve cell; this is *(K+?* *Na+?).* *Na+*

b. Potassium is *28-30* times more concentrated *(outside? inside?)* the neuron.

c. Label the major cation in each site. Also label the major anion inside the neuron. (Values for all of these ions are shown graphically on Figure 27.4, page 909 in the text).

d. Color arrows across membranes using color code ovals to indicate types of transport processes for each ion. (Differences in thickness of arrows indicate relative rates of processes.)

e. Close to the neuron plasma membrane, there is a small buildup of *(positive? negative?)* charges. Use Figure LG 12.2a to explain why this difference exists.

f. Since a difference in electrical charge exists between inside and outside of the neuron, the resting plasma membrane is said to be *(polarized? depolarized?).*

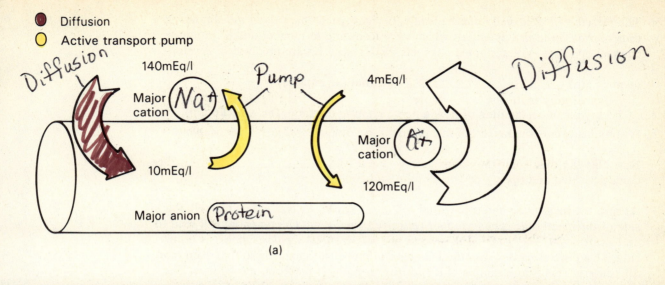

Diffusion
Active transport pump

Diffusion

140mEq/l

Major cation Na+

Pump

4mEq/l

Major cation K+

Diffusion

10mEq/l

120mEq/l

Major anion *Protein*

(a)

(b)

Figure LG 12.2 Diagrams of a nerve cell. (a) Values for ions inside and outside of cell are shown, as well as arrows showing directions of movement of ions. Label and color as directed in Checkpoint C1. (b) Complete as directed in Checkpoint C5.

g. This difference in charge (between inside and outside of the membrane) is said to be

the resting membrane ___*potential*___ (RMP). This is a form of poten-

tial energy which is measured in ___*milivolts*___ (mV). In a resting

membrane this value is typically about ___−70___ mV.

■ **C2.** Do this exercise about roles of ion channels in development of membrane potentials.

a. There are two kinds of ion channels: *(gated? (leakage?)* channels that are always

open and ___*gated*___ channels that require a stimulus to activate opening or closure. The plasma cell of a membrane is more permeable to *((K+) Na+?)* since there are more leakage channels for *((K+) Na+?)*. (See C1(e) above.)

b. Identify the type of gated channel that is responsive in each case described here. Choose from these answers: *(voltage? chemically? mechanically? light?)* gated chan-

nels. Channels in receptors for hearing in the ear: ___*MECHANICAL*___; chan-

nels responsive to the neurotransmitter acetylcholine: ___*CHEMICAL*___;

channels in neurons responsive to vibrations: ___*MECHANICAL*___.

c. The presence of chemically, mechanically, or light gated channels in membranes permits *(action? graded?)* potentials which vary according to number and duration of channel openings. Voltage gated channels for Na^+ and K^+ are involved in development of a(n) _____action_____ potential (or impulse).

■ **C3.** Check your understanding of how a nerve impulse occurs in this Checkpoint.

a. A stimulus causes the nerve cell membrane to become *(more? less?)* permeable to

Na^+. Na^+ can then enter the cell as voltage gated Na^+ _____ become activated and open.

b. At rest, the membrane had a potential of _____ mV. As Na^+ enters the cell, the inside of the membrane becomes more *(positive? negative?)*. The potential will tend to go toward *(−80? −60?)* mV. The process of *(polarization? depolarization?)* is occurring. This process causes structural changes in more Na^+ channels so that even more Na^+ enters. This is an example of a *(positive? negative?)* feedback mechanism. The

result is a nerve impulse or nerve _____.

c. The membrane is completely depolarized at exactly *(−50? 0? +30?)* mV. Na^+ channels stay open until the inside of the membrane potential is *(reversed? repolarized?)* at +30 mV.

d. After a fraction of a second, K^+ voltage gated channels at the site of the original stimulus open. K^+ is more concentrated *(outside? inside?)* the cell (as you showed on Figure 12.2a); therefore K^+ diffuses *(in? out?)*. This causes the inside of the membrane to become more negative and return to its resting potential of _____ mV. The process is known as *(de? re?)*-polarization. In fact, outflow of K^+ may be so

great that _____-polarization occurs in which membrane potential becomes closer to *(−50? −90?)* mv.

e. During depolarization, the nerve cannot be stimulated at all. This period is known as the *(absolute? relative?)* refractory period. *(Large? Small?)*-diameter axons have a longer absolute refractory period with this about *(0.4? 4? 40?)* msec. In other words, slower impulses are likely to occur along *(large? small?)*-diameter neurons. Only a stronger-than-normal stimulus will result in an action potential during the

_____ refractory period, which corresponds roughly with *(de? re?)*-polarization.

f. The nerve impulse is essentially a wave of negativity along the *(inside? outside?)* of

the nerve cell membrane. The impulse is propagated (or _____) along the nerve, as adjacent areas are depolarized, causing more channels to be activated and more *(Na+? K+?)* to enter.

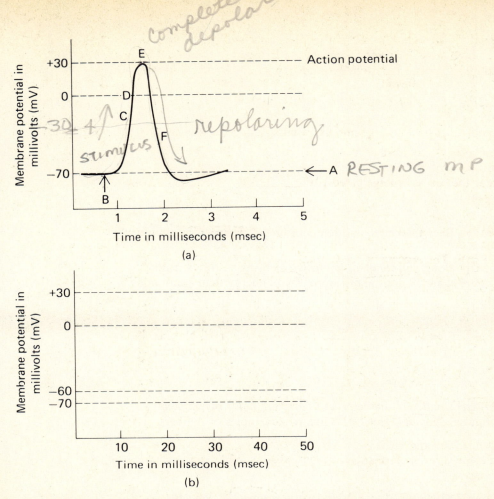

Figure LG 12.3 Diagrams for showing nerve action potentials. (a) Identify letter labels in Checkpoint C4. (b) Complete as directed in Checkpoint D2(b) and (d).

■ **C4.** Write the correct letter label of Figure LG 12.3a next to each description.

B _____ a. The stimulus is applied at this point.

A _____ b. Resting membrane potential is at this level.

F _____ c. Membrane becomes so permeable to K^+ that K^+ diffuses rapidly out of the cell.

C _____ d. The membrane is becoming more positive inside as Na^+ enters; its potential is -30 mV. The process of depolarization is occurring.

D _____ e. The membrane is completely depolarized at this point.

F _____ f. The membrane is repolarizing at this point.

E _____ g. Reversed polarization occurs; enough Na^+ has entered so that this part of the cell is more positive inside than outside.

C5. *For extra review.* Show the events that occur during initiation and transmission of a nerve impulse on Figure LG 12.2b. Compare your diagram to Figures 12.11 and 12.13a, pages 360 and 363 in the text.

C6. Define these terms.

a. Threshold stimulus

b. All-or-none principle

■ **C7.** Describe how myelination, fiber thickness, and temperature affect speed of impulse propagation.

a. Saltatory conduction occurs along (*myelinated?* *unmyelinated?*) nerve fibers. Saltatory transmission is (*faster?* *slower?*) and takes (*more?* *less?*) energy than continuous conduction. Explain why.

less ATP is used because only at the gaps are the NA/K pumps used

b. Type A fibers have (*largest?* *smallest?*) diameter and (*are?* *are not?*) myelinated. These fibers conduct impulses (*rapidly?* *slowly?*). Give two examples of A fibers.

skeletal muscle   touch, pressure

c. Type C fibers have (*largest?* *smallest?*) diameter and (*are?* *are not?*) myelinated. These fibers conduct impulses (*rapidly?* *slowly?*). Give two examples of C fibers.

pain

d. *A clinical challenge.* (*Warm?* *Cool?*) nerve fibers conduct impulses faster. How can this information be applied clinically?

Cool slows pain impulse

■ **C8.** Contrast action potentials of nerve and muscle by completing this table.

Tissue	Typical Resting Membrane Potential (RMP)	Duration of Nerve Impulse
a. Neuron	*higher (-70 mV)*	*Shorter .5-2 msec*
b. Muscle	*lower (-90 mV)*	*longer 1-5 msec. skeletal*

D. Transmission at synapses   (pages 364–370)

■ **D1.** Do this activity about types of synapses.

a. In Chapter 10, you studied the point at which a neuron comes close to contacting a
muscle. This is known as a _neuromuscular_ junction, shown in Figure
LG 10.2, page 173. The minute space between two neurons is known as a
synapse.

b. There are two types of synapses. A(n) *(chemical? electrical?)* synapse is designed to
allow spread of an ionic current from one neuron to the next. These synapses consist

of _gap_ junctions made of hundreds of proteinaceous tunnels

called _connexons_. Gap junctions are common in *(skeletal? smooth
and cardiac?)* muscle where a group of muscle fibers works in synchrony.

c. In electrical synapses membranes of the two cells *(do? do not?)* touch. Does a nerve
impulse actually "jump" across the synaptic cleft? Explain. _no_
_The nerve impulse travels along the presynaptic
axon to the end of the axon, where synaptic vesicles are
triggered to release neurotransmitters. These chemicals enter the
synaptic cleft + then affect the post synaptic neuron (either excite
or inhibit it)_

d. Now summarize the sequence of events at an electrical synapse by placing these
events in order from first to last:

___B___ → ___C___ → ___A___

   A. Electrical signal: postsynaptic potential (nerve impulse)
   B. Electrical signal: presynaptic potential (nerve impulse)
   C. Chemical signal: release of neurotransmitter into synaptic cleft

e. Write these structures in order to summarize the anatomical pathway of a chemical
synapse.

1 EB. End bulb of presynaptic neuron	4 SC. Synaptic cleft
3 N. Neurotransmitter	2 SV. Synaptic vesicle
5 NR. Neurotransmitter receptor	

___EB___ → ___SV___ → ___N___ → ___SC___ → ___NR___
First   Last

f. Explain the role of calcium ions (Ca^{2+}) in nerve transmission at chemical synapses.
_when impulse arrives a end bulb the depolarizing
phase opens voltaged gated Ca+ channels. Because its
more concentrated in the extracellular fluid Ca+ flows
inward triggering exocytosis of synaptic vessicles_

g. The rationale for *one-way information transfer* at chemical synapses is that only synaptic end bulbs of *(pre? post?)*-synaptic neurons release neurotransmitters, and only
(pre? post?)-synaptic neurons have receptors.

■ **D2.** Complete the following exercise about postsynaptic potentials.

a. If a neurotransmitter causes depolarization of a postsynaptic neuron, for example, from −70 to −65 mV, the postsynaptic potential (PSP) is called *(excitatory? inhibitory?)*; in other words, it is called an *(EPSP? IPSP?)*. EPSPs result from inflow of *(K^+? Na^+? Ca^{2+}?)* through chemically gated channels.

b. Usually a single EPSP within a single neuron *(is? is not?)* sufficient to cause a threshold potential and initiate a nerve impulse. Instead, a single EPSP can cause *(partial? total?)* depolarization. Diagram a partial depolarization from −70 to −65 mV in green on Figure LG 12.3b.

c. If neurotransmitters are released from a number of presynaptic end bulbs at one time, their combined effect may produce threshold EPSP of about ____mV. This phenomenon is known as *(spatial? temporal?)* summation. Temporal summation is that due to accumulation of transmitters from *(one? many?)* presynaptic end bulb(s) over a period of time.

d. If a neurotransmitter causes inhibition of the postsynaptic neuron, the process is known as *(de? hyper?)*-polarization, and the PSP is *(excitatory? inhibitory?)*. IPSPs often result from influx of *(Cl^-? K^+? Na^+?)* and/or outflow of *(Cl^-? K^+? Na^+?)* through gated channels. An example of an IPSP would be change in the membrane PSP from −70 to *(−60? −80?)* mV. Diagram such an IPSP in red on Figure 12.3b.

e. In other words, a neuron with an IPSP is *(closer? farther?)* from threshold than a neuron at resting membrane potential (RMP); so a cell with an IPSP is less likely to have an action potential. An example of use of an IPSP is inhibition of your triceps brachii muscle as the _____ is stimulated (with an EPSP) to contract, or innervation of the heart by the vagus nerve (cranial nerve X) which *(in? de?)*-creases heart rate.

■ **D3.** *A clinical challenge.* Once a neurotransmitter completes its job, it must be removed from the synaptic cleft. Describe two mechanisms for getting rid of these chemicals. Notice the consequences of alterations of these mechanisms.

a. Inactivation of a neurotransmitter may occur such as breakdown of _acetylcholine_ (ACh) by the enzyme named _acetylcholinesterase)_ . Certain drugs (such as physostigmine) destroy this enzyme, so postsynaptic neurons (or muscles) would *(remain? be less?)* activated. Predict the effects of such a drug.

b. A neurotransmitter such as _norepnephrine_ may be recycled by the presynaptic neuron that had released it. Cocaine blocks reuptake of two neurotransmitters, namely, _dopamine_ and _NE_ . Describe the physiological consequences of cocaine use:

Short-term effects _increased heart rate blood pressure blood sugar & temp_

Long-term effects _____

D4. After reading the Wellness essays on addiction (pages 28–29 in *Essays on Wellness*), do the following exercise.

a. Discuss factors that may be involved in causing addictions. Include these terms: *physiological, genetic, metabolic, dysfunctional, situational,* and *personality.*

b. List eight questions that may help to identify presence of an addiction in yourself or a friend.

1.   5.

2.   6.

3.   7.

4.   8.

c. Give three examples of how you could enhance *wellness* in your lifestyle.

■ **D5.** In this Checkpoint, describe another mechanism besides IPSP that results in inhibition of nerve impulses.

a. An inhibitory neuron releases its inhibitory transmitter at a synapse with the synaptic end bulb of a(n) ___excitatory___ neuron. The result is *(in? de?)* -crease in release of excitatory transmitter.

b. This type of inhibition is known as ___presynaptic___ inhibition. It may last for *(milliseconds? minutes or hours?).*

■ **D6.** Do this exercise about neurotransmitters.

a. We have already discussed acetylcholine (ACh), a neurotransmitter released at synapses and neuromuscular junctions. ACh is *(excitatory? inhibitory?)* toward skeletal muscle, but ACh released from the vagus nerve is *(excitatory? inhibitory?)* toward cardiac muscle. So when the vagus nerve sends impulses to your heart, your pulse (heart rate) becomes *(faster? slower?)*.

b. GABA and glycine are both *(excitatory? inhibitory?)* neurotransmitters. They act by opening *(Cl⁻? Na⁺?)* channels, leading to IPSPs. The most common inhibitory transmitter in the brain is *(GABA? glycine?)*, whereas _____ is more commonly released by neurons in the spinal cord.

c. Strychnine is a chemical that blocks *(GABA? glycine?)* receptors so that muscles are not properly inhibited (relaxed). Strychnine poisoning is likely to lead to death because the muscles of the _diaphragm_ cannot relax so that air that is high in carbon dioxide cannot be exhaled.

c. List three neurotransmitters that are classified as catecholamines.

■ **D7.** *A clinical challenge.* Alkalosis tends to *(stimulate? depress?)* the neurons of the central nervous system (CNS), leading to tingling, spasms, and possibly convulsions. Acidosis tends to *(stimulate? depress?)* the CNS, leading to _____. With this information in mind, write one sign/symptom of a person whose (arterial) blood pH is 7.28.

■ **D8.** Check your understanding of chemicals that affect transmission at synapses and neuromuscular or neuroglandular junctions by completing this activity. Write E if the effect is excitatory or I if the effect is inhibitory. The first one is done for you. (Lines following descriptions are for the *clinical challenge* activity below.)

___I___ a. A chemical that inhibits release of ACh _3_ Botulism

___I___ b. A chemical that competes for the ACh receptor sites on muscle cells _4_ curare muscle re

___E___ c. A chemical that inactivates acetylcholinesterase _5_ nerve gas

___I___ d. A chemical that increases threshold (for example, from –60 to –40) of a Hypnotics neuron _1_

___E___ e. A chemical that decreases threshold _2_ caffeine, nicotine, benzedrine

A clinical challenge. Match the following chemicals with related mechanisms of actions. Write the number of the chemicals on lines to the right of the above descriptions. One is done for you.

1. Hypnotics, tranquilizers, anesthetics
2. Caffeine, benzedrine, nicotine
3. Botulism toxin, inhibiting muscle contraction
4. Curare, a muscle relaxant
5. Nerve gas such as diisopropyl fluorophosphate; physostigmine

■ **D9.** Match the types of circuits in the box with related descriptions. Answers may be used more than once.

> C. Converging   P. Parallel after-discharge
> D. Diverging   R. Reverberating

_____ a. Impulse from a single presynaptic neuron causes stimulation of increasing numbers of cells along the circuit.

_____ b. An example is a single motor neuron in the brain that stimulates many motor neurons in the spinal cord, therefore activating many muscle fibers.

_____ c. Branches from a second and third neuron in a pathway may send impulses back to the first, so the signal may last for hours, as in coordinated muscle activities.

_____ d. One postsynaptic neuron receives impulses from several nerve fibers.

_____ e. A single presynaptic neuron stimulates intermediate neurons which synapse with a common postsynaptic neuron, allowing this neuron to send out a stream of impulses, as in precise mathematical calculations.

E. Regeneration and repair of nervous tissue; disorders (pages 370–372)

■ **E1.** Which of the following can regenerate if destroyed? Briefly explain why in each case.

a. Neuron cell body

b. CNS nerve fiber

c. PNS nerve fiber

E2. Discuss methods currently being investigated to promote regrowth of damaged neurons.

■ **E3.** Answer these questions about nerve regeneration.

a. In order for a damaged neuron to be repaired, it must have an intact cell body and

also a _____.

b. Can axons in the CNS regenerate? Explain.

c. When a nerve fiber (axon or dendrite) is injured, the changes that follow in the cell body are called *(chromatolysis? Wallerian degeneration?)*. Those that occur in the

portion of the fiber distal to the injury are known as _____.

E4. Explain how peripheral nerves regenerate by describing each of these events.

a. Chromatolysis

b. Wallerian degeneration

c. Retrograde degeneration

d. Accelerated protein synthesis

ANSWERS TO SELECTED CHECKPOINTS

A2. (a) CNS. (b) Aff. (c) SNS or Eff (*Hint:* remember S A M E: Sensory = *A*fferent; *Mo*tor = *E*fferent). (d) ANS.

B1. (a) Neurons. (b) Neuroglia. (c) Neuroglia. (d) Neuroglia.

B3. (a) By neurolemmocytes (Schwann cells) wrapping their cell membranes around nerve fibers. (b) By a similar process involving oligodendroglial cells.

B4.

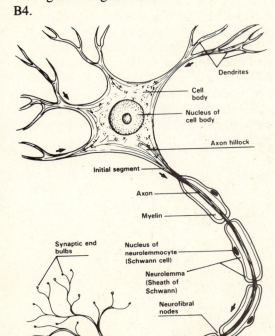

Figure LG 12.1A Structure of a typical neuron as exemplified by an efferent (motor) neuron.

B5. (a) CB. (b) L. (c) M. (d) NF. (e) CS. (f) D. (g) A.

B6. (a) Neuron. (b) Nerve. (c) Nerve.

B9. (a) GVE. (b) GSA. (c) GVA. (d) GSA. (e) GVE. (f) GSE.

B10. (a) Consists of myelinated fibers; myelin is white. (b) Consists of unmyelinated fibers, cell bodies, and neuroglia. (c) Speeds conduction by saltatory effect.

B11.

Structure	Color (Gray or White)	Composition (Cell Bodies or Nerve Fibers)	Location (CNS or PNS)
a. Nerve	White	**Nerve fibers**	**PNS**
b. Tract	**White**	**Nerve fibers**	CNS
c. Nucleus	Gray	**Cell bodies**	**CNS**

C1.

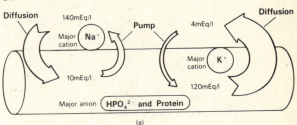

Figure LG 12.2A Diagrams of a nerve cell.

(a) Na^+. (b) 28–30, inside. (c, d) See Figure LG 12.2A. (e) Negative. Three reasons: more protein (negative ions) inside the cell, since protein is synthesized in the cell and is too large to cross the plasma membrane; 50–100× more K^+ exits by diffusion than Na^+ enters by diffusion; 3 Na^+ are electrogenically pumped out for every 2 K^+ pumped in. See Figure LG 12.2A. (f) Polarized. (g) Potential; millivolts; −70.

C2. (a) Leakage, gated; K^+, K^+. (b) Mechanically; chemically; mechanically. (c) Graded; action.

C3. (a) More; channels. (b) −70; positive; −60; depolarization; positive; action potential. (c) 0; reversed. (d) Inside; out; −70; re; hyper, −90. (e) Absolute; small, 4; small; relative, re. (f) Outside; transmitted; Na^+.

C4. (a) B. (b) A. (c) F. (d) C. (e) D. (f) F. (g) E.

C7. (a) Myelinated; faster since the neurofibral nodes (of Ranvier) have a high density of voltage gated Na^+ channels; less energy since only small regions (the nodes) have inflow of Na^+, which then must be pumped out by Na^+/K^+ pumps. (b) Largest, are; rapidly; large, sensory neurons, such as for touch, position of joints, and temperature, as well as nerves to skeletal muscles for quick reactions. (c) Smallest, are not; slowly; nerves that carry impulses to and from viscera. (d) Warm; ice or other cold applications can slow conduction of pain impulses.

C8.

Tissue	Typical Resting Membrane Potential (RMP)	Duration of Nerve Impulse
a. Neuron	Higher (−70 mV)	Shorter (0.5–2 msec)
b. Muscle	Lower (−90 mV)	Longer (1.0–5.0 msec: skeletal muscle; 10–300 msec: cardiac and smooth muscle)

D1. (a) Neuromuscular; synapse. (b) Electrical; gap, connexons; smooth and cardiac. (c) Do not. No. The nerve impulse travels along the presynaptic axon to the end of the axon, where synaptic vesicles are triggered to release neurotransmitters. These chemicals enter the synaptic cleft and then affect the post-synaptic neuron (either excite it or inhibit it). (d) B C A. (e) EB SV N SC NR. (f) The nerve impulse [described in (c) above] opens Ca^{2+} channels, creating an influx of Ca^{2+} into the presynaptic end bulbs. The Ca^{2+} here triggers exocytosis of synaptic vesicles with release of neurotransmitters. (g) Pre, post.

D2.

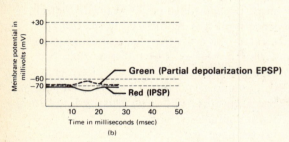

Figure LG 12.3bA Diagram of a facilitation EPSP and an IPSP.

(a) Excitatory; EPSP; Na^+. (b) Is not; partial; see Figure LG 12.3bA. (c) –55; spatial; one. (d) Hyper, inhibitory; Cl^-, K^+; –80; see Figure LG 12.3bA. (e) Farther; biceps; de.

D3. (a) Acetylcholine, acetylcholinesterase (ACHase); remain. Such drugs may help activate muscles of persons with myasthenia gravis since many of their neurotransmitter receptors are destroyed, and the extra activation of ACh can enhance response of remaining receptors. On the other hand, use of some powerful anticholinesterase drugs (as "nerve gases") cause profound and lethal effects. See page 369 of the text. (b) Norepinephrine (NE); NE and dopamine (DA); short-term: increased heart rate, blood pressure, blood sugar, and body temperature, along with euphoria; long-term: depletion of dopamine (since it is not recycled) with harmful effects on heart and/or other organs.

D5. (a) Excitatory; de. (b) Presynaptic; minutes or hours.

D6. (a) ACh; excitatory; inhibitory; slower. (b) Inhibitory; Cl^-; GABA, glycine. (c) Glycine; diaphragm. (c) Norepinephrine (NE), epinephrine (epi), and dopamine (DA).

D7. Stimulate; depress, decreased level of consciousness (LOC), such as lethargy leading to possible coma and death; signs of acidosis are present at pH 7.28 (decreased LOC).

D8. (b) I (4). (c) E (5). (d) I (1). (e) E (2).

D9. (a) D. (b) D. (c) R. (d) C. (e) P.

E1. (a) No. About six months after birth, the mitotic apparatus is lost. (b) No. CNS nerve fibers lack the neurolemma necessary for regeneration, and scar tissue builds up by proliferation of astroglia cells. (c) Yes. PNS fibers do have a neurolemma.

E3. (a) Neurolemma. (b) No. They lack a neurolemma. (c) Chromatolysis, Wallerian degeneration.

MASTERY TEST: Chapter 12

Questions 1–2: Arrange the answers in correct sequence.

<u>A</u> <u>C</u> <u>B</u> 1. In order of transmission across synapse, from first structure to last:
- A. Presynaptic end bulb
- B. Postsynaptic neuron
- C. Synaptic cleft

<u>A</u> <u>C</u> <u>B</u> 2. Membrane potential values, from most negative to zero:
- A. Resting membrane potential (RMP)
- B. Depolarized membrane potential
- C. Threshold potential

Questions 3–9: Circle the letter preceding the one best answer to each question.

3. Choose the one *false* statement.
   - A. The membrane of a resting neuron has a membrane potential of –70 mV.
   - B. In a resting membrane, permeability to K+ ions is about 100 times less than permeability to Na+ ions.
   - C. C fibers are thin, unmyelinated fibers with relatively slow rate of nerve transmission.
   - D. C fibers are more likely to innervate the heart and bladder than structures (such as skeletal muscles) that must make instantaneous responses.

4. All of the following are listed with a correct description *except:*
   - A. Neurolemmocyte—myelination of neurons in the PNS
   - B. Oligodendrocytes—myelination of neurons in the CNS
   - C. Astrocytes—form an epithelial lining of the ventricles of the brain
   - D. Microglia—phagocytic

5. Synaptic end bulbs are located:
   - A. At ends of axon terminals
   - B. On axon hillocks
   - C. On neuron cell bodies
   - D. At ends of dendrites
   - E. At ends of both axons and dendrites

6. Which of these is equivalent to a nerve fiber?
   - A. A neuron
   - B. A neurofibril
   - C. An axon or dendrite
   - D. A nerve, such as sciatic nerve

7. *ACh* is an abbreviation for:
   - A. Acetylcholine
   - B. Norepinephrine
   - C. Acetylcholinesterase
   - D. Serotonin
   - E. Inhibitory postsynaptic potential

8. A term that means the same thing as *afferent* is:
   - A. Autonomic
   - B. Somatic
   - C. Peripheral
   - D. Motor
   - E. Sensory

9. An excitatory transmitter substance that changes the membrane potential from –70 to –65 mV causes:
   - A. Impulse conduction
   - B. Partial depolarization
   - C. Inhibition
   - D. Hyperpolarization

Questions 10–20: Circle T (true) or F (false). If the statement is false, change the underlined word or phrase so that the statement is correct.

T F 10. The concentration of potassium ions (K^+) is considerably <u>greater</u> inside a resting cell than outside of it.

T **F** 11. Since CNS fibers contain no neurolemma, and the neurolemma produces myelin, CNS <u>fibers are all unmyelinated.</u>

T F 12. Neurotransmitter substances are released at <u>synapses and also at neuromuscular junctions.</u>

T **F** 13. Generally, release of excitatory transmitter by <u>a single presynaptic end bulb</u> is sufficient to develop an action potential in the postsynaptic neuron.

T F 14. <u>Epilepsy</u> is a condition that involves abnormal electrical discharges within neurons of the brain.

T **F** 15. In the <u>converging circuit,</u> *diverging* a single presynaptic neuron influences several postsynaptic neurons (or muscle or gland cells) at the same time.

T F 16. Action potentials are measured in <u>milliseconds, which are thousandths of a second.</u> *mili*

T F 17. Nerve fibers with a short absolute refractory period can respond to <u>more rapid</u> stimuli than nerve fibers with a long absolute refractory period.

T **F** 18. Most neurons in the central nervous system (CNS) are classified as <u>unipolar.</u> *multi*

T F 19. A stimulus that is adequate will temporarily <u>increase permeability of the nerve membrane to Na^+.</u>

T **F** 20. The <u>brain and spinal nerves</u> are parts of the peripheral nervous system (PNS).

Questions 21–25: fill-ins. Complete each sentence with the word or phrase that best fits.

AUTONOMIC 21. The ____ nervous system consists of the sympathetic and parasympathetic divisions.

Cooling slows speed of transmission 22. Application of cold to a painful area can decrease pain in that area because ____.

end bulbs 23. One-way nerve impulse transmission can be explained on the basis of release of transmitters only from the ____ of neurons.

_P_ 24. The neurolemma is found only around nerve fibers of the ____ nervous system, so that these fibers can regenerate.

Soma 25. In an axosomatic synapse, an axon's synaptic end bulb transmits nerve impulses to the ____ of a postsynaptic neuron.

ANSWERS TO MASTERY TEST: ■ Chapter 12

Arrange

1. A C B
2. A C B

Multiple Choice

3. B	7. A
4. C	8. E
5. A	9. B
6. C	

True-False

10. T
11. F. May be myelinated since oligodendrocytes myelinate CNS fibers
12. T
13. F. A number of presynaptic end bulbs
14. T
15. F. Diverging
16. T
17. T
18. F. Multipolar
19. T
20. F. Spinal nerves (as well as some other structures; but not the brain)

Fill-ins

21. Autonomic
22. Cooling of neurons slows down the speed of nerve transmission, for example, of pain impulses
23. End bulbs of axons
24. Peripheral
25. Cell body (soma)

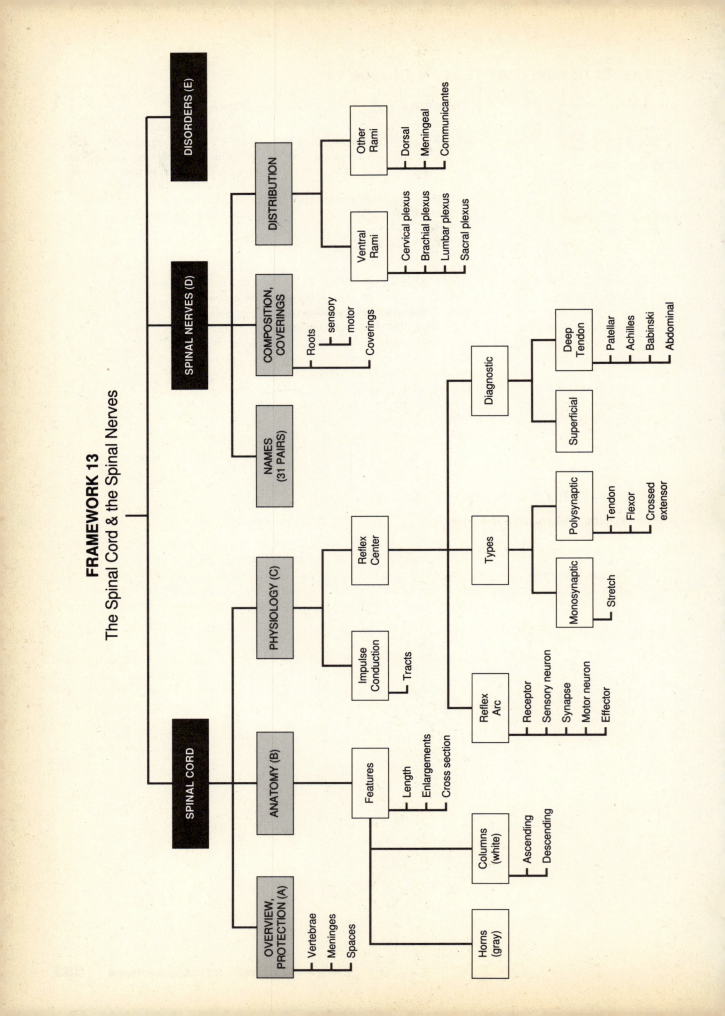

FRAMEWORK 13
The Spinal Cord & the Spinal Nerves

The Spinal Cord and Spinal Nerves

The spinal cord and spinal nerves serve as the major links in the communication pathways between the brain and all other parts of the body. Nerve impulses are conveyed along routes (or tracts) in the spinal cord, laid out much as train tracks: some head north to regions of the brain, and others carry nerve messages south from the brain toward specific body parts. Spinal nerves branch off from the spinal cord, perhaps like a series of bus lines that pick up passengers (nerve messages) to or from train depots (points along the spinal cord) en route to their final destinations. Organization is critical in this nerve impulse transportation network. Any structural breakdowns—by trauma, disease, or other disorders—can lead to interruption in service with resultant chaos (such as spasticity) or standstill (like sensory loss or paralysis).

Welcome aboard, and make your first stop the Chapter 13 Framework with key terms in each section.

TOPIC OUTLINE AND OBJECTIVES

A. Spinal cord: overview; protection

B. Spinal cord: anatomy

1. Describe the protection, gross anatomical features, and cross-sectional structure of the spinal cord.

C. Spinal cord: physiology

2. Describe the functions of the principal sensory and motor tracts of the spinal cord.
3. Describe the components of a reflex arc and its relationship to homeostasis.
4. List and describe several clinically important reflexes.

D. Spinal nerves

5. Describe the composition and coverings of a spinal nerve.
6. Define a plexus and describe the composition and distribution of nerves of the cervical, brachial, lumbar, and sacral plexuses.

E. Disorders

7. Describe spinal cord injury and list the immediate and long-range effects.
8. Explain the causes and symptoms of neuritis, sciatica, shingles, and poliomyelitis.

WORDBYTES

Now study the following parts of words that may help you better understand terminology in this chapter.

Wordbyte	Meaning	Example	Wordbyte	Meaning	Example
arachn-	spider	*arachn*oid	pia	tender	*pia* mater
cauda-	tail	*cauda*te	rhiz-	root	*rhiz*otomy
dura	hard	*dura* mater	soma-	body	*soma*tic

CHECKPOINTS

A. Spinal cord: overview; protection   (page 376)

■ **A1.** List three general functions of the spinal cord.

■ **A2.** The spinal cord is part of the (*central?* *peripheral?*) nervous system.

■ **A3.** Refer to Figure LG 13.1 and do the following exercise.

a. Color the meninges to match color code ovals.
b. Label these spaces: *epidural space, subarachnoid space,* and *subdural space.*
c. Write next to each label of a space the contents of that space. Choose these answers: *cerebrospinal (CSF); lymphatic fluid, fat and connective tissue.*

d. Inflammation of meninges is a condition called _____.

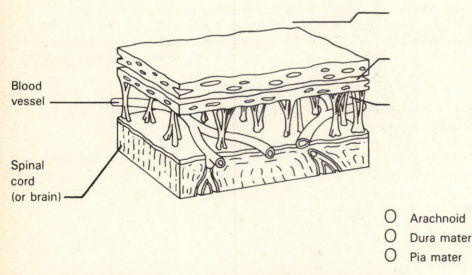

Blood vessel

Spinal cord (or brain)

O Arachnoid
O Dura mater
O Pia mater

Figure LG 13.1   Meninges. Color and label as indicated in Checkpoint A3.

e. The denticulate ligaments are extensions of *(dura mater? arachnoid? pia mater?)* that

attach laterally to _____ along the length of the cord. State functions of these ligaments.

B. Spinal cord: anatomy (pages 376–380)

■ **B1.** Do this activity about your own spinal cord.

a. Identify the location of your own spinal cord. It lies within the vertebral canal, extending from just inferior to the cranium to about the level of your *(waist? sacrum?)*. This level corresponds with about the level of *(L1–L2? L4–L5? S4–S5?)* vertebrae. In other words, the *(spinal cord? vertebral column?)* reaches a lower (more inferior) level in your body. The spinal cord is about 42 to 45 cm (____ inches) in length.

b. Circle the two regions of your spinal cord that have notable enlargements where nerves exit to your upper and lower extremities:
cervical   thoracic   lumbar   sacral   coccygeal

■ **B2.** *A clinical challenge.* Answer these questions about meninges and related clinical procedures.

a. State two or more purposes of a spinal tap (or lumbar puncture).

check for blood

or bacteria

Into which space does the needle enter? *(Epidural? Subdural? Subarachnoid?)*

b. At what level of the vertebral column (not spinal cord) is the needle for a spinal tap

(or lumbar puncture) inserted? Between _____ vertebrae. Explain why this location is a relatively safe site for this procedure.

c. Identify level L4 of the vertebral column on yourself by listing two other landmarks

at this level. _____ _____

d. In some cases anesthetics are introduced into the epidural space, rather than into the subarachnoid space (as in a spinal tap). Explain why an *epidural* is likely to exert its effects on nerves with slower and more prolonged action than a spinal tap.

State one example of a procedure for which an epidural may be used.

■ **B3.** Match the names of the structures listed in the box with descriptions given.

Ca. Cauda equina	F. Filum terminale
Co. Conus medullaris	S. Spinal segment

__Co__ a. Tapering inferior end of spinal cord

__S__ b. Any region of spinal cord from which one pair of spinal nerves arises

__F__ c. Nonnervous extension of pia mater; anchors cord in place

__Ca__ d. "Horse's tail"; extension of spinal nerves in lumbar and sacral regions within subarachnoid space

■ **B4.** Identify the following structures on Figure LG 13.2.

a. On the right side of the diagram, write labels next to leader lines for the following structures: *anterior median fissure, anterior gray horn, lateral gray horn, posterior gray horn, gray commissure,* and *lateral white column.*

b. Tracts conduct nerve impulses in the *(central? peripheral?)* nervous system. Functionally they are comparable to _____ in the peripheral nervous system.

c. Tracts are located in *(gray horns? white columns?).* They appear white since they consist of bundles of *(myelinated? unmyelinated?)* nerve fibers.

Figure LG 13.2 Outline of the spinal cord, roots, and nerves. Color and label according to Checkpoints B4, C1-C3, and D1.

d. Ascending tracts are all *(sensory? motor?)*, conveying impulses between the spinal

cord and the _____.
e. All motor tracts in the cord are *(ascending? descending?)*.

C. Spinal cord: physiology (pages 380–389)

■ **C1.** In Checkpoint A1, you listed main functions of the spinal cord. Expand that description in this Checkpoint.

a. One primary function of the spinal cord is to permit *Communication* between nerves in the periphery, such as arms and legs, and the brain by means of the *(tracts? nerves? ganglia?)* located in the white columns of the cord.

b. Another function of the cord is to serve as a _____ center by

means of spinal nerves. These are attached by _____ roots. The *(anterior?*

posterior?) root contains sensory nerve fibers, and the *anterior* root contains motor fibers. Color the sensory and motor roots on Figure LG 13.2. Select colors according to the color code ovals there.

■ **C2.** Refer to Figure LG 13.2 and do this activity on the conduction function of the spinal cord.

a. Color the five tracts, using color code ovals to demonstrate sensory or motor function of tracts. (*Note:* Tracts are actually present on both sides of the cord, but shown on only one side here.)
b. Write next to the name of each tract its correct functions. Use these answers:
Pain, temperature, crude touch, pressure
Precise voluntary movements
Proprioception, pressure, vibration, and two-point discriminative touch
c. The name *lateral corticospinal* indicates that the tract is located in the

_____*lateral*_____ white column, that it originates in the *(cerebral cortex?*

thalamus? spinal cord?), and that it ends in the _____.
d. The lateral corticospinal tract is *(ascending, sensory? descending, motor?)*, and it is a(n) *(extrapyramidal? pyramidal?)* tract.
e. The rubrospinal, tectospinal, and vestibulospinal tracts are *(extrapyramidal? pyramidal?)* tracts. These are involved with control of:
A. Precise, voluntary movements
B. Automatic movements such as maintenance of muscle tone, posture, and equilibrium

C3. *For extra review.* Draw and label the following tracts on Figure LG 13.2. Then color them according to the color code ovals on the figure: *anterior spinocerebellar, posterior spinocerebellar, rubrospinal, tectospinal,* and *vestibulospinal.*

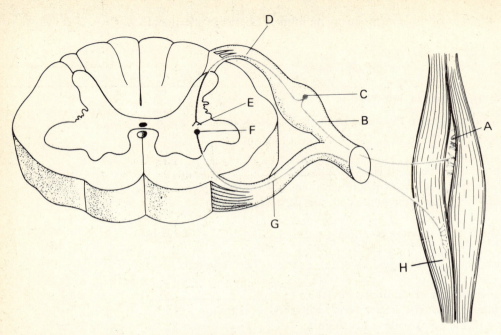

Figure LG 13.3 Reflex arc: stretch reflex. Label as directed in Checkpoints C4 and C5.

■ **C4.** Complete this activity on the function of the spinal cord as a reflex center. Label structures *A–H* on Figure LG 13.3, using the following terms: *effector, motor neuron axon, motor neuron cell body, receptor, sensory neuron axon, sensory neuron cell body, sensory neuron dendrite,* and *synapse* (integrating center). Note that these structures are lettered in alphabetical order along the conduction pathway of a reflex arc. Add arrows showing the direction of nerve transmission in the arc.

■ **C5.** Answer these questions about the reflex arc in Figure LG 13.3.

a. How many neurons does this reflex contain? _____ The neuron that conveys the

impulse toward the spinal cord is a _____ neuron; the one that

carries the impulses toward the effector is a _____ neuron.
b. This is a *(monosynaptic? polysynaptic?)* reflex arc. The synapse, like all somatic synapses, is located in the *(CNS? PNS?)*.
c. Receptors, in this case located in skeletal muscle, are called

_____. They are sensitive to changes in

_____. This type of reflex might therefore be called a

_____ reflex.
d. Since sensory impulses enter the cord on the same side as motor impulses leave, the reflex is called *(ipsilateral? contralateral? intersegmental?)*.

e. What structure is the effector? _____

f. One example of a stretch reflex is the _____, in which stretching of the patellar tendon initiates the reflex.

■ **C6.** Explain how the brain may get the message that a stretch reflex (or other type of reflex) has occurred.

■ **C7.** Do this exercise describing how tendon reflexes protect tendons.
   a. Receptors located in tendons are named *(muscle spindles?* (tendon organs?)*.* They are sensitive to changes in muscle *(length?* (tension?)*,* as when the hamstring muscles are contracted excessively, pulling on tendons.
   b. When this occurs, association neurons cause *((excitation?* (inhibition?)*)* of this same muscle, so that the hamstring fibers *(contract further?* (relax?)*.*
   c. Simultaneously, other association neurons fire impulses which stimulate *(synergistic?* (antagonistic?)* muscles (such as the quadriceps in this case). These muscles then *(con-tract?* (relax?)*.*
   d. The net effect of such a *(mono?* (poly?)* synaptic tendon reflex is that the tendons are *((protected?* (injured?)*.*

■ **C8.** Contrast stretch, flexor, and crossed extensor reflexes in this learning activity.
   a. A flexor reflex *(does? does not?)* involve association neurons, and so it is *(more? less?)* complex than a stretch reflex.
   b. A flexor reflex sends impulses to *(one? several?)* muscle(s), whereas a stretch reflex, such as the knee jerk, activates *(one? several?)* muscle(s), such as the quadriceps.

   A flexor reflex is also known as a _____ reflex, and it is

   _____-lateral.
   c. If you simultaneously contract the flexor and extensor (biceps and triceps) muscles of your forearm with equal effort, what action occurs? (Try it.)

   _____ In order for movement to occur, it is necessary for extensors (triceps) to be inhibited while flexors (biceps) are stimulated. The nervous

   system exhibits such control by a phenomenon known as _____ innervation.
   d. Reciprocal innervation also occurs in the following instance. Suppose you step on a tack under your right foot. You quickly withdraw that foot by *(flexing? extending?)* your right leg using your hamstring muscles. (Stand up and try it.) What happens to your left leg? You *(flex? extend?)* it. This is an example of reciprocal innervation

   involving a _____ reflex.
   e. A crossed extensor reflex is *(ipsilateral? contralateral?)*. It *(may? may not?)* be intersegmental. Therefore, many muscles may be contracted to extend your left thigh and leg to shift weight to your left side and provide balance during the tack episode.

■ **C9.** Why are deep tendon reflexes (that involve stretching a tendon such as that of the quadriceps femoris muscle in the patellar reflex) particularly helpful diagnostically?

■ **C10.** Complete the table about reflexes that are of clinical significance.

	Patellar Reflex	Achilles Reflex	Babinski Sign
a. Procedure used to demonstrate reflex	Tap patellar ligament		
b. Nature of positive response		Plantar flexion	
c. Spinal nerves and muscles evaluated by procedure in reflex			Determines if corticospinal tract is myelinated yet
d. Cause of negative response			Plantar flexion reflex: all toes curl under; normal after age 1½
e. Cause of exaggerated response		Damage to motor tracts in S1 or S2 segments of cord	

D. Spinal nerves   (pages 389–397)

■ **D1.** Complete this exercise about spinal nerves.

a. There are ____ pairs of spinal nerves. Write the number of pairs in each region.

_____ Cervical _____ Thoracic _____ Lumbar _____ Sacral _____ Coccygeal

b. Which of these spinal nerves form the cauda equina?

c. Spinal nerves are attached by two roots. The posterior root is *(sensory? motor?*

mixed?), the anterior root is _____, whereas the spinal nerve is

_____.

d. Individual nerve fibers are wrapped in a connective tissue covering known as *(endo-? epi-? peri-?)* neurium. Groups of nerve fibers are held in bundles (fascicles) by

_____-neurium. The entire nerve is wrapped with _____-neurium.

e. Spinal nerves branch when they leave the intervertebral foramen. These branches are

called _____. Since they are extensions of spinal nerves, rami are *(sensory? motor? mixed?).*

f. Which ramus is larger? *(ventral? dorsal?)* What areas does it supply?

g. What area does the dorsal ramus innervate? Label it on Figure LG 13.2.

h. Name two other branches (rami) and state their functions.

■ **D2.** Match the plexus names in the box with descriptions. Refer to Figure 13.2a (page 378 in your text) for help.

> B. Brachial   L. Lumbar
> C. Cervical   S. Sacral
> I. Intercostal

_____ a. Provides the entire nerve supply for the arm

_____ b. Contains origin of phrenic nerve (nerve that supplies diaphragm)

_____ c. Forms median, radial, and axillary nerves

_____ d. Not a plexus at all, but rather segmentally arranged nerves

_____ e. Supplies nerves to scalp, neck, and part of shoulder and chest

_____ f. Supplies fibers to the femoral nerve which innervates the quadriceps, so injury to this plexus would interfere with actions such as touching the toes

_____ g. Forms the largest nerve in the body (the sciatic) which supplies posterior of thigh and the leg

■ **D3.** *For extra review.* Match names of spinal nerves in the box with their descriptions. On the line following the description, write the site of origin of the nerve. The first one is done for you.

> Axillary   Pudendal
> Inferior gluteal   Radial
> Musculocutaneous   Sciatic

_____**Sciatic**_____ a. Consists of two nerves, the tibial and common peroneal; supplies the hamstrings, adductors, and all muscles distal to the knee. __**L4–S3**__

_____ b. Supplies the deltoid muscle. _____

_____ c. Innervates the major flexors of the arm. _____

_____ d. Supplies most extensor muscles of the forearm, wrist, and fingers; may be damaged by extensive use of crutches. _____

_____ e. Supplies the gluteus maximus muscle. _____

_____ f. May be anesthetized in childbirth since it innervates external genitalia and lower part of the vagina. _____

■ **D4.** *A clinical challenge.* If the cord were completely transected (severed) just below the C7 spinal nerves, how would the functions listed below be affected? (Remember that nerves that originate below this point would not communicate with the brain, so would lose much of their function.) Explain your reasons in each case. (*Hint:* Refer to Figure 13.2a, page 378 in the text.)

a. Breathing via diaphragm

b. Movement and sensation of thigh and leg

c. Movement and sensation of the arm

d. Use of muscles of facial expression and muscles that move jaw, tongue, eyeballs

D5. Describe the general pattern of dermatomes:

a. In the trunk

b. In the extremities

E. Disorders (page 397)

E1. List several causes of spinal cord injury.

■ **E2.** Match the terms in the box with descriptions below.

> H. Hemiplegia   P. Paraplegia
> M. Monoplegia   Q. Quadriplegia

_____ a. Paralysis of one extremity only

_____ b. Paralysis of both legs

_____ c. Paralysis of both arms and both legs

_____ d. Paralysis of the arm, leg, and trunk on one side of the body

■ **E3.** Describe changes associated with transection of the spinal cord in this activity.

a. Hemisection of the cord refers to transection of *(half? all?)* of the spinal cord.

b. If the posterior columns (cuneatus and gracilis) and lateral corticospinal tracts on the right side of the cord are severed at level T11–T12, symptoms of loss of awareness of muscle sensations (proprioception), loss of touch sensations, and paralysis are likely to occur in the *(right arm? left arm? right leg? left leg?)*. (Circle all that apply.)

c. The period of spinal shock is likely to last for several *(days to weeks? years?)*. During this time, the person is likely to experience *(areflexia? hyperreflexia?)*, meaning *(exaggerated? no?)* reflexes.

d. When reflexes gradually return to the injured person, they are likely to return in a specific sequence. Arrange these reflexes in order from first to last, according to their return.

____ ____ ____

C. Crossed extensor reflex
F. Flexion reflex
S. Stretch reflex

■ **E4.** Shingles is an infection of the *(central? peripheral?)* nervous system. The causative virus is also the agent of *(chickenpox? measles? cold sores?)*. Following recovery from chickenpox, the virus remains in the body in the *(spinal cord? dorsal root ganglia?)*. At times it is activated and travels along *(sensory? motor?)* neurons, causing *(pain? paralysis?)*.

■ **E5.** Match names of disorders in the box with definitions below.

> A. Areflexia   Sc. Sciatica
> N. Neuritis   Sh. Shingles
> P. Poliomyelitis

__N__ a. Inflammation of a single nerve

__A__ b. Lack of reflex activity

__Sh__ c. Acute inflammation of the nervous system by *Herpes zoster* virus

__Sc__ d. Neuritis of a nerve in the posterior of hip and thigh; often due to a slipped disc in the lower lumbar region

__P__ e. Also known as infantile paralysis; caused by a virus that may destroy motor cell bodies in the brainstem or in the anterior gray horn of the spinal cord

E6. After reading pages 30–31 in *Essays on Wellness*, do this activity about wellness.

a. How would you define *wellness lifestyle* as it pertains to your own life.

eatting right
getting enough rest + exercise
handling stress correctly

b. Contrast what the terms *handicapped person* and *person with special needs* mean to you.

handicap infers a less perfect person
person with special needs infers that the person is
capable if special accomodations are given.

c. List several categories of persons with special needs, and list names of friends or family members who are included in one or more of those categories.

Category	Friend or Family Member
emphesema	*father*
diabetes	*mother*

d. Choose one or two persons whom you named in (c) and write several suggestions about how the person's special needs in the following areas may be met:
Activities of daily living such as mobility, eating, and bathing

Recreational, leisure, emotional, and/or spiritual needs

Education

Health care

ANSWERS TO SELECTED CHECKPOINTS

A1. Processing center for (reflexes), integration (summing) of afferent or efferent nerve impulses, and pathway for those impulses.

A2. Central (CNS).

A3.

Figure LG 13.1A Meninges.

(d) Meningitis. (e) Pia mater, dura mater; anchor the cord and protect it from displacement.

B1. (a) Waist; L1–2; vertebral column; 16–18.
(b) Cervical, lumbar.

B2. (a) To insert anesthetics, antibiotics, chemotherapy or contrast media; to withdraw cerebrospinal fluid (CSF) for diagnostic purposes such as for analysis for blood or microorganisms; subarachnoid. (b) L3–L4 or L4–L5; the spinal cord ends at about L1–L2, so the cord is not likely to be injured at this lower level. (c) The iliac crest and umbilicus (navel) are both at about this level. (d) The anesthetic must penetrate epidural tissues and then all three layers of meninges before reaching nerve fibers, whereas an anesthetic in the subarachnoid space needs to penetrate only the pia mater to reach nerve tissue; childbirth (labor and delivery).

B3. (a) Co. (b) S. (c) F. (d) Ca.

B4. (a) See Figure LG 13.2A.

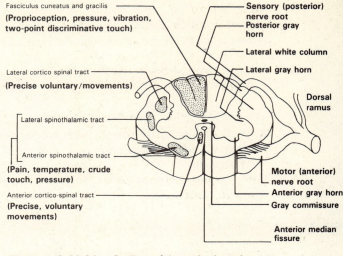

Figure LG 13.2A Outline of the spinal cord, roots, and nerves.

(b) Central; nerves. (c) White columns; myelinated. (d) Sensory, brain. (e) Descending.

C1. (a) Conduction, tracts. (b) Reflex; 2; posterior (dorsal), anterior (ventral). See Figure LG 13.2A.

C2. (a–b) See Figure LG 13.2A. (c) Lateral, cerebral cortex, spinal cord (in anterior gray horn). (d) Descending, motor, pyramidal. (e) Extrapyramidal; B.

C4.

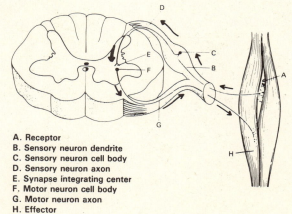

A. Receptor
B. Sensory neuron dendrite
C. Sensory neuron cell body
D. Sensory neuron axon
E. Synapse integrating center
F. Motor neuron cell body
G. Motor neuron axon
H. Effector

Figure LG 13.3A Reflex arc: stretch reflex.

C5. (a) 2; sensory; motor. (b) Monosynaptic; CNS. (c) Muscle spindles; length (or stretch); stretch. (d) Ipsilateral. (e) Skeletal muscle. (f) Knee jerk (patellar reflex).

C6. Branches from axons of sensory or association neurons travel through tracts to the brain.

C7. (a) Tendon organs; tension. (b) Inhibition, relax. (c) Antagonistic; contract. (d) Poly, protected.

C8. (a) Does, more. (b) Several, one withdrawal, ipsi. (c) No action; reciprocal. (d) Flexing; extend; crossed extensor. (e) Contralateral; may.

C9. They can readily pinpoint a disorder of a specific spinal nerve or plexus, or portion of the cord, since they do not involve the brain.

C10.

	Patellar Reflex	Achilles Reflex	Babinski Sign
a. Procedure used to demonstrate reflex	Tap patellar ligament	Tap on calcaneal (Achilles) tendon	Light stimulation of outer margin of sole of foot
b. Nature of positive response	Extension of leg	Plantar flexion	Great toe extends, with or without fanning of other toes. Abnormal after age 1½; shows incomplete myelination.
c. Spinal nerves and muscles evaluated by procedure in reflex	L2-4 (quadriceps muscles)	L4-5, S1-3 (gastrocnemius and soleus muscles)	Determines if corticospinal tract is myelinated yet
d. Cause of negative response	Chronic diabetes mellitus, neurosyphilis	Diabetes, neurosyphilis, alcoholism	Plantar flexion reflex: all toes curl under; normal after age 1½
e. Cause of exaggerated response	Injury to corticospinal tracts	Damage to motor tracts in S1 or S2 segments of cord	Positive Babinski after age 1½ indicates interruption of corticospinal tracts

D1. (a) 31; 8, 12, 5, 5, 1. (b) Lumbar, sacral and coccygeal. (c) Sensory, motor, mixed. (d) Endo-; peri-; epi-. (e) Rami; mixed. (f) Ventral: all of the extremities and the ventral and lateral portions of the trunk. (g) Muscles and skin of the back; see Figure LG 13.2A. (h) Meningeal branch supplies primarily vertebrae and meninges; rami communicantes have autonomic functions.

D2. (a) B. (b) C. (c) B. (d) I. (e) C. (f) L. (g) S.

D3. (b) Axillary, C5–C6. (c) Musculocutaneous, C5–C7. (d) Radial, C5–C8, T-1. (e) Inferior gluteal, L5–S2. (f) Pudendal, S2–S4.

D4. (a) Not affected since (phrenic) nerve to the diaphragm originates from the cervical plexus (at C3–C5), higher than the transection. So this nerve continues to receive nerve impulses from the brain. (b) Complete loss of sensation and paralysis since lumbar and sacral plexuses originate below the injury, and therefore no longer communicate with the brain. (c) Most arm functions are not affected. As shown on Figure 13.2a, page 378 of the text, the brachial plexus originates from C-5 through T-1, so most nerves to the arm (those from C-5 through T-1) still communicate with the brain. (d) Not affected since all are supplied by cranial nerves which originate from the brain.

E2. (a) M. (b) P. (c) Q. (d) H.

E3. (a) Half. (b) Right leg (same side as cord is severed, and in areas served by nerves inferior to the transection such as the right leg. (c) Days to weeks; areflexia, no. (d) S F C.

E4. Peripheral; chickenpox; dorsal root ganglia; sensory, pain.

E5. (a) N. (b) A. (c) Sh. (d) Sc. (e) P.

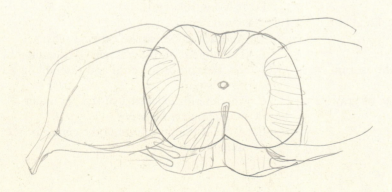

Questions 1–4: Arrange the answers in correct sequence.

<u>B</u>   <u>C</u>   <u>A</u>

1. From superficial to deep:
   A. Subarachnoid space
   B. Epidural space
   C. Dura mater

<u>B</u>   <u>C</u>   <u>A</u>

2. From anterior to posterior in the spinal cord:
   A. Fasciculus gracilis and cuneatus
   B. Anterior spinothalamic tract
   C. Central canal of the spinal cord

<u>C</u>   <u>B</u>   <u>A</u>   <u>D</u>

3. The plexuses, from superior to inferior:
   A. Lumbar
   B. Brachial
   C. Cervical
   D. Sacral

<u>D</u>   <u>B</u>   <u>C</u>   <u>A</u>   <u>E</u>
Receptor Sensory Integrative Motor Effector

4. Order of structures in a conduction pathway, from origin to termination:
   A. Motor neuron
   B. Sensory neuron
   C. Integrative center
   D. Receptor
   E. Effector

Questions 5–10: Circle the letter preceding the one best answer to each question.

5. Herniation (or "slipping" of the disc between L4 and L5 vertebrae is most likely to result in damage to the ___ nerve.
   A. Femoral
   B. Sciatic
   C. Radial
   D. Musculocutaneous

6. All of these tracts are sensory *except:*
   A. Anterior spinothalamic
   B. Lateral spinothalamic
   C. Fasciculus cuneatus
   D. Lateral corticospinal
   E. Posterior spinocerebellar

7. Choose the *false* statement about the spinal cord.
   A. It has enlargements in the cervical and lumbar areas.
   B. It lies in the vertebral foramen.
   C. It extends from the medulla to the sacrum.
   D. It is surrounded by meninges.
   E. In cross section an H-shaped area of gray matter can be found.

8. All of these structures are composed of white matter *except:*
   A. Posterior root (spinal) ganglia
   B. Tracts
   C. Lumbar plexus
   D. Sciatic nerve
   E. Ventral ramus of a spinal nerve

9. Which is a *false* statement about the patellar reflex?
   A. It is also called the knee jerk.
   B. It involves a two-neuron, monosynaptic reflex arc.
   C. It results in extension of the leg by contraction of the quadriceps femoris.
   D. It is contralateral.

10. The cauda equina is:
    A. Another name for the cervical plexus
    B. The lumbar and sacral nerves extending below the end of the cord and resembling a horse's tail
    C. The inferior extension of the pia mater
    D. A denticulate ligament
    E. A canal running through the center of the spinal cord

(T) F 11. The layer of the meninges that gets its name from its delicate structure, which is much like a spider's web, is the arachnoid.

T (F) 12. The two main functions of the spinal cord are that it serves as a reflex center and it is ~~the site where sensations are felt.~~ *conduction site*

(T) F 13. Dorsal roots of spinal nerves are sensory, ventral roots are motor, and spinal nerves are mixed.

(T) F 14. A tract is a bundle of nerve fibers inside the central nervous system (CNS).

T (F) 15. Synapses are *not* present in posterior (dorsal) root ganglia.

(T) F 16. After a person reaches 18 months, the Babinski sign should be negative, as indicated by plantar flexion (curling under of toes and foot).

T (F) 17. Visceral reflexes are used diagnostically ~~more~~ *less* often than somatic ones since it is ~~easy~~ *difficult* to stimulate most visceral receptors.

(T) F 18. Transection of the spinal cord at level C-6 will result in greater loss of function than transection at level T-6.

T (F) 19. The ventral root of a spinal nerve contains axons ~~and dendrites~~ of ~~both~~ motor ~~and sensory~~ neurons.

T F 20. A lumbar puncture (spinal tap) is usually performed at about the level of vertebrae ~~LX to X2~~ *L3–L4* since the cord ends between about ~~L3 and L4.~~ *L1–L2*

_____ 21. Write one or more suggestions that address how health care providers can make health care more accessible to persons with special needs.

menengitis 22. An inflammation of the dura mater, arachnoid, and/or pia mater is known as ___.

poly ipsi 23. Tendon reflexes are ___-synaptic and ___-lateral.

thoracic upper lumbar sacral 24. Lateral gray horns are found only in ___ regions of the spinal cord.

pia 25. The filum terminale and denticulate ligaments are both composed of ___ mater.

ANSWERS TO MASTERY TEST: ■ Chapter 13

Arrange

1. B C A
2. B C A
3. C B A D
4. D B C A E

Multiple Choice

5. B   8. A
6. D   9. D
7. C   10. B

True-False

11. T
12. F. Reflex center, conduction site
13. T
14. T
15. F. Are not
16. T
17. F. Less often than somatic ones since it is difficult to stimulate most visceral receptors
18. T
19. F. Axons of motor
20. F. L3–L4, L1–L2.

Fill-ins

21. Some suggestions: make health educational materials available in large print and on tape; design more accessible offices; legislate to make health care (particularly health promotion and disease prevention) available to all persons
22. Meningitis
23. Poly, ipsi
24. Thoracic, upper lumbar, sacral (further explanation in Chapter 17)
25. Pia

FRAMEWORK 14
Brain & Cranial Nerves

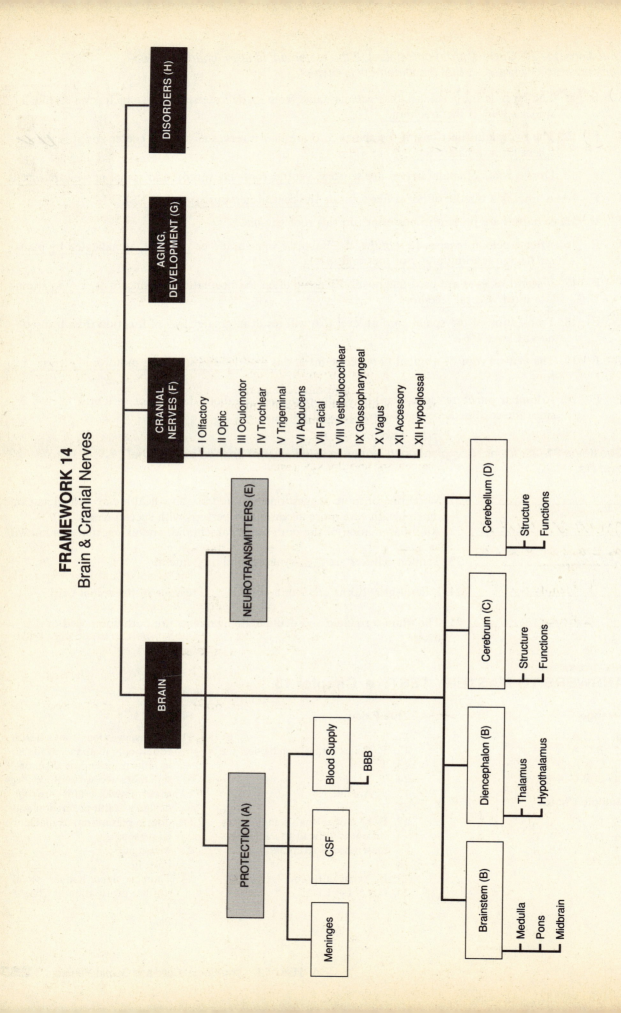

The Brain and Cranial Nerves

The brain is the major control center for the global communication network of the body: the nervous system. The brain requires round-the-clock protection and maintenance afforded by bones, meninges, cerebrospinal fluid, a special blood-brain barrier, along with a fail-safe blood supply. This vital control center consists of four major substructures: the brain stem, diencephalon, cerebrum, and cerebellum. Each brain part carries out specific functions and each releases specific chemical neurotransmitters. Twelve pairs of cranial nerves convey information to and from the brain.

Structural defects may occur in construction (brain development) and also with the normal wear and tear that accompanies aging. Just as a giant computer network may experience minor disruptions in service or major shutdowns, disorders within the brain or cranial nerves may lead to minor, temporary changes in nerve functions or profound and fatal outcomes.

First study the Chapter 14 Framework and the key terms associated with each section.

TOPIC OUTLINE AND OBJECTIVES

A. Brain: introduction; protection and coverings

1. Identify the principal parts of the brain, and describe how the brain is protected.
2. Explain the formation and circulation of cerebrospinal fluid (CSF).
3. Describe the blood supply to the brain and the concept of the blood-brain barrier (BBB).

B. Brain: brain stem and diencephalon

C. Brain: cerebrum

D. Brain: cerebellum

4. Compare the structure and functions of the brain stem, diencephalon, cerebrum, and cerebellum.

E. Neurotransmitters in the brain

5. Discuss the various neurotransmitters found in the brain, as well as the different types of neuropeptides and their functions.

F. Cranial nerves

6. Define a cranial nerve and identify the 12 pairs of cranial nerves by name, number, type, location, and function.

G. Aging and developmental anatomy of the nervous system

7. Describe the effects of aging on the nervous system.
8. Describe the development of the nervous system.

H. Disorders, medical terminology

9. List the clinical symptoms of these disorders of the nervous system: cerebrovascular accidents (CVAs), transient ischemic attacks (TIAs), Alzheimer's disease (AD), brain tumors, cerebral palsy (CP), Parkinson's disease (PD), multiple sclerosis (MS), dyslexia, headache, and Reye's syndrome (RS).
10. Define medical terminology associated with the central nervous system.

WORDBYTES

Now study the following parts of words that may help you better understand terminology in this chapter.

Wordbyte	Meaning	Example	Wordbyte	Meaning	Example
cephalo-	head	hydro*cephal*ic	ophthalm-	eye	*ophthalm*ic
cortico-	bark	cerebral *cortex*	pons	bridge	*pons*
enceph-	brain	di*enceph*alon	quadri-	four	corpora
falx	sickle	*falx* cerebri			*quadri*gemina
glossi-	tongue	hypo*gloss*al	vita	life	arbor *vita*e
hemi-	half	*hemi*sphere			

CHECKPOINTS

A. Brain: introduction; protection and coverings (pages 405–411)

■ **A1.** Identify numbered parts of the brain on Figure 14.1. Then complete this exercise.

a. Structures 1–3 are parts of the <u>brain stem</u>.

b. Structures 4 and 5 together form the <u>diencephalon</u>.

c. Structure 6 is the largest part of the brain, the <u>cerebrum</u>.

d. The second largest part is structure 7, the <u>cerebellum</u>.

■ **A2.** List three ways in which the brain is protected.

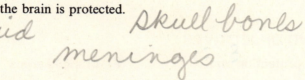

spinal fluid   *skull bones*   *meninges*

A3. Review the layers of the meninges covering the brain and spinal cord by listing them here. *For extra review.* Label the layers on Figure LG 14.1 and review Chapter 13 Checkpoint A3 (pages LG 240–241).

dura mater
arachnoid
pia mater

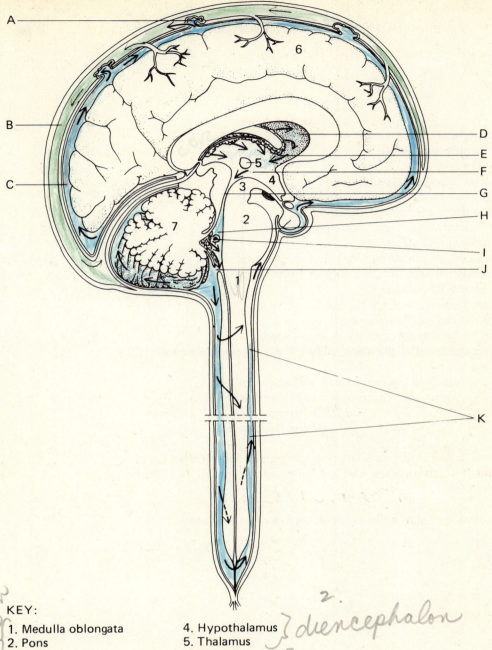

KEY:

1. Medulla oblongata
2. Pons
3. Midbrain

4. Hypothalamus
5. Thalamus
6. Cerebrum
7. Cerebellum

2.
} diencephalon
— 3
— 4

A. Arachnoid villus
B. Cranial venous sinus
C. Subarachnoid space of brain
D. Lateral venticle
E. Interventricular foramen
F. Third ventricle

G. Cerebral aqueduct
H. Fourth ventricle
I. Lateral aperture
J. Median aperture
K. Subarachnoid space of spinal cord

Figure LG 14.1 Brain and meninges seen in sagittal section. Parts of the brain are numbered; refer to Checkpoint A1. Letters indicate pathway of cerebrospinal fluid (CSF); refer to Checkpoint A5.

■ **A4.** Circle all correct answers about CSF.

a. The entire nervous system contains 80–150 ml of CSF. This amount is equal to approximately:
   A. 1 to 2 tablespoons   C. 1 to 2 cups
   (B.) ⅓ to ⅔ cup   D. 1 quart

b. The color of CSF is:
   A. Yellow   C. Red
   (B.) Clear, colorless   D. Green

c. Choose the function(s) of CSF.
   (A.) Serves as a shock absorber for brain and cord
   B. Contains red blood cells
   (C.) Contains white blood cells called lymphocytes
   (D.) Contains nutrients

d. Which statement(s) describe its formation?
   A. It is formed by diffusion of substances from blood.
   (B.) It is formed by filtration and secretion.
   (C.) It is formed from blood in capillaries called choroid plexuses.
   (D.) It is formed by ependymal cells that line all four ventricles.

e. Which statement(s) describe its pathway?
   A. It circulates around the brain, but not the cord.
   (B.) It flows inferior to the end of the spinal cord.
   C. It bathes the brain by flowing through the epidural space.
   (D.) It passes via projections (villi) of the arachnoid into blood vessels (venous sinuses) surrounding the brain.
   (E.) It is formed initially from blood, and finally flows back to blood.

f. An accumulation of CSF within the ventricles is a condition known as:
   A. Hydrarthrosis   (C.) Hydrocephalus
   B. Hydrophobia   D. Hydrocholecystitis

■ **A5.** To check your understanding of the pathway of cerebrospinal fluid (CSF), list in order the structures through which it passes. Use key letters on Figure LG 14.1. Start at the site of formation of CSF. D E F G H I J K C A B

A6. Briefly state results of oxygen starvation of the brain for about 4 minutes.

Explain the role of lysosomes in this process.

Now list effects of glucose deprivation of the brain.

■ **A7.** Describe the blood-brain barrier (BBB) in this exercise.

a. Blood capillaries supplying the brain are *(more?* (*less?*)) leaky than most capillaries of the body. This fact is related to a large number of *(gap?* (*tight?*)) junctions, as well as an abundance of neuroglia named __*astrocytes*__ pressed against capillaries.

b. What advantages are provided by this membrane?

foreign substances can't enter brain

c. List two or more substances needed by the brain that do normally cross the BBB.

glucose oxygen water ions: Na+ K+
anesthetics

d. What problems may result from the fact that some chemicals cannot cross this barrier?

helpful medications can't get in

e. List several substances that may harm the brain that *are* able to cross this barrier.

alcohol, caffeine, heroin, nicotine

■ **A8.** Describe *circumventricular organs* (CVOs) in this exercise.

a. Name three organs that are CVOs. *hypothalamus pineal gland, pituitary*

b. CVOs are unique in the brain region in that they *(have?* (*lack?*)) the blood-brain barrier. State one advantage and one disadvantage of this structural feature.

advantage
then can monitor activities such as fluid balance + hunger

disadvantage permeable to disease

B. Brain: brain stem, diencephalon (pages 411–418)

■ **B1.** Describe the principal functions of the medulla in this exercise.

a. The medulla serves as a ___*Conduction*___ pathway for all ascending and descending tracts. Its white matter therefore transmits *(sensory? motor? (both sensory and motor?))* impulses.

b. Included among these tracts are the triangular ___*pyramidal*___ tracts which are the principal *(sensory? (motor?))* pathways. The main fibers that pass through the pyramids are the *(spinothalamic? (corticospinal?))* tracts.

c. Crossing (or ___*decussation*___) of fibers occurs in the medulla. This explains why movements of your right hand are initiated by motor neurons that originate in the *(right? (left?))* side of your cerebrum. The *((axons?) dendrites?)* of these motor neurons decussate in the medulla to proceed down the right lateral

___*cortico*___-spinal tract.

d. The medulla contains gray areas as well as white matter. Several important nuclei lie in the medulla. Two of these are synapse points for sensory axons that have ascended in the posterior columns of the cord. These two nuclei, named

___*cuneatus)*___ and ___*gracilis*___, transmit impulses for

sensations of _____, _____, and

_____.

e. A hard blow to the base of the skull can be fatal since the medulla is the site of

three vital centers: the _____ center, regulating the heart; the

_____ center, adjusting the rhythm of breathing; and the

_____ center, regulating blood pressure by altering diameter of blood vessels.

f. Some input to the medulla arrives by means of cranial nerves; these nerves may serve motor functions also. Which cranial nerves are attached to the medulla?

_____

(Functions of these nerves will be discussed later in this chapter.)

■ **B2.** Summarize important aspects of the pons in this learning activity.

a. The name *pons* means _____. It serves as a bridge in two ways.

It contains longitudinally arranged fibers that connect the _____

and _____ with the upper parts of the brain. It has transverse

fibers that connect the two sides of the _____.

b. Cell bodies associated with fibers in cranial nerves numbered

_____ lie in nuclei in the pons.

c. *(Respiration? Heartbeat? Blood pressure?)* is controlled by the pneumotaxic and apneustic areas of the pons.

■ **B3.** You maintain a conscious state, or you wake up from sleep, thanks to the regions

of the brain known as the RAS, or R_____ A_____

S_____. These areas are especially sensitive to sensations such as

_____ from ears and _____ or

_____ from skin. The RAS is part of the reticular formation.
Where is the reticular formation located?

■ **B4.** Relate the midbrain to the pons and medulla in this exercise.
   a. Like the pons and medulla, the midbrain is about 1 inch (____ cm) long.
   b. The midbrain is more *(anterior and superior? posterior and inferior?)* compared to the
      pons and medulla.

■ **B5.** Describe the thalamus in this exercise.
   a. If you look for the thalamus in most figures of the brain, you will not find it. It is lo-
      cated *(deep within? on the surface of?)* the cerebrum.
   b. The thalamus is H-shaped. The crossbar of the H is known as the

      _____ mass. It passes through the center of the slitlike

      _____ ventricle. (See Figure LG 14.1.) The two side bars of the
      H form the lateral walls of the third ventricle.
   c. The thalamus is the principal relay station for *(motor? sensory?)* impulses. For exam-
      ple, spinothalamic and lemniscal tracts convey general sensations such as pain,

      _____, _____, _____,

      and temperature to the thalamus where they are relayed to the cerebral cortex. Spe-
      cial sense impulses (for vision and hearing) are relayed through the *(geniculate? re-
      ticular? ventral posterior?)* nuclei of the thalamus.
   d. The thalamus *(does? does not?)* contain nuclei controlling motor functions. However,
      its principal role is conveying *(motor? sensory?)* impulses.

■ **B6.** The name hypothalamus indicates that this structure lies *(above? below?)* the
   thalamus, forming the floor and part of the lateral walls of the

   _____ ventricle.

B7. Expanding on the key words listed below, write a sentence describing major hypothalamic functions.

a. Regulator of visceral activities

b. Psychosomatic

c. Regulating factors to anterior pituitary

d. Feelings (rage)

e. Temperature

f. Thirst

g. Feeding and satiety center

h. Reticular formation: arousal and consciousness

■ **B8.** *For extra review.* Check your understanding of these parts of the brain stem and diencephalon by matching them with the descriptions given below.

H. Hypothalamus	P. Pons
Med. Medulla	T. Thalamus
Mid. Midbrain	

_____ a. It is the principal regulator of visceral activities since it acts as a liaison between cerebral cortex and autonomic nerves that control viscera.

_____ b. It is the site of the red nucleus, the origin of rubrospinal tracts concerned with muscle tone and posture.

_____ c. Cranial nerves III–IV attach to this brain part.

_____ d. Cranial nerves V–VIII attach to this brain part.

_____ e. Cranial nerves VIII–XII attach to this brain part.

_____ f. Feelings of hunger, fullness, and thirst stimulate centers here so that you can respond accordingly.

_____ g. All sensations except smell are relayed through here.

_____ h. Regulation of heart, blood pressure, and respiration occurs by centers located here.

_____ i. It constitutes four-fifths of the diencephalon.

_____ j. It lies under the third ventricle, forming its floor.

_____ k. It forms most of side walls of the third ventricle.

_____ l. Tumor in this region could compress cerebral aqueduct and cause internal hydrocephalus.

_____ m. The olivary and vestibular nuclei associated with equilibrium and posture control are located here.

_____ n. Mammillary bodies and supraoptic and preoptic regions are located here.

C. Brain: cerebrum   (pages 418–426)

■ **C1.** Complete this exercise about cerebral structure.

a. The outer layer of the cerebrum is called _____. It is composed of *(white? gray?)* matter. This means that it contains mainly *(cell bodies? tracts?)*.

b. In the margin, draw a line the same length as the thickness of the cerebral cortex. Use a metric ruler. Note how thin the cortex is.

c. The surface of the cerebrum looks much like a view of tightly packed mountains or

ridges, called _____. The parts where the cerebral cortex dips

down into valleys are called _____ (deep valleys) or

_____ (shallow valleys).

d. The cerebrum is divided into halves called _____. Connecting

them is a band of *(white? gray?)* matter called the _____. Notice this structure in Figure LG 14.1 and in Figure 14.8, page 416 of your text.

e. The falx cerebri is composed of *(nerve fibers? dura mater?)*. Where is it located?

_____ At its superior and inferior margins, the falx is dilated to form channels for venous blood flowing from the brain; these enclosures are

called _____.

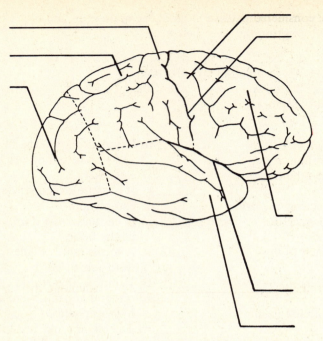

○ Motor speech (Broca's) area
○ Premotor area
○ Primary auditory area
○ Primary motor area

○ Primary somatosensory (general sensory) area
○ Primary visual area
○ Auditory association area

Figure LG 14.2 Right lateral view of lobes and fissures of the cerebrum. Label and color as directed in Checkpoints C2 and C8.

■ **C2.** Label the following structures using leader lines on Figure LG 14.2: *frontal lobe, occipital lobe, parietal lobe, temporal lobe, central sulcus, lateral cerebral sulcus, precentral gyrus, postcentral gyrus.*

■ **C3.** Match the three types of white matter fibers with these descriptions.

A. Association	C. Commissural	P. Projection

_____ a. The corpus callosum contains these fibers and connects the two cerebral hemispheres.

_____ b. Sensory and motor fibers passing between cerebrum and other parts of the CNS are this type of fiber; the internal capsule is an example.

_____ c. These fibers transmit impulses among different areas of the same hemisphere.

C4. List names of structures that are considered parts of the basal ganglia.

Now list two functions of the basal ganglia.

C5. Describe the limbic system in this exercise.

a. This system consists of portions of the cerebrum, thalamus, and hypothalamus. List several component structures.

b. The limbic system is shaped much like a _____ surrounding the brain stem.

c. Explain why the limbic system is sometimes called the "visceral" or "emotional" brain.

d. One other function of the limbic system is _____. Forgetfulness, such as inability to recall recent events, results partly from impairment of this system.

■ **C6.** Which condition can be described as visible bruising of the brain with extended loss of consciousness? *Concussion? Contusion?*

■ **C7.** Draw two important generalizations about brain functions in this activity.

a. In general, the anterior of the cerebrum is more involved with *(motor? sensory?)* control, whereas the posterior of the cerebrum is more involved with *(motor? sensory?)* functions. (Take a moment to visualize those activities taking place in the front and back of your own brain.)

b. As a general rule, *(primary? association?)* sensory areas receive sensations and *(primary? association?)* sensory areas are involved with interpretation and memory of sensations. For example, your ability to see the outline of the cerebrum, as in Figure LG 14.2, depends on your *(primary? association?)* visual areas. The fact that you can distinguish this diagram as a cerebrum (not a hand or heart), along with your memory of its structure for your next test, is based on the health of the neurons in your *(primary? association?)* visual areas.

■ **C8.** Color functional areas of the cerebral cortex listed with color code ovals on Figure 14.2.

■ **C9.** Check your understanding of functional areas of the cerebral cortex by doing this matching exercise. Answers may be used more than once.

> F. Frontal eye field
> G. Gnostic area
> M. Motor speech (Broca's) area
> PA. Primary auditory area
>
> PM. Primary motor area
> PO. Primary olfactory area
> PS. Primary somesthetic area
> PV. Primary visual area

_____ a. In the occipital lobe

_____ b. In the postcentral gyrus

_____ c. Receives sensations of pain, touch, pressure, and temperature

_____ d. In the parietal lobe

_____ e. In the temporal lobe; permits hearing

_____ f. Integrates general and special sensations to form a common thought about them

_____ g. In precentral gyrus of the frontal lobe; controls specific muscles or groups of muscles

_____ h. In the frontal lobe; translates thoughts into speech

_____ i. Controls smell

_____ j. Controls scanning movements of eyes, such as searching for a name in a telephone book

■ **C10.** Do this exercise about control of speech.

a. Language areas are located in the (_left?_ right?) cerebral hemisphere. This is true for most (_persons who are right-handed?_ (persons, regardless of handedness?))

b. The term _aphasia_ refers to inability to _speak_. Fluent aphasia means inability to ((articulate or form) understand?) words. _Word blindness_ refers to inability to understand (spoken? (written?)) words, whereas _word deafness_ is inability

to understand _spoken_.

■ **C11.** Identify the type of brain waves associated with each of the following situations.

> A. Alpha   D. Delta
> B. Beta   T. Theta

__T__ a. Occur in persons experiencing stress and in certain brain disorders.

__D__ b. Lowest frequency brain waves, normally occurring when adult is in deep sleep; presence in awake adult indicates brain damage.

__B__ c. Highest frequency waves, noted during periods of mental activity.

__A__ d. Present when awake but resting, these waves are intermediate in frequency between alpha and delta waves.

■ **C12.** _A clinical challenge._ You are helping to care for Laura, a hospital patient with a contusion of the right side of the brain. Which of the following would you be most likely to observe in Laura? Circle letters of correct answers.

A. Her right leg is paralyzed.
(B) Her left arm is paralyzed.
C. She cannot speak out loud to you nor write a note to you.
(D) She used to sing well, but cannot seem to stay on tune now.
E. She has difficulty with concepts involving numbers.

D. Brain: cerebellum (pages 427–429)

D1. Describe the cerebellum.

a. Where is it located? *posterior to the medulla & pons & inferior to the occipital lobes of the cerebrum*

b. Describe its structure. *somewhat like a butterfly*

c. Describe its functions. Use these key terms: *coordinated movements, posture, equilibrium,* and *emotions.* *it compares the intended movement determined by the motor units of the brain (cerebrum basal ganglia) with what is actually happening. Main brain region that regulates posture & balance.*

■ **D2.** Identify locations of each of the tracts known as *peduncles* that permit communication between the cerebellum and other brain parts. Fill in lines with the terms in the box.

inferior	middle	superior

a. Cerebellum ← _*superior*_ cerebullar peduncle → Midbrain

b. Cerebellum ← _*middle*_ cerebullar peduncle → Pons

c. Cerebellum ← _*inferior*_ cerebullar peduncle → Medulla

E. Neurotransmitters of the brain (pages 429–431)

■ **E1.** Answer these questions about neurotransmitters of the brain.

a. The number of neurotransmitters known (or strongly suspected) to be released by brain neurons is about *(5? 20?* (50?)). Neurotransmitters that cause opening or closing of ion channels in neuron plasma membranes result in *(*(fast)*? slow?)* synaptic transmission. Other neurotransmitters that utilize second messenger systems result in *(fast?* (slow?)) synaptic transmission.

b. Chemically, most neurotransmitters are *(large lipid?* (small, water-soluble?)) molecules, such as those described in Checkpoints E2 and E3.

amino acids
biogenic amines
neuropeptides

■ **E2.** Match the neurotransmitter or neuromodulator with the description that best fits. The first one is done for you. Lines at the end of each description are for Checkpoint E3.

> ACh. Acetylcholine
> DA. Dopamine
> EED. Enkephalins, endorphins, and dynorphins
> GABA. Gamma aminobutyric acid
> Ser. Serotonin
> SP. Substance P

__ACh__ a. Released at neuromuscular junctions and by many neurons, including those of pyramidal tracts **ACh**

_DA_ b. Produced by neurons that degenerate in Parkinson's disease _B_

GABA c. The most common inhibitory neurotransmitter in the brain _AA_

_DA_ d. Made by neurons involved with emotional responses and automatic movements of skeletal muscles _B_

EED e. Morphinelike chemicals that are the body's "natural painkillers" _N_

Ser f. Concentrated in brain stem neurons, this chemical helps control sleep, temperature, and mood _B_

ACh _GABA_

_DA_ g. Produced by neurons that are destroyed in Alzheimer's disease _ACh_

_SP_ h. Transmit pain-related impulses _N_

■ **E3.** Select answers in the box below to identify the correct category of each of the neurotransmitters and neuromodulators listed in Checkpoint E2. Write answers on lines after each of those descriptions. The first one is done for you.

> ACh. Acetylcholine
> AA. Amino acids
> B. Biogenic amines
> N. Neuropeptides

■ **E4.** Identify the neurotransmitters destroyed by the following enzymes.

a. Acetylcholinesterase (AChE) destroys the neurotransmitter named

ACh .

b. Catechol-*o*-methyltransferase (COMT) and monoamine oxidase (MAO) destroy neurotransmitters in the *catecholamine* category. List three such neurotransmitters.

norepinephrine   _epinephrine_   _dopamine_

E5. Discuss the relationship between endogenous opioids (such as endorphins) and biogenic amines (such as norepinephrine and serotonin) in the production of a "high" and/or decrease in depression associated with active exercise.

active exercise increases the release of endorphins Endorphins inhibit pain by blocking release of substance P Serotonin is thought

F. Cranial nerves (pages 431–435)

■ **F1.** Answer these questions about cranial nerves.

a. There are ____ pairs of cranial nerves. They are all attached to the

_____; they leave the _____ via foramina.

b. They are numbered by Roman numerals in the order that they leave the cranium. Which is most anterior? *(I? XII?)* Which is most posterior *(I? XII?)*

c. All spinal nerves are *(purely sensory? purely motor? mixed?)*. Are all cranial nerves mixed *(Yes? No?)*

■ **F2.** Complete the table about cranial nerves. (*For extra review* on locations of cranial nerves, refer back to Chapter 7, Checkpoint B7, page LG 126.)

Number	Name	Functions
a. I	Olfactory	*Smell*
b.	*optic*	Vision (not pain or temperature of the eye)
c. III	*Oculomotor*	
d.	Trochlear	
e. V	*trigeminal*	
f.	*a*	Stimulates lateral rectus muscle to abduct eye; proprioception of the lateral rectus
g.	Facial	
h.	*vest*	Hearing; equilibrium
i. IX	*g*	
j.	Vagus	
k. XI	*a*	
l.	*h*	Supplies muscles of the tongue with motor and sensory fibers

■ **F3.** Check your understanding of cranial nerves by completing this exercise. Write the name of the correct cranial nerve following the related description.

    a. Differs from all other cranial nerves in that it originates from the brainstem and

       from the spinal cord: _____

    b. Eighth cranial nerve (VIII): _____

    c. Is widely distributed into neck, thorax, and abdomen: _____

    d. Senses toothache, pain under a contact lens, wind on the face:

    _____

    e. The largest cranial nerve; has three parts (ophthalmic, maxillary, and mandibular):

    _____

    f. Controls contraction of muscle of the iris, causing constriction of pupil:

    _____

    g. Innervates muscles of facial expression: _____

    h. Two nerves that contain taste fibers and autonomic fibers to salivary glands:

    _____, _____

    i. Three purely sensory cranial nerves: _____,

    _____, _____

■ **F4.** Write the number of the cranial nerve related to each of the following disorders.

_____ a. Bell's palsy   _____ e. Blindness

_____ b. Inability to shrug shoulders or turn   _____ f. Trigeminal neuralgia (tic douloureux)
          head

_____ c. Anosmia   _____ g. Paralysis of vocal cords; loss of sensa-
                                              tion of many organs

_____ d. Strabismus and diplopia   _____ h. Vertigo and nystagmus

G. Aging and developmental anatomy of the nervous system   (pages 431–437)

■ **G1.** Do this exercise describing effects of aging on the nervous system.

    a. The total number of nerve cells ____-creases with age. Since conduction velocity *(in-*

      creases? slows down?), the time required for a typical reflex is ____-creased.

    b. Parkinson's disease is the most common *(motor? sensory?)* disorder in the elderly.

    c. The sense of touch is likely to be *(impaired? heightened?)* in old age, placing the
       older person at *(higher? lower?)* risk for burns or other skin injury.

■ **G2.** Check your understanding of the early development of the nervous system by completing this exercise.

   a. The nervous system begins to develop during the third week of gestation when the

      _____-derm forms a thickening called the

      _____ plate. Soon a longitudinal depression is found in the

      plate; this is the neural _____. As neural folds on the sides of

      the groove grow and meet, they form the neural _____.

   b. Three types of cells form the walls of this tube. Two of these are the marginal layer which forms *(gray? white?)* matter and the mantle layer which becomes *(gray? white?)* matter.

   c. The neural crest forms *(peripheral? central?)* nervous system structures such as

      ganglia as well as spinal and cranial _____.

   d. The anterior portion of the neural plate and tube develops into three enlarged

      _____ by the fourth week. These fluid-filled cavities eventually

      become the _____ which are filled with

      _____ fluid.

   e. By week five the primary vesicles have flexed (bent) to form a total of *(five? ten?)* secondary vesicles. (More about these in the next exercise.)

■ **G3.** Fill in the blanks in this table outlining brain development.

a. Primary Vesicles (3)	b. Secondary Vesicles (5)	c. Principal Parts of Brain Formed (7)
Prosencephalon		1A. _____
	1. Diencephalon →	1B. _____
(_____-brain) →		1C. _____
		2A. Cerebrum
	2. _____ →	2B. _____
_____-encephalon → (Midbrain)	3. Mesencephalon →	3A. _____
_____-encephalon (_____) →	4. _____-encephalon →	4A. _____
	5. Met-_____ →	5A. Pons
		5B. _____

H. Disorders, medical terminology (pages 438–440)

H1. CVA refers to _____.
CVAs are *(rare? common?)* brain disorders. Describe three causes of CVAs.

H2. Contrast CVA with TIA.

H3. Describe the following aspects of Alzheimer's disease (AD):
a. Progressive changes in behavior

b. Pathological findings of brain tissue

c. The possible roles of amyloid and A68 in the causation of AD and Down's syndrome.

H4. Brain tumors are more likely to result from *(neurons? neuroglia?)*. Write three signs or symptoms of brain tumors.

■ **H5.** Complete this Checkpoint about Parkinson's disease.

a. This condition involves a decrease in the neurotransmitter

_____. It is normally produced by cell bodies located in the

substantia nigra, which is part of the _____. Axons lead from

here to the _____ , where dopamine (DA) is released. DA is an
(excitatory? inhibitory?) transmitter.

b. The most common sign of parkinsonism is _____. Explain how
both tremor and muscle rigidity are related to low DA level.

c. Voluntary movements may be slower than normal; this is the condition of *(brady?
tachy?)*-kinesia.

d. Since persons with parkinsonism have a deficit in dopamine, it seems logical to treat
the condition by administration of this neurotransmitter. Does this work? *(Yes?
No?)* Explain.

e. Why do levodopa and/or anticholinergic medications provide some relief to these patients?

■ **H6.** Do this exercise describing some main points about multiple sclerosis (MS).

a. MS affects *(cell bodies? myelin sheaths?)* within the *(central? peripheral?)* nervous
system. At sites along the nerves where myelin is destroyed, many scars (or

_____) form; this is the basis for the name MS.

b. Initial symptoms are most likely to occur around age *(13? 33? 53? 73?)*. MS affects
(motor? sensory? both motor and sensory?) neurons. List signs or symptoms that may
suggest MS.

c. How do immunosuppressants such as ACTH or prednisone help MS?

H7. Contrast headaches of intracranial and extracranial origins.

■ **H8.** *For extra review.* Match the name of the disorder with the related description.

Cerebral palsy	Parkinsonism
Cerebrovascular accident (CVA)	Poliomyelitis
Dyslexia	Reye's syndrome
Multiple sclerosis	Tay-Sachs disease
Neuralgia	Transient ischemic attack (TIA)

a. Degeneration of myelin sheath to form hard plaques in many regions causing loss of motor and sensory function:

_____

b. Degeneration of dopamine-releasing neurons in basal ganglia; characterized by tremor or rigidity of muscles: _____

c. Temporary cerebral dysfunction caused by interference of blood supply to the brain.

_____

d. Most common brain disorder; also called stroke: _____

e. Characterized by difficulty in handling words and symbols, for example, reversal of letters *(b* for *d)*; cause unknown:

_____

f. Attack of pain along an entire nerve, for example, tic douloureux:

_____

g. A disease of children causing swelling of brain cells and fatty infiltration of liver; usually follows viral infection, especially if aspirin is

taken: _____

h. Inherited disease especially among Jewish populations due to excessive lipids in brain cells; affects infants:

_____

i. Of viral origin; often affects only the respiratory system; when nervous system involved, affects movement, but not sensation:

_____

j. Motor disorder caused by damage during fetal life, birth, or infancy; not progressive; apparent mental retardation often actually only speaking or hearing disability:

_____

A1. (a) Brain stem. (b) Diencephalon. (c) Cerebrum. (d) Cerebellum.

A2. Skull bones, meninges, cerebrospinal fluid (CSF).

A4. (a) B. (b) B. (c) A, C, D. (d) B, C, D. (e) B, D, E. (f) C.

A5. D E F G H I J K C A B.

A7. (a) Less; tight; astrocytes. (b) The brain is protected from many substances which could harm the brain, but are kept from the brain by this barrier. (c) Glucose, oxygen, water, ions such as Na$^+$ and K$^+$, and anesthetics. (d) Certain helpful medications, such as most antibiotic, and chemotherapeutic chemicals, cannot cross this barrier. (Fortunately, at the time they are most needed—as in a brain infection—the BBB is more permeable, and so may permit passage of these drugs.) (e) Alcohol, caffeine, nicotine, and heroin.

A8. (a) Hypothalamus, pineal gland, and pituitary. (b) Lack; advantage: since their capillaries are permeable to most substances, they can monitor homeostatic activities such as fluid balance and hunger; disadvantage: they are permeable to harmful substances, for example, possibly the AIDS virus.

B1. (a) Conduction; both sensory and motor. (b) Pyramidal, motor; corticospinal. (c) Decussation; left; axons, cortico. (d) Cuneatus, gracilis; two-point discriminatory touch, proprioception, vibrations. (e) Cardiac, respiratory, vasomotor (vasoconstrictor). (f) Part of VIII, as well as IX–XII.

B2. (a) Bridge; spinal cord, medulla; cerebellum. (b) V–VII and part of VIII. (c) Respiration.

B3. Reticular activating system; sound, temperature, and pain; medulla, pons, and midbrain, as well as portions of the diencephalon (thalamus and hypothalamus) and spinal cord.

B4. (a) 2.5. (b) Anterior and superior.

B5. (a) Deep within. (b) Intermediate, third. (c) Sensory; touch, pressure, proprioception; geniculate. (d) Does, sensory.

B6. Below, third.

B8. (a) H. (b) Mid. (c) Mid. (d) P. (e) Med. (f) H. (g) T. (h) Med. (i) T. (j) H. (k) T. (l) Mid. (m) Mid. (n) H.

C1. (a) Cerebral cortex; gray; cell bodies. (b) 2 to 4 mm. (c) Gyri; fissures, sulci. (d) Hemispheres; white, corpus callosum. (e) Dura mater; between cerebral hemispheres; superior and inferior sagittal sinuses (see Figure 14.2, page 407 in the text).

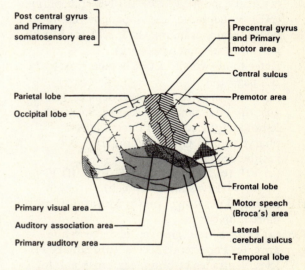

Figure LG 14.2A   Right lateral view of lobes and fissures of the cerebrum.

C2. See Figure LG 14.2A.

C3. (a) C. (b) P. (c) A.

C6. Contusion.

C7. (a) Motor, sensory. (b) Primary, association; primary; association.

C8. See Figure 14.2A.
C9. (a) PV. (b) PS. (c) PS. (d) PS. (e) PA. (f) G. (g) PM. (h) M. (i) PO. (j) F.
C10. (a) Left; persons, regardless of handedness. (b) Speak; understand; written; spoken.
C11. (a) T. (b) D. (c) B. (d) A.
C12. B D
D2. (a) Superior. (b) Middle. (c) Inferior. (Note that the names of these tracts are logical, based on locations of the midbrain, pons, and medulla.)
E1. (a) 50; fast; slow. (b) Small, water-soluble.
E2. (b) DA, B. (c) GABA, AA. (d) DA, B. (e) EED, N. (f) Ser, B. (g) ACh, ACh. (h) SP, N.
E3. See E2.
E4. (a) Acetylcholine (ACh). (b) Norepinephrine (NE), epinephrine, and dopamine (DA).
F1. (a) 12; brain; cranium. (b) I; XII. (c) Mixed; no (three are purely sensory).
F2. Refer to answers to LG Chapter 7, Activity B7, page 133.
F3. (a) Accessory. (b) Vestibulocochlear. (c) Vagus. (d) Trigeminal. (e) Trigeminal. (f) Oculomotor. (g) Facial. (h) Facial, glossopharyngeal. (i) Olfactory, optic, vestibulocochlear.
F4. (a) VII. (b) XI. (c) I. (d) III. (e) II. (f) V. (g) X. (h) VIII (vestibular branch).
G1. (a) De; slows down; in. (b) Motor. (c) Impaired, higher.
G2. (a) Ecto, neural; groove; tube. (b) White, gray. (c) Peripheral, nerves. (d) Primary vesicles; ventricles of brain, cerebrospinal. (e) Five.

G3.

a. Primary Vesicles (3)	b. Secondary Vesicles (5)	c. Principal Parts of Brain Formed (7)
Prosencephalon (__Fore__ -brain) →	1. Diencephalon →	1A. __Thalamus__ 1B. __Hypothalamus__ 1C. __Pineal gland__
	2. __Telencephalon__ →	2A. Cerebrum 2B. __Basal ganglia__
__Mes__ -encephalon → (Midbrain)	3. Mesencephalon →	3A. __Midbrain__
__Rhomb__ -encephalon (__Hindbrain__) →	4. __Myel__ -encephalon →	4A. __Medulla oblongata__
	5. Met-__encephalon__ →	5A. Pons 5B. __Cerebellum__

H5. (a) Dopamine; midbrain; basal ganglia; inhibitory. (b) Tremor; lack of inhibitory DA causes a double-negative effect: extra unnecessary movements occur (tremor), which may then interfere with normal movements (rigidity) (c) Brady. (d) No. DA cannot cross the blood-brain barrier. (e) Levodopa (L-dopa) is a precursor (leads to formation) of DA; parkinsonism involves an excessive amount of acetylcholine (ACh) relative to the amount of dopamine (DA).
H6. (a) Myelin sheaths, central; multiple scleroses. (b) 33; both motor and sensory; muscle weakness, double vision, bladder infections. (c) Suppress autoimmune response in which the body is making antibodies which destroy the oligodendrocytes necessary for CNS myelin formation.
H8. (a) Multiple sclerosis. (b) Parkinsonism. (c) TIA. (d) CVA. (e) Dyslexia. (f) Neuralgia. (g) Reye's syndrome. (h) Tay-Sachs disease. (i) Poliomyelitis. (j) Cerebral palsy.

Questions 1–9: Circle the letter preceding the one best answer to each question.

1. All of these are located in the medulla *except:*
   A. Nucleus cuneatus and nucleus gracilis
   B. Cardiac center, which regulates heart
   C. Site of decussation (crossing) of pyramidal tracts
   D. The olive
   E. Origin of cranial nerves V to VIII

2. Which of these is a function of the postcentral gyrus?
   A. Controls specific groups of muscles, causing their contraction
   B. Receives general sensations from skin, muscles, and viscera
   C. Receives olfactory impulses
   D. Primary visual reception area
   E. Somesthetic association area

3. All of these structures contain cerebrospinal fluid *except:*
   A. Subdural space
   B. Ventricles of the brain
   C. Central canal of the spinal cord
   D. Subarachnoid space

4. Damage to the accessory nerve would result in:
   A. Inability to turn the head or shrug shoulders
   B. Loss of normal speech function
   C. Hearing loss
   D. Changes in heart rate
   E. Anosmia

5. Which statement about the trigeminal nerve is *false?*
   A. It sends motor fibers to the muscles used for chewing.
   B. Pain in this nerve is called trigeminal neuralgia (tic douloureux).
   C. It carries sensory fibers for pain, temperature, and touch from the face, including eyes, lips, and teeth area.
   D. It is the smallest cranial nerve.

6. All of these are functions of the hypothalamus *except:*
   A. Control of body temperature
   B. Release of chemicals (regulating factors) that affect release or inhibition of hormones
   C. Principal relay station for sensory impulses
   D. Center for mind-over-body (psychosomatic) phenomena
   E. Involved in maintaining sleeping or waking state

7. Damage to the occipital lobe of the cerebrum would most likely cause:
   A. Loss of hearing
   B. Loss of vision
   C. Loss of ability to smell
   D. Paralysis
   E. Loss of feeling in muscles (proprioception)

8. Choose the one *false* statement.
   A. Multiple sclerosis is a progressive disorder that worsens as time lapses.
   B. Cerebral palsy (CP) is a progressive disorder that gets worse with time.
   C. About 70 percent of CP victims appear mentally retarded, but, in fact, this may be a reflection of inability to speak or walk well.
   D. Poliomyelitis may affect motor nerves, but does not affect sensations.

9. All of the following neurotransmitters are classified as catecholamines *except:*
   A. Acetylcholine
   B. Epinephrine
   C. Norepinephrine
   D. Dopamine

_____ _____ _____ 10. From superior to inferior:
    A. Thalamus
    B. Hypothalamus
    C. Corpus callosum

_____ _____ _____ 11. Order in which impulses are relayed in conduction pathway for vision:
    A. Optic nerve
    B. Optic tract
    C. Optic chiasma

_____ _____ _____ _____ 12. Pathway of cerebrospinal fluid, from formation to final destination:
    A. Choroid plexus in ventricle
    B. Subarachnoid space
    C. Cranial venous sinus
    D. Arachnoid villi

_____ _____ _____ 13. From anterior to posterior:
    A. Fourth ventricle
    B. Pons and medulla
    C. Cerebellum

Questions 14–20: Circle T (true) or F (false). If the statement is false, change the underlined word or phrase so that the statement is correct.

T   F   14. The thalamus, hypothalamus, and cerebrum are all developed from the <u>forebrain (prosencephalon)</u>.

T   F   15. <u>Endorphins, enkephalins, and dopamine</u> are all chemicals which are considered the "body's own pain killers."

T   F   16. The language areas are located in the <u>cerebellar cortex.</u>

T   F   17. Cranial nerves I, III, and VIII are all <u>purely sensory</u> nerves.

T   F   18. The limbic system functions in control of <u>emotional aspects of behavior.</u>

T   F   19. <u>GABA, serotonin, and acetylcholine may</u> all function as excitatory transmitter substances.

T   F   20. <u>Delta</u> brain waves have the lowest frequency of the four kinds of waves produced by normal individuals.

Questions 21–25: fill-ins. Complete each sentence with the word or phrase that best fits.

_____ 21. The ____ is the extension of dura mater that separates the cerebral hemispheres, whereas the ____ separates the cerebrum from the cerebellum.

_____ 22. The superior and inferior colliculi, associated with movements of eyeballs and head in response to visual stimuli, are located in the ____.

_____ 23. If the brain is deprived of oxygen for more than 4 minutes, ____ of brain cells release enzymes that cause these cells to self-destruct.

_____ 24. The ____ controls wakefulness.

_____ 25. The corpus striatum, including the caudate and lentiform nuclei, is part of the ____.

Multiple Choice

1. E   6. C
2. B   7. B
3. A   8. B
4. A   9. C
5. D

Arrange

10. C A B
11. A C B
12. A B D C
13. B A C

True-False

14. T
15. F. Endorphins, enkephalins, and dynorphin
16. F. Cerebral cortex
17. F. I, II, and VIII
18. T
19. F. Serotonin and acetylcholine
20. T

Fill-ins

21. Falx cerebri; tentorium cerebelli
22. Midbrain
23. Lysosomes
24. Reticular formation or reticular activating system (RAS)
25. Basal ganglia

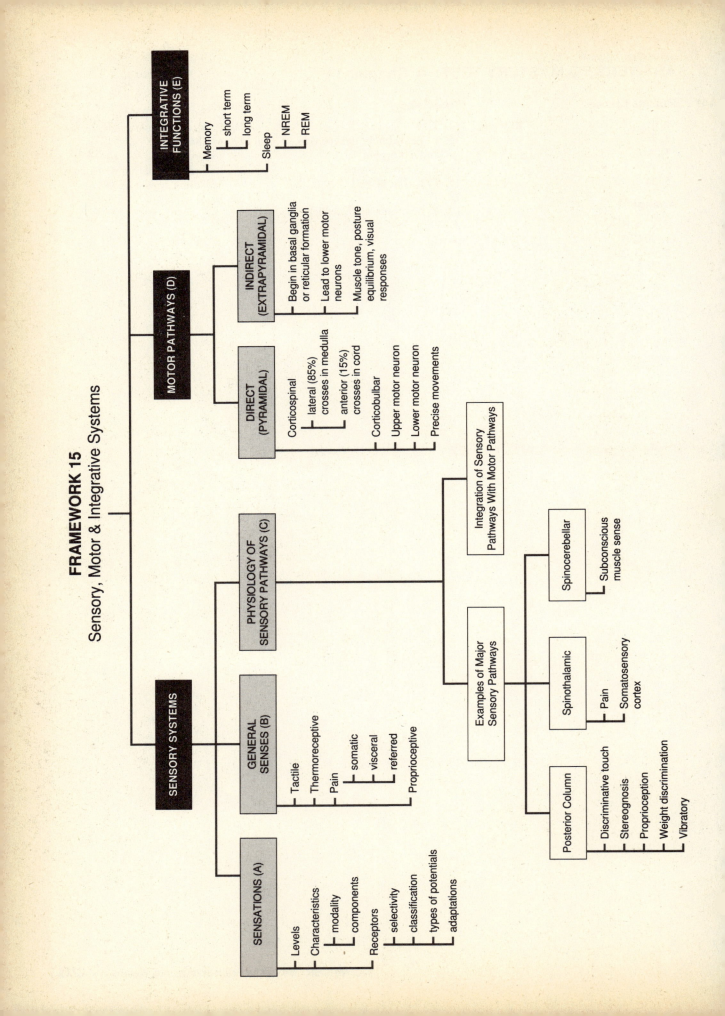

FRAMEWORK 15
Sensory, Motor & Integrative Systems

SENSORY SYSTEMS

SENSATIONS (A)
- Levels
- Characteristics
  - modality
  - components
- Receptors
  - selectivity
  - classification
  - types of potentials
  - adaptations

GENERAL SENSES (B)
- Tactile
- Thermoreceptive
- Pain
  - somatic
  - visceral
  - referred
- Proprioceptive

PHYSIOLOGY OF SENSORY PATHWAYS (C)

Integration of Sensory Pathways With Motor Pathways

Examples of Major Sensory Pathways

Posterior Column
- Discriminative touch
- Stereognosis
- Proprioception
- Weight discrimination
- Vibratory

Spinothalamic
- Pain
- Somatosensory cortex

Spinocerebellar
- Subconscious muscle sense

MOTOR PATHWAYS (D)

DIRECT (PYRAMIDAL)
- Corticospinal
  - lateral (85%) crosses in medulla
  - anterior (15%) crosses in cord
- Corticobulbar
- Upper motor neuron
- Lower motor neuron
- Precise movements

INDIRECT (EXTRAPYRAMIDAL)
- Begin in basal ganglia or reticular formation
- Lead to lower motor neurons
- Muscle tone, posture equilibrium, visual responses

INTEGRATIVE FUNCTIONS (E)
- Memory
  - short term
  - long term
- Sleep
  - NREM
  - REM

Sensory, Motor, and Integrative Systems

The past three chapters have introduced the components of the nervous system and demonstrated the arrangement of neurons in the spinal cord, brain, spinal nerves, and cranial nerves. Chapter 15 integrates this information in the study of major nerve pathways including those for (a) general senses (such as touch, pain, and temperature), (b) motor control routes for precise, conscious movements and for unconscious movements, and (c) complex integrative functions such as memory and sleep.

To begin this chapter, consider the organization of concepts in the Chapter 15 Framework and note key terms associated with each section.

TOPIC OUTLINE AND OBJECTIVES

A. Sensations

1. Define a sensation and list the characteristics of sensations.
2. Describe the classification of receptors.

B. General senses

3. List the location and function of the receptors for tactile sensations (touch, pressure, vibration), thermoreceptive sensations (heat and cold), and pain.
4. Distinguish somatic, visceral, referred, and phantom pain.
5. Identify the proprioceptive receptors and indicate their functions.

C. Physiology of sensory pathways

6. Discuss the neuronal components and functions of the posterior column–medial lemniscus and the spinothalamic and spinocerebellar pathways.

7. Describe the integration of sensory input and motor output.

D. Motor pathways

8. Compare the location and functions of the direct and indirect motor pathways.
9. Explain how the basal ganglia and cerebellum are related to motor responses.

E. Integrative functions

10. Compare integrative functions such as memory, wakefulness, and sleep.

WORDBYTES

Now study the following parts of words that may help you better understand terminology in this chapter.

Wordbyte	Meaning	Example	Wordbyte	Meaning	Example
a-, an	without	*an*esthesia	-esthesia	sensation	an*esthesia*, par*esthesia*
-algia	pain	neur*algia*, an*alge*sia	ipsi-	same	*ipsi*lateral
-ceptor	receiver	proprio*ceptor*	para-	abnormal	*par*esthesia
contr-	against, opposite	*contra*lateral	proprio-	one's own	*proprio*ception
			rhiz-	root	*rhiz*otomy

CHECKPOINTS

A. Sensations (pages 444–446)

■ **A1.** Review the three basic functions of the nervous system by listing them in sequence here:

a. Receiving _____ input

b. _____ information

c. Transmitting _____ impulses to _____ or

_____

A2. For a moment, visualize what your life would be like if you were unable to experience any sensations. List the three types of sensations that you believe you would miss most.

_____ _____ _____

Now write a sentence describing how your health and safety might be endangered by lack of ability to perceive sensations.

■ **A3.** Do this activity about levels of sensation.

a. *(Perception? Sensation?)* is the awareness (either conscious or unconscious) of stimuli, such as the act of seeing a bird, whereas _____ is the conscious awareness and interpretation of sensation, such as the recognition of that bird as a bluebird and not a robin.

b. Identify levels of sensation by matching parts of the nervous system listed in the box with descriptions below.

B. Brain stem	S. Spinal cord
C. Cerebral cortex	T. Thalamus

_____ 1. Immediate reflex action is possible without involvement of the brain.

_____ 2. Involves subconscious motor reactions, such as those that facilitate balance and equilibrium.

_____ 3. Offers a general (or "crude") awareness of locale and type of sensations, such as awareness of pressure within the leg or that something is being heard; offers no specifics.

_____ 4. Gives precise information about sensations, such as distinguishing Bach from Baez from Buffett, as well as memories of hearing a particular musical work.

■ **A4.** *A clinical challenge.* Mr. Hamilton was in a motorcycle accident that resulted in his left leg being severed above the knee. Months later he is referred to the university pain clinic for an assessment. He states, "I feel this pain in my left foot, and sometimes my toes itch like crazy, but they're not there to scratch!" How might you interpret his experience?

■ **A5.** Do this exercise on the characteristics of sensations.

a. The quality that makes the sensation of pain different from the sense of touch or

 which makes hearing different from vision is called the _____
 of sensation. Any single sensory neuron carries sensations of *(one modality? many modalities?)*. Different modalities of sensations are distinguished precisely at the level of the *(spinal cord? thalamus? cerebral cortex?)*.

b. Recall getting dressed this morning. The fact that you felt the shirt touching your back just after you put on that shirt, but that the awareness of that touch has dissi-

 pated over time is known as the characteristic of _____. Circle the type of receptors that adapt most slowly. *(Pain? Pressure? Smell? Touch?)* Explain how such slow adaptation is advantageous to you.

c. Describe what is meant by the *selectivity* of receptors.

 Is selectivity absolute? *(Yes? No?)* Explain.

■ **A6.** Arrange in correct order the components in the pathway of sensation.

C. Conduction	Td. Transduction
S. Stimulation	Tl. Translation

1._____ → 2. _____ → 3. _____ → 4. _____
Now match each of these components of sensation with the correct description below.

_____ a. Activation of a neuron

_____ b. Stimulus → generator potential

_____ c. Generator potential → nerve impulse; impulse is conveyed along a first-order neuron into the spinal cord or brainstem up to a higher level of the brain.

_____ d. Nerve impulse → sensation; usually occurs in the cerebral cortex

■ **A7.** Identify the class of receptor that fits each description.

| E. Exteroceptor | I. Interoceptor | P. Proprioceptor |

_____ a. These receptors inform you that you are hungry or thirsty; these also inform your brain of need for adjustment of blood pressure.

_____ b. Your ears, eyes, and receptors in your skin for pain, touch, hot, and cold are of this type.

_____ c. With your eyes closed, you can tell your exact body position, thanks to these receptors.

■ **A8.** Complete these sentences describing types of receptors based on nature of stimulus detected.

a. You can maintain balance by means of inner ear _____-receptors sensitive to change in position.

b. Different tastes and smells are distinguished with the help of

_____-receptors in nose and mouth.

c. _____-receptors inform you of the temperature around you.

■ **A9.** Choose the descriptions associated with general (G) or special (S) receptors.

_____ a. These are numerous and widespread in the body.

_____ b. Relatively complex receptors are of this type.

_____ c. Sight, hearing, smell, and taste involve receptors of this type.

_____ d. Sensations of the skin (cutaneous) are of this type.

■ **A10.** Contrast generator potential (GP) with receptor potential (RP) by completing this table.

	GP	RP
a. Triggers action potential. *(Yes? No?)*		
b. Most receptors for *(general? special?)* senses are of this type.		
c. *(Always excitatory? Always inhibitory? Excitatory or inhibitory?)*		

B. General senses   (pages 446–451)

■ **B1.** List six examples of cutaneous sensations.

■ **B2.** Do the following exercise about tactile sensations.

a. Name the tactile sensations.

These sensations are all sensed by _____-receptors. Receptors
for these sensations are *(evenly? unevenly?)* distributed throughout the skin of the
body.

b. *(Discriminative? Light?)* touch refers to ability to determine that something has
touched the skin, but its precise location, shape, size, or texture cannot be distinguished.

c. In general, pressure receptors are more *(superficial? deep?)* in location than touch receptors.

d. Sensations of _____ result from rapid repetition of sensory impulses from tactile receptors.

■ **B3.** Match names of receptors with their descriptions. Answers may be used more than
once.

C. Corpuscles of touch (Meissner's corpuscles)	N. Nociceptors
H. Hair cells of the inner ear	T. Tendon organs
J. Joint kinesthetic receptors	Type I. Type I cutaneous mechanoreceptors (tactile or Merkel discs)
L. Lamellated (Pacinian) corpuscles	Type II. Type II cutaneous mechanoreceptors (end organs of Ruffini)
M. Muscle spindles	

_____ a. Egg-shaped masses located in dermal
papillae, especially in fingertips, palms
of hands, and soles of feet

_____ b. Onion-shaped structures sensitive to
pressure and high-frequency vibration

_____ c. Free nerve endings that sense pain

_____ d. Slowly adapting touch receptors (two
answers)

_____ e. Proprioreceptors (four answers)

_____ f. May respond to any type of stimulus
if stimulus is strong enough to cause
tissue damage

_____ g. Respond to chemicals released from
injured tissue, such as prostaglandins
and kinins

_____ h. Type Ia and type II fibers that are sensitive to stretch

_____ i. Sensitive to tension; monitor the force
of muscle contraction

B4. *For extra review.* Refer to Figure LG 5.1 (page 87). Identify and label types of receptors on that figure. Add your own drawings of three other types of receptors in appropriate layers of skin. Label those receptors also.

B5. Contrast terms in each pair:

a. Visceral pain/somatic pain

b. General anesthesia/spinal anesthesia

■ **B6.** *A clinical challenge.* Explain why patients experiencing visceral pain (as during a heart attack or gallbladder attack) may feel pain in locations quite distant from those two organs.

   a. Pain impulses that originate in the heart, for example, during a "heart attack," enter the spinal cord at the same level as do sensory fibers from skin covering the

   _____. This level of the cord is about _____ to _____.

   b. Refer to Figure 15.2 (page 449 in your text). Notice that some liver and gallbladder pain is felt in a region quite distant from these organs, that is, to the

   _____ region. These organs lie just inferior to the dome-shaped

   _____, which is supplied by the phrenic nerve from the *(cervical? brachial? lumbar?)* plexus. A painful gallbladder can send impulses to the cord

   via the phrenic nerve, which enters the cord at the same level (about C _____ or

   C _____) as nerves from the neck and shoulder. So gallbladder pain is said to be

   _____ to the neck and shoulder area.

C. Physiology of sensory pathways   (pages 451–454)

■ **C1.** Arrange these levels of sensations from first to last in the pathway of sensation. Sensory impulses:

   ___ ___ ___

A. Cross to the opposite side of the spinal cord or brainstem and ascend to the thalamus
B. Reach the spinal cord or brainstem
C. Pass to the somatosensory region of the cerebral cortex and also travel (via collaterals) to the cerebellum and reticular formation

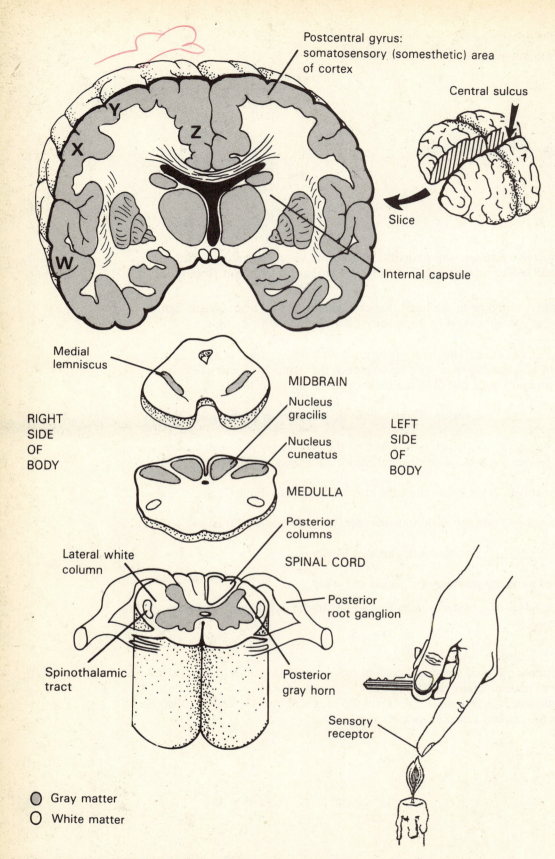

Postcentral gyrus:
somatosensory (somesthetic) area
of cortex

Central sulcus

Slice

Internal capsule

Medial
lemniscus

MIDBRAIN

RIGHT
SIDE
OF
BODY

Nucleus
gracilis

Nucleus
cuneatus

LEFT
SIDE
OF
BODY

MEDULLA

Posterior
columns

SPINAL CORD

Lateral white
column

Posterior
root ganglion

Spinothalamic
tract

Posterior
gray horn

Sensory
receptor

○ Gray matter
○ White matter

Figure LG 15.1 Diagram of the central nervous system. Refer to Checkpoints C2 and C3.

■ **C2.** Do this learning activity about sensory reception in the cerebral cortex.

a. Picture the brain as roughly similar in shape to a small, rounded loaf of bread. The crust would be comparable to the cerebral _____*cortex*_____.

b. Suppose you cut a slice of bread in midloaf that represents the postcentral gyrus. It looks somewhat like the "slice" or section of brain shown at the top of Figure LG 15.1. Locate the Z on that figure. This portion (next to the _____*lateral*_____ fissure) receives sensations from the *(face? hand? hip? foot?)*. (Refer to Figure 15.6, page 453 of your text, for help.)

c. Suppose a person suffered a loss of blood supply (CVA or "stroke") to the area of the brain marked X in Figure LG 15.1. Loss of sensation in the *(face? hand? leg?)* area would result. Damage to the area marked *W* would be likely to lead to loss of the sense of *(vision? hearing? taste?)* since the W is located on the temporal lobe.

d. The letters W, X, Y, and Z are marked on the *(right? left?)* side of the brain. (Note that this is an anterior view of this brain section. If a friend stands and faces you, the *right* side of your friend's brain will be in your *left* field of view.) What effects would brain damage to area *Y* of the postcentral gyrus have? Loss of *(sensation? movement?)* to the *(face? hand? foot?)* on the *(right? left?)* side of the body.

e. Certain regions of the body are represented by large areas of the somesthetic or _____ cortex, indicating that these areas *(are? are not?)* very sensitive. Name two of these areas.

■ **C3.** Check your understanding of pathways by drawing them according to the directions below. Try to draw them from memory; do not refer to the text figures or the table until you finish. Label the *first-, second-,* and *third-order neurons* as I, II, and III.

a. Define *stereognosis.*

Draw in *blue* on Figure LG 15.1 the pathway that allows you to accomplish stereognosis so that you can recognize the shape of a key in your left hand. (*Hint:* The correct recognition of the key involves discriminative touch, proprioception, and weight discrimination.)

b. Now draw in *red* the pathway for sensing pain and temperature as the index finger of the left hand comes too close to a flame. For simplification, draw only the lateral (not anterior) component.

C4. Contrast: *acute pain/chronic pain.*

Which type of pain impulses travel along myelinated A nerve fibers? *(Acute? Chronic?)*

■ **C5.** *A clinical challenge.* Match methods of pain relief listed in the box with the following descriptions.

A. Anesthetic (Novocaine)	O. Opioids such as morphine,
AAl. Aspirin, acetaminophen (Tylenol),	codeine, or meperidine (Demerol)
ibuprofen (Motrin, Advil)	R. Rhizotomy
C. Cordotomy	

_____ a. Surgical technique that severs spinal cord pathways for pain, such as spinothalamic tracts.

_____ b. Surgical technique involving cutting of posterior (sensory) root of spinal nerve(s)

_____ c. Medication that alters the quality of pain perception, so that pain feels less noxious

_____ d. Medication that inhibits formation of chemicals (prostaglandins) that stimulate pain receptors

_____ e. Medication that blocks conduction of nerve impulses in first-order neurons

■ **C6.** Answer these questions about the spinocerebellar tracts.

a. They are concerned with *(conscious? subconscious?)* muscle sense.

b. The spinocerebellar tracts permit reflex adjustments for _____

and _____ .

c. The left posterior spinocerebellar tract conveys impulses from muscles on *(the left side? the right side? both sides?)* of the body.

d. The left anterior spinocerebellar tract conveys impulses from muscles on *(the left side? the right side? both sides?)* of the body.

C7. Does every bit of sensory input that reaches the CNS elicit a motor response? *(Yes? No?)* Discuss integration of sensory input and motor output.

■ **C8.** The linking of sensory to motor response occurs at different levels in the central nervous system. Indicate which is the highest level required for each of these responses.

BG. Basal ganglia	CC. Cerebral cortex
C. Cerebellum	S. Spinal cord

_____ a. Quick withdrawal of a hand when it touches a hot object

_____ b. Unconscious responses to proprioceptive impulses to permit smooth, coordinated movements (two answers)

_____ c. Playing piano or writing a letter

D. Physiology of motor pathways (pages 454–461)

■ **D1.** Refer to Figure LG 15.2 and do the following exercise about motor control.

   a. This "slice" or section of the brain containing the primary motor cortex is located in the *(frontal? parietal?)* lobe.

   b. Area P in this figure controls *(sensation? movement?)* to the *(left? right?)* *(arm? leg? side of the face?)*. Motor neurons in area P would send impulses to the pons to

   activate cranial nerve _____ to facial muscles.

   c. Describe effects of damage to area R.

■ **D2.** Show the route of impulses along the principal pyramidal pathway by listing in correct sequence the structures that comprise the pathway. Refer to Figure 15.8 (page 455 in the text) if you have difficulty with this activity.

___ ___ ___ ___ ___ ___ ___ ___

   A. Anterior gray horn (lower motor neuron)
   B. Midbrain and pons
   C. Effector (skeletal muscle)
   D. Internal capsule

   E. Lateral corticospinal tract
   F. Medulla, decussation site
   G. Precentral gyrus (upper motor neuron)
   H. Ventral root of spinal nerve

■ **D3.** Now draw this pathway on Figure LG 15.2. Show the control of muscles in the left hand by means of neurons in the lateral corticospinal pathway (pyramidal tracts proper). Label the *upper* and *lower motor neurons*.

■ **D4.** Complete this exercise contrasting anterior with lateral corticospinal tracts.

   a. Only about ____ percent of the upper motor neurons pass through anterior corticospinal tracts; ____ percent pass through the large _____ corticospinal (pyramidal) tracts.

   b. Left anterior corticospinal tracts consist of axons that originated in upper motor neurons on the *(right? left?)* side of the motor cortex.

■ **D5.** Give an anatomical explanation for the fact that a 6-month-old infant cannot be expected to walk.

■ **D6.** State the function of *corticobulbar tracts*.

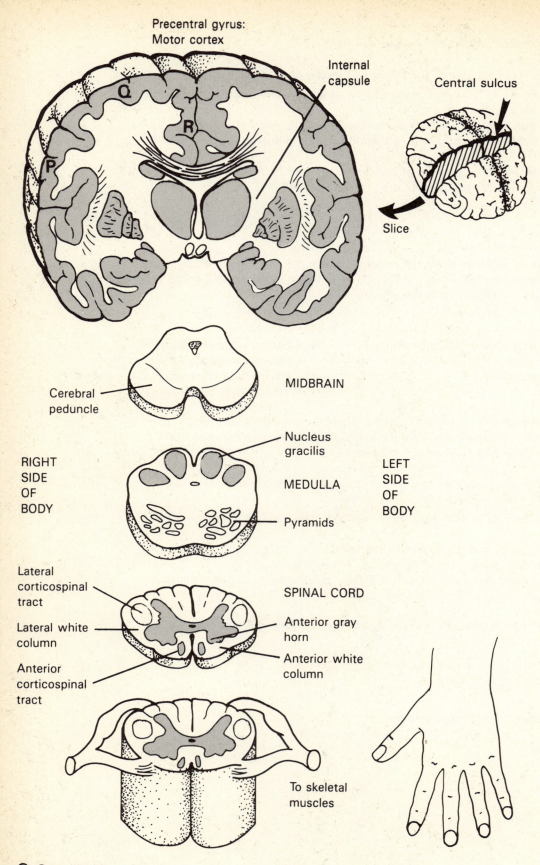

Precentral gyrus: Motor cortex

Internal capsule

Central sulcus

Slice

Q

R

P

Cerebral peduncle

MIDBRAIN

Nucleus gracilis

RIGHT SIDE OF BODY

MEDULLA

LEFT SIDE OF BODY

Pyramids

Lateral corticospinal tract

SPINAL CORD

Lateral white column

Anterior gray horn

Anterior corticospinal tract

Anterior white column

To skeletal muscles

⬤ Gray matter

◯ White matter

Figure LG 15.2 Diagram of the central nervous system. Refer to Checkpoints D1 and D3.

■ **D7.** Do this exercise on *upper motor neurons* and *lower motor neurons.*

   a. Upper motor neurons are located in the *(brain stem? basal ganglia? motor cortex?)*,

   whereas lower motor neurons are in _____ or

   _____.

   b. Destruction of *(upper? lower?)* motor neurons results in spastic paralysis since the brain is not controlling movements. Flaccid paralysis results from damage of *(upper? lower?)* motor neurons.

   c. Why are lower motor neurons called the *final common pathway?*

■ **D8.** *A clinical challenge.* Describe two conditions that involve damage to basal ganglia in this exercise.

   a. *(Huntington's? Parkinson's?)* disease involves changes in basal ganglia associated with decrease in the neurotransmitter *(acetylcholine? dopamine? GABA?)*. Signs of this condition are *(jerky movements and facial twitches? muscle rigidity and tremors?)*.

   b. A predominant sign of Huntington's disease (HD) is jerky movements (or

   "dance") known as _____ related to loss of neurons that make

   the inhibitory transmitter named _____. Signs of this progressive condition are most likely to be first noted at about age *(4 years? 40 years? 70 years?)*. A child of a parent with HD has *(almost no? a 50–50?)* chance of having this condition.

D9. Summarize how the basal ganglia and cerebellum, together with associated neurons, help you to carry out coordinated, precise movements, such as those involved in a game of tennis.

D10. Assess a study partner for cerebellar damage. *Note:* Signs of such damage can be expected to be *(contra-? ipsi?)*-lateral.

■ **D11.** Describe locations of indirect (extrapyramidal) tracts in this exercise:

a. All of these tracts lie *(within? outside of?)* corticospinal tracts. All of them originate in the *(brainstem? basal ganglia? cerebral cortex?)* and end in the

_____.

b. Fill in names of extrapyramidal tracts.

1. Midbrain → _____-spinal → spinal cord

2. Midbrain → _____-spinal → spinal cord

3. Pons → _____-spinal → spinal cord

4. Medulla → _____-spinal → spinal cord

5. Medulla → _____-spinal → spinal cord

E. Integrative functions   (pages 461–465)

■ **E1.** List three types of activities that require complex integration processes by the brain.

E2. Define these terms: *learning, memory, engram.*

E3. List several parts of the brain that are considered to assist with memory.

■ **E4.** Do this exercise on memory.

a. Looking up a phone number, dialing it, and then quickly forgetting it is an example of *(short-term? long-term?)* memory. One theory of such memory is that the phone

number is remembered only as long as a _____ neuronal circuit is active.

b. Repeated use of a telephone number can commit it to long-term memory. Such

reinforcement is known as memory _____. *(Most? Very little?)* of the information that comes to conscious attention goes into long-term memory, since the brain is selective about what it retains.

c. Some evidence indicates that *(short-term? long-term?)* memory involves electrical and chemical events rather than anatomical changes. Supporting this idea is the fact that chemicals used for anesthesia and shock treatments interfere with *(recent? long-term?)* memory.

E5. Discuss the following anatomical or biochemical changes that may be associated with enhanced long-term memory:

a. Presynaptic neurons and neurotransmitter

b. RNA

c. NMDA receptors

■ **E6.** Complete this exercise about sleep and wakefulness.

a. _____ *rhythm* is a term given to the usual daily pattern of sleep and wakefulness in humans.

b. Whether you are awake or asleep depends upon whether the reticular formation, also known as the _____ system (RAS) is active. The RAS has two principal parts; the *(thalamic? pons and midbrain?)* portion is responsible for your waking from a deep sleep, and the _____ portion helps you to maintain a conscious state.

c. List three types of sensory input that can stimulate the RAS.

d. Stimulation of the reticular formation leads to *(increased? decreased?)* activity of the cerebral cortex, as indicated by _____ recordings of brain waves.

e. Continued feedback between the RAS, the _____, and the _____ maintains the state called *consciousness*. Inactivation of the RAS produces a state known as _____.

■ **E7.** Do the following learning activity about sleep.

a. Name two kinds of sleep.

b. *(REM? NREM?)* sleep is required first in order for *(REM? NREM?)* sleep to occur.

c. In NREM sleep, a person gradually progresses from stage *(1 to 4? 4 to 1?)* into deep sleep. Alpha waves are present in stage(s) *(1? 3 and 4?)*, whereas slow delta waves characterize stage(s) *(1? 3 and 4?)* sleep.

d. REM sleep is also called _____ sleep since it follows, but is much more active than, the deep stage of sleep. Respirations and pulse are much

____-creased over their levels in stage 4; alpha waves are present as in stage ____

sleep. The name REM indicates that _____ movements can be observed. Dreaming occurs during *(stages 1–4 of NREM? REM?)* sleep.

e. Periods of REM and NREM sleep alternate throughout the night in about ____-minute cycles. During the early part of an 8-hour sleep period REM periods last about *(5–10? 30? 50?)* minutes; they gradually increase in length until the final REM

period lasts about ____ minutes.

■ **E8.** *A clinical challenge.* Do this exercise on insomnia.

a. Mrs. Douglas reports to the clinical nurse specialist (CNS) that she is having difficulty sleeping. What questions do you think the CNS might ask Mrs. Douglas to assess the insomnia?

b. Mrs. Douglas states that she has been trying to sleep by using either "a sleeping pill, a couple of glasses of wine, or both." What suggestions do you imagine that the CNS may make to this client?

ANSWERS TO SELECTED CHECKPOINTS

A1. (a) Sensory. (b) Integrating, associating, and storing. (c) Motor (efferent), muscles or glands.

A3. (a) Sensation, perception. (b) 1. S. 2. B. 3. T. 4. C.

A4. Phantom pain (phantom limb sensation). This results from irritation of nerves in the stump. Pathways from the stump to the cord and brain are still intact and may be stimulated even though the original sensory receptors in his foot are missing. Remember that he feels pain or itching in his *brain,* not in his leg or foot.

A5. (a) Modality; one modality; cerebral cortex. (b) Adaptation; pain; protects you by continuing to provide warning signals of painful stimuli. (c) Vigorous response of each type of receptor to a specific type of stimulus, such as touch receptor response to mechanical energy. No. Receptors may respond somewhat to other types of stimuli.

A6. 1. S → 2. Td → 3. C → 4. Tl. (a) S. (b) Td. (c) C. (d) Tl.

A7. (a) I (b) E. (c) P.

A8. (a) Mechano-. (b) Chemo-. (c) Thermo-.

A9. (a) G. (b) S. (c) S. (d) G.

A10.

	GP	RP
a. Triggers action potential. *(Yes? No?)*	**Yes**	**No**
b. Most receptors for *(general? special?)* senses are of this type.	**General (such as pain) (*Hint:* remember G as in "generator for general senses.")**	**Special (such as vision)**
c. *(Always excitatory? Always inhibitory? Excitatory or inhibitory?)*	**Always excitatory**	**Excitatory or inhibitory**

B1. Touch, pressure, vibration, cold, heat, and pain.

B2. (a) Touch, pressure, vibration, as well as itch and tickle; mechano; unevenly. (See Mastery Test Question 5.) (b) Light. (c) Deep. (d) Vibration.

B3. (a) C. (b) L. (c) N. (d) Type I and Type II. (e) H, J, M, T. (f) N. (g) N. (h) M. (i) T.

B6. (a) Medial aspects of left arm; T1, T4. (b) Shoulder and neck (right side); diaphragm, cervical; 3, 4; referred.

C1. B A C

C2. (a) Cortex. (b) Longitudinal; foot. (c) Face; hearing. (d) Right; sensation, hand, left. (e) General sensory or somatosensory, are; lips, face, thumb and other fingers (see Figure 15.6, page 453 in the text).

C3. (a) Stereognosis is the ability to recognize an object such as a key in the hand by its size (including weight), shape, and texture. This information is provided by receptors for touch and by proprioceptors in joints and muscles and their pathways shown in Figure LG 15.1A. (b) See Figure LG 15.1A.

C5. (a) C. (b) R. (c) O. (d) AA1. (e) A.

C6. (a) Subconscious. (b) Posture, muscle tone. (c) The left side. (d) Both sides.

C8. (a) S. (b) C and BG. (c) CC.

D1. (a) Frontal. (b) Movement, left side of the face; VII. (c) Movement of left foot affected (spasticity).

D2. G D B F E A H C.

D3. See Figure 15.2A.

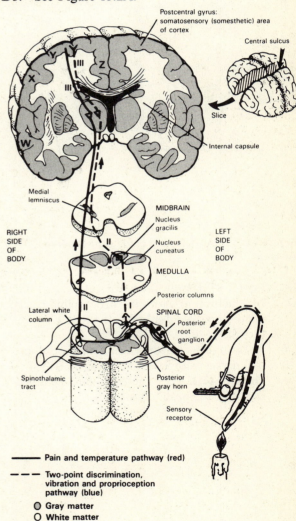

— Pain and temperature pathway (red)

--- Two-point discrimination, vibration and proprioception pathway (blue)

● Gray matter
○ White matter

Figure LG 15.1A Diagram of the central nervous system.

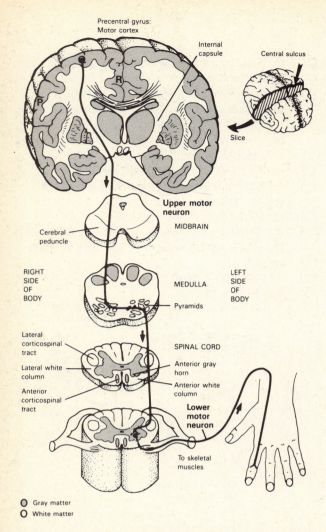

Precentral gyrus:
Motor cortex

Internal
capsule

Central sulcus

Slice

**Upper motor
neuron**

MIDBRAIN

Cerebral
peduncle

RIGHT
SIDE
OF
BODY

MEDULLA

LEFT
SIDE
OF
BODY

Pyramids

Lateral
corticospinal
tract

SPINAL CORD

Lateral white
column

Anterior gray
horn

Anterior white
column

Anterior
corticospinal
tract

**Lower
motor
neuron**

To skeletal
muscles

◎ Gray matter
○ White matter

Figure LG 15.2A   Diagram of the central nervous
system.

D4.   (a) 15; 85, lateral. (b) Left.
D5.   Myelination of axons in corticospinal tracts
is not complete until after the first 12
months of life.
D6.   These tracts synapse with neurons of all cranial nerves with motor functions (III–VII
and IX–XII). They control voluntary movements of the head and neck.

D7.   (a) Motor cortex; brainstem or anterior gray
of the spinal cord. (b) Upper; lower. (c)
Since these neurons which are activated by
input from a variety of sources (brain or spinal nerves) must function for *any* movement
to occur.
D8.   (a) Parkinson's, dopamine; muscle rigidity
and tremors. (b) Chorea, GABA; 40 years;
50–50.
D11.   (a) Outside of; brainstem, spinal cord. (b) 1,
2. Rubro and tecto; 3. Anterior reticulo; 4–
5. Vestibulo and lateral reticulo.
E1.   Memory, sleep/wakefulness, emotions.
E4.   (a) Short-term; reverberating. (b) Consolida-
tion; very little. (c) Short-term, recent.
E6.   (a) Circadian. (b) Reticular activating; thalamic; pons and midbrain (mesencephalic).
(c) Examples; light, sounds, touch, impulses
from cerebral cortex. (d) Increased, electro-
encephalograph (EEG). (e) Cerebral cortex;
spinal cord, sleep.
E7.   (a) Rapid eye movement (REM) and non-
rapid eye movement (NREM). (b) NREM,
REM. (c) 1 to 4; 1; 3 and 4. (d) Paradoxi-
cal; in, 1; rapid eye; REM. (e) 90; 5–10, 50.
E8.   (a) Questions about how long the insomnia
has persisted, sleep patterns (how long it
takes her to fall asleep, how often she wakes
up at night, what awakens her, how early
she awakens in the morning), how lack of
sleep affects her day, what she has tried to
improve her sleep, and life changes/stressors
since the insomnia began. (b) Sleeping pills
lose their effectiveness, may become addic-
tive, and may even exacerbate (make worse)
the condition. Although alcohol may help
her get to sleep, alcohol produces a light,
restless sleep. Suggestions to improve sleep:
alter the sleep environment (noise, temperature, light), avoid drinking fluids or eating a
large meal before bedtime to reduce needs
to go to the bathroom during the night, exercise more, try stress reduction techniques,
and evaluate her progress in coping with the
problem by keeping a sleep diary for a week
and then discussing again with the CNS.

Questions 1–4: Circle the letter preceding the one best answer to each question.

1. All of these sensations are conveyed by the posterior column-medial lemniscus pathway *except:*
   A. Pain and temperature
   B. Proprioception
   C. Fine touch, two-point discrimination
   D. Vibration
   E. Stereognosis

2. Choose the *false* statement about REM sleep.
   A. Infant sleep consists of a higher percentage of REM sleep than does adult sleep.
   B. Dreaming occurs during this type of sleep.
   C. The eyes move rapidly behind closed lids during REM sleep.
   D. EEG readings are similar to those of stage 4 of NREM sleep.
   E. REM sleep occurs periodically throughout a typical 8-hour sleep period.

3. The hallmark of cerebellar disease is known as:
   A. Ataxia
   B. Parkinson's disease
   C. Huntington's disease
   D. Paresthesia (tingling sensations)

4. Choose the *false* statement about pain receptors.
   A. They may be stimulated by any type of stimulus, such as heat, cold, or pressure.
   B. They have a simple structure with no capsule.
   C. They are characterized by a high level of adaptation.
   D. They are found in almost every tissue of the body.
   E. They are important in helping to maintain homeostasis.

Questions 5–8: Arrange the answers in correct sequence.

_____ _____ _____ 5. Density of sense receptors, from most dense to least:
   A. In tip of tongue
   B. In tip of finger
   C. In back of neck

_____ _____ _____ _____ 6. In order of occurrence in a sensation:
   A. Transduction of a stimulus to a generator potential
   B. Translation of the impulse into a sensation
   C. Conduction along a nervous pathway
   D. A stimulus

_____ _____ _____ _____ 7. Levels of sensation, from those causing simplest, least precise reflexes to those causing most complex and precise responses:
   A. Thalamus
   B. Brain stem
   C. Cerebral cortex
   D. Spinal cord

_____ _____ _____ _____ _____ 8. Pathway for conduction of most of the impulses for voluntary movement of muscles:
   A. Anterior gray horn of the spinal cord
   B. Precentral gyrus
   C. Internal capsule
   D. Location where decussation occurs
   E. Lateral corticospinal tract

(T) F   9. In general, the <u>left</u> side of the brain controls the right side of the body.

(T) F   10. Stimulation of pressure receptors <u>can</u> result in the sensation of pain.

T **(F)**   11. The final common pathway consists of <u>upper motor</u> neurons.   *lower*

T **(F)**   12. Circadian rhythm pertains to the <u>complex feedback circuits involved in producing coordinated movements.</u>   *24hr*

T **(F)**   13. Sight, hearing, smell, and <u>pressure</u> are all special senses.   *taste*

(T) F   14. Pain in the diaphragm, liver, or gallbladder may seem to be felt in the <u>shoulder and neck region</u> since the phrenic nerve enters the cord at the same levels (C3 to C5) as cutaneous nerves of the shoulder and neck.

(T) F   15. The neuron that crosses to the opposite side in sensory pathways is usually the <u>second-order</u> neuron.

(T) F   16. Muscle spindles, tendon organs, joint kinesthetic receptors are all examples of <u>proprioceptive</u> receptors.

T **(F)**   17. Damage to the left lateral spinothalamic tract would be most likely to result in loss of awareness of <u>pain and vibration</u> sensations in the <u>left</u> side of the body.   *temp* *right*

(T) F   18. Conscious sensations, such as those of sight and touch, can occur only in the <u>cerebral cortex.</u>

(T) F   19. Pain experienced by an amputee as if the amputated limb were still there is an example of <u>phantom pain.</u>

(T) F   20. Damage to the final common pathway will result in <u>flaccid paralysis.</u>

Proprioception   21. ____ refers to the awareness of muscles, tendons, joints, balance, and equilibrium.

RAS   22. Inactivation of the ____ results in the state of sleep.

Extrapyramidal   23. Rubrospinal, tectospinal, and vestibulospinal tracts are all classified as ____ tracts.

_____   24. A CVA affecting the medial portion of the postcentral gyrus of the right hemisphere (area Z in Figure LG 15.1, page 290) is most likely to result in symptoms such as ____.

_____   25. Voluntary motor impulses are conveyed from motor cortex to neurons in the spinal cord by ____ pathways.

Multiple Choice

1. A   3. A
2. D   4. C

Arrange

5. A B C
6. D A C B
7. D B A C
8. B C D E A

True-False

9. T
10. T
11. F. Lower motor
12. F. Events that occur at approximately 24-hour intervals, such as sleeping and waking
13. F. Sight, hearing, smell, and taste (not pressure)
14. T
15. T
16. T
17. F. Pain and temperature sensations in the right
18. T
19. T
20. T

Fill-ins

21. Proprioception
22. Reticular activating system (RAS)
23. Extrapyramidal
24. Loss of sensation of left foot
25. Direct (pyramidal or corticospinal) pathways

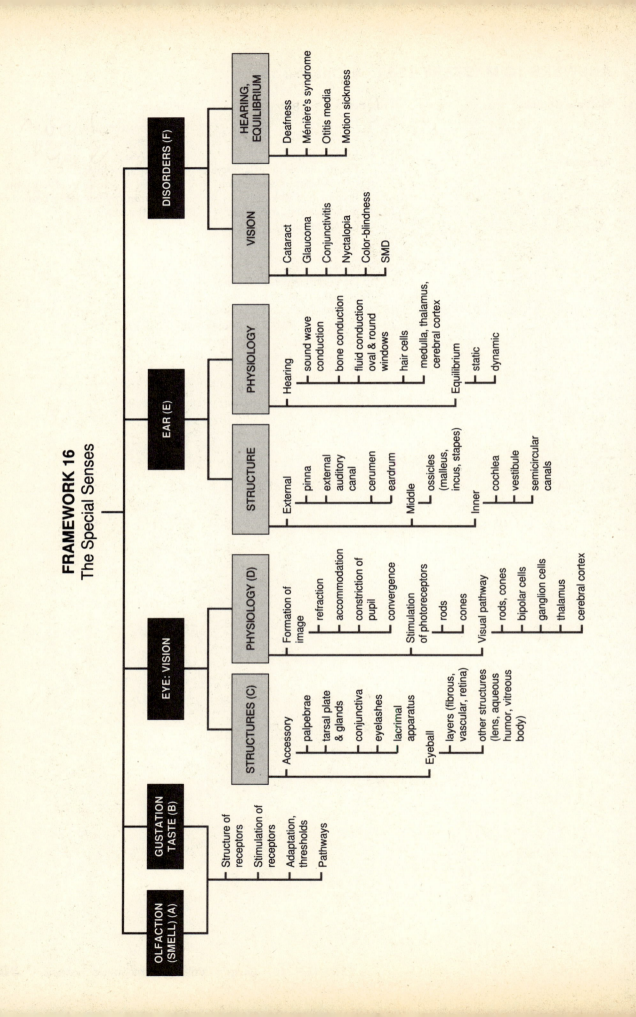

FRAMEWORK 16
The Special Senses

OLFACTION (SMELL) (A)

GUSTATION TASTE (B)
- Structure of receptors
- Stimulation of receptors
- Adaptation, thresholds
- Pathways

EYE: VISION

STRUCTURES (C)
- Accessory
  - palpebrae
  - tarsal plate & glands
  - conjunctiva
  - eyelashes
  - lacrimal apparatus
- Eyeball
  - layers (fibrous, vascular, retina)
  - other structures (lens, aqueous humor, vitreous body)

PHYSIOLOGY (D)
- Formation of image
  - refraction
  - accommodation
  - constriction of pupil
  - convergence
- Stimulation of photoreceptors
  - rods
  - cones
- Visual pathway
  - rods, cones
  - bipolar cells
  - ganglion cells
  - thalamus
  - cerebral cortex

EAR (E)

STRUCTURE
- External
  - pinna
  - external auditory canal
  - cerumen
  - eardrum
- Middle
  - ossicles (malleus, incus, stapes)
- Inner
  - cochlea
  - vestibule
  - semicircular canals

PHYSIOLOGY
- Hearing
  - sound wave conduction
  - bone conduction
  - fluid conduction oval & round windows
  - hair cells
  - medulla, thalamus, cerebral cortex
- Equilibrium
  - static
  - dynamic

DISORDERS (F)

VISION
- Cataract
- Glaucoma
- Conjunctivitis
- Nyctalopia
- Color-blindness
- SMD

HEARING, EQUILIBRIUM
- Deafness
- Ménière's syndrome
- Otitis media
- Motion sickness

The Special Senses

All sensory organs contain receptors for increasing sensitivity to the environment. In Chapter 15 you considered the relatively simple receptors and pathways for the general senses of touch, pressure, temperature, pain, and vibration. The special senses include smell, taste, vision, hearing, and equilibrium. Special afferent pathways in many ways resemble general afferent pathways. However, a major point of differentiation is the arrangement of special sense receptors in complex sensory organs, specifically the nose, tongue, eyes, and ears.

First study carefully the Chapter 16 Framework and note the key terms associated with each section.

TOPIC OUTLINE AND OBJECTIVES

A. Olfactory sensations

1. Locate the receptors for olfaction and describe the neural pathway for smell.

B. Gustatory sensations

2. Identify the gustatory receptors and describe the neural pathway for taste.

C. Visual sensations: anatomy

3. List and describe the accessory structures of the eye and the structural divisions of the eyeball.
4. Discuss image formation by describing refraction, accommodation, and constriction of the pupil.

D. Visual sensations: physiology

5. Describe how photoreceptors and photopigments function in vision.

6. Describe the retinal processing of visual input and the neural pathway of light impulses to the brain.

E. Auditory sensations and equilibrium

7. Describe the anatomical subdivisions of the ear.
8. List the principal events in the physiology of hearing.
9. Identify the receptor organs for equilibrium and how they function.

F. Disorders, medical terminology

10. Contrast the causes and symptoms of glaucoma, senile macular degeneration, deafness, Ménière's syndrome, otitis media, and motion sickness.
11. Define medical terminology associated with the sense organs.

Now study the following parts of words that may help you better understand terminology in this chapter.

Wordbyte	Meaning	Example	Wordbyte	Meaning	Example
aqua-	water	*aqu*eous humor	orbit-	eye socket	supra*orbit*al
blephar-	eyelid	*blephar*oplasty	ossi-	bone	*ossi*cle
kerat-	cornea	radial *kerat*otomy	ot(o)-	ear	*ot*olith
macula	spot	*macula* lutea	presby-	old	*presby*opia
med-	middle	otitis *med*ia	sclera	hard	*sclera*, oto*scler*osis
ocul-	eye	*ocul*omotor nerve	tympano-	drum	*tympan*ic
olfact-	smell	*olfact*ory			membrane
ophthalm-	eye	*ophthalm*ologist	vitr-	glassy	*vitr*eous humor
opt-	eye	*opt*ic nerve, *opt*itian			

CHECKPOINTS

A. Olfactory sensations (pages 468–469)

■ **A1.** Describe olfactory receptors in this exercise.

a. Receptors for smell are located in the *(superior? inferior?)* portion of the nasal cavity. Receptors consist of ____-polar neurons which are located between

_____ cells. The distal end of each olfactory receptor cell

consists of dendrites that have cilia known as olfactory _____.

b. What is the function of olfactory (Bowman's) glands in the nose?

c. Since smell is a _____ sense, receptors in olfactory hair membranes respond to different chemical molecules, leading to a

_____ and _____.

d. Adaptation to smell occurs *(slowly? rapidly?)* at first and then happens at a much slower rate.

■ **A2.** Complete this exercise about the nerves associated with the nose.

a. Upon stimulation of olfactory hairs, impulses pass to cell bodies and axons; the axons of these olfactory cells form cranial nerves *(I? II? III?)*, the olfactory nerves. These pass from the nasal cavity to the cranium to terminate in the olfactory

_____ located just inferior to the _____
lobes of the cerebrum.

b. Neurons in the olfactory bulb then convey impulses along the olfactory

_____ directly to the _____. Note that olfaction is one sense that *(does? does not?)* involve the thalamus. What is the effect of projection of sensations of smell and taste to the limbic system?

c. The olfactory pathway just described transmits sensations for the *(general? special?)* sense of smell. General senses, such as irritation of the nose by pepper leading to sniffles or tears, are carried primarily by cranial nerve ____, the facial nerve.

B. Gustatory sensations: taste (pages 469–471)

■ **B1.** Describe receptors for taste in this exercise.

a. Receptors for taste, or _____ sensation, are located in taste

buds. These receptors are protected by a _____ formed of

supporting cells, as well as the _____ cells that produce the supporting cells. Within each taste bud are about *(5? 50?)* gustatory receptor cells with

gustatory _____ that project from a taste bud pore. The life span of each gustatory receptor cell is about *(10 days? 10 years?)*.

b. Taste buds are located on elevated projections of the tongue called

_____. The largest of these are located in a V-formation at the

back of the tongue. These are _____ papillae.

_____ papillae are mushroom-shaped and located on the sides

and tip of the tongue. Pointed _____ papillae cover the

_____ of the tongue, and these *(do? do not?)* contain many
taste buds.

■ **B2.** Do the following exercise about taste.

a. There are only four primary taste sensations: _salty_,
sweet, _sour_, and
bitter. The taste buds at the tip of the tongue are most sensitive to _sweet_ and _salty_ tastes, whereas
the posterior portion is most sensitive to _bitter_ taste.

b. The four primary tastes are modified by _olfactory_ sensations to
produce the wide variety of "tastes" experienced. If you have a cold (with a "stuffy
nose"), you may not be able to discriminate the usual variety of tastes, largely
because of loss of the sense of _smell_.

c. The threshold for smell is quite (high? low?), meaning that (a large? only a small?)
amount of odor must be present in order for smell to occur. The threshold varies for
the four primary tastes; the threshold for (sour? sweet and salt? bitter?) is lowest,
while those for _____ are highest.

d. Taste impulses are conveyed to the brainstem by cranial nerves _____, _____,
and _____; there nerve messages travel to lower parts of the brain, including the
_____, _____, and
_____, and finally stimulate the cerebral cortex.

C. Visual sensations: anatomy (pages 471–482)

C1. Examine your own eye structure in a mirror. With the help of Figure 16-3 (page
471 in your text), identify each of the accessory structures of the eye listed below.

Conjunctiva	Palpebrae
Eyebrow	Tarsal glands
Eyelashes	Tarsal plate
Lacrimal glands: identify site	

■ **C2.** *For extra review.* Match the structures listed above with the following descriptions. Write the name of the structure on the line provided.

a. Eyelid: _palpebra_

b. Thick fold of connective tissue that forms much of the eyelid: _tarsal plate_

c. Covers the orbicularis oculis muscle; provides protection: _eyebrow_

d. Short hairs; infection of glands at the base of these hairs is *sty:* _eyelashes_

e. Located in superolateral region of orbit; secrete tears: _lacrimal glands_

f. Secrete oil; infection is *chalazion:* _tarsal glands_

g. Mucous membrane lining the eyelids and covering anterior surface of the eye: _conjunctiva_

■ **C3.** Do this exercise on accessory eye structures.

a. Lacrimal glands produce ___tears___ as a result of *(sympathetic?* (parasympathetic?)) stimulation. Glands normally produce _1_ ml per day of tears. State two or more functions of tears. _antibacterial_ _flush dust_

b. Each eye is moved by *(3? 4?* (6?)) extrinsic eye muscles; these are innervated by cranial nerves _III_, _IV_, and _VI_. *(For extra review,* see Chapter 11, Activity C4, page LG 198).

c. The eye sits in the bony socket known as the ___orbit___ with only the anterior ((one-sixth?) *one-half?)* of the eyeball exposed.

■ **C4.** On Figure LG 16.1, color the three layers (or tunics) of the eye using color code ovals. Next to the ovals write letters of labeled structures that form each layer. One is done for you. Then label all structures in the eye and check your answers against the key.

■ **C5.** *A clinical challenge.* A cornea that is surgically transplanted *(is?* (is not?)) likely to be rejected. State your rationale. _has no blood supply thus has no antibodies to attack foreign body_

■ **C6.** Complete this exercise about blood vessels of the eye.

a. Retinal blood vessels ((can?) *cannot?)* be viewed through an ophthalmoscope. Of what advantage is this? _detect disease_

b. The central retinal artery enters the eye at the ((optic disc?) *central fovea?).*

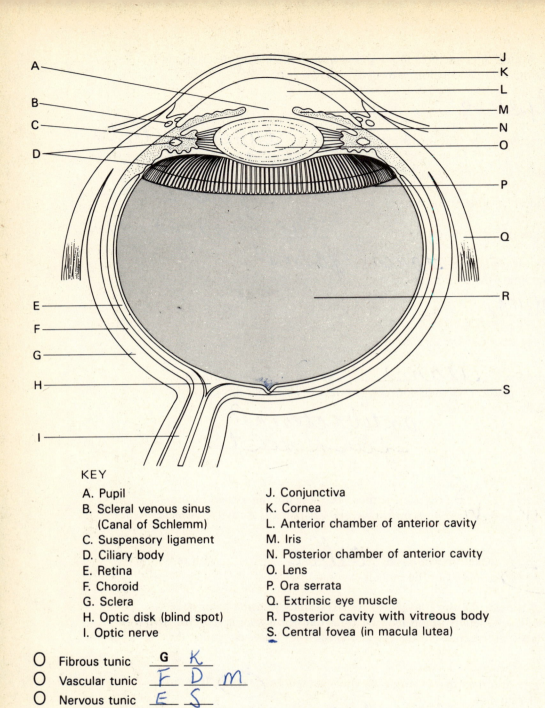

A
B
C
D
E
F
G
H
I

J
K
L
M
N
O
P
Q
R
S

KEY

A. Pupil
B. Scleral venous sinus
   (Canal of Schlemm)
C. Suspensory ligament
D. Ciliary body
E. Retina
F. Choroid
G. Sclera
H. Optic disk (blind spot)
I. Optic nerve

J. Conjunctiva
K. Cornea
L. Anterior chamber of anterior cavity
M. Iris
N. Posterior chamber of anterior cavity
O. Lens
P. Ora serrata
Q. Extrinsic eye muscle
R. Posterior cavity with vitreous body
S. Central fovea (in macula lutea)

uvea = choroid + cilliary body

O Fibrous tunic   G K
O Vascular tunic   F D M
O Nervous tunic   E S

Figure LG 16.1 Structure of the eyeball in horizontal section. Color as directed in Checkpoint C4.

■ **C7.** Contrast the two types of photoreceptor cells by writing R for rods or C for cones before the related descriptions.

__C__ a. About 6 million in each eye; most concentrated in the central fovea of the macula lutea.

__R__ b. Over 120 million in each eye; located mainly in peripheral regions of the eye.

__C__ c. Sense color and acute (sharp) vision.

__R__ d. Used for night vision.

■ **C8.** Do this exercise that describes the retina.

   a. After light strikes the back of the retina, nerve impulses pass anteriorly through three zones of neurons. First is the region of rods and cones, or

   photoreceptor zone; next is the _bipolar_ layer; most

   anterior is the _ganglion_ layer.

   b. Axons of the ganglion layer converge to form the optic _nerve_.

   This nerve exits from the eye at the point known as the _optic_

   disc, or _blind spot_. The name indicates that no image formation can occur here since no rods or cones are present.

   c. What function(s) do *amacrine cells* and *horizontal cells* serve?

■ **C9.** Describe the lens and the cavities of the eye in this exercise.

   a. The lens is composed of *(lipid? protein?)* arranged in layers, much like an onion. Normally the lens is *(clear? cloudy?)*. A loss of transparency of the lens occurs in

   the condition called _____.

   b. The lens divides the eye into anterior and posterior *(chambers? cavities?)*. The

   posterior cavity is filled with _vitreous_ body, which has a *(watery?*

   jellylike?) consistency. This body helps to hold the _lens_ in place. Vitreous body is formed during embryonic life and *(is? is not?)* replaced in later life.

   c. The anterior cavity is subdivided by the _____ into two chambers. Which is larger? *(Anterior chamber? Posterior chamber?)* A watery fluid called

   aqueous humor is present in the anterior cavity. This fluid is

   formed by the _cillary_ processes, and it is normally replaced about every 90 *(minutes? days?)*.

   d. How is the formation and final destination of aqueous fluid similar to that of cerebrospinal fluid (CSF)?

   it has a distinct flow

   e. State two functions of aqueous humor. _nourish the lens & cornea_

   f. Aqueous fluid creates pressure within the eye; in glaucoma this pressure *(increases?*

   decreases?). What effects may occur?

   blindness

■ **C10.** *For extra review.* Check your understanding by matching eye structures in the box with descriptions below. Use each answer once.

CF. Central fovea	OD. Optic disc
Cho. Choroid	P. Pupil
CM. Ciliary muscle	R. Retina
Cor. Cornea	S. Sclera
I. Iris	SVS. Scleral venous sinus
L. Lens	(Canal of Schlemm)

_____ a. "White of the eye."

_____ b. Normally clear; if cloudy, it is a cataract.

_____ c. Blind spot; area in which there are no cones or rods.

_____ d. Area of sharpest vision; area of densest concentration of cones.

_____ e. Anteriorly located, avascular coat responsible for about 75 percent of all focusing done by the eye.

_____ f. Consists of a pigment epithelium and a neural portion; if these two portions separate (detach), blindness may result.

_____ g. Brown-black layers prevent reflection of light rays; also nourishes eyeball since it is vascular.

_____ h. A hole; appears black, like a circular doorway leading into a dark room.

_____ i. Regulates the amount of light entering the eye; colored part of the eye.

_____ j. Attaches to the lens by means of radially arranged fibers called the suspensory ligaments.

_____ k. Located at the junction of iris and cornea; drains aqueous humor.

C11. Define each of these processes involved in *image formation* on the retina.

a. *Refraction* of light

b. *Accommodation* of the lens

c. *Constriction* of the pupil

d. *Convergence* of the eyes

■ **C12.** Answer these questions about formation of an image on the retina.

a. Bending of light rays so that images focus exactly upon the retina is a process

known as _____. The *(cornea? lens?)* accomplishes most (75 percent) of the refraction within the eye. Explain why you think this is so.

b. Draw how a letter "e" would look as it is focused on the retina? ____
c. Explain why you see things right side up even though refraction produces inverted images on your retina.

■ **C13.** Explain how *accommodation* enables your eyes to focus clearly.

a. A *(concave? convex?)* lens will bend light rays so that they converge and finally intersect. The lens in each of your eyes is *(biconcave? biconvex?)*.
b. When you bring your finger toward your eye, light rays from your finger must converge *(more? less?)* so that they will focus on your retina. Thus for near vision, the anterior surface of your lens must become *(more? less?)* convex. This occurs by a thickening and bulging forward of the lens caused by *(contraction? relaxation?)* of

the _____ muscle. In fact, after long periods of close work (such as reading), you may experience eye strain.

c. What is the *near point of vision* for a young adult? _____ cm (_____ inches) For

a 70-year-old? _____ cm (_____ inches) What factor explains this difference?

■ **C14.** *A clinical challenge.* Describe errors in refraction and accommodation in this exercise.

a. The normal (or _____ eye) can refract rays from a distance of

_____ meters (_____ feet) and form a clear image on the retina.

b. In the nearsighted (or _____) eye, images focus *(in front of? behind?)* the retina. This may be due to an eyeball that is too *(long? short?)*. Corrective lenses should be slightly *(concave? convex?)* so they refract rays less and allow them to focus farther back on the retina.
c. The farsighted person can see *(near? far?)* well, but has difficulty seeing *(near? far?)*

without the aid of corrective lenses. The farsighted (or _____) person requires *(concave? convex?)* lenses. After about age 40, most persons lose the ability to see near objects clearly.

■ **C15.** Choose the correct answers to complete each statement.

a. Both accommodation of the lens and constriction of the pupil involve *(extrinsic? intrinsic?)* muscles, whereas convergence of the eyes involves *(extrinsic? intrinsic?)* muscles.
b. The pupil *(dilates? constricts?)* in the presence of bright light; the pupil *(dilates? constricts?)* in stressful situations when sympathetic nerves stimulate the iris.

D. Physiology of vision (pages 482–487)

■ **D1.** List three principal processes that are necessary for vision to occur.

a.

b.

c.

■ **D2.** Do this exercise on photoreceptors and photopigments.

a. The basis for the names *rods* and *cones* rests in the shape of the *(inner? outer?)* segments of these cells, that is, the segment closest to the *(choroid? vitreous body?)*. In which segment are photopigments such as rhodopsin located? *(Inner? Outer?)* In which segment can the nucleus, Golgi complex, and mitochondria be found? *(Inner? Outer?)*

b. How many types of photopigment are found in the human retina? (____) Rhodopsin is the photopigment found in *(rods? cones?)*. How many types of photopigments are found in cones?

c. Any photopigment consists of two parts: *retinal* and *opsin*. What is the function of retinal?

d. Retinal is a vitamin ____ derivative and is formed from carotenoids. List several carotenoid-rich foods.

Tell how such foods may help to prevent *nyctalopia* (night blindness)?

■ **D3.** Complete this exercise about what happens in the rods of your eyes after you walk into a dark movie theatre. Refer to Figure 16.12 (page 483) in your text.

a. In order to see in dim light (as in the dark theatre), you use *(rods? cones?)*.

b. Rods produce a pigment called _____ (or visual purple). But in bright light (as in the lobby) this chemical is broken down very *(rapidly? slowly?)*. So as you enter the theatre your rods lack sufficient rhodopsin to see in the dark.

c. During the short time that you are stumbling around in the darkness, rhodopsin can form in the following manner. An enzyme converts retinal from its trans form to its

_____ form, which fits well with a glycoprotein named

_____ to produce rhodopsin.

d. As soon as you have sufficient rhodopsin present in the rods, stimulation of the rods by the dim light in the theatre will begin breaking down rhodopsin. This is a rapid but stepwise process ultimately resulting in the cleavage of rhodopsin into two parts:

again the glycoprotein _____ and the *(cis? trans?)* form of

retinal. As a result of this process, a _____ impulse develops, and a nerve impulse then passes to your brain informing you of what you have seen in dim light, such as images of persons in seats near you. What you are experiencing in the dark theatre is *(dark? light?)* adaptation in which visual sensitivity increases very slowly.

e. Your vision becomes increasingly better after a few more minutes in the dim light since you continue to form rhodopsin. [See item (c) above.] But should you step out into the lobby to buy popcorn, the bright light will *(produce? destroy?)* rhodopsin so rapidly and completely that you will have to start the dark-induced synthesis of rhodopsin all over again once you reenter the dimly lit theatre.

■ **D4.** Explain what is meant by the "dark current." Describe the roles of *cyclic GMP* and *recoverin* in this process.

■ **D5.** Explain how cones help you to see in "living color" by completing this paragraph.

Cones contain pigments which *(do? do not?)* require bright light for breakdown (and therefore for a generator potential). Three different pigments are present, sensitive

to three colors: _____, _____, and

_____. A person who is red-green color-blind lacks some of the

cones receptive to two colors (_____ and

_____) and so cannot distinguish between these colors. This condition occurs more often in *(males? females?)*.

■ **D6.** Describe the conduction pathway for vision by arranging these structures in sequence. Write the letters in correct order on the lines provided. Also, indicate the four points where synapsing occurs by placing an asterisk (*) between the letters.

> B. Bipolar cells
> C. Cerebral cortex (visual areas)
> G. Ganglion cells
> OC. Optic chiasma
> ON. Optic nerve
> OT. Optic tract
> P. Photoreceptor cells (rods, cones)
> T. Thalamus

___ ___ ___ ___ ___ ___ ___ ___

■ **D7.** Answer these questions about the visual pathway to the brain. It may be helpful to refer to Figure 16.15 (page 486 in the text).

a. Hold your left hand up high and to the left so you can still see it. Your hand is in the *(temporal? nasal?)* visual field of your left eye, and in the

_____ visual field of your right eye.

b. Due to refraction, the image of your hand will be projected onto the *(left? right?)* *(upper? lower?)* portion of the retinas of your eyes.

c. All nerve fibers from these areas of your retinas reach the *(left? right?)* side of your thalamus and cerebral cortex.

d. Damage to the right optic tract, right side of thalamus, or right visual cortex would result in loss of sight of the *(left? right?)* visual fields of each eye.

E. Auditory sensations and equilibrium (pages 487–498)

■ **E1.** Refer to Figure LG 16.2 and do the following exercise.

a. Color the three parts of the ear using color code ovals. Then write on lines next to ovals the letters of structures that are located in each part of the ear. One is done for you.

b. Label each lettered structure using leader lines on the figure.

■ **E2.** *For extra review.* Select the ear structures in the box that fit the descriptions below. Not all answers will be used.

> A. Auricle (pinna)
> AT. Auditory tube
> I. Incus
> M. Malleus
> OW. Oval window
> RW. Round window
> S. Stapes
> TM. Tympanic membrane

_____ a. Tube used to equalize pressure on either side of tympanic membrane

_____ b. Eardrum

_____ c. Structure on which stapes exerts pistonlike action

_____ d. Ossicle adjacent to eardrum

_____ e. Anvil-shaped ear bone

_____ f. Portion of the external ear shaped like the flared end of a trumpet

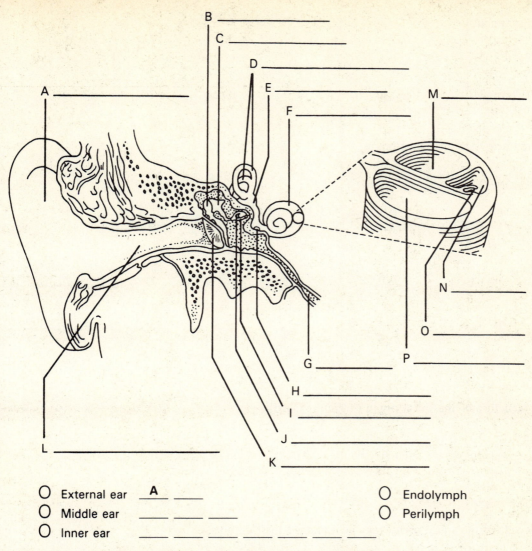

B	_____	
C	_____	
D	_____	
E		
F	_____	
A	_____	M _____

G _____   P _____

H _____

I _____

J _____

K _____

L _____

N _____

O _____

○ External ear   **A** ____   ○ Endolymph
○ Middle ear   ____ ____ ____   ○ Perilymph
○ Inner ear   ____ ____ ____ ____ ____ ____

Figure LG 16.2 Diagram of the ear in frontal section. Color and label as directed in Checkpoints E1 and E5.

E3. Contrast locations, anatomy, and physiology of the following structures:

a. Tensor tympani muscle—stapedius muscle

b. Bony labyrinth—membranous labyrinth

c. Modiolus—helicotrema

d. Vestibular membrane—tectorial membrane

e. Ossicles—otoliths

E4. Describe the structure of the spiral organ (organ of Corti).

Explain how hair cells of the spiral organ may be damaged by loud noises.

■ **E5.** Color *endolymph* and *perilymph* and related color code ovals in Figure LG 16.2.

■ **E6.** Fill in the blanks and circle correct answers about sound waves.
   a. Sound waves heard by humans range from frequencies of 20 to 20,000 cycles/sec

   (Hz). Humans can best hear sounds in the range of _____ Hz.
   b. A musical high note has a *(higher? lower?)* frequency than a low note. So frequency
      is *(directly? indirectly?)* related to pitch.

   c. Sound intensity (loudness) is measured in units called _____.

   Normal conversation is at a level of about _____ dB, whereas sounds at

   _____ dB can cause pain.

■ **E7.** Summarize events in the process of hearing in this activity. It may help to refer to Figure 16.20 (page 493 in the text).

a. Sound waves travel through the _____ and strike the

_____ membrane. Sound waves are magnified by the action of

the three _____ in the middle ear.

b. The ear bone named *(malleus? incus? stapes?)* strikes the *(round? oval?)* window, setting up waves in *(endo-? peri-?)* lymph. This pushes on the floor of the upper *scala (vestibuli? tympani?)*. As a result the cochlear duct is moved, and so is the perilymph in the lower canal, the *scala (vestibuli? tympani?)*. The pressure of the perilymph is finally expended by bulging out the *(round? oval?)* window.

c. As the cochlear duct moves, tiny hair cells embedded in the floor of the duct are

stimulated. These hair cells are part of the _____ organ; its

name is based on its spiral arrangement on the _____ membrane all the way around the 2¾ coils of the cochlear duct.

d. As spiral organ hair cells are moved by waves in endolymph, hairs move against the

gelatinous _____ membrane. This movement generates receptor potentials in hair cells that excite nearby sensory neurons of the

_____ branch of cranial nerve _____. The pathway continues

to the brainstem, _____ (relay center), and finally to the

_____ lobe of the cerebral cortex.

■ **E8.** Check your understanding of the pathway of fluid conduction by placing the following structures in correct sequence. Write the letters on the lines provided.

B. Basilar membrane	ST. Scala tympani (perilymph)
C. Cochlear duct (endolymph)	SV. Scala vestibuli (perilymph)
O. Oval window	VM. Vestibular membrane
RW. Round window	

___ ___ ___ ___ ___ ___ ___

E9. Describe the roles of K^+ and Ca^{2+} in stimulating hair cells and generating nerve impulses related to hearing.

E10. *A clinical challenge.* After you read pages 36–37 of *Essays in Wellness,* discuss health practices that can help you to maximize wellness of your eyes and ears.

■ **E11.** Contrast receptors for hearing and for equilibrium in this summary of the inner ear.

    a. Receptors for hearing and equilibrium are all located in the *(middle? inner?)* ear. All

       consist of supporting cells and _____ cells that are covered by

       a _____ membrane.

    b. In the spiral organ, which senses _____, the gelatinous mem-

       brane is called the _____ membrane. Hair cells move against this membrane as a result of *(sound waves? change in body position?)*.

    c. In the macula, located in the *(semicircular canals? vestibule?)*, the gelatinous membrane is embedded with calcium carbonate crystals called

       _____. These respond to gravity in such a way that the macula is the main receptor for *(static? dynamic?)* equilibrium. An example of such equilibrium occurs as you are aware of your *(position while lying down? change in position on a careening roller coaster?)*.

    d. In the semicircular canals the gelatinous membrane is called the

       _____. Its shape is *(flat? like an inverted cup?)*. The cupula is part of the *(crista? saccule?)* located in the ampulla. Change in direction (as in a

       roller coaster) causes _____ to bend hairs in the cupula. Cristae in semicircular canals are therefore receptors primarily for *(static? dynamic?)* equilibrium.

E12. Describe vestibular pathways, including the role of the cerebellum in maintaining equilibrium.

F. Disorders, medical terminology (pages 498–499)

■ **F1.** Match the name of the disorder with the description.

B.	Blepharitis	Myo.	Myopia
Cat.	Cataract	Myr.	Myringitis
Con.	Conjunctivitis	P.	Presbyopia
G.	Glaucoma	Str.	Strabismus
H.	Hyperacusia	T.	Tinnitus
K.	Keratitis	V.	Vertigo
Men.	Ménière's syndrome		

_____ a. Condition requiring corrective lenses to focus distant objects

_____ b. Excessive intraocular pressure resulting in blindness; second most common cause of blindness

_____ c. Ringing in the ears

_____ d. Pinkeye

_____ e. Abnormally sensitive hearing

_____ f. Inflammation of the eardrum

_____ g. Inflammation of the eyelid

_____ h. Disturbance of the inner ear with excessive endolymph

_____ i. Loss of transparency of the lens

_____ j. Farsightedness due to loss of elasticity of lens, especially after age 40

_____ k. Inflamation of the cornea

_____ l. Sense of spinning or whirling

F2. Discuss causes and treatments of:

a. Otitis media

b. Motion sickness

F3. Contrast sensorineural deafness with conduction deafness.

A1. (a) Superior; bi-, supporting; hairs. (b) Produce mucus that acts as a solvent for odoriferous substances. (c) Chemical, generator potential and nerve impulse. (d) Rapidly.

A2. (a) I; bulb, frontal. (b) Tracts, cerebral cortex; does not; awareness of certain smells, such as putrid odors or fragrance of roses, may lead to emotional responses. (c) Special; VII.

B1. (a) Gustatory; capsule, basal; 50, hairs; 10 days. (b) Papillae; circumvallate; fungiform; filiform, anterior, do not.

B2. (a) Sweet, sour, salt, and bitter; sweet and salty, bitter. (b) Olfactory; smell. (c) Low, only a small; bitter; sweet and salt. (d) VII, IX, and X; thalamus, hypothalamus and limbic system.

C2. (a) Palpebra. (b) Tarsal plate. (c) Eyebrow. (d) Eyelashes. (e) Lacrimal glands. (f) Tarsal glands. (g) Conjunctiva.

C3. (a) Tears (lacrimal fluid), parasympathetic; 1.0; tears protect against infection via the bactericidal enzyme lysozyme and by flushing away irritating substances; tears are signs of emotions. (b) 6; III, IV and VI. (c) Orbit, one-sixth.

C4. Fibrous tunic; G, K; vascular tunic: D, F, M; nervous tunic: E, S.

C5. Is not. Since the cornea is avascular, antibodies that cause rejection do not circulate there.

C6. (a) Can; health of blood vessels can be readily assessed, for example, in persons with diabetes. (b) Optic disc.

C7. (a) C. (b) R. (c) C. (d) R.

C8. (a) Photoreceptor; bipolar; ganglion. (b) Nerve; optic, blind spot. (c) They modify signals transmitted along optic pathways.

C9. (a) Protein; clear; cataract. (b) Cavities; vitreous, jellylike; retina; is not. (c) Iris, anterior chamber; aqueous; ciliary, minutes. (d) Both are formed from blood vessels (choroid plexuses) and the fluid finally returns to venous blood. (e) Provides nutrients and oxygen and removes wastes from the cornea and lens; also provides pressure to separate the cornea from lens. (f) Increases; damage to the retina with possible blindness.

C10. (a) S. (b) L. (c) OD. (d) CF. (e) Cor. (f) R. (g) Cho. (h) P. (i) I. (j) CM. (k) SVS.

C12. (a) Refraction; cornea; the density of the cornea differs considerably from the air anterior to it. (b) "∂" ("e" inverted 180° and much smaller). (c) Your brain learned early in your life how to "turn" images so that what you see corresponds with locations of objects you touch.

C13. (a) Convex; biconvex. (b) More; more; contraction, ciliary. (c) 10, 4; 20; 8. The lens loses elasticity (presbyopia) with aging.

C14. (a) Emmetropic, 6 (20 feet). (b) Myopic, in front of; long; concave. (c) Far, near; hypermetropic, convex.

C15. (a) Intrinsic, extrinsic. (b) Constricts; dilates.

D1. (a) Formation of an image on the retina. (b) Stimulation of photoreceptors so that a light stimulus is converted into an electrical stimulus (receptor potential and nerve impulse). (c) Transmission of the impulse along neural pathways to the thalamus and visual cortex.

D2. (a) Outer, choroid; outer; inner. (b) 4; rods; one in each of three types of cones that are sensitive to different colors. (c) Absorbs light. (d) A; carrots, yellow squash, broccoli, and spinach; these foods lead to production of retinal, which is a component of the retinal in rods and cones.

D3. (a) Rods. (b) Rhodopsin; rapidly. (c) Cis, opsin. (d) Opsin, trans; generator; dark. (e) Destroy.

D5. Do; yellow to red, green, blue; red, green; males.

D6. P * B * G ON OC OT * T * C. (Note that some crossing but no synapsing occurs in the optic chiasma.)

D7. (a) Temporal, nasal. (b) Right, lower. (c) Right. (d) Left.

E1.

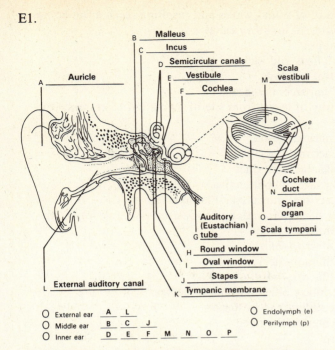

Malleus
B
C Incus
D Semicircular canals
Auricle E Vestibule
F Cochlea
A

Scala
M vestibuli

p

e

p

Cochlear
N duct

Spiral
organ

Auditory
(Eustachian) O
G tube P Scala tympani

H Round window
I Oval window
J Stapes
K Tympanic membrane

L External auditory canal

O External ear   A   L
O Middle ear   B   C   J
O Inner ear   D   E   F   M   N   O   P

O Endolymph (e)
O Perilymph (p)

Figure LG 16.2A   Diagram of the ear in frontal section.

E2.   (a) AT. (b) TM. (c) OW. (d) M. (e) I. (f) A.
E5.   See Figure LG 16.2A
E6.   (a) 1,000–4,000. (b) Higher; directly. (c) Decibels (dB); 45, 115–120.
E7.   (a) External auditory canal, tympanic; ossicles. (b) Stapes, oval, peri-; vestibuli; tympani; round. (c) Spiral; basilar. (d) Tectorial; cochlear, VIII; thalamus, temporal.
E8.   O SV VM C B ST RW.
E11. (a) Inner; hair, gelatinous. (b) Hearing, tectorial; sound waves. (c) Vestibule, otoliths; static; position while lying down. (d) Cupula; like an inverted cup; crista; endolymph; dynamic.
F1.   (a) Myo. (b) G. (c) T. (d) Con. (e) H. (f) Myr. (g) B. (h) Men. (i) Cat. (j) P. (k) K. (l) V.

MASTERY TEST: Chapter 16

Questions 1–5: Circle T (true) or F (false). If the statement is false, change the underlined word or phrase so that the statement is correct.

T   F   1. Parasympathetic nerves stimulate the circular iris muscle (constrictor pupillae) to constrict the pupil, whereas sympathetic nerves cause dilation of the pupil.

T   F   2. Convergence and accommodation are both results of contraction of smooth muscle of the eye.

T   F   3. Crista, macula, otolith, and spiral organ are all structures located in the inner ear.

T   F   4. Both the aqueous body and vitreous humor are replaced constantly throughout your life.

T   F   5. The receptor organs for special senses are less complex structurally than those for general sense.

Questions 6–12: Arrange the answers in correct sequence.

_____ _____ _____   6. Layers of the eye, from superficial to deep:
A. Sclera
B. Retina
C. Choroid

_____ _____ _____ _____   7. From anterior to posterior:
A. Vitreous body
B. Optic nerve
C. Cornea
D. Lens

_____ _____ _____ _____   8. Pathway of aqueous humor, from site of formation to destination:
A. Anterior chamber
B. Scleral venous sinus
C. Ciliary body
D. Posterior chamber

_____ _____ _____ _____ _____ 9. Pathway of sound waves and resulting mechanical action:
   A. External auditory canal
   B. Stapes
   C. Malleus and incus
   D. Oval window
   E. Tympanic membrane

_____ _____ _____ _____ _____ 10. Pathway of tears, from site of formation to entrance to nose:
   A. Lacrimal gland
   B. Lacrimal duct
   C. Nasolacrimal duct
   D. Surface of conjunctiva
   E. Lacrimal punctae and lacrimal canals

_____ _____ _____ _____ _____ 11. Order of impulses along conduction pathway for smell:
   A. Olfactory bulb
   B. Olfactory hairs
   C. Olfactory nerves
   D. Olfactory tract
   E. Primary olfactory area of cortex

_____ _____ _____ _____ _____ 12. From anterior to posterior:
   A. Anterior chamber
   B. Iris
   C. Lens
   D. Posterior cavity
   E. Posterior chamber

Questions 13–20: Circle the letter preceding the one best answer to each question.

13. In darkness, Na^+ channels are held open by a molecule called:
    A. Glutamate
    B. Cyclic GMP
    C. PDE
    D. Transducin

14. Infections in the throat (pharynx) are most likely to lead to ear infections in the following manner. Bacteria spread through the:
    A. External auditory meatus to the external ear
    B. Auditory (eustachian) tube to the middle ear
    C. Oval window to the inner ear
    D. Round window to the inner ear

15. Choose the *false* statement about rods.
    A. There are more rods than cones in the eye.
    B. Rods are concentrated in the fovea and are less dense around the periphery.
    C. Rods enable you to see in dim (not bright) light.
    D. Rods contain rhodopsin.
    E. No rods are present at the optic disc.

16. Choose the *false* statement about the lens of the eye.
    A. It is biconvex.
    B. It is avascular.
    C. It becomes more rounded (convex) as you look at distant objects.
    D. Its shape is changed by contraction of the ciliary muscle.
    E. Change in the curvature of the lens is called accommodation.

17. Choose the *false* statement about the middle ear.
    A. It contains three ear bones called ossicles.
    B. Infection in the middle ear is called otitis media.
    C. It functions in conduction of sound from the external ear to the inner ear.
    D. The cochlea is located here.

18. Choose the *false* statement about the semicircular canals.
    A. They are located in the inner ear.
    B. They sense acceleration or changes in position.
    C. Nerve impulses begun here are conveyed to the brain by the vestibular branch of cranial nerve VIII.
    D. There are four semicircular canals in each ear.
    E. Each canal has an enlarged portion called an ampulla.

19. Destruction of the left optic tract would result in:
    A. Blindness in the left eye
    B. Loss of left visual field of each eye
    C. Loss of right visual field of each eye
    D. Loss of lateral field of view of each eye ("tunnel vision")
    E. No effects on the eye

20. Mr. Frederick has a detached retina of the lower right portion of one eye. As a result he is unable to see objects in which area in the visual field of that eye?
    A. High in the left   C. Low in the left
    B. High in the right   D. Low in the right

Questions 21–25: fill-ins. Complete each sentence with the word or phrase that best fits.

_____ 21. ____ is the most common disorder leading to blindness.

_____ 22. Name the four processes necessary for formation of an image on the retina: ____ of light rays, ____ of the lens, ____ of the pupil, and ____ of the eyes.

_____ 23. ____ is the study of the structure, functions, and diseases of the eye.

_____ 24. The names of three ossicles are ____, ____, and ____.

_____ 25. Two functions of the ciliary body are ____ and ____.

ANSWERS TO MASTERY TEST: ■ Chapter 16

True-False

1. T
2. F. Accommodation, but not convergence, is a result.
3. T
4. F. Aqueous humor, but not vitreous humor, is
5. F. More

Arrange

6. A C B
7. C D A B
8. C D A B
9. A E C B D
10. A B D E C
11. B C A D E
12. A B E C D

Multiple Choice

13. B   17. D
14. B   18. D
15. B   19. C
16. C   20. A

Fill-ins

21. Cataract
22. Refraction, accommodation, constriction, convergence
23. Ophthalmology
24. Malleus, incus, and stapes
25. Production of aqueous humor and alteration of lens shape for accommodation

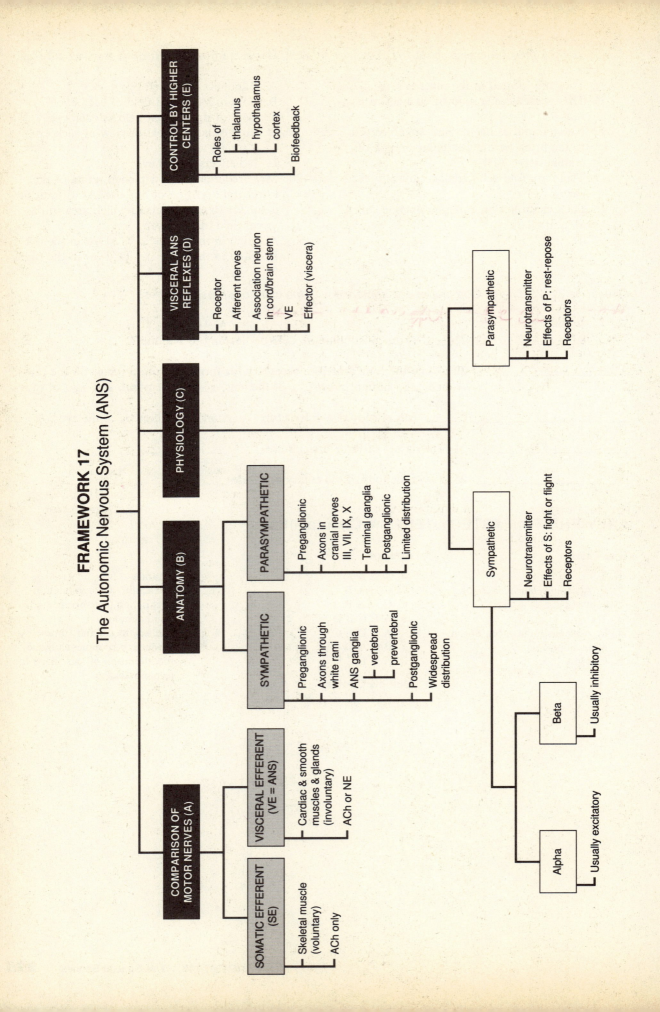

FRAMEWORK 17
The Autonomic Nervous System (ANS)

COMPARISON OF MOTOR NERVES (A)

SOMATIC EFFERENT (SE)
- Skeletal muscle (voluntary)
- ACh only

VISCERAL EFFERENT (VE = ANS)
- Cardiac & smooth muscles & glands (involuntary)
- ACh or NE

ANATOMY (B)

SYMPATHETIC
- Preganglionic
- Axons through white rami
- ANS ganglia
  - vertebral
  - prevertebral
- Postganglionic
- Widespread distribution

PARASYMPATHETIC
- Preganglionic
- Axons in cranial nerves III, VII, IX, X
- Terminal ganglia
- Postganglionic
- Limited distribution

PHYSIOLOGY (C)

Sympathetic
- Neurotransmitter
- Effects of S: fight or flight
- Receptors
  - Alpha
    - Usually excitatory
  - Beta
    - Usually inhibitory

Parasympathetic
- Neurotransmitter
- Effects of P: rest-repose
- Receptors

VISCERAL ANS REFLEXES (D)
- Receptor
- Afferent nerves
- Association neuron in cord/brain stem
- VE
- Effector (viscera)

CONTROL BY HIGHER CENTERS (E)
- Roles of
  - thalamus
  - hypothalamus
  - cortex
- Biofeedback

The Autonomic Nervous System (ANS)

The autonomic nervous system (ANS) consists of the special branch of the nervous system that exerts unconscious control over viscera. The ANS regulates activities such as heart rate and blood pressure, glandular secretion, and digestion. The two divisions of the ANS—the sympathetic and parasympathetic—carry out a continual balancing act, readying the body for response to stress or facilitating rest and relaxation, according to momentary demands. ANS neutrotransmitters hold particular significance clinically since they are often mimicked or inhibited by medications.

First study the Chapter 17 Framework and the key terms associated with each section.

TOPIC OUTLINE AND OBJECTIVES

A. Comparison of somatic and autonomic nervous systems

1. Compare the structural and functional differences between the somatic and autonomic nervous system.

B. Anatomy

2. Identify the principal structural features of the autonomic nervous system.
3. Compare the sympathetic and parasympathetic divisions of the autonomic nervous system in terms of anatomy physiology, and neurotransmitters released.

C. Physiological effects of the ANS

4. Describe the various postsynaptic receptors involved in automatic responses.

D. Visceral autonomic reflexes

5. Describe the components of a visceral autonomic reflex.

E. Control by higher centers

6. Explain the relationship of the hypothalamus to the autonomic nervous system.

WORDBYTES

Now study the following parts of words that may help you better understand terminology in this chapter.

Wordbyte	Meaning	Example	Wordbyte	Meaning	Example
auto-	self	*auto*nomic	post-	after	*post*ganglionic
nomos	law, governing	auto*nomic*	pre-	before	*pre*ganglionic
para-	near, beside	*para*vertebral, *para*thyroid	ram(i)-	branch	white *ram*us
			splanchn-	viscera	*splanchn*ic nerve

CHECKPOINTS

A. Comparison of somatic and autonomic nervous systems   (page 503)

■ **A1.** Why is the autonomic nervous system (ANS) so named? Is the ANS entirely independent of higher control centers? Explain. *It was thought to be automatic w/o any conscious control. It is regulated by the hypothalamus the medulla with input from the limbic system*

■ **A2.** Contrast the somatic and autonomic nervous systems in this table.

	Somatic	Autonomic
a. Sensations and movements are mostly (*conscious? unconscious/automatic?*)	*Conscious*	*unconscious/ automatic*
b. Types of tissue innervated by efferent nerves	Skeletal muscle	*visceral cardiac*
c. Efferent neurons are excitatory (E), inhibitory (I), or both (E + I)	*E*	*E + I*
d. Examples of sensations (input)	*special senses proprioception*	Mostly visceral sensory (such as changes in CO_2 level of blood or stretching of visceral wall)
e. Number of neurons in efferent pathway	*1*	*2*
f. Neurotransmitter(s) released by motor neurons	Acetylcholine (ACh)	*ACh + NE*

■ **A3.** Name the two divisions of the ANS.

Sympathetic parasympathetic

■ **A4.** Many visceral organs have dual innervation by the autonomic system.
  a. What does dual innervation mean? *has both afferent + efferent*

  b. How do the sympathetic and parasympathetic divisions work in harmony to control viscera? *sympathetic speeds up & activates parasympathetic rests + restores*

B. Anatomy (pages 503–509)

■ **B1.** Complete this exercise describing structural differences between visceral efferent and somatic efferent pathways.

  a. Between the spinal cord and effector, somatic pathways include __1__ neuron(s), whereas visceral pathways require __2__ neuron(s), known as the pre- *ganglion* and *post ganglion* neurons.
  b. Somatic pathways begin at (all? only certain?) levels of the cord, whereas visceral routes begin at *only certain* levels of the cord.

■ **B2.** Contrast preganglionic with postganglionic fibers in this exercise.

  a. Consider the sympathetic division first. Its preganglionic cell bodies lie in the lateral gray of segments __T1__ to __L2__ of the cord. For this reason, the sympathetic division is also known as the *thoracolumbar* outflow.
  b. Preganglionic axons (are? are not?) myelinated, and therefore they appear white.

  These axons may branch to reach a number of different *ganglia* where they synapse. These ganglia contain (pre-? post-?) ganglionic neuron cell bodies whose axons proceed to the appropriate organ. Postganglionic axons are (gray? white?) since they (do? do not?) have a myelin covering.
  c. Parasympathetic preganglionic cell bodies are located in two areas. One is the *Brainstem* from which axons pass out in cranial nerves __III__, __VII__, __IX__, and __X__. A second location is the (anterior? lateral?) gray horns of segments __S2__ through __S4__ of the cord.
  d. Based on location of (pre? post?) -ganglionic neuron cell bodies, the parasympathetic division is also known as the *craniosacral* outflow. The axons of these neurons extend to (the same? different?) autonomic ganglia from those in sympathetic pathways (see Checkpoint B4). There postganglionic cell bodies send out short (gray? white?) axons to innervate viscera.
  e. Most viscera (do? do not?) receive fibers from both the sympathetic and the parasympathetic divisions. However, the origin of these divisions in the CNS and the pathways taken to reach viscera (are the same? differ?).

■ **B3.** Contrast these two types of ganglia.

	Posterior Root Ganglia	Autonomic Ganglia
a. Contains neurons that are *(afferent? efferent?)*.		
b. Contains neurons in *(somatic? visceral? both?)* pathways.		
c. Synapsing *(does? does not?)* occur here.		

■ **B4.** Complete the table contrasting types of autonomic ganglia.

	Sympathetic Trunk	Prevertebral	Terminal
a. Sympathetic or para-sympathetic		Sympathetic	
b. Alternate name	Paravertebral ganglia or vertebral chain		
c. General location			Close to or in walls of effectors

■ **B5.** Refer to Figure 17.2 (page 507 in your text). Trace with your finger the route of a preganglionic neuron. Now describe the pathway common to *all* sympathetic pregan-glionic neurons by listing the structures in correct sequence. ___ ___ ___ ___

> A. Ventral root of spinal nerve
> B. Sympathetic trunk ganglion
> C. Lateral gray of spinal cord (between T1 and L2)
> D. White ramus communicans

■ **B6.** Suppose you could walk along the route of nerve impulses in sympathetic pathways. Start at the sympathetic trunk ganglion. It may be helpful to refer to the "map" provided by Figure 17.2, page 507 in your text.

    a. What is the shortest possible path you could take to reach a synapse to a postganglionic neuron cell body?

    b. You could then ascend and/or descend the sympathetic chain to reach trunk ganglia at other levels. This is important since sympathetic preganglionic cell bodies are

       limited in location to _____ to _____ levels of the cord, yet the sympathetic

       trunk extends from _____ to _____ levels of the vertebral column. Sympathetic fibers must be able to reach these distant areas to provide the entire body with sympathetic innervation.

    c. Now "walk" the sympathetic nerve pathway to sweat glands, blood vessels, or hair muscles in skin or extremities. Again trace this route on Figure 17.2, page 507 in the text. Preganglionic neurons synapse with postganglionic cell bodies in sympathetic trunk ganglia (at the level of entry or after ascending or descending). Postganglionic

       fibers then pass through _____ which connect to

       _____, and then convey impulses to skin or extremities.

    d. Some preganglionic fibers do not synapse as described in (b) or (c), but pass on through trunk ganglia without synapsing there. They course through

       _____ nerves to _____ ganglia. Follow their route as they synapse with postganglionic neurons whose axons form

       _____ en route to viscera.

    e. Prevertebral ganglia are located only in the _____ . In the neck, thorax, and pelvis the only sympathetic ganglia are those of the trunk (see Figure 17.1, page 506 in the text). As a result, cardiac nerves, for example, contain only *(preganglionic? postganglionic?)* fibers, as indicated by broken lines in the figure.

    f. A given sympathetic preganglionic neuron is likely to have *(few? many?)* branches; these may take any of the paths you have just "walked." Once a branch synapses, it

       (can? cannot?) synapse again, since any autonomic pathway consists of just _____ neurons (preganglionic and postganglionic).

■ **B7.** Test your understanding of these routes by completing the sympathetic pathway shown in Figure LG 17.1. A preganglionic neuron located at level T5 of the cord has its axon drawn as far as the white ramus. Finish this pathway by showing how the axon may branch and synapse to eventually innervate these three organs: the heart, a sweat gland in skin of the thoracic region, and the stomach. Draw the preganglionic fibers in solid lines and the postganglionic fibers in broken lines.

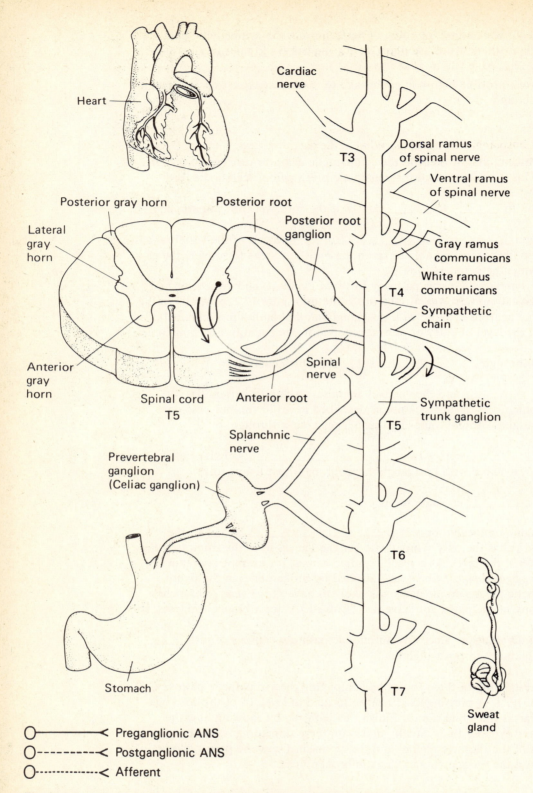

Heart

Cardiac
nerve

T3

Dorsal ramus
of spinal nerve

Ventral ramus
of spinal nerve

Posterior gray horn

Posterior root

Posterior root
ganglion

Lateral
gray
horn

Gray ramus
communicans

White ramus
communicans

T4

Sympathetic
chain

Anterior
gray
horn

Spinal cord
T5

Anterior root

Spinal
nerve

Sympathetic
trunk ganglion

T5

Splanchnic
nerve

Prevertebral
ganglion
(Celiac ganglion)

T6

Stomach

T7

Sweat
gland

O————< Preganglionic ANS

O----------< Postganglionic ANS

O·············< Afferent

Figure LG 17.1 Typical sympathetic pathway beginning at level T5 of the cord. Complete
pathways according to directions in Checkpoints B7 and D2.

■ **B8.** In the past several Checkpoints, you have been focusing on the *(sympathetic? para-sympathetic?)* division of the ANS. Now that you are somewhat familiar with the rather complex sympathetic pathways, you will notice that parasympathetic pathways are much simpler. This is due to the fact that there is no parasympathetic chain of

ganglia; the only parasympathetic ganglia are _____, which are in or close to the organs innervated. Also parasympathetic preganglionic fibers synapse with *(more? fewer?)* postganglionic neurons. What is the significance of the structural simplicity of the parasympathetic system?

■ **B9.** Complete the table about the cranial portion of the parasympathetic system.

Cranial Nerve		Name of Terminal Ganglion	Structures Innervated
Number	**Name**		
a.			Iris and ciliary muscle of eye
b.		1. Pterygopalatine	1.
		2. Submandibular	2.
c.	Glossopharyngeal		
d.		Ganglia in cardiac and pulmonary plexuses and in abdominal plexuses	

■ **B10.** *For extra review.* Write S if the description applies to the sympathetic division of the ANS, P if it applies to the parasympathetic division, and P, S if it applies to both

_____ a. Also called thoracolumbar outflow

_____ b. Has long preganglionic fibers leading to terminal ganglia and very short postganglionic fibers

_____ c. Celiac and superior mesenteric ganglia are sites of postganglionic neuron cell bodies

_____ d. Sends some preganglionic fibers through cranial nerves

_____ e. Has some preganglionic fibers synapsing in vertebral chain (trunk)

_____ f. Has more widespread effect in the body, affecting more organs

_____ g. Has some fibers running in gray rami to supply sweat glands, hair muscles, and blood vessels

_____ h. Has fibers in white rami (connecting spinal nerve with vertebral chain)

_____ i. Contains fibers that supply viscera with motor impulses

_____ j. The only division to supply kidneys and adrenal glands

■ **B11.** In Activity B1, we emphasized that autonomic pathways require two neurons.

One exception to this rule is the pathway to the _____ glands. Since this pair of glands function like modified *(sympathetic? parasympathetic?)* ganglia, no "postganglionic neurons" lead out to any other organ.

C. Physiology (pages 509–512)

■ **C1.** *(All? Most? No?)* viscera receive both sympathetic and parasympathetic innerva-

tion, known as _____ innervation of the ANS. List exceptions here:

a. Sympathetic only: _____

b. Parasympathetic only: _____

■ **C2.** Fill in the blanks to indicate which neurotransmitters are released by sympathetic (S) or parasympathetic (P) neurons or by somatic neurons. (It may help to refer to Figure LG 17.2.) Use these answers:

> ACh. Acetylcholine (cholinergic)
> NE. Norepinephrine = noradrenalin, or possibly epinephrine = adrenalin (adrenergic)

_____ a. All preganglionic neurons (both S and P) release this neurotransmitter

_____ b. Most S postganglionic neurons

_____ c. A few S postganglionic neurons, namely those to sweat glands and some blood vessels

_____ d. All P postganglionic neurons

_____ e. All somatic neurons

For extra review, color the neurotransmitters on Figure LG 17.2, matching colors you use in the key with those you use in the figure.

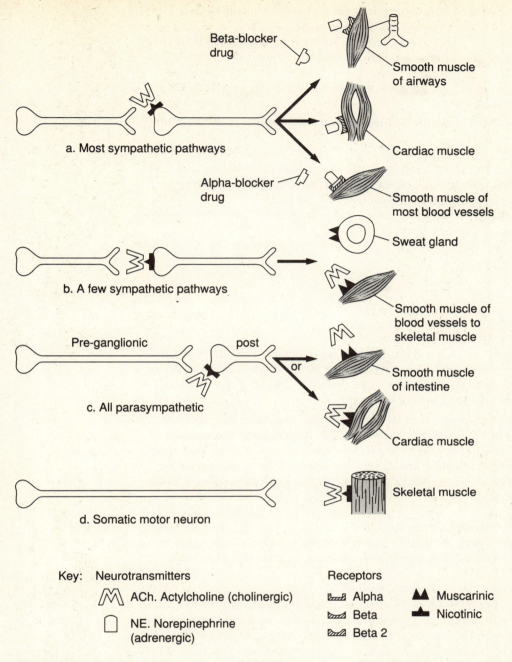

Beta-blocker drug

Smooth muscle of airways

a. Most sympathetic pathways

Cardiac muscle

Alpha-blocker drug

Smooth muscle of most blood vessels

Sweat gland

b. A few sympathetic pathways

Smooth muscle of blood vessels to skeletal muscle

Pre-ganglionic   post

or

Smooth muscle of intestine

c. All parasympathetic

Cardiac muscle

d. Somatic motor neuron

Skeletal muscle

Key: Neurotransmitters   Receptors

ACh. Actylcholine (cholinergic)   Alpha   Muscarinic

NE. Norepinephrine   Beta   Nicotinic
(adrenergic)   Beta 2

Figure LG 17.2   Diagram of autonomic and somatic neurons, neurotransmitters, and receptors. Note that neurotransmitters and receptors are relatively enlarged. Refer to Checkpoints C2 and C5 to C7.

■ **C3.** Do this activity summarizing the effects of autonomic neurotransmitters.

a. Axons that release the transmitter acetylcholine (ACh) are known as *(adrenergic? cholinergic?)*. Those that release norepinephrine (NE, also called noradrenalin) are

called _____. Most sympathetic nerves are said to be *(choliner-*

gic? adrenergic?), whereas all parasympathetic nerves are _____.

b. During stress, the *(sympathetic? parasympathetic?)* division of the ANS prevails, so stress responses are primarily *(adrenergic? cholinergic?)* responses.

c. Another source of NE as well as epinephrine is the gland known as the

_____. Therefore, chemicals released by this gland (as in stress) will mimic the action of the *(sympathetic? parasympathetic?)* division of the ANS.

■ **C4.** Do this exercise relating the fate of ANS neurotransmitters and overall effects of the two divisions of the ANS.

a. On the arrows below, write the initials of the enzymes that destroy each transmitter:

ACh ⟶ A̶C̶h̶   NE ⟶ N̶E̶

b. The ANS neurotransmitter that "hangs around" longer (has more lasting effect) is *(ACh? NE?)*. This fact provides one explanation for the longer-lasting effects of *(sympathetic/adrenergic? parasympathetic/cholinergic?)* neurons.

c. State two other rationales for the more lasting and widespread effects of the sympathetic system:

1.

2.

■ **C5.** Again refer to Figure LG 17.2 and do this exercise on neurotransmitters and related receptors. Be sure to match the color you use for receptor symbols in the key to the color you use for receptors in the figure. (The symbol S denotes *sympathetic* and P indicates *parasympathetic* in the following statements.)

a. First identify all nicotinic receptors in the figure. Notice that these receptors are found on cell bodies of *(P postganglionic neurons? S postganglionic neurons? autonomic effector cells? somatic effector cells [skeletal muscle]?)*. (Circle *all* that apply.) In other words, nicotinic receptors are all sites for action of the neurotransmitter *(ACh? NE? either ACh or NE?)*. Activation of nicotinic receptors is *(always? sometimes? never?)* excitatory.

b. On what other types of receptors can ACh act? _____. Identify these receptors on the figure. As you do, notice that these receptors are found on *all effectors* stimulated by *(P? S?)* postganglionic neurons. In addition, muscarinic

receptors are found on a few S effectors, such as _____. Activation of muscarinic receptors is *(always? sometimes? never?)* excitatory. For example, as P nerves (vagus) release ACh to a muscarinic receptor in cardiac muscle, heart activity *(increases? decreases?)*, but when P nerves (vagus) release ACh to a muscarinic receptor in the wall of the stomach or intestine, gastrointestinal activity *(increases? decreases?)*.

c. Explain what accounts for the names *nicotinic* and *muscarinic* for ACh receptors.

The names are based on the fact that the action of _____ on

these receptors is mimicked by action of _____ on nicotinic

receptors and by a poison named _____ from mushrooms on muscarinic receptors.

d. *Alpha* and *beta* receptors are found only on effectors innervated by *(P? S?)* nerves; these effectors are stimulated by *(ACh? NE and epinephrine?)*. NE and epinephrine

are categorized as _____ neurotransmitters.

e. *A clinical challenge.* Alex receives word that his brother has been seriously injured in an automobile accident. Among other immediate physical responses, his skin "goes pale." Explain this response. (Refer to Exhibit 17.3, page 512 in your text, for help.) Which type of receptor is found in smooth muscle of blood vessels of skin? *(Alpha? Beta$_1$? Beta$_2$?)* Alpha receptors are usually *(excitatory? inhibitory?)*. As sympathetic (the "stress system") nerves release NE, smooth muscle of blood vessels are excited, causing them to *(constrict/narrow? dilate/widen?)*. As a result, blood flows *(into? out of?)* skin, and the skin pales. How may this response help Alex?

■ **C6.** *A clinical challenge.* Certain medications can mimic the effects of the body's own transmitters. Again refer to Figure LG 17.2 and do this exercise.

a. A beta$_1$-stimulator (or beta-exciter) mimics *(ACh? NE?)*, causing the heart rate and strength of contraction to ____-crease. Name one such drug. (For help, refer to a

pharamacology text.) _____ Such a drug has a structure that is close enough to the shape of NE that it can "sit on" the beta$_1$ receptor as NE would and activate it.

b. A beta-blocker (or beta-inhibitor) also has a shape similar to that of NE and so can

sit on _____ receptors. Name one such drug.

_____. However, this drug cannot activate the receptor, but simply prevents NE from doing so. So a beta-blocker ____-creases heart rate and force of contraction.

c. Other drugs, known as anticholinergics (or ACh-blockers), can take up residence on ACh receptors so that ACh being released from vagal nerve stimulation cannot exert its normal effects. A person taking such a medication may exhibit a(n) *(increased? decreased?)* heart rate.

■ **C7.** *For extra review.* Do this additional exercise on more effects of medications on ANS nerves.

a. What types of receptors are located on smooth muscle of bronchi and bronchioles (airways)? *(Alpha? Beta$_1$? Beta$_2$?)* (For help, refer to Exhibit 17.3, page 512 in your text.) The effect of the sympathetic stimulation of lungs is to cause *(dilation? constriction?)* of airways, making breathing *(easier? more difficult?)* during stressful times.

b. The effect of NE on the smooth muscle of the stomach, intestines, bladder, and uterus is also *(contraction? relaxation?)*, so that during stress, the body ____-creases activity of these organs and can focus on more vital activities such as heart contractions.

c. From Activity C5e, you may recognize that the sympathetic response during stress causes many blood vessels, such as those in skin, to *(constrict? dilate?)*. Prolonged stress may therefore lead to *(high? low?)* blood pressure. One type of medication used to lower blood pressure is an alpha-*(blocker? stimulator?)*, since it tends to dilate these vessels and "pool" blood in nonessential areas like skin, thereby avoiding overloading the heart with blood flow.

■ **C8.** Use arrows to show whether parasympathetic (P) or sympathetic (S) fibers stimulate (↑) or inhibit (↓) each of the following activities. Use a dash (–) to indicate that there is no parasympathetic innervation. The first one is done for you.

a. P ___↓___   S ___↑___   Dilation of pupil

b. P _____   S _____   Heart rate and blood flow to coronary (heart muscle) blood vessels

c. P _____   S _____   Constriction of skin blood vessels

d. P _____   S _____   Salivation and digestive organ contractions

e. P _____   S _____   Erection of genitalia

f. P _____   S _____   Dilation of bronchioles for easier breathing

g. P _____   S _____   Contraction of bladder and relaxation of internal urethral sphincter causing urination

h. P _____   S _____   Contraction of pili of hair follicles causing "goose bumps"

i. P _____   S _____   Contraction of spleen which transfers some of its blood to general circulation, causing increase in blood pressure

j. P _____   S _____   Release of epinephrine and norepinephrine from adrenal medulla

k. P _____   S _____   Coping with stress, fight-or-flight response

C9. Write a paragraph outlining the changes effected by the ANS during a fight-or-flight response, such as running in fear.

D. Visceral autonomic reflexes   (pages 512–513)

■ **D1.** Autonomic neurons can be stimulated by *(somatic afferents? visceral afferents? either somatic or visceral afferents?)*.

■ **D2.** On Figure LG 17.1 draw an afferent neuron (in contrasting color or with a dotted line) to show how the sympathetic neuron could be stimulated by stomach pain.

■ **D3.** Arrange in correct sequence structures in the pathway for a painful stimulus at your fingertip to cause a visceral (autonomic) reflex, such as sweating. Write the letters of the structures in order on the lines provided.

___ ___ ___ ___ ___ ___ ___ ___ ___ ___

> A. Association neuron in spinal cord
> B. Nerve fiber in gray ramus
> C. Pain receptor in skin
> D. Cell body of postganglionic neuron in trunk ganglion
> E. Cell body of preganglionic neuron in lateral gray of cord
>
> F. Nerve fiber in anterior root of spinal nerve
> G. Nerve fiber in white ramus
> H. Fiber in spinal nerve in brachial plexus
> I. Sweat gland
> J. Sensory neuron

■ **D4.** Most visceral sensations *(do? do not?)* reach the cerebral cortex, so most visceral sensations are at *(conscious? subconscious?)* levels. Hunger and nausea are exceptions.

E. Control by higher centers   (pages 513–514)

E1. Explain the role of each of these structures in control of the autonomic system.

a. Hypothalamus (Which part controls sympathetic nerves? Which controls parasympathetic?)

b. Cerebral cortex (When is it most involved in ANS control? In stress or nonstress situations?)

E2. Explain how biofeedback is used to help in each of the following cases.

a. Tension headache

b. Childbirth

ANSWERS TO SELECTED CHECKPOINTS

A1. It was thought to be autonomous (self-governing). However, it is regulated by brain centers such as the hypothalamus and medulla, with input from the limbic system and other parts of the cerebrum.

A2.

	Somatic	Autonomic
a. Sensations and movements are mostly *(conscious? unconscious/automatic?)*	**Conscious**	**Unconscious/ automatic**
b. Types of tissue innervated by efferent nerves	Skeletal muscle	**Cardiac muscle, smooth muscle, and most glands**
c. Efferent neurons are excitatory (E), inhibitory (I), or both (E or I)	**E**	**E or I**
d. Examples of sensations (input)	**Special senses, general somatic senses, or proprioceptors, and possibly visceral sensory, as when pain of appendicitis or menstrual cramping causes a person to curl up (flexion of thighs, legs)**	Mostly visceral sensory (such as changes in CO_2 level of blood or stretching of visceral wall)
e. Number of neurons in efferent pathway	**One (from spinal cord to effector)**	**Two (pre- and postganglionic)**
f. Neurotransmitter(s) released by motor neurons	Acetylcholine (ACh)	**ACh or norepinephrine (NE) and epinephrine**

A3. Sympathetic and parasympathetic.

A4. (a) Most viscera are innervated by both sympathetic and parasympathetic divisions. (b) One division excites, and the other division inhibits the organ's activity.

B1. (a) 1, 2, ganglionic, postganglionic. (b) All, only certain (see Checkpoint B2).

B2. (a) T1, L2; thoracolumbar. (b) Are; ganglia; post-; gray, do not. (c) Brainstem; III, VII, IX, X; lateral; S2, S4. (d) Pre-, craniosacral; different; gray. (e) Do; differ.

B3. (a) Afferent, efferent. (b) Both, visceral. (c) Does not, does.

B4.

	Sympathetic Trunk	Prevertebral	Terminal
a. Sympathetic or para-sympathetic	**Sympathetic**	Sympathetic	**Para-sympathetic**
b. Alternate name	Paravertebral ganglia or vertebral chain	**Collateral or prevertebral (celiac, superior and inferior mesenteric)**	**Intramural (in walls of viscera)**
c. General location	**In vertical chain along both sides of vertebral bodies from base of skull to coccyx**	**In three sites anterior to spinal cord and close to major abdominal arteries**	Close to or in walls of effectors

B5. C A D B.

B6. (a) Immediately synapse in trunk ganglion. (b) T1, L2; C3, sacral. (c) Gray rami, spinal nerves. (d) Splanchnic, prevertebral; plexuses. (e) Abdomen; postganglionic. (f) Many; cannot, 2.

B7.

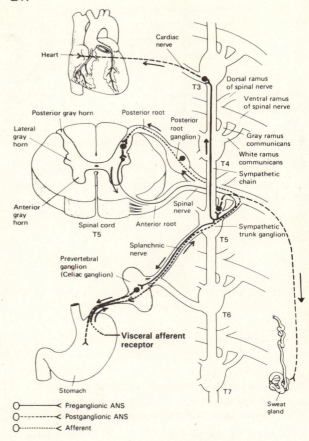

Figure LG 17.1A   Typical sympathetic pathway beginning at level T5 of the cord.

B8. Sympathetic; terminal ganglia; fewer. Parasympathetic effects are much less widespread than sympathetic, distributed to fewer areas of the body, and characterized by a response of perhaps just one organ, rather than an integrated response in which many organs respond *in sympathy* with one another.

B9.

| Cranial Nerve | | Name of Terminal Ganglion | Structures Innervated |
Number	Name		
a. III	Oculomotor	Ciliary	Iris and ciliary muscle of eye
b. VII	Facial	1. Pterygopalatine 2. Submandibular	1. **Nasal mucosa, palate, pharynx, lacrimal gland** 2. **Submandibular and submaxillary salivary glands**
c. IX	Glossopharyngeal	Otic	**Parotid salivary gland**
d. X	Vagus	Ganglia in cardiac and pulmonary plexuses and in abdominal plexuses	**Thoracic and abdominal viscera**

B10. (a) S. (b) P. (c) S. (d) P. (e) S. (f) S. (g) S. (h) S. (i) P, S. (j) S.

B11. Adrenal (medulla); sympathetic.

C1. Most, dual. (a) Sweat glands, arrector pili (hair) muscles, fat cells, kidneys, adrenal gland, and most blood vessels. (b) Lacrimal (tear) glands. (Refer to Exhibit 17.3 in your text, pages 512–513)

C2. (a) ACh. (b) NE. (c–e) ACh. (Refer to Figure LG 17.2.)

C3. (a) Cholinergic; adrenergic; adrenergic, cholinergic. (b) Sympathetic, adrenergic. (c) Adrenal medulla; sympathetic.

C4. (a)

ACh $\xrightarrow{\text{AChE}}$ ~~ACh~~

NE $\xrightarrow{\text{COMT or MAO}}$ ~~NE~~

(b) NE; sympathetic/adrenergic. (c) 1. Greater divergence of sympathetic neurons, for example, through the extensive sympathetic trunk chain and through gray rami to all spinal nerves. 2. Mimicking of sympathetic responses via release of catecholamines (epinephrine and NE) from the adrenal medulla.

C5. (a) Refer to key for Figure LG 17.2; both P and S postganglionic neurons and somatic effector cells (skeletal muscle); ACh; always. (b) Muscarinic (*Hint:* Take time to notice that in this figure and in Figure 17.3, page 510 of your text, the *m*uscarinic receptors are ones shaped so that ACh "sits down" in an "M" [for muscarinic] position, whereas nicotinic receptors have ACh attach in the "W" position); P (notice this at top of Exhibit 17.3, page 512 in text); blood vessels that supply skeletal muscles and many sweat glands; sometimes; decreases; increases. (c) ACh, nicotine, muscarine. (d) S; NE, and epinephrine; catecholamines. (e) Alpha; excitatory; constrict/narrow; out of; blood flows into major vessels to increase blood pressure, which may assist in a stress response.

C6. (a) NE, in; isoproterinol (Isoprel) or epinephrine. (b) Beta; propranolol (Inderal); de. (c) Increased (a double negative: the inhibiting vagus is itself inhibited).

C7. (a) Beta$_2$; dilation, easier. (b) Relaxation, de. (c) Constrict; high; blocker.

C8. (b) P ↓, S ↑. (c) P–, S ↑. (d) P ↑, S ↓. (e) P ↑, S ↓. (f) P ↓, S ↑. (g) P ↑, S ↓. (h) P–, S ↑. (i) P–, S ↑. (j) P–, S ↑. (k) P ↓, S ↑.

D1. Either somatic or visceral afferents.

D2. See Figure LG 17.1A above.

D3. C J A E F G D B H I.

D4. Do not, subconscious.

Questions 1–6: Circle letters preceding all correct answers to each question.

1. Choose all *true* statements about the vagus nerve.
   A. Its autonomic fibers are sympathetic.
   B. It supplies ANS fibers to viscera in the thorax and abdomen, but not in the pelvis.
   C. It causes the salivary glands and other digestive glands to increase secretions.
   D. Its ANS fibers are mainly preganglionic.
   E. It is a cranial nerve originating from the medulla.

2. Which activities are characteristic of the stress response, or fight-or-flight reaction?
   A. The liver breaks down glycogen to glucose.
   B. The heart rate decreases.
   C. Kidneys increase urine production since blood is shunted to kidneys.
   D. There is increased blood flow to genitalia, causing erect state.
   E. Hairs stand on end ("goose pimples") due to contraction of arrector pili muscles.
   F. In general, the sympathetic system is active.

3. Which fibers are classified as autonomic?
   A. Any visceral efferent nerve fiber
   B. Any visceral afferent nerve fiber
   C. Nerves to salivary glands and sweat glands
   D. Pain fibers from ulcer in stomach wall
   E. Sympathetic fibers carrying impulses to blood vessels
   F. All nerve fibers within cranial nerves
   G. All parasympathetic nerve fibers within cranial nerves

4. Choose all *true* statements about gray rami.
   A. They contain only sympathetic nerve fibers.
   B. They contain only postganglionic nerve fibers.
   C. They carry impulses from trunk ganglia to spinal nerves.
   D. They are located at all levels of the vertebral column (from C1 to coccyx).
   E. They carry impulses between lateral ganglia and collateral ganglia.
   F. They carry preganglionic neurons from anterior ramus of spinal nerve to trunk ganglion.

5. Which of the following structures contain some sympathetic preganglionic nerve fibers?
   A. Splanchnic nerves
   B. White rami
   C. Sciatic nerve
   D. Cardiac nerves
   E. Ventral roots of spinal nerves
   F. The sympathetic chains

6. Which are structural features of the parasympathetic system?
   A. Ganglia close to the CNS and distant from the effector
   B. Forms the craniosacral outflow
   C. Distributed throughout the body, including extremities
   D. Supplies nerves to blood vessels, sweat glands, and adrenal gland
   E. Has some of its nerve fibers passing through lateral (paravertebral) ganglia

Questions 7–10: Circle the letter preceding the one best answer to each question.

7. All of the following axons are cholinergic *except:*
   A. Parasympathetic preganglionic
   B. Parasympathetic postganglionic
   C. Sympathetic preganglionic
   D. Sympathetic postganglionic to sweat glands
   E. Sympathetic postganglionic to heart muscle

8. All of the following are collateral ganglia *except:*
   A. Superior cervical
   B. Prevertebral ganglia
   C. Celiac ganglion
   D. Superior mesenteric ganglion
   E. Inferior mesenteric ganglion

9. Which region of the cord contains no preganglionic cell bodies at all?
   A. Sacral
   B. Lumbar
   C. Cervical
   D. Thoracic

10. Which statement about postganglionic neurons is *false?*
    A. They all lie entirely outside of the CNS.
    B. Their axons are unmyelinated.
    C. They terminate in visceral effectors.
    D. Their cell bodies lie in the lateral gray matter of the cord.
    E. They are very short in the parasympathetic system.

Questions 11–20: Circle T (true) or F (false). If the statement is false, change the underlined word or phrase so that the statement is correct.

T F 11. Control of the ANS by the cerebral cortex occurs primarily during times when a person is <u>relaxed (nonstressed)</u>.

T F 12. Synapsing occurs in both <u>sympathetic and parasympathetic</u> ganglia.

T F 13. The sciatic, brachial, and femoral nerves all contain <u>sympathetic postganglionic</u> nerve fibers.

T F 14. Biofeedback and meditation confirm that the ANS is <u>independent of</u> higher control centers.

T F 15. Under stress conditions the <u>sympathetic system</u> dominates over the parasympathetic.

T F 16. When one side of the body is deprived of its sympathetic nerve supply, as in Horner's syndrome, the following symptoms can be expected: <u>constricted pupil (miosis) and lack of sweating (anhidrosis)</u>.

T F 17. Sympathetic cardiac nerves <u>stimulate</u> heart rate, and the vagus <u>slows down</u> heart rate.

T F 18. Viscera <u>do have sensory nerve fibers, but they are not included in the autonomic nervous system</u>.

T F 19. <u>All</u> viscera have dual innervation by sympathetic and parasympathetic divisions of the ANS.

T F 20. The sympathetic system has a <u>more</u> widespread effect in the body than the parasympathetic does.

Questions 21–25: fill-ins. Complete each sentence with the word or phrase that best fits.

_____ 21. The three types of tissue (effectors) innervated by the ANS nerves are ___.

_____ 22. ___ nerves convey impulses from the sympathetic trunk ganglia to collateral ganglia.

_____ 23. About 80 percent of the craniosacral outflow (parasympathetic nerves) is located in the ___ nerves.

_____ 24. Alpha (α) and beta (β) receptors are stimulated by the transmitter ___.

_____ 25. Drugs that block (inhibit) beta receptors in the heart will cause ___-crease in heart rate and blood pressure.

ANSWERS TO MASTERY TEST: ■ Chapter 17

Multiple Answers

1. B, C, D, E
2. A, E, F
3. A, C, E, G
4. A, B, C, D
5. A, B, E, F
6. B

Multiple Choice

7. E   9. C
8. A   10. D

True-False

11. F. Stressed
12. T
13. T
14. F. Dependent on
15. T
16. T
17. T
18. T
19. F. Most (or some)
20. T

Fill-ins

21. Cardiac muscle, smooth muscle, and glandular epithelium
22. Splanchnic
23. Vagus
24. NE (and also epinephrine)
25. De

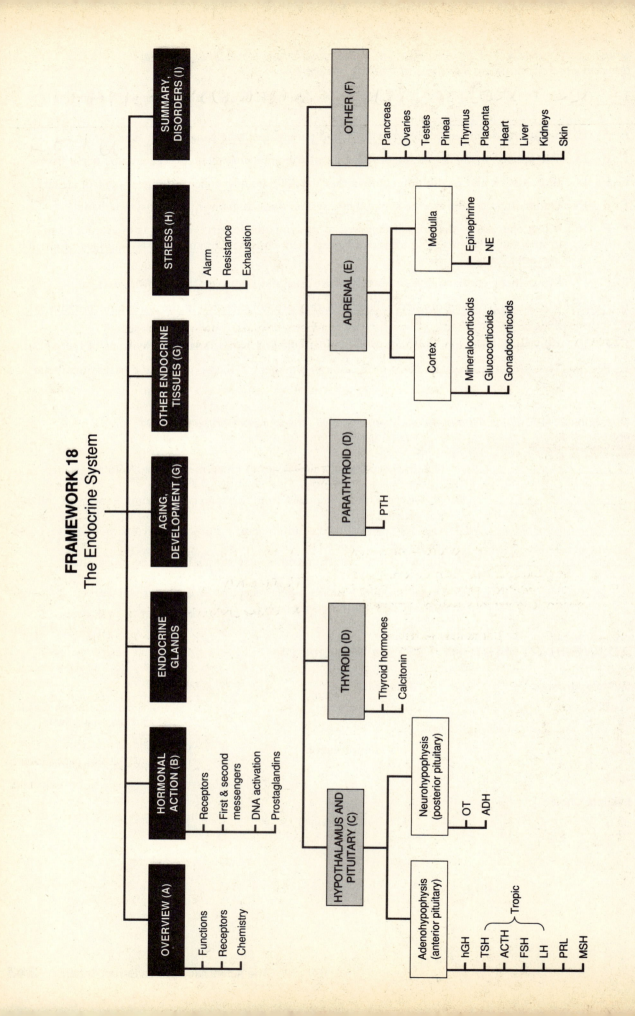

FRAMEWORK 18
The Endocrine System

OVERVIEW (A)
- Functions
- Receptors
- Chemistry

HORMONAL ACTION (B)
- Receptors
- First & second messengers
- DNA activation
- Prostaglandins

ENDOCRINE GLANDS

AGING, DEVELOPMENT (G)

OTHER ENDOCRINE TISSUES (G)

STRESS (H)
- Alarm
- Resistance
- Exhaustion

SUMMARY, DISORDERS (I)

HYPOTHALAMUS AND PITUITARY (C)
- Adenohypophysis (anterior pituitary)
  - hGH
  - TSH
  - ACTH
  - FSH
  - LH
  - PRL
  - MSH
  - Tropic (ACTH, FSH, LH)
- Neurohypophysis (posterior pituitary)
  - OT
  - ADH

THYROID (D)
- Thyroid hormones
- Calcitonin

PARATHYROID (D)
- PTH

ADRENAL (E)
- Cortex
  - Mineralocorticoids
  - Glucocorticoids
  - Gonadocorticoids
- Medulla
  - Epinephrine
  - NE

OTHER (F)
- Pancreas
- Ovaries
- Testes
- Pineal
- Thymus
- Placenta
- Heart
- Liver
- Kidneys
- Skin

The Endocrine System

CHAPTER
18

Hormones produced by at least 18 organs exert widespread effects on just about every body tissue. Hormones are released into the bloodstream, which serves as the vehicle for distribution throughout the body. In fact, analysis of blood levels of hormones can provide information about the function of specific endocrine glands. How do hormones "know" which cells to affect? And how does the body "know" when to release more or less of a hormone? Mechanisms of action and regulation of hormone levels are discussed in this chapter. The roles of all major hormones and effects of excesses or deficiencies in those hormones are also included. The human body is continually exposed to stressors, such as rapid environmental temperature change, a piercing sound, or serious viral infection. Adaptations to stressors are vital to survival; however, certain adaptations lead to negative consequences. A discussion of stress and adaptation concludes Chapter 18.

One way to alleviate the stress associated with learning about stress is to consult the Chapter 18 Framework and to study the key terms for each section.

TOPIC OUTLINE AND OBJECTIVES

A. Overview: functions, receptors, chemistry

1. Define the components of the endocrine system and discuss the functions of the endocrine and nervous system in maintaining homeostasis.
2. Describe how hormones are transported in the blood and how they interact with target cell receptors.
3. Compare the four chemical classes of hormones.

B. Mechanisms of hormonal action and regulation

4. Explain the two general mechanisms of hormonal action.
5. Describe the control of hormonal secretions via feedback cycles and give several examples.

C. Hypothalamus and pituitary (hypophysis)

6. Describe the release of hormones stored in the posterior pituitary gland.

D. Thyroid and parathyroids

E. Adrenals

F. Other endocrine glands: pancreas, gonads, pineal (epiphysis cerebri), thymus

7. Describe the location, histology, hormones, and functions of the following endocrine glands: pituitary, thyroid, parathyroids, adrenals, pancreas, ovaries, testes, pineal, and thymus.

G. Aging and developmental anatomy of the endocrine system; other endocrine tissues

8. Describe the development of the endocrine system.
9. Describe the effects of aging on the endocrine system.

H. Stress and homeostasis

10. Define the general adaptation syndrome (GAS) and compare homeostatic responses and stress responses.

I. Summary of hormones, disorders

11. Discuss the symptoms of pituitary dwarfism, giantism, acromegaly, diabetes insipidus, cretinism, myxedema, goiter, tetany, osteitis fibrosa cystica, aldosteronism, Addison's disease, Cushing's syndrome, adrenogenital syndrome, pheochromocytoma, diabetes mellitus, and hyperinsulinism.

WORDBYTES

Now study the following parts of words that may help you better understand the terminology in this chapter.

Wordbyte	Meaning	Example	Wordbyte	Meaning	Example
adeno-	gland	*adeno*hypophysis	insipid-	without taste	diabetes *insipidus*
auto-	self	*auto*crine			
crin-	to secrete	endo*crine*	mellit-	sweet	diabetes *mellitus*
endo-	within	*endo*crine	oxy(s)-	swift	*oxy*tocin
exo-	outside	*exo*crine	para-	around	*para*thyroid, *para*crine
gen-	to create	diabeto*gen*ic			
horm-	excite, urge on	*horm*one	-tocin	childbirth	pi*tocin*
			trop-	turn	thyro*trop*in

CHECKPOINTS

A. Overview: functions, receptors, chemistry (pages 518–522)

■ **A1.** Complete the table contrasting exocrine and endocrine glands.

	Secretions Transported In	Examples
a. Exocrine	*ducts to body cavities*	*sweat, oil, mucous & digestive*
b. Endocrine	*blood stream*	*pituitary, adrenal, pineal, thyroid, pancreas* *Hormones*

■ **A2.** Refer to Figure LG 18.1 and do this exercise.

a. Color and label each endocrine gland next to letters A–H.
b. Next to numbers 1–10 label each organ containing endocrine tissue. *For extra review,* identify the location of each organ in your own body.

■ **A3.** Compare the ways in which the nervous and endocrine systems exert control over the body. Write N (nervous) or E (endocrine) next to descriptions that fit each system.

_N_ a. Sends messages to muscles, glands, and neurons only.

_E_ b. Sends messages to virtually any part of the body.

_N_ c. Effects are generally faster and shorter-lived.

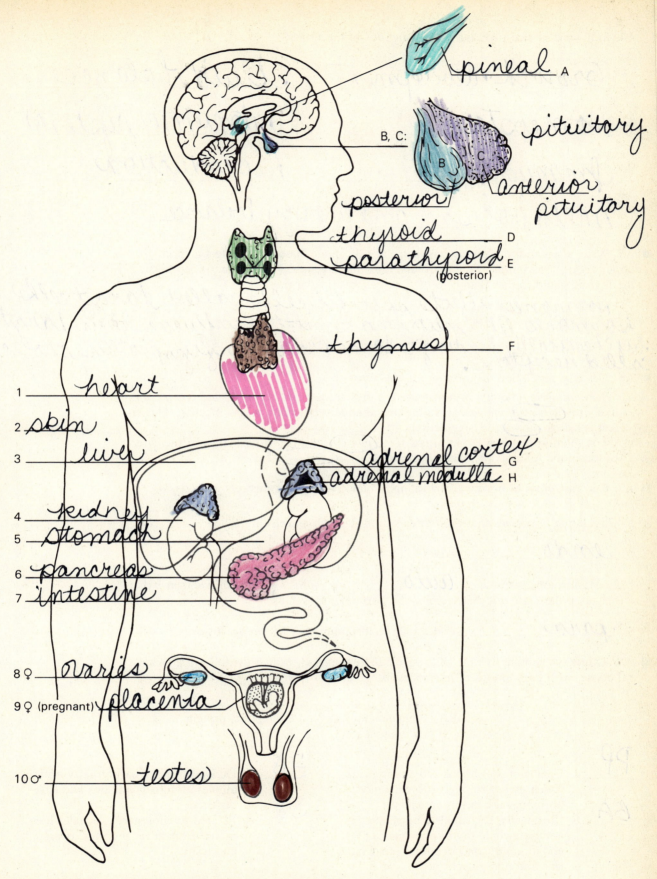

pineal A

pituitary

B, C:

posterior

anterior
pituitary

thyroid D

parathyroid E
(posterior)

thymus F

1 heart

2 skin

3 liver

adrenal cortex G
adrenal medulla H

4 kidney

5 stomach

6 pancreas

7 intestine

8 ♀ ovaries

9 ♀ (pregnant) placenta

10 ♂ testes

Figure LG 18.1   Diagram of endocrine glands in black (A–H) and organs containing endocrine tissue (1–10) in gray. Label and color as directed in Checkpoints A2 and C1.

A4. In a word or two each, list seven functions of hormones.

1. _Growth & development_
2. _Homeostasis_
3. _Immune_
4. _metabolism and energy balance_
5. _fluid balance_
6. _autonomic system_
7. _Reproduction_

■ **A5.** Write a sentence explaining how a specific hormone "knows" to affect a specific target organ, for example, how thyroid-stimulating hormone (TSH) "knows" to affect the thyroid gland, rather than cells of pancreas, ovaries, or muscles.

Hormones affects specific cells called target cells. Hormones like neurotransmitters influence their target cell by chemically binding to large protein or glycoprotein molecule called receptors.

■ **A6.** Describe regulation of hormones in this exercise.

a. When excessive amounts of hormones are present, the number of related receptors is likely to (decrease? increase?), so the overabundant hormone is less effective. This effect is called (down? up?)-regulation.

b. Up-regulation makes a target organ (less? more?) sensitive to hormones by increase of receptors, for example, when hormone level is (deficient? excessive?).

■ **A7.** Contrast types of hormones by filling in blanks. Use these answers: *auto, endo, para*.

a. Circulating hormones that exert effects on distant target cells are called

endo -crines.

b. Local hormones include _auto_ -crines such as interleuken-2 that act upon the same cell that secreted the hormone, as well as

para -crines such as histamine that act on nearby cells.

■ **A8.** Carefully examine Exhibit 18.1, page 521 in your text. Match the following chemical classes with descriptions of hormones listed below.

BA. Biogenic amines	PP. Peptide, protein
E. Eicosanoid	S. Steroid

_PP_ a. Chains of 3 to about 200 amino acids which may form glycoproteins; most hormones are in this category.

_BA_ b. Chemically, the simplest hormones; examples include thyroid hormones (T_3 and T_4) and catecholamines (epinephrine and norepinephrine).

_E_ c. Derived from 20-carbon fatty acid; examples are prostaglandins and leukotrienes.

_S_ d. Related to cholesterol; includes hormones from adrenal cortex, ovary, and testis.

■ **A9.** *A clinical challenge.* Give a rationale for *parenteral* (by injection) rather than *p.o.* (by mouth) administration of insulin.

goes directly into blood stream doesn't get digested in stomach destroying their peptide bonds.

■ **A10.** Contrast transport mechanisms of hormones. Use these answers:

> FF. Water-soluble hormone that circulates in *free form* in blood
> TP. Lipid-soluble hormone that circulates attached to *transport protein*

The lines following each hormone group are for Activity B1.

TP a. Thyroid hormone *I*

TP b. Steroid, such as hydrocortisone or progesterone *I*

FF c. Catecholamine *P*
NE & Epi

FF d. Protein or peptide, including all hypothalamic and pituitary hormones
P

B. Mechanisms of hormonal action and regulation   (pages 522–524)

■ **B1.** Look back at Activity A10. On the lines following descriptions of chemical groups of hormones (a–d), write letters indicating the expected mechanism of hormonal action. (*Hint:* Remember that each action depends on the chemistry of the hormone.) Use these answers:

I. Since a lipid-soluble hormone, it can enter cell to use *intracellular receptors* that activate DNA to direct synthesis of new proteins

P. Since not lipid-soluble, the hormone uses *plasma membrane receptors* to activate second messengers such as cyclic AMP.

■ **B2.** Study the cyclic AMP mechanism in Figure 18.4 (page 523 in your text). Then complete this exercise.

a. A hormone such as antidiuretic hormone (ADH) acts as the

_____*first*_____ messenger, as it carries a message from the

_____ where it is secreted (in this case the hypothalamus) to

the _____ where it acts (in this case kidney cells).

b. The hormones then bind to a receptor on the *(inner? outer?)* surface of the plasma membrane, causing activation of ____-proteins.

c. Such activation increases activity of the enzyme _____, located on the *(inner? outer?)* surface of the plasma membrane. This enzyme catalyzes

conversion of ATP to _____, which is known as the

_____ messenger.

d. Cyclic AMP then activates one or more enzymes known as

_____ to help transfer *(calcium? phosphate?)* from ATP to a protein, usually an enzyme. The resulting chemical can then set off the target cell's response. In the case of ADH, this is an increase in permeability of kidney cells,

with a resulting decrease in _____ production.

e. Effects of cyclic AMP are *(short? long?)*-lived. This chemical is rapidly degraded by

an enzyme named _____.

f. A number of hormones are known to act by means of the cyclic AMP mechanism.

Included are most of the _____-soluble hormones. Name several.

g. Name two or more chemicals besides cyclic AMP that may act as second messengers.

B3. Explain what is meant by amplification of hormone effects.

■ **B4.** Describe hormonal interactions in this activity.

a. The interaction by which effects of progesterone on the uterus in preparation for pregnancy are enhanced by earlier exposure to estrogen is known as a(n) *(antagonistic? permissive? synergistic?)* effect.

b. Insulin and glucagon exert _____ effects on blood glucose levels.

■ **B5.** Precise regulation of hormone levels is critical to homeostasis. Briefly describe control mechanisms for each hormone listed below. Use these answers:

> C. Blood level of a chemical that is controlled by the hormone
> H. Blood level of another hormone
> N. Nerve impulses

_____ a. Epinephrine   _____ c. Cortisol

_____ b. Parathyroid hormone

■ **B6.** *(Negative? Positive?)* feedback mechanisms are utilized for most hormonal regulation. For example, a low level of Ca^{2+} circulating through the parathyroid gland causes a(n) *(increase? decrease?)* in release of parathyroid hormone (PTH). State one or more example(s) of hormone control by *positive* feedback.

C. Hypothalamus and pituitary (hypophysis)   (pages 524-535)

■ **C1.** Complete this exercise about the pituitary gland.

a. The pituitary is also known as the _____.

b. Defend or dispute this statement, "The pituitary is the master gland."

c. Where is it located?

d. Seventy-five percent of the gland consists of the *(anterior? posterior?)* lobe, called

the _____-hypophysis. The posterior lobe (or

_____-hypophysis) is somewhat smaller.

e. The anterior pituitary is known to secrete *(2? 5? 7?)* different hormones. Write abbreviations for each of these hormones next to the diagram of the anterior pituitary on Figure LG 18.1.

f. What advantage is afforded by the special arrangement of blood vessels that link the hypothalamus with the pituitary (hypophysis)?

g. What are functions of these three types of anterior pituitary cells?

   1. Lactotrophs:_____

   2. Somatotrophs:_____

   3. Gonadotrophs:_____

h. Define *tropic* hormones.

Which four of the anterior pituitary hormones are tropic hormones?

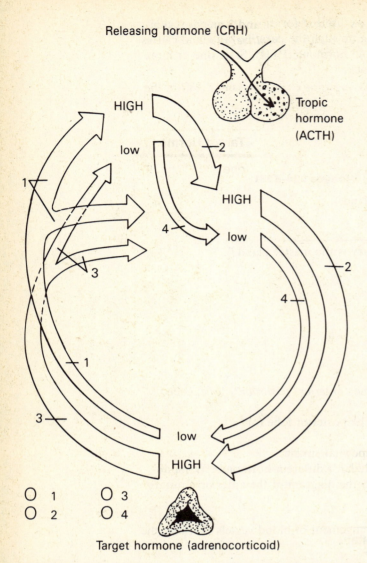

Releasing hormone (CRH)

HIGH

low

Tropic
hormone
(ACTH)

2

HIGH

low

4

2

4

1

1

3

3

low

HIGH

low

HIGH

○ 1   ○ 3
○ 2   ○ 4

Target hormone (adrenocorticoid)

Figure LG 18.2   Control of hormone secretion: example of releasing, tropic, and target
hormones. Thickness of arrows indicates amounts of hormones. Color arrows as directed in
Checkpoint C2. The branching of arrows 1 and 3 indicates that target hormones affect both the
hypothalamus (releasing hormones) and the anterior pituitary (tropic hormones).

■ **C2.** Tropic hormones are involved in feedback mechanisms summarized in steps 1 to 4
in Figure LG 18.2. Trace the pathway of those four steps as you do this exercise.

a. Color the arrow for step 1, in which a low blood level of target hormone, in this example, adrenocorticoid such as cortisone produced by the target gland

_____, triggers the hypothalamus to secrete a *(high? low?)* level
of releasing hormone (CRH). Simultaneously, the low level of target hormones directly stimulates the anterior pituitary to produce a *(high? low?)* level of ACTH.
b. Color the two arrows for step 2. First the high level of releasing hormone (CRH)

passes through blood vessels to the _____ and stimulates more
release of ACTH. A high level of this *(tropic? target?)* hormone in blood passing
through the adrenal cortex stimulates secretion of a *(high? low?)* level of target hormone.

c. Color steps 3 and 4, and note thickness of arrows. Both steps 1 and 3 are *(positive? negative?)* feedback mechanisms, whereas steps 2 and 4 are *(positive? negative?)* feedback mechanisms.

■ **C3.** Complete this table listing releasing-tropic-target hormone relationships. Fill in the name of a hormone next to each letter and number.

Hypothalamic Hormone	→ Anterior Pituitary Hormone	→ Target Hormone
a. GHRH (growth hormone releasing hormone or somatocrinin)		None
b. CRH (corticotropin releasing hormone)	1. 2.	1. 2. None
c.	1. TSH (thyrotropin) 2.	1. 2. None
d. GnRH (gonadotropic releasing hormone)	1. FSH (follicle stimulating hormone) 2. LH (luteinizing hormone)	1. 2.

■ **C4.** Fill in this table naming the three known hypothalamic inhibiting hormones and their related anterior pituitary hormones. Hypothalamic inhibiting hormones provide additional regulation for the anterior pituitary hormones that *(are? are not?)* tropic hormones. (Notice this in the right column in the table below.)

Hypothalamic Hormone	→ Anterior Pituitary Hormone	→ Target Hormone
a. GHIH (growth hormone inhibiting hormone or somatostatin)		None
b.	MSH (melanocyte stimulating hormone)	None
c. PIH (prolactin inhibiting hormone)		None

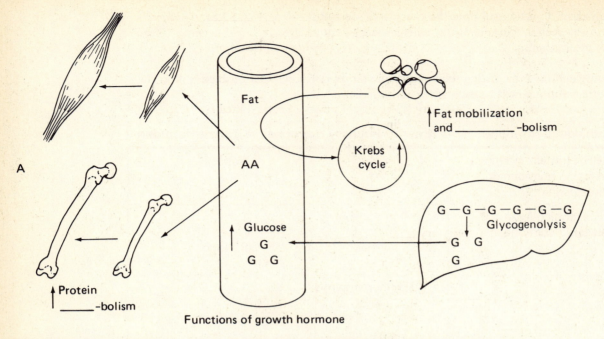

Fat

Krebs
cycle

AA

A

↑ Fat mobilization
and _____ -bolism

Glucose
G
G   G

G—G—G—G—G—G
Glycogenolysis
G   G
G

↑ Protein
_____ -bolism

Functions of growth hormone

Figure LG 18.3 Functions of growth hormone in regulating metabolism. AA, amino acid; G-G-G-G-G-G, glycogen; G, glucose. Complete as directed in Checkpoint C5.

■ **C5.** Refer to Figure LG 18.3 as you do this checkpoint about growth hormone.

a. A main function of human growth hormone (_____) is to stimulate growth and

maintain size of _____ and _____. The

alternative name for hGH, _____ indicates its function: to
"turn" *(trop-)* the body *(soma-)* to growth. Note that somatotropic hormone *(is? is not?)* classified as a true tropic hormone since it *(does? does not?)* stimulate another endocrine gland to produce a hormone.

b. hGH stimulates growth by *(accelerating? inhibiting?)* entrance of amino acids into

cells where they can be used for _____ synthesis. Therefore, growth hormone stimulates protein *(anabolism? catabolism?)*. Indicate this by filling in the blank on the left side of Figure LG 18.2.

c. hGH promotes breakdown of glycogen in liver to _____. It also promotes release of stored body fats, as well as stimulates

_____ -bolism of fats. Show this by filling in the blank on the right side of the figure.

d. Since cells now turn to fats for energy (by means of the Krebs cycle), and liver

glycogen is broken down, the level of glucose in blood _____ -creases, a condition known as *(hyper? hypo?)* -glycemia. Since hyperglycemia may overstimulate and "burn out" beta cells of the pancreas, hGH may be considered

_____ -genic.

e. hGH secretion is controlled by two regulating factors: _____

and _____. One stimulus that promotes release of hGH is low
blood sugar (for example, during growth when cells require much energy). This condition, *(hyper-? hypo-?)* glycemia, stimulates secretion of *(GHRH? GHIH?)*, which
then causes release of *(high? low?)* levels of hGH.

f. Somatomedins are made in the *(liver? pituitary?)* when it is stimulated by *(CRH?
hGH?)*. What are effects of somatomedins?

g. Hypersecretion of GH in early years leads to *(dwarfism? giantism? acromegaly?),*

whereas deficiency causes pituitary _____. Excess hGH after

closure of epiphyseal plates causes _____, characterized by

bones in three areas: _____, _____, and

_____.

C6. *For extra review.* Complete the table describing factors that affect production of
each of the following hormones. The first one is done for you.

Hormone	How Affected	Factors
a. hGH	Decreased by	HGIH, hGH (by negative feedback), REM sleep (stages 3 or 4), obesity, increased fatty acids and decreased amino acids in blood; emotional deprivation; low thyroid
b. hGH	Increased by	
c. ACTH	Increased by	
d. ADH	Increased by	

■ **C7.** Do this exercise about hormonal control of the mammary glands.

a. Milk is produced in mammary glands following stimulation by the hormone

_____ (_____) which is secreted by the *(anterior? posterior?)* pituitary. A different hormone causes ejection of milk from the glands at the time of

the baby's suckling. This hormone, named _____, is released by the *(anterior? posterior?)* pituitary.

b. During pregnancy, prolactin levels *(increase? decrease?)* due to increased levels of *(PIH? PRH?)*. During most of a woman's lifetime prolactin levels are low as a result of high levels of *(PIH? PRH? PRF?)*. An exception is the time just prior to each menstrual flow when PIH levels *(increase? decrease?)*. Due to this lack of inhibition prolactin secretion increases, activating breast tissues (and causing tenderness).

■ **C8.** Describe posterior pituitary hormone function by writing the correct terms on the lines provided on Figure LG 18.4. Use the list of terms on the figure.

■ **C9.** Answer these questions about posterior pituitary hormones.

a. The time period from the start of a baby's suckling at the breast to the delivery of

milk to the baby is about _____-seconds. Although afferent impulses (from breast to hypothalamus) require only *(a fraction of a second? 45 seconds?)*, passage of the

oxytocin through the _____ to the breasts requires about

_____ seconds.

b. On a day when your body becomes dehydrated (by loss of sweat), your ADH

production is likely to _____-crease.

c. Effects of diuretic medications are *(similar to? opposite of?)* ADH. In other words,

diuretics _____-crease urine production and _____-crease blood volume and blood pressure.

d. Inadequate ADH production, as occurs in diabetes *(mellitus? insipidus?)* results in *(enormous? minute?)* daily volumes of urine. Unlike the urine produced in diabetes mellitus, this urine *(does? does not?)* contain sugar. It is bland or *insipid.*

e. Alcohol *(stimulates? inhibits?)* ADH secretion, which contributes to _____-creased urine production after alcohol intake. *For extra review* of other factors affecting ADH production, refer to Checkpoint C6.

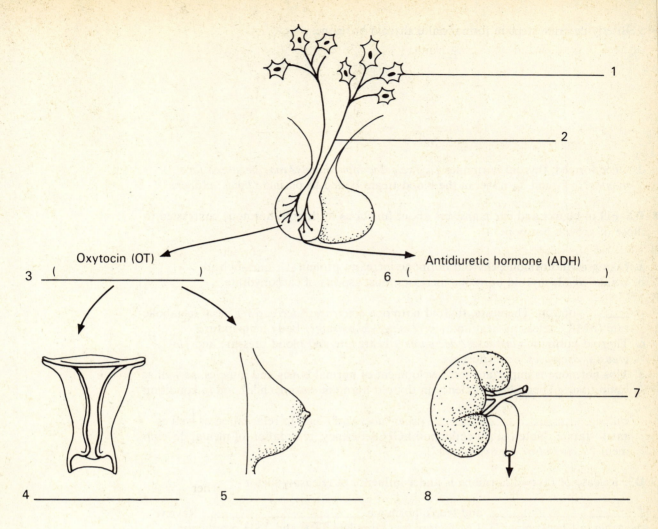

Oxytocin (OT)

Antidiuretic hormone (ADH)

3 (_____)

6 (_____)

4 _____

5 _____

7 _____

8 _____

LIST OF TERMS

Axons to posterior pituitary   Pitocin
Blood volume ↑   Urine ↓
Breast milk let-down   Uterine contraction
Neurosecretory cells in   Vasopressin
  hypothalamus

Figure LG 18.4 Posterior pituitary hormones. Label as directed in Checkpoint C8.

D. Thyroid and parathyroids (pages 535–542)

D1. Describe the histology and hormone secretion of the thyroid gland in this exercise.

a. The thyroid is composed of two types of glandular cells. *(Follicular? Parafollicular?)*

cells produce the two hormones _____ and

_____. Parafollicular cells manufacture the hormone

_____.

b. The ion *(Cl⁻? I⁻? Br⁻?)* is highly concentrated in the thyroid since it is an essential component of T_3 and T_4. Each molecule of thyroxine (T_4) contains four of the

I⁻ ions, whereas T_3 contains _____ ions of I⁻.

c. Briefly describe steps in formation of thyroid hormone.

d. Once formed, thyroid hormones *(are released immediately? may be stored for months?)*. T_3 and T_4 travel in the bloodstream bound to plasma *(lipids? proteins?)*.

■ **D2.** Fill in blanks and circle answers about functions of thyroid hormone contrasted to those of growth hormone.

a. Like growth hormone, thyroid hormone increases protein _____-bolism.
b. Unlike hGH, thyroid hormone increases most aspects of carbohydrate

_____-bolism. Therefore, thyroid hormone *(increases? decreases?)* basal metabolic rate (BMR), releasing heat and *(increasing? decreasing?)* body temperature.
c. Thyroid hormone *(increases? decreases?)* heart rate and blood pressure and *(increases? decreases?)* nervousness.
d. This hormone is important for development of normal bones and muscles, as well as brain tissue. Therefore, deficiency in thyroid hormone during childhood (a condition

called _____) *(does? does not?)* lead to retardation as well as small stature. Note that in childhood hGH deficiency, retardation of mental development *(occurs also? does not occur?)*.

D3. Release of thyroid hormone is under influence of releasing factor

_____ and tropic hormone _____. Review the table in Checkpoint C3. List factors that stimulate TRF and TSH production.

■ **D4.** A simple goiter is *(an enlargement? a decrease in size?)* of the thyroid gland. It results from the futile attempts of the gland to produce thyroid hormone when iodine is present in *(excessive? deficient?)* quantities.

■ **D5.** Color arrows on Figure LG 18.5 to show hormonal control of calcium. This figure emphasizes that CT and PTH are *(antagonists? synergists?)* with regard to effects on calcium. One way that PTH increases blood calcium level is by stimulating kidney pro-

duction of the active form of vitamin ____, (also known as _____),

which facilitates _____
absorption in the intestine. Both CT and PTH act synergistically to *(lower? raise?)* blood phosphate levels.

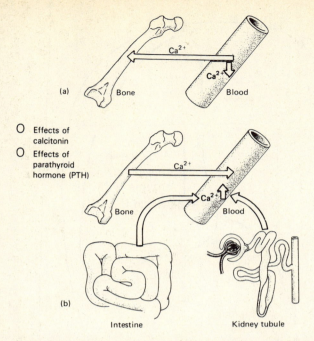

Figure LG 18.5 Hormonal control of calcium ion (Ca²⁺). (a) Increased calcium storage in bones lowers blood calcium (hypocalcemic effect). (b) Calcium from three sources increases blood calcium level (hypercalcemic effect). Arrows show direction of calcium flow. Color arrows as directed in Checkpoint D5.

■ **D6.** *A clinical challenge.* Consider calcium imbalances in this exercise.

a. Hypocalcemia can result from *(hypo? hyper?)*-parathyroid conditions. Decreased

blood calcium results in abnormal _____-crease in nerve impulses to muscles, and

the condition known as _____.

b. Parathyroid tumors or certain types of cancer cells may lead to hyperparathyroidism. Explain why persons with those conditions are more subject to fractures.

E. Adrenals (pages 542–549)

E1. Name the three zones of the adrenal cortex and the primary hormone(s) secreted by each zone.

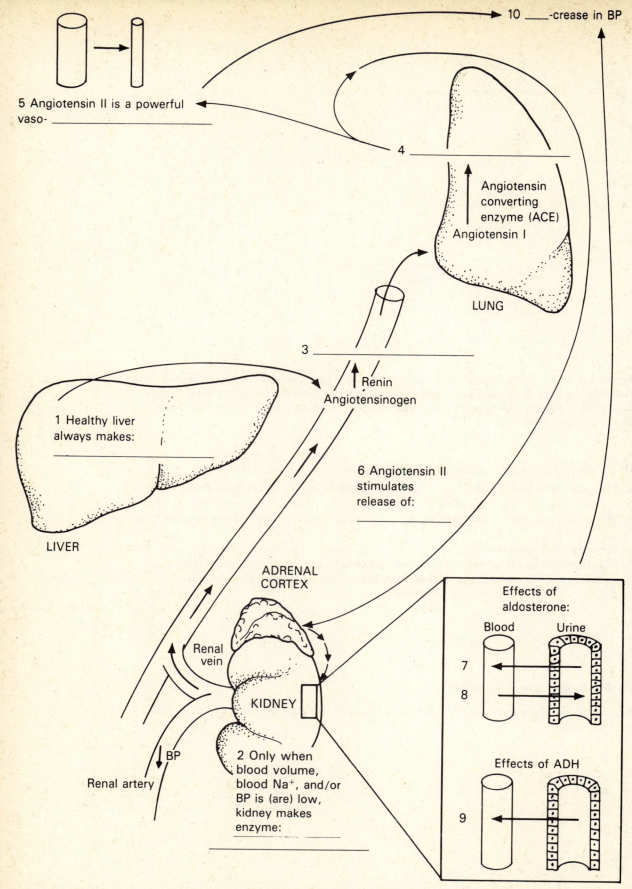

10 _____-crease in BP

5 Angiotensin II is a powerful vaso-_____

4 _____

Angiotensin converting enzyme (ACE)

Angiotensin I

LUNG

3 _____

↑ Renin
Angiotensinogen

1 Healthy liver always makes:
_____

LIVER

6 Angiotensin II stimulates release of:
_____

ADRENAL CORTEX

Renal vein

KIDNEY

↓ BP

Renal artery

2 Only when blood volume, blood Na⁺, and/or BP is (are) low, kidney makes enzyme:
_____

Effects of aldosterone:

Blood   Urine

7

8

Effects of ADH

9

Figure LG 18.6 Hormonal control of fluid and electrolyte balance and blood pressure (BP) by renin, angiotensin, aldosterone and ADH. Fill in blanks 1–10 as directed in Checkpoint E2.

■ **E2.** Fill in blank lines on Figure LG 18.6 to show mechanisms that control fluid balance and blood pressure.

■ **E3.** *A clinical challenge.* Determine whether each of the following conditions is likely to *increase* or *decrease* blood pressure.

a. Excessive renin production: _____

b. Use of an angiotensin converting enzyme (A.C.E.)-inhibitor medication:

_____

c. Aldosteronism (for example, due to a tumor of the zona glomerulosa):

_____

■ **E4.** State two factors that stimulate aldosterone production.

E5. Name three glucocorticoids. Circle the one that is most abundant in the body.

■ **E6.** In general, glucocorticoids are involved with metabolism and resistance to stress. Describe their effects in this exercise.

a. These hormones promote _____-bolism of proteins. In this respect their effects are *(similar? opposite?)* to those of both growth hormone and thyroid hormone.

b. However, glucocorticoids may stimulate conversion of amino acids and fats to

glucose, a process known as _____. In this way glucocorticoids (like hGH) *(increase? decrease?)* blood glucose.

c. Refer to Figure 18.26 (page 558 in your text). Note that glucocorticoid (and also hGH) production *(increases? decreases?)* during times of stress. By raising blood glucose levels, these hormones make energy available to cells most vital for a response to stress. During stress glucocorticoids also act to *(increase? decrease?)* blood pressure and to *(enhance? limit?)* the inflammatory process. (They are *anti-inflammatory.*)

d. *For extra review* of regulation of glucocorticoids, refer to Activities C2, C3, and C6.

■ **E7.** A third category of adrenal cortex hormones is the _____ hormones. Name two.

E8. Discuss these three disorders related to the adrenal cortex, including definition, causes, and signs or symptoms.

a. Cushing's syndrome

b. Congenital adrenal hyperplasia (CAH)

c. Addison's disease

E9. Do this exercise on the adrenal medulla.

a. Name the two principal hormones secreted by the adrenal medulla.

   Circle the one that accounts for 80 percent of adrenal medulla secretions. These

   hormones are classified chemically as _____.

b. In general, effects of these hormones mimic the *(sympathetic? parasympathetic?)* nervous system. List three or more effects.

E10. Read Chapter 18 of *Essays in Wellness*. Contrast *eustress* with *distress*, and give an example of each in your own life.

Now write one or more ways that you can convert distress in your life to eustress.

Finally, write two ways that you can change your response to distress to improve your wellness and enjoyment of life.

F. Other endocrine glands: pancreas, gonads, pineal (epiphysis cerebri), thymus (pages 549–554)

■ **F1.** The islets of Langerhans are located in the _____. Do this exercise describing hormones produced there.

a. The name *islets of Langerhans* suggests that these clusters of

_____-crine cells lie amidst a "sea" of exocrine cells within the pancreas. Name the four types of islet cells.

b. Refer to Figure LG 18.7a. Glucagon, produced by *(alpha? beta? delta?)* cells, *(increases? decreases?)* blood sugar in two ways. First, it stimulates the breakdown of

_____ to glucose, a process known as *(glycogenesis? glyco-genolysis? gluconeogenesis?)*. Second, it stimulates the conversion of amino acids and

other compounds to glucose. This process is called _____.
c. Is glucagon controlled directly by an anterior pituitary *tropic* hormone? *(Yes? No?)*

In fact, control is by effect of blood _____ level directly upon the pancreas. When blood glucose is low, then a *(high? low?)* level of glucagon will be produced. This will raise blood sugar.
d. *For extra review.* Define the terms *glycogen* and *glucagon*.

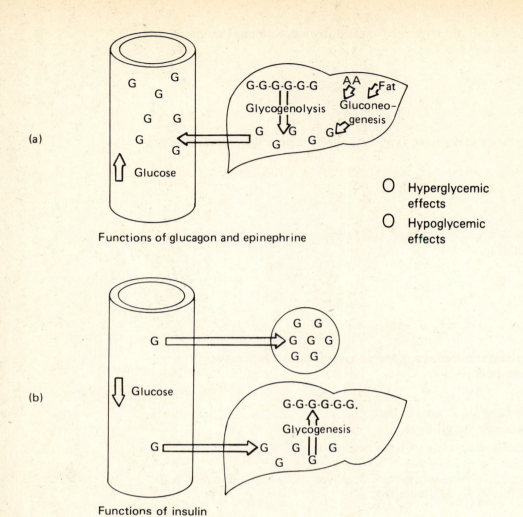

(a)

Functions of glucagon and epinephrine

O Hyperglycemic effects

O Hypoglycemic effects

(b)

Functions of insulin

Figure LG 18.7 Hormones that regulate blood glucose level. AA, amino acid; G, glucose; G-G-G-G-G-G, glycogen. Color according to directions in Checkpoint F1.

e. Now look at Figure LG 18.7b. The action of insulin is *(the same as? opposite?)* that of glucagon. In other words, *insulin decreases blood sugar.* (Repeat that statement three times; it is important!) Insulin acts by several mechanisms. Two are shown in the figure. First, insulin *(helps? hinders?)* transport of glucose from blood into cells.

Second, it accelerates conversion of glucose to _____, the

process called _____.

f. Hormones that raise blood glucose levels are called *(hypo? hyper?)*-glycemic hor-mones. Insulin, the one hormone that lowers blood glucose level is said to be

_____-glycemic. To show this, color arrows in Figure 18.7.

■ **F2.** Contrast type I and type II diabetes mellitus (DM) by writing I or II next to the related descriptions.

_____ a. Also known as maturity-onset DM

_____ b. The more common type of DM

_____ c. Related to insensitivity of body cells to insulin, rather than to absolute insulin deficiency

_____ d. Also known as insulin-dependent diabetes mellitus (IDDM)

_____ e. More likely to lead to serious complications such as ketoacidosis

■ **F3.** List three or more factors that increase insulin secretion.

■ **F4.** *A clinical challenge.* Mrs. Jefferson has diabetes mellitus. In the absence of sufficient insulin, her cells are deprived of glucose. Explain why she experiences the following symptoms.

a. Hyperglycemia and glycosuria

b. Increased urine production (poly-_____) and increased water

   intake (poly-_____)

c. Ketoacidosis, a form of _____-osis

d. Atherosclerosis

■ **F5.** Sex hormones are produced in several locations in the body. List four. (More about these in Chapter 28.)

F6. Where is the pineal gland located?

Secretion of one pineal hormone, melatonin, increases in *(light? darkness?)*. How may this hormone be related to the form of depression known as seasonal affect disorder (SAD)?

■ **F7.** What is the general function of the thymus? (More about this in Chapter 22.)

G. Aging and developmental anatomy of the endocrine system; other endocrine tissues (pages 554–556)

■ **G1.** Name endocrine organs that often cause problems among the elderly.

_____ _____

■ **G2.** Now match the endocrine gland with the correct description of its embryological origin.

AC. Adrenal cortex	P. Pancreas
AM. Adrenal medulla	PT. Parathyroid
AP. Anterior pituitary	T. Thymus

_____ a. Develops from the foregut area that later becomes part of small intestine

_____ b. Derived from the roof of the mouth; called the hypophyseal (Rathke's) pouch

_____ c. Originates from the neural crest which also produces sympathetic ganglia

_____ d. Derived from tissue from the same region that forms gonads

_____ e. Arise from pharyngeal pouches (two answers)

■ **G3.** As you continue your study of anatomy and physiology, you will consider hormones produced in other parts of the body: those from the

_____ tract (Chapter 24), hormones made in the

_____ during pregnancy (Chapter 29), and also the hormone

erythropoietin, which affects _____ production (Chapter 19). The

heart produces a hormone called ANP (a_____

n_____ p_____) that helps to ____-crease
blood pressure.

G4. Write a brief description of functions of each of the following categories of chemicals produced by the body.

a. Leukotrienes (LTs)

b. Prostaglandins (PGs)

c. Thromboxane (TX)

■ **G5.** *A clinical challenge.* Aspirin and nonsteroidal anti-inflammatories (NSAIDs) inhibit synthesis of one of the categories of chemicals listed in Checkpoint G4. Which one?

_____. Write three therapeutic effects of NSAIDs.

■ **G6.** Match growth factors with descriptions.

C. Cytokines	TAF. Tumor angiogenesis factors
IGF. Insulinlike growth factor	

_____ a. Produced by normal and tumor cells; stimulate growth of new blood vessels.

_____ b. Made by certain white blood cells, these help to provide immunity.

_____ c. Made in the liver (and also in some cancer cells), this factor stimulates growth of cartilage and fibroblasts.

H. Stress and homeostasis   (pages 556–560)

H1. Define and give two examples of *stressors*.

■ **H2.** Summarize the body's response to stressors in this exercise.

a. The part of the brain that senses the stress and initiates response is the

_____. It responds by two main mechanisms.

b. First is the *(alarm? resistance?)* reaction involving the adrenal *(medulla? cortex?)*

and the _____ division of the autonomic nervous system.

c. Second is the _____ reaction. Playing key roles in this response

are three anterior pituitary hormones: _____,

_____, and _____ and their target hor-
mones. Therefore, the adrenal *(medulla? cortex?)* hormones are activated.

■ **H3.** Now describe events that characterize the first (or alarm) stage of response to
stress. Complete this exercise.

a. The alarm reaction is sometimes called the _____ response.

b. During this stage blood glucose *(increases? decreases?)* by a number of hormonal
mechanisms, including those of adrenal medulla hormones

_____ and _____. Glucose must be avail-
able for cells to have energy for the stress response.

c. Oxygen must also be available to tissues; the respiratory system *(increases? decreases?)* its activity. Heart rate and blood pressure *(increase? decrease?)*. Blood

is shunted to vital tissues such as the _____,

_____, and _____ and is directed away

from reservoir organs such as the _____,

_____, and _____.

d. Nonessential activities such as digestion *(increase? decrease?)*. Sweating *(increases?
decreases?)* in order to control body temperature and eliminate wastes.

H4. Describe the effects of each of these hormones in the resistance stage.

a. Mineralocorticoids

b. Glucocorticoids

c. Thyroxin

d. hGH

H5. When the resistance stage fails to combat the stressor, the body moves into the

_____ stage. Explain how this is related to:

a. Loss of potassium

b. Depletion of glucocorticoids

c. Weakening of organs

I. Summary of hormones, disorders (pages 516–560)

■ **I1.** *For extra review.* Test your understanding of these hormones by writing the name (or abbreviation where possible) of the correct hormone after each of the following descriptions.

a. Stimulates release of growth hormone: _____

b. Stimulates testes to produce testosterone: _____

c. Promotes protein anabolism, especially of bones and muscles:

   _____

d. Tropic hormone for thyroxin: _____

e. Stimulates development of ova in females and sperm in males:

   _____

f. Present in bloodstream of nonpregnant women to prevent lactation:

   _____

g. Stimulates ovary to release egg and change follicle cells into corpus luteum:

   _____

h. Stimulates uterine contractions and also breast milk let-down:

   _____

i. Inhibited by MIH: _____

j. Stimulates kidney tubules to produce small volume of concentrated urine:

   _____

k. Target hormone of ACTH: _____

l. Increases blood calcium: _____

m. Decreases blood calcium and phosphate: _____

n. Contains iodine as an important component: _____

o. Raises blood glucose (three answers): _____,

   _____, _____

p. Lowers blood glucose: _____

q. Serves anti-inflammatory functions: _____

r. Mimics many effects of sympathetic nerves: _____

■ **I2.** *For extra review.* Match the disorder with the hormonal imbalance.

> Ac. Acromegaly   M. Myxedema
> Ad. Addison's disease   PC. Pheochromocytoma
> C. Cretinism   PD. Pituitary dwarfism
> DI. Diabetes insipidus   S. Simple goiter
> DM. Diabetes mellitus   T. Tetany
> G. Graves' disease

_____ a. Deficiency of hGH in child; slow bone growth

_____ b. Excess of hGH in adult; enlargement of hands, feet, and jawbones

_____ c. Deficiency of ADH; production of enormous quantities of "insipid" (non-sugary) urine

_____ d. Deficiency of effective insulin; hyperglycemia and glycosuria (sugary urine)

_____ e. Deficiency of thyroxin in child; short stature and mental retardation

_____ f. Deficiency of thyroxin in adult; edematous facial tissues, lethargy

_____ g. Excess of thyroxin; protruding eyes, "nervousness," weight loss

_____ h. Deficiency of thyroxin due to lack of iodine; most common reason for enlarged thyroid gland

_____ i. Result of deficiency of PTH; decreased calcium in blood and fluids around muscles resulting in abnormal muscle contraction

_____ j. Deficiency of adrenocorticoids; increased K^+ and decreased Na^+ resulting in low blood pressure and dehydration

_____ k. Tumor of adrenal medulla causing sympathetic-type responses, such as increased pulse and blood pressure, hyperglycemia, and sweating

ANSWERS TO SELECTED CHECKPOINTS

A1.

	Secretions Transported In	Examples
a. Exocrine	Ducts	Sweat, oil, mucus, digestive juices
b. Endocrine	Bloodstream	Hormones

A2. (a) A, pineal; B, posterior pituitary; C, anterior pituitary; D, thyroid; E, parathyroid; F, thymus; G, adrenal cortex; H, adrenal medulla. (b) 1, heart; 2, skin; 3, liver; 4, kidney; 5, stomach; 6, pancreas; 7, intestine; 8, ovary; 9, placenta; 10, testis.

A3. (a) N. (b) E. (c) N.

A5. Hormones affect only cells with specific receptors that "fit" the hormone.

A6. (a) Decrease; down. (b) More, deficient.

A7. (a) Endo. (b) Auto, para.

A8. (a) PP. (b) BA. (c) E. (d) S.

A9. Insulin is a protein that would be destroyed by digestive enzymes.

A10. (a–b) TP. (c–d) FF.

B1. (a–b) I. (c–d) P.

B2. (a) First, endocrine gland, target cell. (b) Outer, G. (c) Adenylate cyclase, inner; cyclic AMP, second messenger. (d) Protein kinases, phosphate; urine. (e) Short; phosphodiesterase. (f) Water; ADH, OT, FSH, LH, TSH, ACTH, CT, PTH, glucagon, hypothalamic regulating hormones, epinephrine, and NE. (g) Ca^{2+}, cGMP, IP_3, or DAG.

B4. (a) Permissive. (b) Antagonistic.

B5. (a) N. (b) C. (c) H.

B6. Negative; increase. Examples of positive feedback: uterine contractions stimulate more oxytocin, or high estrogen level stimulates high LH level and ovulation.

C1. (a) Hypophysis. (b) The pituitary regulates many body activities, yet this gland is "ruled" by releasing and inhibiting hormones from the hypothalamus. (c) In the sella turcica just inferior to the hypothalamus. (d) Anterior, adeno; neuro. (e) 7; hGH, PRL, MSH, ACTH, TSH, FSH, LH. (f) Hypothalamic hormones reach the pituitary quickly and without dilution since they do not have to traverse systemic circulation. (g) Lactotrophs: cells that produce PRL; somatotrophs: cells that produce hGH; gonadotrophs: cells that produce FSH and LH. (h) Hormones that stimulate release of other hormones: ACTH, TSH, FSH, and LH.

C2. (a) Adrenal cortex, high; high. (b) Anterior pituitary; tropic, high. (c) Negative, positive.

C3.

Hypothalamic Hormone →	Anterior Pituitary Hormone →	Target Hormone
a. GHRH (growth hormone releasing hormone or somatocrinin)	hGH (human growth hormone)	None
b. CRH (corticotropin releasing hormone)	1. ACTH (corticotropin or adrenocorticotropic hormone) 2. MSH (melanocyte stimulating hormone)	1. Cortisol 2. None
c. TRH (thyrotropin releasing hormone)	1. TSH (thyrotropin or thyroid stimulating hormone) 2. PRL (prolactin)	1. Thyroxine 2. None
d. GnRH (gonadotropic releasing hormone)	1. FSH (follicle stimulating hormone) 2. LH (luteinizing hormone)	1. Estrogen 2. Estrogen, progesterone

C4. Are not. (a) hGH (human growth hormone). (b) MIH (MSH inhibiting hormone). (c) PRL (prolactin).

C5. (a) hGH, bones, skeletal muscles; somatotropin; is not, does not. (b) Accelerating, protein; anabolism. (c) Glucose; cata. (d) In, hyper; diabeto. (e) GHRH, GHIH; hypo, GHRH, high. (f) Liver, hGH; they regulate effects of hGH, such as raising blood glucose level. (g) Giantism, dwarfism; acromegaly; hands, feet, and face.

C7. (a) Prolactin (or lactogenic hormone) (PRL), anterior; oxytocin (OT), posterior. (b) Increase, PRH; PIH; decrease.

C8. 1, neurosecretory cells in hypothalamus; 2, axons to posterior pituitary; 3, pitocin; 4, uterine contraction; 5, breast milk let-down; 6, vasopressin; 7, blood volume ↑ (by increased reabsorption of water from urine in kidneys); 8, urine ↓.

C9. (a) 30–60; a fraction of a second, bloodstream, 30–60. (b) In. (c) Opposite of; in, de. (d) Insipidus, enormous; does not. (e) Inhibits; in.

D2. (a) Ana. (b) Cata; increases, increasing. (c) Increases, increases. (d) Cretinism, does; does not occur.

D4. An enlargement; deficient.

D5. Labels on Figure LG 18.5: (a) Calcitonin (CT). (b) Parathyroid hormone (PT). Antagonists; D, calcitriol or 1, 25-dihydroxy vitamin D_3, calcium, phosphate and magnesium; lower.

D6. (a) Hypo; in, tetany. (b) Demineralization of bones occurs, since PTH stimulates osteoclasts to destroy bone, leading to osteitis fibrosa cystica.

E2. 1, The protein angiotensinogen; 2, renin; 3, angiotensin I; 4, angiotensin II; 5, constrictor; 6, aldosterone; 7, Na^+ (and H_2O by osmosis, and Cl^- and HCO_3^- to achieve electrochemical balance); 8, K^+ and H^+; 9, H_2O; 10, In.

E3. (a) Increase (known as *high renin hypertension*). (b) Decrease, such as the antihypertensive captopril (Capoten). (c) Increase.

E4. Angiotensin II (resulting from renin-angiotensin pathway) and increased blood level of K^+ (hyperkalemia).

E6. (a) Cata; opposite. (b) Gluconeogenesis; increase. (c) Increases; increase, limit.

E7. Gonadocorticoids or sex hormones; estrogens and androgens.

F1. Pancreas. (a) Endo; alpha, beta, delta, and F. (b) Alpha, increases; glycogen, glycogenolysis; gluconeogenesis. (c) No; glucose; high. (d) *Glycogen* is a storage form of glucose; *glucagon* is a hormone that breaks down glycogen to glucose. (e) Opposite; helps; glycogen, glycogenesis. (f) Hyper; hypo; glucagon and epinephrine are hyperglycemic; insulin is hypoglycemic.

F2. (a–c) II. (d–e) I.

F3. Hyperglycemia, as well as hormones that lead to hyperglycemia (hGH, ACTH, cortisol, glucagon, epinephrine, or NE), increased blood level of amino acids, and parasympathetic (vagal) nerve impulses.

F4. (a) Lack of insulin prevents glucose from entering cells, so more is left in blood and spills into urine. (b) -Uria, -dipsia; glucose in urine draws water (osmotically), and as result of water loss, patient is thirsty. (c) Acid; since cells have little glucose to use, they turn to excessive fat catabolism which leads to keto-acid formation (discussed in Chapter 25). (d) As fats are transported in extra amounts (for catabolism), some fats are deposited in walls of vessels; fatty (= *athero*) thickening (= *sclerosis*) results.

F5. Ovaries, testes, adrenal cortex and placenta.
F7. Immunity.
G1. Pancreas and thyroid, as well as testes and ovaries.
G2. (a) P. (b) AP. (c) AM. (d) AC. (e) PT, T.
G3. Gastrointestinal, placenta, red blood cell; atrial natriuretic peptide.
G5. Prostaglandins (PGs); reduce fever, pain, and inflammation.
G6. (a) TAF. (b) C. (c) IGF.
H2. (a) Hypothalamus. (b) Alarm, medulla, sympathetic. (c) Resistance; ACTH, hGH; TSH; cortex.
H3. (a) Fight-or-flight. (b) Increases; norepinephrine, epinephrine. (c) Increases; increase; skeletal muscles, heart, or brain; spleen, GI tract, skin. (d) Decrease; increases.

I1. (a) GHRH. (b) LH. (c) hGH (also testosterone). (d) TSH. (e) FSH. (f) PIH. (g) LH. (h) OT. (i) MSH. (j) ADH. (k) Adrenocorticoid (or glucocorticoid, such as cortisol). (l) PTH. (m) CT. (n) Thyroid hormone. (o) hGH, glucagon, epinephrine, norepinephrine, adrenocorticoids (or ACTH indirectly). (p) Insulin. (q) Adrenocorticoid. (r) Epinephrine, norepinephrine.
I2. (a) PD. (b) Ac. (c) DI. (d) DM. (e) C. (f) M. (g) G. (h) S. (i) T. (j) Ad. (k) PC.

MASTERY TEST: Chapter 18

Questions 1–2: Arrange the answers in correct sequence.

_____ _____ _____ _____ _____ 1. Steps in action of TSH upon a target cell:
  A. Adenylate cyclase breaks down ATP.
  B. TSH attaches to receptor site.
  C. Cyclic AMP is produced.
  D. Protein kinase is activated to cause the specific effect mediated by TSH, namely, stimulation of thyroid hormone production.
  E. G-protein is activated.

_____ _____ _____ _____ 2. Steps in renin-angiotensin mechanism to increase blood pressure:
  A. Angiotensin I is converted to angiotensin II.
  B. Renin converts angiotensinogen, a plasma protein, to angiotensin I.
  C. Angiotensin II causes vasoconstriction and stimulates aldosterone production which causes water conservation and raises blood pressure.
  D. Low blood pressure or low blood Na^+ level causes secretion of enzyme renin from cells in kidneys.

3. All of these compounds are synthesized in the hypothalamus *except:*
   A. ADH
   B. GHRH
   C. CT
   D. PIH
   E. Oxytocin

4. Choose the *false* statement about endocrine glands.
   A. They secrete chemicals called hormones.
   B. Their secretions enter extracellular spaces and then pass into blood.
   C. Sweat and sebaceous glands are endocrine glands.
   D. Endocrine glands are ductless.

5. All of the following correctly match hormonal imbalance with related signs or symptoms *except:*
   A. Deficiency of ADH—excessive urinary output
   B. Excessive aldosterone—high blood pressure
   C. Deficiency of aldosterone—low blood level of potassium and muscle weakness
   D. Excess of parathyroid production—demineralization of bones; high blood level of calcium and tetany

6. All of the following hormones are secreted by the anterior pituitary *except:*
   A. ACTH
   B. FSH
   C. Prolactin
   D. Oxytocin
   E. hGH

7. Which one of the following is a function of adrenocorticoid hormones?
   A. Lower blood pressure
   B. Raise blood level of calcium
   C. Convert glucose to amino acids
   D. Lower blood level of sodium
   E. Anti-inflammatory

8. All of these hormones lead to increased blood glucose *except:*
   A. ACTH
   B. Insulin
   C. Glucagon
   D. Growth hormone
   E. Epinephrine

9. Choose the *false* statement about the anterior pituitary.
   A. It secretes at least seven hormones.
   B. It develops from mesoderm.
   C. It secretes tropic hormones.
   D. It is stimulated by releasing factors from the hypothalamus.
   E. It is also known as the adenohypophysis.

10. Choose the *false* statement.
   A. Both ADH and aldosterone tend to lead to retention of water and increase in blood pressure.
   B. PTH activates vitamin D, which enhances calcium absorption.
   C. PTH activates osteoclasts, leading to bone destruction.
   D. Tetany is a sign of hypocalcemia.
   E. ADH and OT pass from hypothalamus to posterior pituitary via pituitary portal veins.

Questions 11–20: Circle T (true) or F (false). If the statement is false, change the underlined word or phrase so that the statement is correct.

T F 11. <u>Polyphagia, polyuria, and hypoglycemia are all</u> symptoms associated with insulin deficiency.

T F 12. Regulating hormones are all secreted by <u>the anterior pituitary and affect the hypothalamus.</u>

T F 13. The hormones of the adrenal medulla mimic the action of <u>parasympathetic</u> nerves.

T F 14. Prostaglandins are classified as <u>eicosanoids</u> and have a <u>broad</u> range of effects.

T F 15. <u>Epinephrine and cortisol are both</u> adrenocorticoids.

T F 16. Secretion of <u>growth hormone, insulin, and glucagon</u> is controlled (directly or indirectly) by blood glucose level.

T F 17. Most of the feedback mechanisms that regulate hormones are <u>positive (rather than negative)</u> feedback mechanisms.

T F 18. The alarm reaction occurs <u>before</u> the resistance reaction to stress.

T F 19. Most hormones studied so far appear to act by a mechanism involving <u>cyclic AMP</u> rather than by the <u>gene activation</u> mechanism.

T F 20. A high level of thyroxine circulating in the blood will tend to lead to a <u>low</u> level of TRH and TSH.

Questions 21–25: fill-ins. Complete each sentence with the word or phrase that best fits.

_____ 21. ____ is a hormone used to induce labor since it stimulates contractions of smooth muscle of the uterus.

_____ 22. ____ and ____ are two hormones that are classified chemically as bioamines.

_____ 23. ____ is a term that refers to the control mechanism by which the number of receptors for a hormone will be decreased when that hormone is present in excessive amounts.

_____ 24. Type ____ diabetes mellitus is a non-insulin-dependent condition in which insulin level is adequate but cells have decreased sensitivity to insulin.

_____ 25. Three factors that stimulate release of ADH are ____.

ANSWERS TO MASTERY TEST: ■ Chapter 18

Arrange

1. B E A C D
2. D B A C

Multiple Choice

3. C	7. E
4. C	8. B
5. C	9. B
6. D	10. E

True-False

11. F. Polyphagia and polyuria are both or Polyphagia, poluria, and hyperglycemia are all
12. F. Hypothalamus and affect the anterior pituitary
13. F. Sympathetic
14. T
15. F. Cortisol is
16. T
17. F. Negative (rather than positive)
18. T
19. T
20. T

Fill-ins

21. Oxytocin (OT or pitocin)
22. Thyroid hormones (T_3 and T_4), epinephrine or norepinephrine
23. Down-regulation
24. II (maturity onset)
25. Decreased extracellular water, pain, stress, trauma, anxiety, nicotine, morphine, tranquilizers

UNIT IV

Maintenance of the Human Body

FRAMEWORK 19
Blood

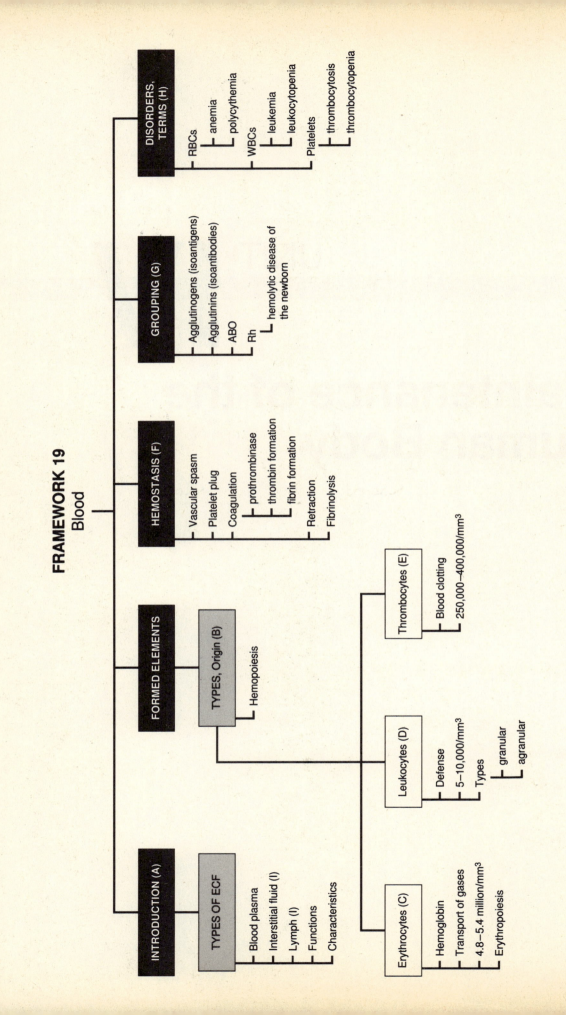

INTRODUCTION (A)

TYPES OF ECF
- Blood plasma
- Interstitial fluid (I)
- Lymph (I)
- Functions
- Characteristics

Erythrocytes (C)
- Hemoglobin
- Transport of gases
- 4.8–5.4 million/mm³
- Erythropoiesis

FORMED ELEMENTS

TYPES, Origin (B)
- Hemopoiesis

Leukocytes (D)
- Defense
- 5–10,000/mm³
- Types
  - granular
  - agranular

Thrombocytes (E)
- Blood clotting
- 250,000–400,000/mm³

HEMOSTASIS (F)
- Vascular spasm
- Platelet plug
- Coagulation
  - prothrombinase
  - thrombin formation
  - fibrin formation
- Retraction
- Fibrinolysis

GROUPING (G)
- Agglutinogens (isoantigens)
- Agglutinins (isoantibodies)
- ABO
- Rh
  - hemolytic disease of the newborn

DISORDERS, TERMS (H)
- RBCs
  - anemia
  - polycythemia
- WBCs
  - leukemia
  - leukocytopenia
- Platelets
  - thrombocytosis
  - thrombocytopenia

The Cardiovascular System: The Blood

The major role of the blood is transportation through an intricate system of channels—blood vessels—that reach virtually every part of the body. Blood constantly courses through the vessels, carrying gases, fluids, nutrients, electrolytes, hormones, and wastes to and from body cells. Most of these chemicals "ride" within the fluid portion of blood (plasma). Some, such as oxygen, piggyback on red blood cells. White blood cell "passengers" take short trips in the blood and depart from the vessels at sites calling for defense. Platelets and other clotting factors congregate to plug holes in the network of vessels.

To explore this marvelous transport system, start by examining the Chapter 19 Framework and the list of key words for each section.

TOPIC OUTLINE AND OBJECTIVES

A. Types of body fluids, functions and characteristics of blood

1. Contrast the general roles of blood, lymph, and interstitial fluid in maintaining homeostasis.
2. Define the functions and physical characteristics of the various components of blood.
3. List the components of plasma and explain their importance.

B. Components of blood: origin of formed elements

4. Compare the origins, histology, and functions of the formed elements in blood.

C. Erythrocytes

D. Leukocytes

E. Thrombocytes

F. Hemostasis

5. Identify the stages involved in blood clotting and explain the various factors that promote and inhibit blood clotting.

G. Grouping (typing)

6. Explain ABO and Rh blood grouping.

H. Disorders, medical terminology

7. Identify the clinical symptoms of several types of anemia, infectious mononucleosis (IM), and leukemia.
8. Define medical terminology associated with blood.

WORDBYTES

Now study the following parts of words that may help you better understand terminology in this chapter.

Wordbyte	Meaning	Example	Wordbyte	Meaning	Example
a-, an-	not, without	*an*emia	leuko-	white	*leuko*cytosis
-crit	to separate	hemato*crit*	mega-	large	*mega*karyocyte
-cyte	cell	leuko*cyte*	-osis	increase	leukocyt*osis*
dia-	through	*dia*phragm	-pedesis	moving	dia*pedesis*
-emia	blood	leuk*emia*	-penia	want, lack	leukocyto*penia*
erythro-	red	*erythro*poiesis	-poiesis	formation	hemo*poiesis*
heme-, hemo-	blood	*hemo*stasis	thromb-	clot	pro*thromb*in

CHECKPOINTS

A. Types of body fluids, functions and characteristics of blood (pages 567–568)

A1. Review relationships among these three body fluids: blood, interstitial fluid, and lymph. (Refer to Figure LG 1.1, page 8 of the Learning Guide.)

■ **A2.** Contrast body fluids in the table.

	Interstitial Fluid and Lymph	Plasma
a. Amount of protein (more or less)	*less*	*more*
b. Formed elements present	*some leukocytes*	Red blood cells, white blood cells, platelets
c. Location	Between cells and in lymph vessels	*w/i blood vessels*

■ **A3.** Name the structures that comprise the:

a. Cardiovascular system

blood   *heart +*   *blood vessels*

b. Lymphatic system

lymph   *lymph vessels*   *lymph nodes tonsils spleen*

■ **A4.** Describe functions of blood in this Checkpoint.

a. Blood transports many substances. List six or more.

_____oxygen_____   _____CO2_____   _____iron_____

_____leukocytes_____   _____hormones_____   _____heat_____

b. List four aspects of homeostasis regulated by blood. One is done for you.

_____pH_____   _____temp_____

_____fluid_____   _____electrolytes_____

c. Explain how your blood protects you.

contains blood clotting proteins
defense proteins (antibodies)
phagocytic cells

■ **A5.** Circle the answers that correctly describe characteristics of blood.

a. Average temperature:
36°C (96.8°C)   37°C (98.6°C)   38°C (100.4°C)

b. pH:
6.8   7.0–7.1   7.35–7.45

c. Volume of blood in average adult:
1.5–3 liters   5–6 liters   8–10 liters

A6. Contrast the following procedures: *venipuncture/arterial stick/finger stick.*

■ **A7.** Match the names of components of plasma with their descriptions.

A. Albumin		GAF. Glucose, amino acids, and fats	
E. Electrolytes		HE. Hormones and enzymes	
F. Fibrinogen		UC. Urea, creatinine	
G. Globulin		W. Water	

___W___ a. Makes up about 92 percent of plasma

___HE___ b. Regulatory substances carried in blood

___E___ c. Cations and anions carried in plasma

___A___ d. Constitutes about 55 percent of plasma protein

___F___ e. Made by liver; a protein used in clotting

___G___ f. Antibody protein

___UC___ g. Wastes carried to kidneys or sweat glands

___GAF___ h. Food substances carried in blood

B. Components of blood: origin of formed elements   (pages 568–570)

■ **B1.** Describe components of blood in this exercise.

a. Blood consists of about _**55**_ percent plasma and _**45**_ percent formed elements

which include _**cells**_ and _**cell fragment**_ .

b. The three types of formed elements (or cells) are: red blood cells (_**erythro**_ -cytes),

white blood cells (_**leuko**_ -cytes), and platelets (_**thrombo**_ -cytes).

■ **B2.** Check your knowledge about blood formation in this activity.

a. Blood formation is a process known as _**hemopoiesis**_ . All blood cells

arise from _**stem**_ cells. Myeloblasts form mature *(red? granular white? agranular white?)* blood cells.

b. Most erythropoiesis occurs in *(lymphoid? red bone marrow?)*. Name six or more bones in which hemopoiesis occurs after birth.

**sternum, ribs, epiphyses of femur +**
**flat bones of skull   pelvis   humerus**

c. Erythropoietin is a hormone made mostly in the _**~~liver~~ kidney**_ . This hormone stimulates production of *(RBCs? WBCs? platelets?)*.

d. Most of the cytokines that stimulate hemopoiesis are known as

_____ (CSFs) or _____ (Ils). Name three or more CSFs.

_____   _____   _____

■ **B3.** *A clinical challenge.* Explain how the following chemicals produced via recombinant DNA may be helpful chemically:

a. *EPO* for patients with kidney failure _**stimulates RBC production**_
**in failing kidneys that don't produce**
**erythropoietin**

b. *Neupogen* for patients with immune deficiency
**stimulates neutrophil production in people**
**whose bone marrow production is inadeq**

C. Erythrocytes   (pages 570–574)

■ **C1.** Refer to Figure LG 19.1. Which diagram represents a mature erythrocyte? _**A**_ Note that it *(does? does not?)* contain a nucleus. The chemical named

**iron** accounts for the color of red blood cells (RBCs). Label and color the RBC.

**hemoglobin**

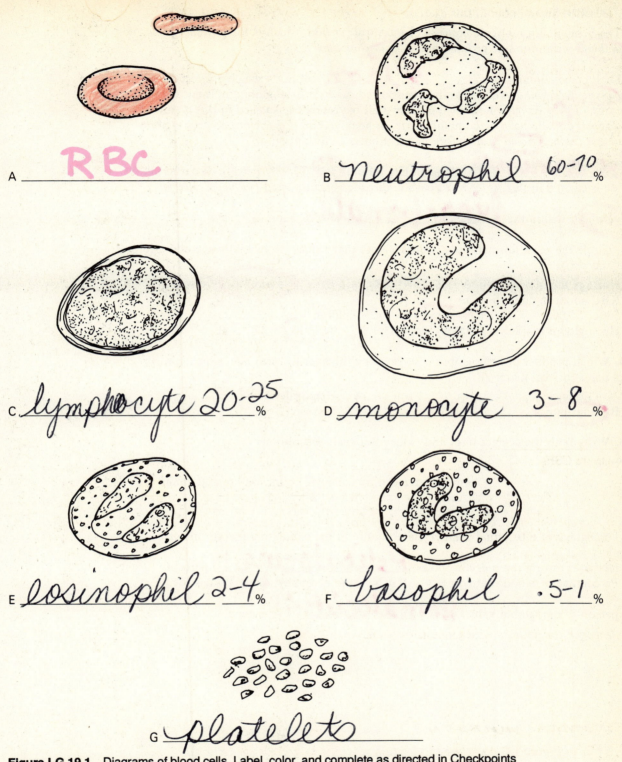

A RBC

B neutrophil 60-70 %

C lymphocyte 20-25 %

D monocyte 3-8 %

E eosinophil 2-4 %

F basophil .5-1 %

G platelets

Figure LG 19.1 Diagrams of blood cells. Label, color, and complete as directed in Checkpoints C1, D1, and E2.

■ **C2.** Describe hemoglobin in this exercise.

a. The hemoglobin molecule consists of a central portion, which is the protein *(heme?* *globin?)* with four side groups called *(hemes? globins?)*.

b. Each heme contains one _____ *Iron* _____ atom on which a molecule of *(oxygen? carbon dioxide?)* can be transported, so each hemoglobin molecule can carry *(1? 4?)* oxygen molecule(s). Almost all oxygen is transported in this manner.

c. Carbon dioxide has one combining site on hemoglobin; it is the amino acid portion of the *(heme? globin?)*; therefore, it contributes to formation of

_____ *carbo amino* _____-hemoglobin. About *(23? 70? 97?)* percent of carbon dioxide is carried in this state. Most carbon dioxide is transported *(in blood cells? in*

plasma?) within the _____ *bicarbonate* _____ (HCO_3^-) ion.

■ **C3.** Circle the most normal blood values. (Note that values vary slightly according to age and sex.)

a. Average life of a red blood cell: 4 hours   4 days   (4 months)   4 years

b. RBC count in cube this size: □ (mm^3)

    500   5,000   250,000   (5 million)   250 million

c. Hemoglobin in adults (g/100 ml blood):   1   8   (15)   27   41

■ **C4.** Which product of hemoglobin is secreted by the liver to form bile, which contributes color to fecal material?

    A. Globin   C. Hemosiderin

    B. (Bilirubin)   D. Ferritin

■ **C5.** Write both the correct term and value after each description.

> Terms: hematocrit, reticulocyte count, differential count
> Values: 1   8   15   27   41

a. Percentage of RBCs that are not quite mature in the circulating blood; this value

increases during rapid erythropoiesis: _____ *Reticulocyte 1* _____ percent

b. Percentage of blood that consists of RBCs; value is usually about 3 times the

person's hemoglobin value: _____ *Hematocrit 41* _____ percent

■ **C6.** Explain the roles of the terms in the box in red blood cell production and destruction by matching terms with descriptions.

> BB. Bilirubin and biliverdin
> B₁₂. Vitamin B_{12}
> E. Erythropoietin
> FH. Ferritin and hemosiderin
>
> H. Hypoxia
> IF. Intrinsic factor
> T. Transferrin

_H_   a. Decrease of oxygen in cells; serves as a signal that erythropoiesis is needed (to help provide more oxygen to tissues)

_E_   b. Hormone produced by kidneys when they are hypoxic; stimulates erythropoiesis in red bone marrow

B₁₂   c. Vitamin necessary for normal erythropoiesis

_IF_   d. Substance produced by the stomach lining and necessary for normal vitamin B_{12} absorption

_T_   e. Plasma protein that transfers iron (Fe) from dead RBCs to new RBCs forming in marrow

_FH_   f. Storage forms of iron (Fe) in muscle and liver cells

_BB_   g. Products of breakdown of the heme portion of hemoglobin

■ **C7.** _A clinical application._ Explain the rationale behind each of the following practices of some athletes in attempts to increase oxygen-carrying capacity of blood, and, ultimately, athletic performance.

a. Training in a high-altitude city such as Mexico City or Denver for a competition to be held in one of those cities.

b. _Induced erythrocytemia (blood doping):_ removal of RBCs a certain period before competition with reinjection of the cells shortly before the competition.

D. Leukocytes   (pages 574–577)

■ **D1.** Refer to Figure LG 19.1 and do this exercise about leukocytes.

a. Leukocytes _(have? lack?)_ hemoglobin, and so these cells are known as _(red blood cells or RBCs? white blood cells or WBCs?)._

b. Label each leukocyte (WBC) on Figure 19.1, and indicate what percentage of the total WBC count is accounted for by each type of WBC. Such a breakdown of white blood cells is known as a _differential_ WBC count.

c. Each white blood cell (WBC) _(has? lacks?)_ a nucleus. Which type of WBC has a large kidney-shaped nucleus? _monocyte_ Which WBC has a nucleus that occupies most of the cell? _lymphocyte._ Which is polymorphonuclear? _neutrophile_

d. Which three WBCs are granulocytes? _N E B_

e. _For extra review._ Color nucleus, cytoplasm, and granules of all WBCs.

■ **D2.** Check your understanding of types of WBCs by matching names of WBCs with descriptions. Answers may be used more than once.

B. Basophils	M. Monocytes
E. Eosinophils	N. Neutrophils
L. Lymphocytes	

__N__ a. Constitute the largest percentage of WBCs

__N__ b. Produce defensins, lysozymes, and oxidants with antibiotic activity

__L__ c. Involved in immunity: develop into plasma cells that produce antibodies

__B__ d. Involved in allergic response: release histamine, heparin, and serotonin

__E__ e. Involved in allergic reactions, combat histamines and provide protection against parasitic worms

__M__ f. Form fixed and wandering macrophages that clean up sites of infection

__MN__ g. Important in phagocytosis (two answers)

__L M__ h. Classified as agranular leukocytes (two answers)

__N__ i. Known as *bands* in immature state

__N__ j. In stained cells lilac-colored granules are visible

__E__ k. In stained cells red-orange granules are visible

■ **D3.** What are MHC antigens and how do they affect the success rate of transplants?

WBC have surface proteins Called major histocompatibility Antigens The more similar the MHC antigens less chance of transplant reject

D4. Write a paragraph summarizing response of leukocytes to pathogenic invasion. Be sure to include the following terms: ~~phagocytosis~~, ~~chemotaxis~~, *kinins*, ~~CSF~~, ~~diapedesis~~ *(emigration)*, *defensins*, and *histamine*.

Neutrophils & macrophages are active in phagocytosis. They ingest bacteria & dead matter. Different chemicals in inflammed tissue attract phagocytes toward the tissue. This is chemotaxis. CSF enhance phagocy activity by neutrophils & macrophages. most leukocytes can squeeze through minut spaces between cells that form capillarywal neutrophils contain defensins that form pept spears that poke holes in microbe membrane Histamines intensify the allergic reaction Kinins act as chemotactic agents for phago cy

■ **D5.** Complete this exercise about antigens and antibodies.

a. Antigens are defined as substances that _stimulate an immune response_
Most are composed of *(carbohydrate? lipid? protein?)*. Most *(are? are not?)* synthesized by the body. Several examples of antigens are: _toxins bacterial enzyme_

b. In response to antigens, *(B? T?)*-lymphocytes become plasma cells that secrete
antibodies. These chemicals are all globulin-type *(lipids? proteins?)*.
They function to *(activate? inactivate?)* antigens.

c. Other lymphocytes are known as *(A? L? T?)*-lymphocytes. One group of T-cells are activated by antigens to kill these antigens either directly or with help of other lymphocytes and macrophages. Such T-cells are appropriately called
cytotoxic T-cells.

■ **D6.** A normal leukocyte count is _5000 – 10000_ mm^3. In other words, a typical ratio of RBCs to WBCs is about:

A. 700 to 1
B. 30:1
C. 2:1
D. 1:1

■ **D7.** *A clinical challenge.* Mrs. Doud arrives at a health clinic with a suspected acute infection. Complete this Checkpoint about her.

a. During infection, it is likely that Mrs. Doud's leukocyte count will *(in? de?)*-crease. A count of *(4000? 8000? 12,000?)* leukocytes/mm^3 blood is most likely. This condition is known as *(leukocytosis? leukopenia?)*.

b. A differential increase in the number of WBCs named _neutrophils_ is most indicative of enhanced phagocytic activity during infection. Neutrophils are most likely to account for *(48? 62? 76?)* percent of the total white count in Mrs. Doud's blood. A sign of a chronic infection is increase in the percent of cells that
can become macrophages; these are _monocytes_

D8. Briefly describe the procedure of *bone marrow transplant.* Describe the process as well as conditions for which this procedure is used.

E. Thrombocytes (pages 577–578)

■ **E1.** Circle correct answers related to thrombocytes.

a. Thrombocytes are also known as:
antibodies   red blood cells
platelets   white blood cells

b. Thrombocytes are formed in:
bone marrow
tonsils and lymph nodes
spleen

c. Platelets are:
entire cells
chips off the old meta-megakaryocytes

d. A normal range for platelet count is ____/mm^3
5000–1000   4.5–5.5 million
250,000–400,000

e. The primary function of platelets is related to:
O_2 and CO_2 transport   blood clotting
defense   blood typing

CHAPTER **19** The Cardiovascular System: The Blood **387**

■ **E2.** Label platelets on Figure LG 19.1.

■ **E3.** List the components of a complete blood count (CBC).

RBC, WBC, WBC dif, hemoglobin + hematocrit platelet count

F. Hemostasis   (pages 579–583)

■ **F1.** List the three basic mechanisms of hemostasis:

a. Vascular *spasm*

b. *Platelet* plug formation

c. *Coagulation* (clotting)

F2. List chemicals present within granules of platelets.

a. In alpha granules:

b. In dense granules:

■ **F3.** Check your understanding of three phases of platelet plug formation in this Check-point.

a. When platelets snag on the inner lining of a damaged blood vessel, their shape is altered. They become *(small, disc-shaped? large irregular-shaped?)*. Additional platelets aggregate on the uneven projections. This phase is known as platelet *(adhesion? release reaction? plug?)*.

b. Platelets then release chemicals. Some chemicals *(dilate? constrict?)* vessels. How does this help hemostasis?

Name two of these chemicals. *serotonin + thromboxane 2*

c. State two roles of the ADP that is released by platelets.

Activates platelets + makes them stickier

d. The resulting platelet plug is useful in preventing blood loss in *(large? small?)* vessels.

■ **F4.** The third step in hemostasis, coagulation, requires clotting factors which are derived from three sources.

   a. Some of these factors are made in the liver and constantly circulate in blood (*plasma?* serum?).

   b. Others are released by (red blood cells? *platelets?*).

   c. One important factor, called ___*tissue*___ factor (TF) or thromboplastin, is released from damaged cells. If these cells are within blood or in the lining of blood vessels, then the (extrinsic? *intrinsic?*) pathway to clotting is initiated. If other tissues are traumatized and release tissue factor, then the (*extrinsic?* intrinsic?) pathway is initiated.

■ **F5.** Summarize the three major steps in the cascade of events in coagulation by filling in blanks A–E in Figure LG 19.2. Choose from the list of terms on the figure.

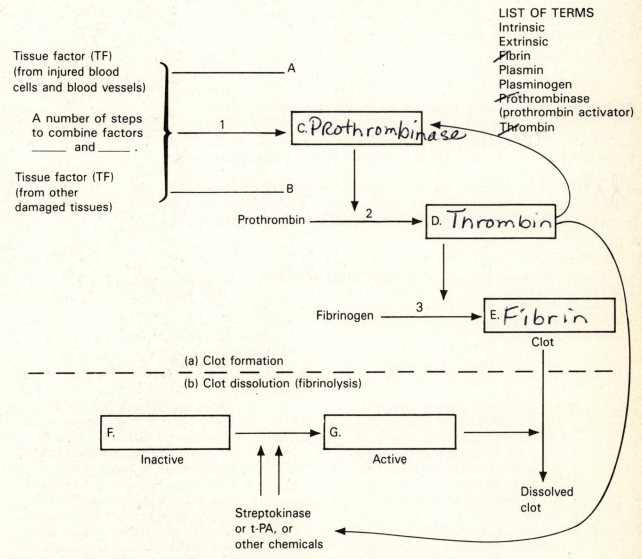

Figure LG 19.2 Summary of steps in clot formation and dissolution. Complete as directed in Checkpoints F5, F6, and F8.

■ **F6.** *For extra review,* refer to Figure LG 19.2 and do this exercise.

a. The figure indicates that thrombin exerts a *(positive? negative?)* feedback effect upon prothrombin, causing the clot to *(enlarge? shrink?)*. Which chemical finally absorbs 90 percent of the thrombin to prevent snowballing of the clot?

_____

b. The ion *(Ca²⁺? K⁺? Na⁺?)*, also known as factor ____, plays a critical role in many aspects of coagulation. Write this ion on the figure next to each of the three major steps.

c. Fill in blanks on Figure LG 19.2 with Roman numerals to indicate which two factors combine to form prothrombinase.

d. Write in factor numbers for other clotting chemicals shown on Figure LG 19.2.

■ **F7.** Although there are many forms of hemophilia, all types involve deficiency of

_____. In the most common form of hemophilia, known as hemophilia *(A? B? C?)*, factor ____ is missing. Write three signs or symptoms of hemophilia.

■ **F8.** Following clot formation, clots undergo two changes. Summarize these.

a. First, clot *retraction* or syneresis occurs.

b. The lower portion of Figure LG 19.2 shows the next step: clot dissolution or *fibrinolysis*. An inactive enzyme (F on the figure) is activated to (G) which dissolves clots. Label F and G on the figure. F is activated by chemicals produced within the body, for example, from damaged blood vessels and also by

(D) _____.

■ **F9.** *A clinical challenge.* Match each chemical with its role in clot formation or dissolution.

H. Heparin	S, t-PA. Streptokinase and t-PA
K. Vitamin K	W. Warfarin (Coumadin)
PGI₂. Prostacyclin	

_____ a. Required for synthesis of factors II, VII, IX, and X in clot formation. Commercial preparations may be required for newborns who lack the intestinal bacteria that synthesize this chemical, and also for persons with malabsorption of fat since this is a fat-soluble chemical.

_____ b. Slow-acting anticoagulant that is antagonistic to vitamin K so decreases synthesis of prothrombin in liver; used to prevent clots in persons with history of clot formation; also the active ingredient in rat poison.

_____ c. Fast-acting anticoagulant that blocks conversion of prothrombin to thrombin; produced in the body by mast cells and basophils, but also available in pharmacologic preparations.

_____ d. Commercial preparations introduced into coronary arteries can limit clot size and prevent severe heart attacks; classified as fibrinolytic agents since they activate plasminogen (shown on Figure LG 19.2) to dissolve clot.

_____ e. A chemical made by white blood cells and cells lining blood vessels; opposes actions of thromboxane A2.

■ **F10.** *A clinical challenge.* Match the correct term with the description.

E. Embolus	T. Thrombus

_____ a. A blood clot

_____ b. A "clot-on-the-run," dislodged from the site at which it formed (usually a deep vein of the leg); also fat from broken bone, bubble of air, or amniotic fluid traveling through blood, possibly to lung (pulmonary) vessels

■ **F11.** *A clinical challenge.* Mr. B, a chronic alcoholic, needs to have a surgical procedure. Explain why he may be at risk for excessive bleeding.

G. Grouping (typing) of blood   (pages 583–586)

■ **G1.** Contrast agglutination with coagulation.

a. Coagulation (or _____) involves coagulation factors found in platelets, plasma, or other tissue fluids. The process *(requires? can occur in absence of?)* red blood cells.

b. Agglutination (or "clumping") of erythrocytes is an

antigen-_____ process that *(does? does not?)* require red blood

cells since these are sites of _____ used in the agglutination reaction.

■ **G2.** Do this exercise about factors responsible for blood groups.

a. Agglutinogens are *(antigens? antibodies?)* located *(in plasma? on surface of red blood cells?)*.

b. Type A blood has *(A? B? both A and B? neither A nor B?)* antigen on RBCs. Genotypes ____ and ____ produce persons with type A blood.

c. Type A blood most likely has *(A? B? both A and B?)* antibody, or

_____. These are also called *anti-B* antibodies since they attack agglutinogen B.

d. If a person with type A blood receives type B blood, the recipient's *(anti-A? anti-B?)* antibodies will attack the *(A? B?)* antigens on the donor's cells. Are the donor's anti-A antibodies likely to attack the recipient's type A cells and cause a major incompatibility reaction?

G3. Complete the table contrasting ABO blood types.

Blood Group	Percent of White Population	Percent of Black Population	Sketch of Blood Showing Correct Agglutinogens and Agglutinins	Can Donate Safely To	Can Receive Blood Safely From
a. Type A					A, O
b.	10				
c. Type AB		7			
d.				A, B, AB, O	

■ **G4.** Type O is known as the universal *(donor? recipient?)* with regard to the ABO group,

since type O blood lacks ___*antigens*___ of the ABO group. Type _____ is known as the universal recipient. Explain why.

■ **G5.** Complete this exercise about the *Rh system.*

a. The Rh *(+? −?)* group is more common. Rh *(+? −?)* blood has Rh agglutinogens on the surfaces of RBCs.

b. Under normal circumstances plasma of *(Rh⁺ blood? Rh⁻ blood? both Rh groups? neither Rh group?)* contains anti-Rh agglutinins.

c. Rh *(+? −?)* persons can develop these agglutinins when they are exposed to Rh *(+? −?)* blood.

d. An example of this occurs in fetal-maternal incompatibility when a mother who is Rh *(+? −?)* has a baby who is Rh *(+? −?)* and some of the baby's blood enters the mother's bloodstream. The mother develops anti-Rh agglutinins which may cross the placenta in future pregnancies and hemolyze the RBCs of Rh *(+? −?)* babies. Such a

condition is known as _____.

e. *A clinical challenge.* Discuss precautions taken with Rh⁻ mothers soon after delivery, miscarriage, or abortion of an Rh⁺ baby to prevent future problems with Rh incompatibility.

G6. Discuss risks associated with blood transfusions and precautions taken to protect recipients.

H. Disorders, medical terminology   (pages 586–588)

■ **H1.** Match names of types of anemia with descriptions below.

> A. Aplastic   N̶. Nutritional
> Hl. Hemolytic   P̶. Pernicious
> H̶r. Hemorrhagic S̶. Sickle cell

N a. Condition resulting from inadequate diet, such as deficiency of iron, vitamin B_{12}, or amino acids

P b. Condition in which intrinsic factor is not produced, so absorption of vitamin B_{12} is inadequate

S c. Inherited condition in which hemoglobin forms stiff rodlike structures causing erythrocytes to assume sickle shape and rupture, reducing oxygen supply to tissues

Hl d. Rupture of red blood cell membranes due to variety of causes, such as parasites, toxins, or antibodies

Hr e. Condition due to excessive bleeding, as from wounds, gastric ulcers, heavy menstrual flow

A f. Inadequate erythropoiesis as a result of destruction or inhibition of red bone marrow

H2. Persons with one gene for sickle cell anemia are said to have *(sickle cell trait?* sickle cell anemia?)*. These individuals have *(greater?* less?)* resistance to malaria. Explain why.

■ **H3.** *A clinical challenge.* Amy (24 years old) has a red blood count of 7 million/mm³. She has the condition known as *(anemia?* polycythemia?)*. Amy has a hemotocrit done. It is more likely to be *(under 32?* over 45?)*. Her blood is *(more?* less?)* viscous than normal, which is likely to cause *(high?* low?)* blood pressure.

■ **H4.** Infectious mononucleosis is caused by the Epstein-Barr *(bacteria?* virus?)*. It affects mostly *(teenagers?* the elderly?)*. The name of this infection is based upon the fact that B lymphocytes become abnormal in appearance, resembling

monocytes . A differential count shows a great increase in

lympho -cytes. Permanent damage usually *(does?* does not?)* result.

■ **H5.** Leukemia is a form of cancer involving abnormally high production of *(erythrocytes? leukocytes?)*. Briefly explain why the following symptoms are likely to occur.

a. Anemia

b. Hemorrhage

c. Infection

H6. Define and explain the clinical significance of each of these terms.

a. Gamma globulin

b. Thrombocytopenia

c. Septicemia

ANSWERS TO SELECTED CHECKPOINTS

A2.

	Interstitial Fluid and Lymph	Plasma
a. Amount of protein (more or less)	Less	More
b. Formed elements present	Contain some leukocytes but no red blood cells or platelets	Red blood cells, white blood cells, platelets
c. Location	Between cells and in lymph vessels	Within blood vessels

A3. (a) Blood, heart and blood vessels. (b) Lymph, lymph vessels, lymph nodes, tonsils, spleen).

A4. (a) Oxygen, carbon dioxide and other wastes, nutrients, hormones, enzymes, heat. (b) pH, temperature, water (fluids) and dissolved chemicals (such as electrolytes). (c) Contains phagocytic cells, blood clotting proteins, and defense proteins such as antibodies, complement, and interferon.

A5. (a) 38°C (100.4°C). (b) 7.35–7.45. (c) 5–6 liters.

A7. (a) W. (b) HE. (c) E. (d) A. (e) F. (f) G. (g) UC. (h) GAF.

B1. (a) 55, 45, cells and cell fragments. (b) Erythro, leuko, thrombo.

B2. (a) Hemopoiesis (or hematopoiesis); stem; granular white. (b) Red bone marrow; proximal epiphyses of femur and humerus, flat bones of skull, sternum, ribs, vertebrae, pelvis. (c) Kidneys; RBCs. (d) Colony-stimulating factors; interleukins; GM-CSF, M-CSF, G-GSF, and multi-CSF (interleukin-3).

B3. EPO stimulates RBC production in persons whose failing kidneys no longer produce erythropoietin. (b) Neupogen (G-CSF) stimulates neutrophil production in persons whose bone marrow function is inadequate, for example, transplant or AIDS patients.

C1. A; does not; hemoglobin; see Figure 19.2, page 571.

C2. (a) Globin, hemes. (b) Iron (Fe), oxygen, 4. (c) Globin, carbamino-; 23; in plasma, bicarbonate.

C3. (a) 4 months. (b) 5 million. (c) 15.

C4. B.

C5. (a) Reticulocyte, 1. (b) Hematocrit, 41.

C6. (a) H. (b) E. (c) B_{12}. (d) IF. (e) T. (g) FH. (h) BB.

C7. Mild hypoxia stimulates RBC formation. Hypoxia is induced in (a) by a move to high altitudes where the air is "thinner" (lower pressure, so lower in oxygen). Residents of high-altitude cities develop higher hematocrits. Removal of RBCs [in (b)] induces hypoxia and erythropoiesis, giving this athlete an unethical boost of hematocrit once original RBCs are reintroduced.

D1. (a) Lack, white blood cells or WBCs. (b) B, neutrophil (60–70); C, lymphocyte (20–25); D, monocyte (3–8); E, eosinophil (2–4); F, basophil (0.5–1.0); differential. (c) Has; monocyte (D); lymphocyte (C); neutrophil (B). (d) B C D. (e) See Figure 19.2, page 571 of your text.

D2. (a) N. (b) N. (c) L. (d) B. (e) E. (f) M. (g) M, N. (h) L, M. (i) N. (j) N. (k) E.

D3. MHC antigens are proteins on cell surfaces; they are unique for each person. The greater the similarity between major histocompatibility (MHC) antigens of donor and recipient, the less likely is rejection of the transplant.

D5. (a) Stimulate an immune response; protein; are not; toxins, bacterial enzymes and structural proteins. (b) B, antibodies; proteins; inactivate. (c) T; cytotoxic.

D6. 5,000–10,000; A.

D7. (a) In; 12,000; leukocytosis. (b) Neutrophils (PMNs or polys); 76; monocytes.

E1. (a) Platelets. (b) Bone marrow. (c) Chips off the old meta-megakaryocytes. (d) 250,000–400,000. (e) Blood clotting.

E2. G.

E3. RBC, WBC, and platelet counts; differential WBC count; hemoglobin and hematocrit (H & H).

F1. (a) Spasm. (b) Platelet. (c) Coagulation.

F3. (a) Large, irregular-shaped; adhesion. (b) Constrict; reduces blood flow into injured area; serotonin, thromboxane A2. (c) Activates platelets and makes them stickier, thereby inviting more platelets to the scene. (d) Small.

F4. (a) Plasma. (b) Platelets. (c) Tissue; intrinsic; extrinsic.

F5. A, intrinsic. B, extrinsic. C, Prothrombinase (prothrombin activator). D, Thrombin. E, Fibrin.

F6. (a) Positive, enlarge; fibrin. (b) Ca^{2+}, IV. (c) X and V. (d) I, Fibrinogen; II, Prothrombin; III, Tissue factor (thromboplastin).

F7. Some clotting factor; A, VIII; hemorrhaging, blood in urine (hematuria), and joint pain and damage.

F8. (a) Retraction. (b) Fibrinolysis; F, plasminogen; G, plasmin; D, thrombin.

F9. (a) K. (b) W. (c) H. (d) S, t-PA. (e) PGI_2.

F10. (a) T. (b) E.

F11. Most coagulation factors are normally synthesized in the liver, which is severely damaged by cirrhosis, a liver disease that can be caused by excessive consumption of alcohol.

G1. (a) Clotting; can occur in absence of. (b) Antibody, does, antigens.

G2. (a) Antigens, on surface of red blood cells. (b) A; AO or AA. (c) b, agglutinin (or isoagglutinin). (d) Anti-B, B; not likely to have major reaction since donor's antibodies are rapidly diluted in the recipient's plasma.

G4. Donor, *a* or *b* agglutinins; AB; persons with type AB blood lacks both *a* and *b* agglutinins.

G5. (a) +; +. (b) Neither Rh group. (c) −, +. (d) −, +; +, hemolytic disease of the newborn (HDN) or erythroblastosis fetalis. (e) The chemical RhoGAM, an agglutinogen, given soon after delivery, binds to any fetal agglutinogens, destroying them. Therefore, the Rh⁻ mother will not produce agglutinins that would attack agglutinogens of future Rh⁺ fetuses.

H1. (a) N. (b) P. (c) S. (d) H1. (e) Hr. (f) A.

H3. Polycythemia; over 45; more, high.

H4. Virus; teenagers; monocytes; lympho; does not.

H5. Focus on production of only immature leukocytes in abnormal bone marrow limits production of RBCs which leads to (a) anemia; platelet production also decreases (b: hemorrhage) and WBC production declines (c: infection).

Questions 1–2: Arrange the answers in correct sequence.

_____ _____ _____   1. Events in the coagulation process:
        A. Retraction or tightening of fibrin clot
        B. Fibrinolysis or clot dissolution by plasma
        C. Clot formation

_____ _____ _____   2. Stages in the clotting process:
        A. Formation of prothrombin activator (prothrombinase)
        B. Conversion of prothrombin to thrombin
        C. Conversion of fibrinogen to fibrin

Questions 3–16: Circle the letter preceding the one best answer to each question.

3. The normal RBC count in healthy females is
   ____ RBCs/mm^3
   A. 4.8 million   D. 250,000
   B. 2 million   E. 8000
   C. 0.5 million

4. All of the following types of formed elements are produced in bone marrow *except:*
   A. Neutrophils   D. Erythrocytes
   B. Basophils   E. Lymphocytes
   C. Platelets

5. Which chemicals are forms in which bilirubin is excreted in urine or feces?
   A. Ferritin and hemosiderin
   B. Biliverdin and transferrin
   C. Urobilinogen and stercobilin
   D. Interleukin 5 and 7

6. Megakaryocytes are involved in formation of:
   A. Red blood cells   D. Platelets
   B. Basophils   E. Neutrophils
   C. Lymphocytes

7. Choose the *false* statement about Norine, who has type O blood.
   A. Norine has neither A nor B agglutinogens on her red blood cells.
   B. She has both anti-A and anti-B agglutinins in her plasma.
   C. She is called the universal donor.
   D. She can receive blood safely from both type O and type AB persons.

8. Choose the *false* statement about blood.
   A. Blood is thicker than water.
   B. It normally has a pH of 7.0.
   C. The human body normally contains about 5 to 6 liters of it.
   D. It normally consists of more plasma than cells.

9. Choose the *false* statement about factors related to erythropoiesis.
   A. Erythropoietin stimulates RBC formation.
   B. Oxygen deficiency in tissues serves as a stimulus for erythropoiesis.
   C. Intrinsic factor, which is necessary for RBC formation, is produced in the kidneys.
   D. A reticulocyte count of over 1.5 percent of circulating RBCs indicates that erythropoiesis is occurring rapidly.

10. Choose the *false* statement about plasma.
    A. It is red in color.
    B. It is composed mainly of water.
    C. Its concentration of protein is greater than that of interstitial fluid.
    D. It contains plasma proteins, primarily albumin.

11. Choose the *false* statement about neutrophils.
    A. They are actively phagocytic.
    B. They are the most abundant type of leukocyte.
    C. Neutrophil count decreases during most infections.
    D. An increase in their number would be a form of leukocytosis.

12. All of the following correctly match parts of blood with principal functions *except:*
    A. RBC: carry oxygen and CO_2
    B. Plasma: carries nutrients, wastes, hormones, enzymes
    C. WBCs: defense
    D. Platelets: determine blood type

13. Which of the following chemicals is an enzyme that converts fibrinogen to fibrin?
   A. Heparin
   B. Thrombin
   C. Prothrombin
   D. Coagulation factor VI
   E. Tissue factor

14. A type of anemia that is due to a deficiency of intrinsic factor production, caused, for example, by alterations in the lining of the stomach, is:
   A. Aplastic
   B. Sickle cell
   C. Hemolytic
   D. Pernicious

15. A person with a hematocrit of 66 and hemoglobin of 22 is most likely to have the condition named:
   A. Anemia
   B. Polycythemia
   C. Infectious mononucleosis
   D. Leukemia

16. Which condition is best described as a malignant disorder of plasma cells of the bone marrow?
   A. Aplastic anemia   C. Porphyria
   B. Multiple myeloma   D. Leukemia

Questions 17–20: Circle T (true) or F (false). If the statement is false, change the underlined word or phrase so that the statement is correct.

T   F   17. Plasmin is an enzyme that facilitates clot <u>formation.</u>

T   F   18. Erythroblastosis fetalis is most likely to occur with an Rh <u>positive mother and her Rh negative babies.</u>

T   F   19. In both black and white populations in the U.S., type O is <u>most</u> common and type AB is <u>least</u> common of the ABO blood groups.

T   F   20. In 100 ml of blood there are usually about <u>15 ml</u> of formed elements, most of which are red blood cells and this value is known as the <u>hematocrit.</u>

Questions 21–25: fill-ins. Answer questions or complete sentences with the word or phrase that best fits.

_____ 21. Name five types of substances transported by blood.

_____ 22. Name three types of plasma proteins.

_____ 23. Name the type of leukocyte that is transformed into plasma cells which then produce antibodies.

_____ 24. Write a value for a normal leukocyte count: ___/mm^3.

_____ 25. All blood cells and platelets are derived from ancestor cells called ____.

ANSWERS TO MASTERY TEST: ■ Chapter 19

Arrange

1. C A B   2. A B C

Multiple Choice

3. A	10. A
4. E	11. C
5. C	12. D
6. D	13. B
7. D	14. D
8. B	15. B
9. C	16. B

True-False

17. F. Dissolution or fibrinolysis
18. F. Negative mother and her Rh positive babies
19. T
20. F. 47 ml (males) to 42 ml (females); hematocrit

Fill-ins

21. Oxygen, carbon dioxide, nutrients, wastes, sweat, hormones, enzymes
22. Albumin, globulins, and fibrinogen
23. Lymphocyte
24. 5,000–10,000
25. Hemocytoblasts

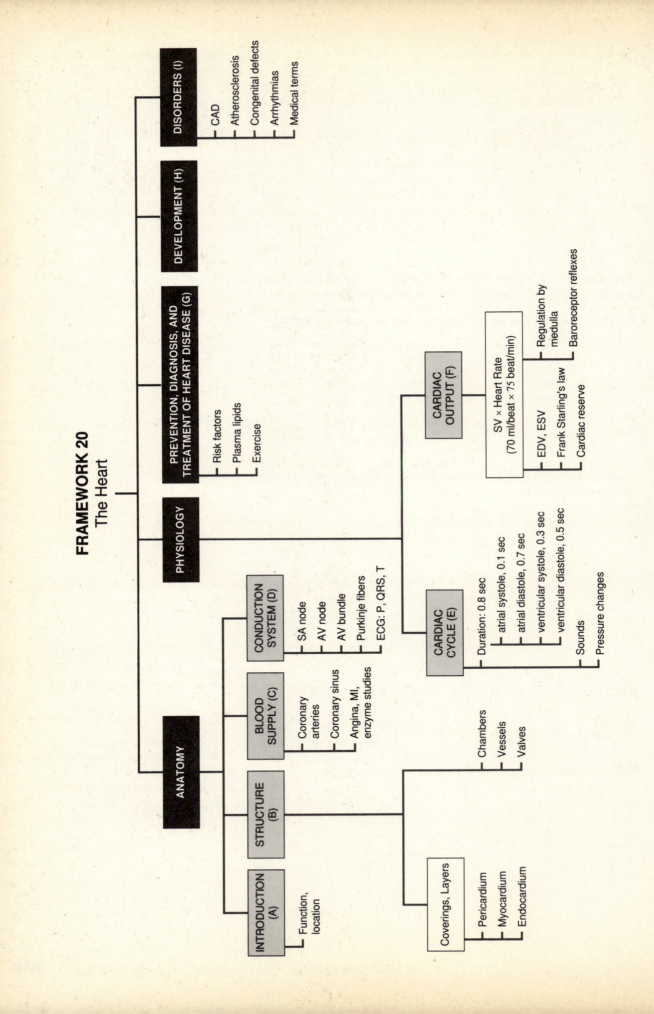

FRAMEWORK 20
The Heart

ANATOMY

INTRODUCTION (A)
— Function, location

STRUCTURE (B)
— Coverings, Layers
  — Pericardium
  — Myocardium
  — Endocardium
— Chambers
— Vessels
— Valves

BLOOD SUPPLY (C)
— Coronary arteries
— Coronary sinus
— Angina, MI, enzyme studies

CONDUCTION SYSTEM (D)
— SA node
— AV node
— AV bundle
— Purkinje fibers
— ECG: P, QRS, T

PHYSIOLOGY

CARDIAC CYCLE (E)
— Duration: 0.8 sec
  — atrial systole, 0.1 sec
  — atrial diastole, 0.7 sec
  — ventricular systole, 0.3 sec
  — ventricular diastole, 0.5 sec
— Sounds
— Pressure changes

CARDIAC OUTPUT (F)
SV × Heart Rate
(70 ml/beat × 75 beat/min)
— EDV, ESV
— Frank Starling's law
— Cardiac reserve
— Regulation by medulla
— Baroreceptor reflexes

PREVENTION, DIAGNOSIS, AND TREATMENT OF HEART DISEASE (G)
— Risk factors
— Plasma lipids
— Exercise

DEVELOPMENT (H)

DISORDERS (I)
— CAD
— Atherosclerosis
— Congenital defects
— Arrhythmias
— Medical terms

The Cardiovascular System: The Heart

CHAPTER 20

Make a fist, then open and squeeze tightly again. Repeat this about once a second as you read the rest of this Overview. Envision your heart as a muscular pump about the size of your fist. Unfailingly, your heart exerts pressure on your blood, moving it onward through the vessels to reach all body parts. The heart has one job; it is simply a pump. But this function is critical. Without the force of the heart, blood would come to a standstill, and tissues would be deprived of fluids, nutrients, and other vital chemicals. To serve as an effective pump, the heart requires a rich blood supply to maintain healthy muscular walls, a specialized nerve conduction system to synchronize actions of the heart, and intact valves to direct blood flow correctly. Heart sounds and ECG recordings as well as a variety of more complex diagnostic tools provide clues to the status of the heart.

Is your fist tired yet? The heart ordinarily pumps 24 hours a day without complaint and seldom reminds us of its presence. As you complete this chapter on the heart, keep in mind the value and indispensability of this organ. Start by studying the Chapter 20 Framework and the key terms for each section.

TOPIC OUTLINE AND OBJECTIVES

A. Introduction

B. Heart structure

1. Describe the location of the heart and the structure and functions of the wall, chambers, great vessels, and valves of the heart.

C. Blood supply

2. Describe the blood supply of the heart.

D. Conduction system, electrocardiogram

3. Explain the structural and functional features of the conduction system of the heart.
4. Describe the physiology of cardiac muscle contraction.
5. Explain the meaning of an electrocardiogram (ECG) and its diagnostic importance.

E. Cardiac cycle

6. Describe the phases, timing, and sounds associated with a cardiac cycle.

F. Cardiac output

7. Define cardiac output (CO) and describe the factors that affect it.
8. Explain how heart rate is regulated.

G. Prevention, diagnosis, and treatment of heart disease

9. List and explain the risk factors involved in heart disease.
10. Explain the relationship between plasma lipids and heart disease.

H. Developmental anatomy

11. Describe the developmental anatomy of the heart.

I. Disorders, medical terminology

12. Explain the benefits of regular exercise on the heart.

13. Define the following disorders: coronary artery disease (CAD), congenital defect, and arrhythmias.

14. Define medical terminology associated with the heart.

WORDBYTES

Now study the following parts of words that may help you better understand terminology in this chapter.

Wordbyte	Meaning	Example	Wordbyte	Meaning	Example
cardi-	heart	*cardi*ologist	myo-	muscle	*myo*cardial infarction
coron-	crown	*coron*ary arteries			
endo-	within	*endo*carditis	peri-	around	*peri*cardium
			vascul-	small vessel	cardio*vascul*ar

CHECKPOINTS

A. Introduction: heart functions and location (page 592)

■ **A1.** Do this exercise about heart function.

a. What is the primary function of the heart? ___*pump blood*___

b. Visualize a coffee cup; now imagine that sitting next to it is a gallon-size milk container. Keep in mind that a gallon holds about __4__ quarts or approximately 4 liters. Now take a moment to imagine a number of containers lined up and holding the amount of blood pumped out by the heart.

   1. With each heartbeat, the heart pumps a stroke volume equal to about ⅓ cup (70 ml), since 1 cup = about _210_ ml.
   2. Each minute, the heart pumps enough blood to fill up 1.3 gallons (____ liters).
   3. Within the 1440 minutes in a day, the heart would fill up about 1980 gallons (over ____ liters).

c. Reflecting on the work your heart is doing for you, what message would you give to your heart?

   thank you

■ **A2.** Closely examine Figure 20.1, page 593 in your text. Consider the location, size, and shape of your heart as you do this exercise. Trace its outline on your body.

a. Your heart lies in the _____ portion of your thorax, between

your two _____. About *(one-third? one-half? two-thirds?)* of the mass of your heart lies to the left of the midline of your body.

b. Your heart is about the size and shape of your _____.

c. The pointed part of your heart, called the _____, lies in the

_____ intercostal space, about _____ cm (_____ inches) from the midline of your body.

d. The region of the heart that lies directly superior to the diaphragm is the right *(atrium? ventricle?)*.

B. Heart structure (pages 592–600)

■ **B1.** Arrange in order from most superficial to deepest. ____ ____ ____ ____ ____ ____

E. Endocardium	PC. Pericardial cavity (site of pericardial fluid)
FP. Fibrous pericardium	PP. Parietal pericardium
M. Myocardium	VP. Visceral pericardium (epicardium)

For extra review of the pericardium, look again at Checkpoint D2 in Chapter 4, page 78 in the Learning Guide.

■ **B2.** Inflammation of the pericardium is known as _____. This

condition may lead to cardiac tamponade, which means _____

_____.

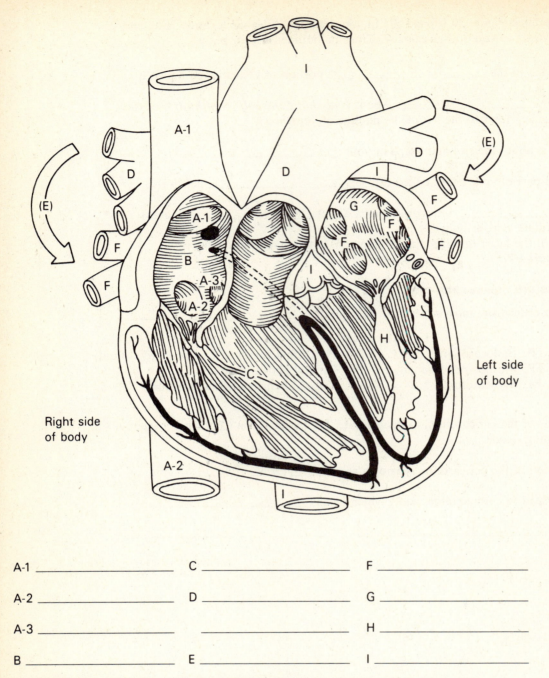

Right side
of body

Left side
of body

A-1 _____ C _____ F _____

A-2 _____ D _____ G _____

A-3 _____   _____ H _____

B _____ E _____ I _____

Figure LG 20.1 Diagram of a frontal section of the heart. Letters follow the path of blood through the heart. Label, color, and draw arrows as directed in Checkpoints B3 and D1.

■ **B3.** Refer to Figures LG 20.1 and LG 20.2 to do the following activity.

a. Identify all structures with letters (A–I) by writing labels on lines A–I.

b. Draw arrows on Figure LG 20.1 to indicate direction of blood flow.

c. Color red the chambers of the heart and vessels that contain highly oxygenated blood; color blue the regions in which blood is low in oxygen and high in carbon dioxide.

d. On Figure LG 20.1, label the four valves that control blood flow through the heart.

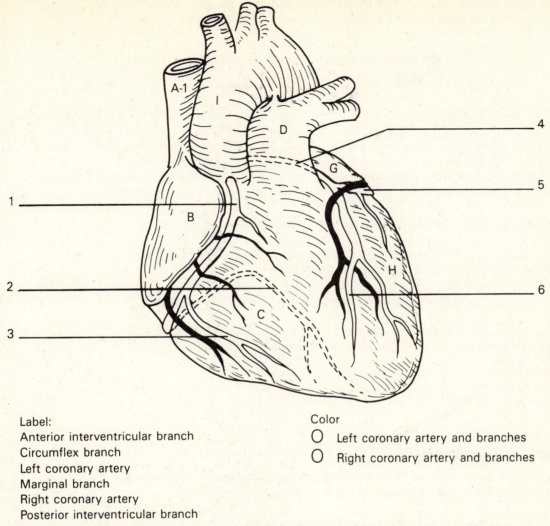

Label:
Anterior interventricular branch
Circumflex branch
Left coronary artery
Marginal branch
Right coronary artery
Posterior interventricular branch

Color
O Left coronary artery and branches
O Right coronary artery and branches

Figure LG 20.2 Anterior view of the heart. Label and color as directed in Checkpoints B3 and C2. Cardiac veins are solid black.

■ **B4.** Check your understanding of heart structure and the pathway of blood through the heart by selecting terms that fit descriptions below. Not all answers will be used.

A. Atria	PV. Pulmonary vein
CT. Chordae tendiniae	S. Septum
PA. Pulmonary artery	V. Ventricles
PM. Papillary muscles	

_____ a. Thin-walled chambers that receive blood from veins

_____ b. Blood vessel that carries blood that is rich in oxygen from lungs to left atrium

_____ c. Wall separating right side from left side of the heart

_____ d. Strong tendons that anchor atrioventricular valves to ventricular muscle

_____ e. Sites of most of the myocardium

■ **B5.** Check your understanding of the valves by doing this matching exercise. More than one answer may be required for each description.

> A. Aortic semilunar   P. Pulmonary semilunar
> B. Bicuspid   T. Tricuspid

_____ a. Also called the mitral valve

_____ b. Prevents backflow of blood from right ventricle to right atrium

_____ c. Prevents backflow from pulmonary trunk to right ventricle

_____ d. Prevents backflow of blood into left atrium

_____ e. Have half-moon-shaped leaflets or cusps (two answers)

_____ f. Also called atrioventricular (AV) valves (two answers)

■ **B6.** *A clinical challenge.* Name the microorganism that causes rheumatic fever (RF).

_____. What condition is likely to signal the presence of this microbe and warn of possible effects (sequelae) upon the heart?

_____ Which parts of the heart are most likely to be affected?

_____ A thought to ponder: as you look at heart physiology, speculate about why these two valves might be more susceptible.

C. Blood supply (pages 600–601)

C1. Defend or dispute this statement: "The myocardium receives all of the oxygen and nutrients it needs from blood that is passing through its four chambers."

■ **C2.** Refer to Figure LG 20.2 and complete this checkpoint.
   a. Using color code ovals on the figure, color the two coronary arteries and their main branches.
   b. Label each of the vessels, using the list of labels on the figure.

■ **C3.** The coronary sinus functions as *(an artery? a vein?)*. It collects blood that has

passed through coronary arteries and capillaries into _____ veins. The coronary sinus finally empties this blood into the *(right? left?) (atrium? ventricle?)*.

C4. Define *ischemia* and explain how it is related to *angina pectoris*.

■ **C5.** *A clinical challenge.* Do this exercise about "heart attacks."

a. "Heart attack" is a common name for a myocardial _____

(MI). What does the term *infarction* mean? _____ In what other areas of the body might infarctions also occur? Cerebral infarction:

_____; pulmonary infarction: _____.

b. List two or more immediate causes of an MI.

c. Define *reperfusion.*

d. Explain possible damage caused by reperfusion of the heart, including the role of *free radicals.*

e. List three or more other diseases or conditions that may be related to free radicals.

D. Conduction system; cardiac muscle physiology; electrocardiogram (pages 601–605)

■ **D1.** In this exercise describe how the heart beats regularly and continuously.

a. In embryonic life about ____ percent of cardiac muscle fibers become *autorhythmic.* State two functions of these specialized cells of the heart.

b. Label the parts of the conduction system on Figure LG 20.1.

c. The normal pacemaker of the heart is the *(SA? AV?)* node. Each time the SA node "fires," impulses travel via the conduction system and also by the

_____ junctions in _____ discs of cardiac muscle.

d. What structural feature of the heart makes the AV node and AV bundle necessary for conduction from atria to ventricles?

e. Through which part of the conduction system do impulses pass most slowly? *(SA node? AV node? AV bundle [of His]?).* Based on the anatomy of this tissue, why

does the rate of impulse conduction slow down? _____ Of what advantage is this slowing?

■ **D2.** Do this exercise about the pacemaker(s) of the heart.

a. The SA node normally fires at about ____ to ____ times per minute (normal pulse).

How might this "normal pacemaker" be damaged? _____. If this occurs, then responsibility for setting the pace of the heart may be passed on to the ____ node, which fires at ____ to ____ times per minute.

b. If both the SA and AV nodes fail, then autorhythmic fibers in

_____ may take over with a rate of only ____ to ____ beats per minute. Pacemakers at "other than the normal site" are known as

_____ pacemakers, and tend to be *(more? less?)* effective than the SA node in pacing the heart.

■ **D3.** Describe the physiology of cardiac muscle contraction in this activity.

a. Contractile fibers of the normal heart have a resting membrane potential of *(−70? −90?)* mV.

b. Arrange in correct sequence the events in an action potential of cardiac muscle. The

blank parentheses are for Activity D3c. One is done for you.

$\underline{\quad C \quad}$ (RD) → _____ ( ) → _____ **or** ( ) → _____ ( ) → _____ ( )

A. *Slow Ca^{2+} channels* open so Ca^{2+} enters muscle fibers. __P__

B. Combined flow of Na^+ and Ca^{2+} maintains depolarization for about 250 msec. ____

C. Na^+ enters through *fast Na^+ channels;* voltage rises rapidly to about +20 mV. ____

D. A second contraction cannot be triggered during this period, which is longer than for skeletal muscle. ____

E. The presence of Ca^{2+} binding to troponin permits myocardial contraction via sliding of actin filaments next to myosin filaments.

F. K^+ channels open so K^+ ions leave the fiber; meanwhile less Na^+ or Ca^{2+} enters as those channels are closing; voltage returns to resting level. ____

c. Now label events A–F above by filling in parentheses. One is done for you. (For help refer to Figure 20.7, page 603 in your text.) Use these answers:

P. Plateau	Rf. Refractory
RD. Rapid depolarization	Rp. Repolarization

d. *A clinical challenge.* Calcium channel blockers [such as verapamil (Procardia)] tend to *(in? de?)*-crease contraction of the cardiac muscle (and of smooth muscle of coronary arteries). Write one or more condition(s) for which such a medication might be prescribed.

Now name one chemical that enhances flow of Ca^{2+} through these channels to

strengthen contraction of the heart. _____

■ **D4.** Describe an ECG (or EKG) in this activity.

a. What do the letters ECG (or EKG) stand for?

An ECG is a recording of *(electrical changes associated with impulse conduction? muscle contractions?)* of the heart.

b. On what parts of the body are leads (electrodes) placed for a 12-lead ECG?

c. In Figure LG 20.3b, an ECG tracing using lead *(I? II? III?)* is shown. Label the following parts of that ECG on the figure: *P wave, P-R interval, QRS wave (complex), S-T segment, T wave.*

d. *A clinical challenge.* Answer the following questions with names of parts of ECG.

1. Indicates atrial depolarization leading to atrial contraction

   _____

2. Prolonged if the AV node is damaged, for example, by rheumatic fever, or in the

   condition described as "bundle (of His) block" _____

3. Elevated in acute MI _____

4. Elevated in hyperkalemia (high blood level of K^+) _____

5. Flatter than normal in coronary artery disease _____

6. Represents ventricular repolarization _____

7. Represents ventricular depolarization _____

8. Represents atrial repolarization _____

E. Cardiac cycle (pages 605–608)

■ **E1.** Complete the following overview of movement of blood through the heart.

a. Blood moves through the heart as a result of _____ and

   _____ of cardiac muscle, as well as _____

   and _____ of valves. Valves open or close due to

   _____ within the heart.

b. Contraction of heart muscle is known as _____, whereas

   relaxation of myocardium is called _____.

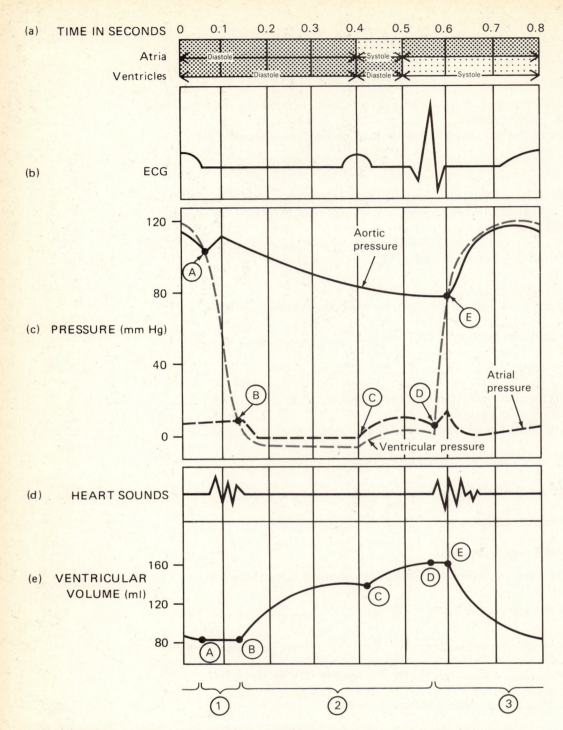

(a) TIME IN SECONDS

Atria

Ventricles

(b) ECG

(c) PRESSURE (mm Hg)

Aortic pressure

Atrial pressure

Ventricular pressure

(d) HEART SOUNDS

(e) VENTRICULAR VOLUME (ml)

Figure LG 20.3 Cardiac cycle. (a) Systole and diastole of atria and ventricles related to time. (b) ECG related to cardiac cycle. (c) Pressures in the aorta, atria, and ventricles. (d) Heart sounds related to cardiac cycle. (e) Volume of blood in ventricles. Label as directed in Checkpoints D4c and E2 to E4.

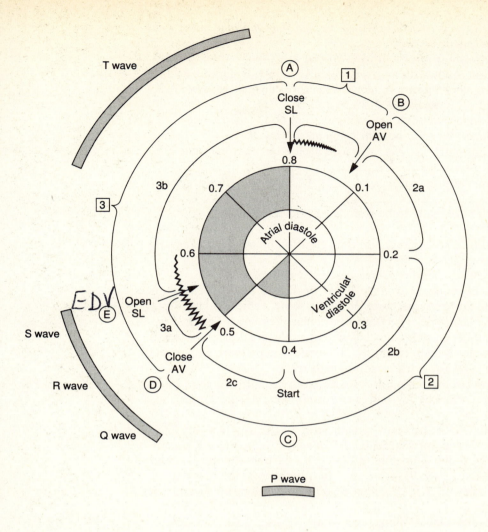

T wave

Close
SL

Open
AV

A

B

1

3b

0.7

0.8

0.1

2a

3

Atrial diastole

0.6

0.2

EDV

Open
SL

E

Ventricular
diastole

S wave

3a

0.5

0.3

R wave

Close
AV

D

2c

0.4

2b

2

Start

Q wave

C

P wave

KEY

1   Isovolumetric relaxation
2a   Rapid ventricular filling
2b   Reduced ventricular filling (diastasis)
2c   Atrial contraction
3a   Isovolumetric contraction
3b   Ventricular ejection

Figure LG 20.4   Cardiac cycle. Events of this cycle are shown on a special 0.8-sec "clock." Sectors are 0.1 sec each. Atrial activity is represented within inner circle and ventricular activity within outer circle. Shaded areas: systole; white areas of circle: diastole. Sounds are represented as ⋀⋁⋀⋁⋀⋁ . Valve openings and closings and ECG correlations are also included. Brackets with key numbers 1 to 3b define important periods within the cycle. Refer to Checkpoints E2 and E3.

■ **E2.** In this exercise, we will consider details of the cardiac cycle. Refer alternately to Figure LG 20.3 and Figure LG 20.4. Figure LG 20.3 will allow you to correlate pressure and volume changes within the heart to an ECG recording and to heart sounds. Figure LG 20.4 will help you to visualize events as they occur in a one-cycle "clock."

a. If the heart beats at 75 beats per minute, then *(54? 60? 75?)* complete cardiac cycles occur per 60-sec period. In other words, if your pulse is 75 (beats per minute), the duration of one cardiac cycle ("heartbeat") within you is 8/10 sec.

b. Now look at Figure LG 20.4, which represents one complete 0.8-sec cardiac cycle on a special 0.8-sec "clock." The clock is divided into eight sectors, or time periods, each of which is .01 sec in duration.

Note: In the following activities, it is critical to remember that the total time elapsed during one cardiac (clock) cycle is only 0.8 sec.

c. We could examine the cycle shown in Figure LG 20.4 at any point on the "clock." We will, however, begin our study with an overview of the cycle. Beginning at the top of the clock, the "12 o'clock" position, identify the three distinct phases—1, 2, and 3—that are bracketed on Figures LG 20.3 and LG 20.4.

   1. Phase 1 is a period in which the atria and ventricles are both *(contracting? relaxing?)*. In other words, they are both in *(systole? diastole?)*.

   2. Throughout phase 2 the ventricles *(eject? fill with?)* blood as they continue in *(systole? diastole?)* while atria go into ___systole___ toward the end of Phase 2.

   3. Now the filled ventricles are ready to be stimulated (indicated by the ___QRS___ wave of the ECG) to contract and then eject the blood. That is exactly what happens in phase ___3___.
   At the bottom of Figure 20.3, label these three bracketed phases.

d. Now let's look at details. Begin again at "12 o'clock" on Figure LG 20.4 and at the left side of LG Figure LG 20.3c. As we come on to the scene, ventricles have just stopped contracting in the last cycle. Therefore ventricular pressure is *(rising? dropping?)*. When pressure within ventricles dips below that in the great arteries, blood fills semilunar valves and closes them (point ___A___). Notice that pressure within ventricles drops at a rapid rate now as ventricles continue to relax.

e. When intraventricular pressure becomes lower than that in atria, the force of blood within atria causes (atrioventricular) AV valves to *(open? close?)*. This happens at point ___B___.

f. Note (Figure LG 20.4) that AV valves now remain open all the way to point D, in other words, throughout the time of ventricular *(ejection? filling?)*. Trace a pencil lightly along the filling curve shown on Figure LG 20.3e, starting at point B. Note that the curve is steepest just *(after B? before C?)*. In other words, the first part of ventricular diastole (just after the AV "floodgates" open) is the time when ventricles fill *(rapidly? slowly?)*, whereas slower filling, known as ___diastasis___, occurs later (just before point C).

g. Which wave of the ECG signals atrial systole? *((P? QRS? T?)*. Duration of atrial systole is about *(0.1? 0.3? 0.5?)* sec. As a result of atrial contraction, pressure within the atria *(increases? decreases?)* (point C in Figure LG 20.3c), forcing the last 30 percent of all of the blood that will enter ventricles into those chambers.

h. Point *D* marks the end of atrial systole, as well as the end of ventricular *(diastole? systole?)*. Ventricles are now filled to their maximum volume (see Figure LG 20.3e), which is about ___130___ ml of blood. This volume is known as the ___end___-___diastolic___ volume (EDV). Write EDV next to point E on Figure LG 20.3e.

i. The *(P? QRS? T?)* wave signals ventricular systole. Note that although the QRS wave itself happens over a very short period, the ventricular contraction that follows occurs over a period of about *(0.1? 0.3? 0.5?)* sec.

j. Trace a pencil lightly along the curve in Figure LG 20.3c showing ventricular pressure changes following the QRS wave. Notice that pressure there *(increases? decreases?)* *(slightly? dramatically?)*. In fact it quickly surpasses atrial pressure (at point ___), forcing the AV valves *(open? closed?)*.

k. A brief time later ventricular pressure becomes so great that it even surpasses pressure in the great arteries (pulmonary artery and aorta). This pressure forces blood against the undersurface of the _____ valves, *(opening? closing?)* them. This occurs at point ___ on the figure.

l. Continued ventricular systole ejects blood from the heart into the great vessels. In the aorta this creates a systolic blood pressure of about *(15? 80? 120?)* mmHg.

m. Repolarization of ventricles (after the _____ wave) causes these chambers to go into *(systole? diastole?)*. Consequently, ventricular pressure begins to *(increase? decrease?)*. This completes one cardiac cycle.

n. The volume remaining in the ventricles at the end of systole is known as

_____-_____ volume (ESV). Write ESV next to point *A* to indicate that this much blood remains in the heart as a new cycle begins. ESV is normally about ____ ml. The amount of blood ejected from the heart

during systole is known as _____ volume (SV), and equals about ____ ml.

o. Label points *B, C, D,* and E on Figure LG 20.3c to indicate which valves open or close at those points. Check your answers by referring to Figure LG 20.4.

p. *For extra review.* Refer to a figure of the heart, such as Figure LG 20.1, as you review this activity.

■ **E3.** Summarize changes in ventricular volume during the cycle by relating events described in the key for Figure LG 20.4 with the curve in Figure LG 20.3e.

a. For example, the fact that the portion of the curve between points A and B is "flat" indicates that the volume of these chambers *(does? does not?)* change during this period. This is logical, since the AV and SL valves are all *(open? closed?)* during this

time. This period (A to B) is known as the period of _____ relaxation.

b. Note that the "flat" portion of the curve in Figure LG 20.3e between points D and E is a time when ventricles are *(relaxing? contracting?)*. Which valves are closed

during this period? _____ In other words, ventricles have exerted enough pressure to close *(AV? SL?)* valves, but not enough pressure yet to

open ____ valves. Name this period: isovolumetric _____.
(*Note:* remember that the reason that periods A–B and D–E are called *isovolumetric* (iso = same) is because no blood enters or leaves ventricles since both sets of valves are *closed.*)

■ **E4.** Do this exercise on heart sounds.

a. The first heart sound *(lubb? dupp?)* is produced by turbulence of blood at the

(opening? closing?) of the _____ valves. What causes the second sound?

b. Write *first* and *second* next to the parts of Figure LG 20.3d showing each of these sounds.

c. *A clinical challenge.* Using Figure 20.11, page 608 in your text, recognize why a stethoscope is placed at several points on the chest in order to best hear different heart sounds. If a stethoscope is placed at the level of about the fifth interspace (between ribs 5 and 6) on the left side of the sternum, the *(first? second?)* sound is better heard, since turbulence from the *(mitral? aortic semilunar?)* valve is readily detected there. To best hear the second sound (related to SL valve closure), the stethoscope must then be moved in a more *(inferior? superior?)* direction on the chest.

E5. Explain what causes heart murmurs.

F. Cardiac output; regulation of the heart (pages 608–612)

■ **F1.** Determine the average cardiac output in a resting adult.
Cardiac output = stroke volume × heart rate

= _____ ml/stroke × _____ strokes/min

= _____ ml/min (_____ liter/min)

■ **F2.** At rest Dave has a cardiac output of 5 liters per minute. During a strenuous cross-country run, Dave's maximal cardiac output is 20 liters per minute. Calculate Dave's cardiac reserve.

$$\text{Cardiac reserve} = \frac{\text{maximal cardiac output}}{\text{cardiac output at rest}} = \frac{20 \text{ lts}}{5 \text{ lts}} \qquad 15 \text{ lts}$$

Note that cardiac reserve is usually (*higher*? *lower*?) in trained athletes than in sedentary persons.

■ **F3.** The two major factors (shown in Activity F1) that control cardiac output are

___stroke V___ and ___heart rate___.

■ **F4.** Do this activity about stroke volume (SV).

a. Which statement is true about stroke volume (SV)? (*Hint:* refer to Checkpoint E2h and n.)

A. SV = ESV–EDV   B. SV = EDV–ESV   C. SV = $\frac{ESV}{EDV}$ D. SV = $\frac{EDV}{ESV}$

b. List the three factors that control stroke volume (SV):

_____   _____   _____

c. During exercise, skeletal muscles surrounding blood vessels squeeze (*more*? *less*?) blood back to the heart. This increase in venous return to the heart (*increases*? *decreases*?) end-diastolic volume (EDV). As a result, muscles of the heart are stretched (*more*? *less*?), so that preload (= stretching of the heart muscle) (*increases*? *decreases*?).

d. Within limits, a stretched muscle contracts with (*greater*? *less*?) force than a muscle that is only slightly stretched. This is a statement of the ___Frank___ ___Starling___ law of the heart. To get an idea of this, blow up a balloon slightly and then let it go. Then blow up a balloon quite full of air (much like the heart when stretched by a large venous return) and let that balloon go. In which stage do the walls of the balloon (heart) compress the air (blood) with greater force?

e. As a result of the Frank-Starling law, during exercise the ventricles of the normal heart contract (*more*? *less*?) forcefully, and stroke volume (*increases*? *decreases*?).

f. State several reasons why venous return might be decreased, leading to reduction in stroke volume and cardiac output.

g. Stroke volume and cardiac output are directly related to the length of ventricular *(systole? diastole?)*. When the heart rate is very rapid, the cardiac cycle is much shorter. Therefore there is *(more? less?)* time for ventricular filling. So a person with a heart rate of 160 beats per minute is likely to have a(n) *(increased? reduced?)* stroke volume and cardiac output.

■ **F5.** Summarize factors affecting heart rate (HR), stroke volume (SV), and cardiac output (CO) by inserting an ↑ (for increase) or ↓ (for decrease) in the blanks provided. The first one is done for you.

a. ↑ exercise → ↑ preload
b. Moderate ↑ in preload → | SV
c. ↑ SV (if heart rate is constant) → | CO
d. ↑ in heart contractility → | SV and CO
e. Positively inotropic medication such as digoxin (Lanoxin) → | SV and CO
f. ↑ afterload (for example, due to hypertension) → | SV and CO
g. Hypothermia → | heart rate and CO
h. ↑ heart rate (if SV is constant) → | SV and CO
i. ↑ stimulation of the cardioaccelerator center → | heart rate
j. ↑ sympathetic nerve impulses → | heart strength (and SV) and | heart rate
k. ↑ vagal nerve impulses → | heart rate
l. Hyperthyroidism → | heart rate
m. ↑ Ca^{2+} → | heart strength (and SV) and | heart rate
n. ↓ SV (as in CHF) → | ESV → excessive EDV and preload → | SV and CO

■ **F6.** *A clinical challenge.* Do this exercise about Ms. S, who has congestive heart failure (CHF).

a. Her weakened heart becomes overstretched much like a balloon (except much thicker!) that has been expanded 5000 times. Now her heart myofibers are stretched beyond the optimum length according to the Frank-Starling law. As a result, the force of her heart is *de*-creased, and its stroke volume *de*-creases (as shown in Activity F5n).

b. Mrs. S has right-sided heart failure, so her blood is likely to back up, distending vessels in *(lungs? systemic regions, such as in neck and ankles?)*; this is

    Systemic _____ edema results, with signs such as swollen hands and feet.

c. Write a sign or symptom of left-sided heart failure.

    lungs fill

d. An intra-aortic balloon pump (IABP) is especially helpful for *(right? left?)*-sided CHF since the IABP ____-creases afterload, and therefore can ____-crease stroke volume.

G. Risk factors in heart disease   (pages 612–615)

G1. List eight risk factors for heart disease. Circle five of those that can be modified by a healthy lifestyle.

a. _____   e. _____

b. _____   f. _____

c. _____   g. _____

d. _____   h. _____

■ **G2.** One risk factor, obesity, is associated with the fact that the heart must work harder to pump blood into an estimated ____ km (____ miles) of blood vessels for each extra pound of fat.

■ **G3.** Distinguish classes of chemicals related to atherosclerosis by doing this exercise.

a. HDL and LDL both consist of complexes of _____ combined

with _____. HDL refers to *(high? low?)*-density lipoprotein,

whereas LDL means _____.

b. *(HDLs? LDLs?)* are considered culprits in causing atherosclerosis since they contain large amounts of cholesterol and deposit some of this lipid in walls of arteries. So a *(high? low?)* blood level of LDL is desirable.

c. High levels of LDL may occur in blood of persons who have too *(many? few?)* LDL receptors on their cells. Consequently, LDL remains in blood and may travel to blood vessel walls and take up residence there.

d. A high level of HDL tends to *(lead to? prevent?)* atherosclerosis. List two ways that HDLs may be increased in your own blood.

G4. List benefits of exercise in improving cardiovascular health.

H. Developmental anatomy   (pages 615–616)

■ **H1.** The heart is derived from _____-derm. The heart begins to develop during the *(third? fifth? seventh?)* week. Its initial formation consists of two

endothelial tubes which unite to form the _____ tube.

■ **H2.** Match the regions of the primitive heart at right with the related parts of a mature heart listed below.

A. Atrium	SV. Sinus venosus
BCTA. Bulbus cordis and truncus arteriosus	V. Ventricle

_____ a. Superior and inferior vena cava

_____ b. Right and left ventricles

_____ c. Parts of the fetal heart connected by foramen ovale

_____ d. Aorta and pulmonary trunk

I. Disorders, medical terminology   (pages 616–620)

■ **I1.** Outline the sequence of changes in atherosclerosis in this learning activity.

a. Atherosclerosis involves damage to the walls of arteries. State several factors that may lead to damage of this lining.

b. The resulting fatty lesion in the arterial lining is known as an atherosclerotic

_____. The rough surface of plaque may snag platelets. These

may then cause _____ formation at the site, possibly leading to unwanted emboli. Platelets may also release a chemical called

_____.

c. Name two other types of cells that release PDGF: _____ and

_____. What is the effect of this chemical?

I2. List several factors that may lead to a coronary artery spasm.

■ **I3.** Match each congenital heart disease in the box with the related description below.

> C. Coarctation of aorta
> IASD. Interatrial septal defect
> IVSD. Interventricular septal defect
>
> PDA. Patent ductus arteriosus
> VS. Valvular stenosis

_____ a. Connection between aorta and pulmonary artery is retained after birth, allowing backflow of blood to right ventricle.

_____ b. Foramen ovale fails to close.

_____ c. Septum between ventricles does not develop properly.

_____ d. Valve is narrowed, limiting forward flow and causing increase in workload of heart.

_____ e. Aorta is abnormally narrow.

■ **I4.** Which of the conditions in Checkpoint I3 result in left-to-right shunts? Explain why this is so.

■ **I5.** Do this exercise on arrhythmias.

a. *(All? Not all?)* arrhythmias are serious.

b. Conduction failure across the AV node results in an arrhythmia known as

_____.

c. Which is more serious? *(Atrial? Ventricular?)* fibrillation. Explain why.

d. Excitation of a part of the heart other than the normal pacemaker (SA node) is

known as a(n) _____ focus. Such excitations, which may be caused by ingestion of caffeine or by lack of sleep, or by more serious problems such as ischemia, may lead to an extra early ventricular systole; these are known as

_____.

I6. Briefly describe each of these cardiac conditions.

a. Congestive heart failure (CHF)

b. Cor pulmonale

c. Cardiomegaly

ANSWERS TO SELECTED CHECKPOINTS

A1. (a) It is a pump. (b) 4 (one gallon = 4 quarts or 3.86 liters). (b1) 237. (b2) 5 (1.3 gallons = 5 liters). (b3) 7200 (1.3 gallons/min × 1440 min/day = 1800 gallons/day = over 7000 liters). (c) Perhaps: "Thanks!" "Great job!" "Amazing!" or "You deserve the best of care, and I intend to see that you get it."

A2. (a) Mediastinal, lungs; two-thirds. (b) Fist. (c) Apex, fifth, 8, 3. (d) Ventricle.

B1. FP PP PC VP M E

B2. Pericarditis; compression of heart due to accumulation of fluid in the pericardial space.

B3.

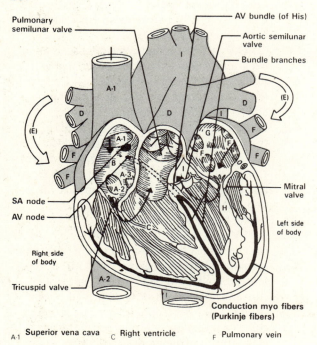

Pulmonary semilunar valve
AV bundle (of His)
Aortic semilunar valve
Bundle branches
A-1
SA node
AV node
Mitral valve
Right side of body
Left side of body
Tricuspid valve
A-2
Conduction myo fibers (Purkinje fibers)

A-1 Superior vena cava   C Right ventricle   F Pulmonary vein

A-2 Inferior vena cava   D Pulmonary trunk   G Left atrium

A-3 Coronary sinus   and arteries   H Left ventricle

B Right atrium   E Vessels in lungs   I Aorta

Figure LG 20.1A Diagram of a frontal section of the heart.

(a, b, d) See Figure LG 20.1A. (c) Blue: A–D and first part of E; red: last part of E; F–I.

B4. (a) A. (b) PV. (c) S. (d) CT. (e) V.

B5. (a) B. (b) T. (c) P. (d) B. (e) A, P. (f) B, T.

B6. Group A, β-hemolytic Streptococcus pyogenes; streptococcal sore throat ("strep throat"); valves, particularly the bicuspid (mitral) and the aortic semilunar. These two valves control blood flow through the left side of the heart, which is a higher-pressure system than the right side; therefore, valve damage is more likely, making these valves more vulnerable to effects of RF.

C2. (a) Left coronary artery: 4–6; right coronary artery: 1–3. (b) 1, right coronary artery; 2, posterior interventricular branch; 3, marginal branch; 4, left coronary artery; 5, circumflex branch; 6, anterior interventricular branch.

C3. Vein; cardiac; right atrium.

C5. Infarction; death of tissue because of lack of blood flow to that tissue; brain, lung. (b) Thrombus (clot), embolus (mobile clot), spasm of coronary artery. (c) Reestablishing blood flow (with oxygen supply) after tissue has been deprived of blood flow (ischemia). (d) An oxygen free radical borrows an electron from another molecule. As a result, a chain reaction of "borrowing electrons" may occur causing damage to heart tissue. (e) Alzheimer's disease, Parkinson's disease, cancer, cataracts, rheumatoid arthritis, and aging.

D1. (a) One; pacemaker and impulse conduction. (b) See Figure LG 20.1A. (c) SA; gap, intercalated. (d) The fibrous ring (including AV valves) that completely separates atria from ventricles. (e) AV node; fibers have smaller diameter; affords time for ventricles to fill more completely before they contract.

D2. (a) 60–100; ischemia as in myocardial infarction; AV, 40–50. (b) Conduction myofibers (Purkinje fibers), 20–40; ectopic; less.

D3. (a) −90. (b,c) C (RD) → (A) P → B or E (both P) → F (Rp) → D (RF) (d) De; coronary artery disease (CAD), high blood pressure (hypertension), rapid heart of rate (tachycardia); epinephrine or other catecholamines.

D4. (a) An electrocardiogram (the recording) or electrocardiograph (the instrument); electrical changes associated with impulse conduction. (b) Arms, legs, and chest. (c) II; refer to Figures 20.8 and 20.9, pages 604 and 606 of the text. (d1) P wave; (d2) P-R interval; (d3) S–T segment; (d4–d6) T wave; (d7) QRS complex; (d8) No wave since masked by QRS complex.

E1. (a) Contraction, relaxation; opening, closing; pressure changes. (b) Systole, diastole.

E2. (a) 75; 0.8 (60 sec/75 cardiac cycles = 0.8 sec/cardiac cycle). (b) 0.1. (c1) Relaxing; diastole; (c2) fill with, diastole, systole; (c3) QRS; 3; phase 1: Atrial and ventricular diastole (or isovolumetric relaxation); phase 2: Ventricular diastole (and filling) with atrial contraction toward the end of this phase; phase 3: ventricular systole and atrial diastole. (d) Dropping; A. (e) Open; B. (f) Filling; after B; rapidly; diastasis. (g) P; 0.1; increases. (h) Diastole; 130; end-diastolic. (i) QRS; 0.3. (j) Increases dramatically; D; closed. (k) Semilunar, opening; E. (l) 120. (m) T, diastole; decrease. (n) End-systolic; 60; stroke, 70. (o) A, closing of SL valves; B, opening of AV valves; D, closing of AV valves, E, opening of SL valves.

E3. (a) Does not; closed; isovolumetric. (b) Contracting; AV and SL; AV, SL; contraction.

E4. (a) Lubb, closing, A–V; turbulence due to closing of SL valves. (b) Refer to Figure 20.9, page 606 in the text. (c) First, mitral; superior.

F1. CO = 70 ml/stroke × 75 strokes (beats)/min = 5250 ml/min = 5.25 liters/min.

F2. Four; 20 liters per min/5 liters per min = 4; higher.

F3. Stroke volume (SV) and heart rate (HR).

F4. (a) B. (b) Preload, contractility, and afterload. (c) More; increases; more, increases. (d) Greater; Frank-Starling; the more expanded balloon (heart) exerts greater force on air (blood). (e) More, increases. (f) Lack of exercise, loss of blood (hemorrhage), or heart attack (myocardial infarction). (g) Diastole; less; reduced.

F5. (a) ↑ exercise → ↑ preload. (b) Moderate ↑ in preload → ↑ SV. (c) ↑ SV (if heart rate is constant) → ↑ CO. (d) ↑ in heart contractility → ↑ SV and CO. (e) Positively inotropic medication such as digoxin (Lanoxin) → ↑ SV and CO. (f) ↑ afterload (for example, due to hypertension) → ↓ SV and CO. (g) Hypothermia → ↓ heart rate and CO. (h) ↑ heart rate (if SV is constant) → ↑ SV and CO. (i) ↑ stimulation of the cardioaccelerator center → ↑ heart rate. (j) ↑ sympathetic nerve impulses → ↑ heart strength (and SV) and ↑ heart rate. (k) ↑ vagal nerve impulses → ↓ heart rate. (l) Hyperthyroidism → ↑ heart rate. (m) ↑ Ca^{2+} → ↑ heart strength (and SV) and ↑ heart rate. (n) ↓ SV (due to CHF) → ↑ ESV → excessive EDV and preload → ↓ SV and CO.

F6. (a) De, de. (b) Systemic regions, such as in neck and ankles; peripheral or systemic. (c) Difficulty breathing or shortness of breath (dyspnea), which is a sign of pulmonary edema. (d) Left, de, in.

G2. 300 km (200 miles) per pound of fat.

G3. (a) Lipids (like cholesterol and triglycerides), proteins; high, low density lipoproteins. (b) LDLs; low. (c) Few. (d) Prevent; exercise and a diet low in total fat, saturated fat, and cholesterol.

H1. Meso; third; primitive heart.

H2. (a) SV. (b) V. (c) A. (d) BCTA.

I1. (a) Cytomegalovirus, carbon monoxide from smoking, diabetes mellitus, prolonged hypertension, and fatty diet. (b) Plaque; clot (thrombus); platelet-derived growth factor (PDGF). (c) Macrophages and endothelial cells; promote growth of smooth muscle cells, thus narrowing arterial lumen.

I3. (a) PDA. (b) IASD. (c) IVSD. (d) VS. (e) C.

I4. IASD, IVSD, and PDA. In all three cases, blood is forced from high pressure area in left side of heart or aorta back to lower pressured right side of heart.

I5. (a) Not all. (b) Heart block. (c) Ventricular; atrial fibrillation reduces effectiveness of the heart by only about 30 percent, but ventricular fibrillation causes the heart to fail as a pump. (d) Ectopic; premature ventricular contractions (PVCs) or ventricular premature contractions (VPCs).

MASTERY TEST: Chapter 20

Questions 1–3: Arrange the answers in correct sequence.

___C___ ___A___ ___B___ ___D___

1. Pathway of the conduction system of the heart:
   - *2* A. AV node
   - *3* B. AV bundle and bundle branches
   - *1* C. SA node
   - *4* D. Conduction myofibers (Purkinje fibers)

___D___ ___E___ ___A___ ___B___ ___C___

2. Route of a red blood cell supplying oxygen to myocardium of left atrium and returning to right atrium:
   - *3* A. Arteriole, capillary, and venule within myocardium
   - *4* B. Branch of great cardiac vein
   - *5* C. Coronary sinus leading to right atrium
   - *1* D. Left coronary artery
   - *2* E. Circumflex artery

___C___ ___D___ ___E___ ___A___ ___B___

3. Route of a red blood cell now in the right atrium:
   - *4* A. Left atrium
   - *5* B. Left ventricle
   - *1* C. Right ventricle
   - *2* D. Pulmonary artery
   - *3* E. Pulmonary vein

Questions 4–15: Circle the letter preceding the one best answer to each question.

4. All of the following are correctly matched *except:*
   - A. Myocardium—heart muscle
   - B. Visceral pericardium—epicardium
   - C. Endocardium—forms heart valves which are especially affected by rheumatic fever
   - **D.** Pericardial cavity—space between fibrous pericardium and parietal layer of serous pericardium

5. Which of the following factors will tend to decrease heart rate?
   - A. Stimulation by cardiac nerves
   - B. Release of the transmitter substance norepinephrine in the heart
   - C. Activation of neurons in the cardioaccelerator center
   - **D.** Increase of vagal nerve impulses
   - E. Increase of sympathetic nerve impulses

6. The average cardiac output for a resting adult is about ___ per minute.
   - A. 1 quart
   - B. 5 pint
   - **C.** 1.25 gallons (5 liters)
   - D. 0.5 liter
   - E. 2000 ml

7. When ventricular pressure exceeds atrial pressure, what event occurs?
   - A. AV valves open
   - **B.** AV valves close
   - C. Semilunar valves open
   - D. Semilunar valves close

8. The second heart sound is due to turbulence of blood flow as a result of what event?
   - A. AV valves opening
   - B. AV valves closing
   - C. Semilunar valves opening
   - **D.** Semilunar valves closing

9. Choose the *false* statement about the circumflex artery.
   - A. It is a branch of the left coronary artery.
   - **B.** It provides the major blood supply to the right ventricle.
   - C. It lies in a groove between the left atrium and left ventricle.
   - D. Damage to this vessel would leave the left atrium with virtually no blood supply.

10. Choose the *false* statement about heart structure.
    - A. The heart chamber with the thickest wall is the left ventricle.
    - **B.** The apex of the heart is more superior in location than the base.
    - C. The heart has four chambers.
    - D. The left ventricle forms the apex and most of the left border of the heart.

11. All of the following are defects involved in Tetralogy of Fallot *except:*
    - A. Ventricular septal defect
    - B. Enlarged right ventricle
    - **C.** Stenosed mitral valve
    - D. Aorta emerging from both ventricles

12. Choose the *true* statement.
    A. The T wave is associated with atrial depolarization.
    B. The normal P–R interval is about 0.4 sec.
    C. Myocardial infarction means strengthening of heart muscle.
    D. Myocardial infarction is commonly known as a "heart attack" or a "coronary."
13. Choose the *false* statement.
    A. Pressure within the atria is known as intraarterial pressure.
    B. During most of ventricular diastole the semilunar valves are closed.
    C. During most of ventricular systole the semilunar valves are open.
    D. Diastole is another name for relaxation of heart muscle.

14. Which of the following structures are located in ventricles?
    A. Papillary muscles
    B. Fossa ovalis
    C. Ligamentum arteriosum
    D. Pectinate muscles
15. The heart is composed mostly of:
    A. Epithelium
    B. Muscle
    C. Dense connective tissue

Questions 16–20: Circle T (true) or F (false). If the statement is false, change the underlined word or phrase so that the statement is correct.

T F 16. The blood in the left chambers of the heart contains <u>higher</u> oxygen content than blood in the right chambers.

T F 17. At the point when intraarterial pressure surpasses ventricular pressure, semilunar valves <u>open</u>.

T F 18. The pulmonary <u>artery carries</u> blood from the lungs to the left atrium.

T F 19. The normal cardiac cycle <u>does not</u> require direct stimulation by the autonomic nervous system.

T F 20. During <u>about half</u> of the cardiac cycle, atria and ventricles are contracting simultaneously.

Questions 21–25: fill-ins. Complete each sentence with the word or phrase that best fits.

_____ 21. Most ventricular filling occurs during atrial ____.

_____ 22. ____ is a cardiovascular disorder that is the leading cause of death in the U.S.

_____ 23. An ECG is a recording of ____ of the heart.

_____ 24. ____ is a term that means abnormality or irregularity of heart rhythm.

_____ 25. Are atrioventricular and semilunar valves ever open at the same time during the cardiac cycle? ____

Arrange

1. C A B D
2. D E A B C
3. C D E A B

Multiple Choice

4. D	10. B
5. D	11. C
6. C	12. D
7. B	13. A
8. D	14. A
9. B	15. B

True-False

16. T
17. F. Close
18. F. Veins carry
19. T
20. F. No part

Fill-ins

21. Diastole
22. Coronary artery disease
23. Electrical changes or currents that precede myocardial contractions
24. Arrhythmia or dysrhythmia
25. No

FRAMEWORK 21
Circulation: Vessels & Routes

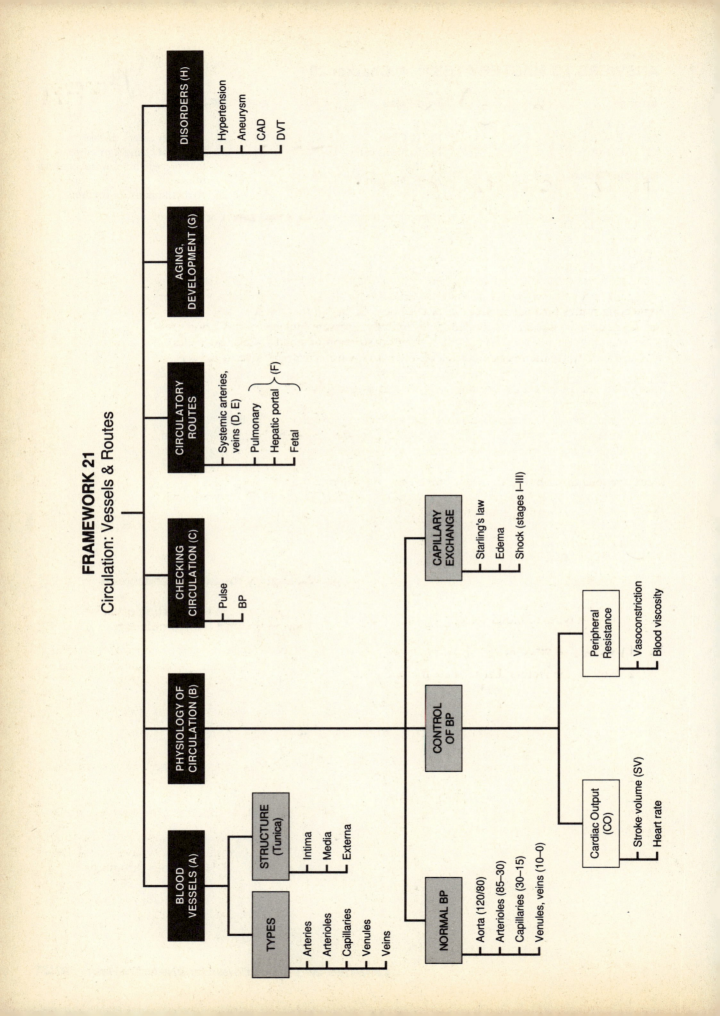

The Cardiovascular System: Blood Vessels and Hemodynamics

CHAPTER
21

Blood is pumped by heart action to all regions of the body. The network of vessels is organized much like highways, major roads, side streets, and tiny alleys to provide interconnecting routes through all sections of the body. In city traffic, tension mounts when traffic is heavy or side streets are blocked. Similarly, pressure builds (arterial blood pressure) when volume of blood increases or when constriction of small vessels (arterioles) occurs. In a city, traffic may be backed up for hours, with vehicles unable to deliver passengers to destinations. A weakened heart may lead to similar and disastrous effects, backing up fluid into tissues (edema) and preventing transport of oxygen to tissues (ischemia, shock, and death of tissues).

Careful map study leads to trouble-free travel along the major arteries of a city. A similar examination of the major arteries and veins of the human body facilitates an understanding of the normal path of blood as well as clinical applications, such as the exact placement of a blood pressure cuff over the brachial artery or the expected route of a deep venous thrombus (pulmonary embolism) to the lungs.

Begin your route through circulatory pathways at the Chapter 21 Framework and familiarize yourself with the names of important routes.

TOPIC OUTLINE AND OBJECTIVES

A. Blood vessels

1. Contrast the structure and function of the various types of blood vessels.

B. Physiology of circulation; shock

2. Explain the factors that regulate the velocity and volume of blood flow.
3. Discuss the various pressures involved in the movement of fluids between capillaries and interstitial spaces.
4. Explain how the return of venous blood to the heart is accomplished.
5. Describe how blood pressure is regulated.
6. Define the three stages of shock.

C. Checking circulation

7. Define pulse and blood pressure (BP) and contrast the clinical significance of systolic, diastolic, and pulse pressure.

D. Circulatory routes: systemic arteries

E. Circulatory routes: systemic veins

F. Circulatory routes: hepatic portal, pulmonary, and fetal circulation

8. Identify the principal arteries and veins of systemic, hepatic portal, pulmonary, and fetal circulation.

G. Aging and developmental anatomy

9. Explain the effects of aging on the cardiovascular system.
10. Describe the development of blood vessels and blood.

H. Disorders and medical terminology

11. List the causes and symptoms of hypertension, aneurysm, coronary artery disease (CAD), and deep-venous thrombosis (DTV).
12. Define medical terminology associated with blood vessals.

Now study the following parts of words that may help you better understand terminology in this chapter

Wordbyte	Meaning	Example	Wordbyte	Meaning	Example
hepat-	liver	*hepat*ic artery	sphygmo-	pulse	*sphygmo*manometer
med-	middle	tunica *med*ia	tunica	sheathe	*tunica* intima
phlebo-	vein	*phleb*itis	vaso-	vessel	*vaso*constriction
port-	carry	*port*al vein	ven-	vein	*veni*puncture
retro-	backward, behind	*retro*peritoneal			

CHECKPOINTS

A. Anatomy of blood vessels (pages 624–629)

■ **A1.** Refer to Figure LG 21.1 and complete this Checkpoint on blood vessels.

a. Label vessels 1–7. Note that numbers are arranged in sequence of the pathway of blood flow. Use these labels:

aorta and other large arteries	small artery
arteriole	small vein
capillary	venule
inferior vena cava	

b. Color vessels according to color code ovals.

c. The human circulatory system is called a(n) *(open? closed?)* system. State the significance of this fact.

■ **A2.** Match the terms to related descriptions:

A. Anastomosis	TI. Tunica intima
L. Lumen	TM. Tunica media
TE. Tunica externa	VV. Vasa vasorum

_____ a. The opening in blood vessels through which blood flows; narrows in atherosclerosis

_____ b. Formed largely of endothelium, it lines lumen; the only layer in capillaries.

_____ c. Junction of two or more vessels supplying one region of the body such as the cerebral arterial circle (of Willis) to the brain

_____ d. Composed of muscle and elastic tissue; contraction permits vasoconstriction

_____ e. Strong tunica adventitia

_____ f. Tiny blood vessels that carry nutrients to walls of blood vessels

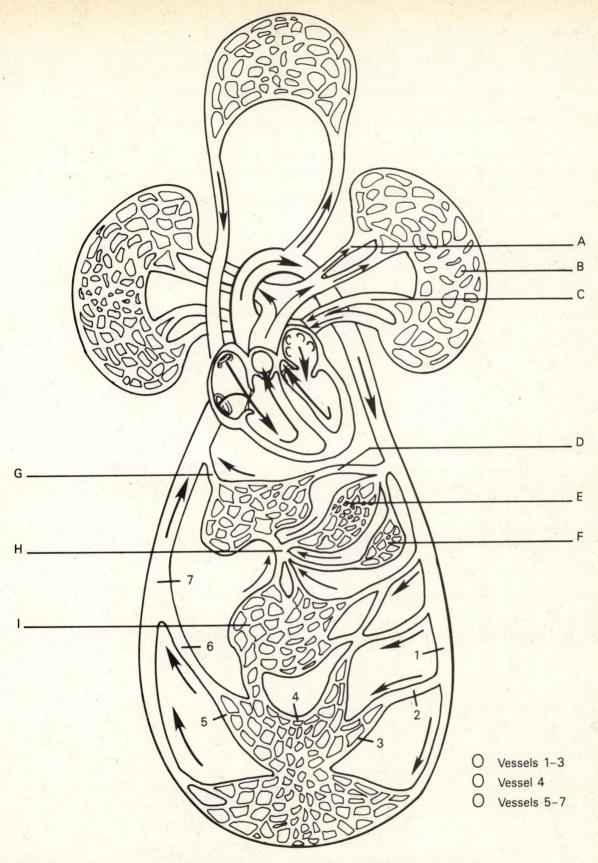

O Vessels 1–3
O Vessel 4
O Vessels 5–7

Figure LG 21.1 Circulatory routes Numbers 1–7 follow the pathway of blood through the systemic blood vessels. Color and label numbered vessels according to directions in Checkpoint A1. Letters A–I refer to Checkpoints A1, D1, and F1.

■ **A3.** Use the names of vessels in A1 to answer the following descriptions.

capillary

_____ a. Site of gas, nutrient, and waste exchange with tissues.

_____ b. Plays primary role in regulating moment-to-moment distribution of blood and in regulating blood pressure.

_____ c. Sinusoids of liver are wider, more leaky versions of this type of vessel.

_____ d. Reservoirs for about 60 percent of the volume of blood in the body; vaso-constriction (due to sympathetic impulses) permits redistribution of blood stored here (three answers).

_____ e. Elastic (conducting) arteries that function collectively as a pressure reservoir.

A4. Contrast terms in the following pairs:

a. Structure of the tunica media of the following vessels:
   conducting arteries/distributing arteries

   Tm walls are thin in proportion to their diameter
   contains more elastic less smooth muscle
   more smooth muscle less elastic fiber

   artery/vein

   artery - lumen small walls thick
   vein lumen large walls thinner

b. Functions of:
   anastomosing arteries/end arteries

   provide alternate routes for blood to reach
   tissue or organ
   directly nourish tissue or organ

c. Number of capillaries in:
   muscles and liver/tendons and ligaments

   many to supply organs + tissue w/ high metab
   metabolic   ra
   activity lower fewer supplied

d. Functions of:
   thoroughfare channel/true capillary _low resistance pathway that ope_
   when constriction of precapillary sphincter reduce bloo
   flow

e. Structure of:
   continuous capillary/fenestrated capillary

fenestrated (pores)

■ **A5.** Complete this exercise about structure of veins.

a. Veins have *(thicker? thinner?)* walls than arteries. This structural feature relates to the fact that the pressure in veins is *(more? less?)* than in arteries. The pressure difference is demonstrated when a vein is cut; blood leaves a cut vein in *(rapid spurts? an even flow?)*.

b. Gravity exerts back pressure on blood in veins located inferior to the heart. To counteract this, veins contain _valves_.

c. When valves weaken, veins become enlarged and twisted. This condition is called _varicose_. This occurs more often in *(superficial? deep?)* veins. Why? _muscle tissue tends to hold deep veins together_

d. A vascular (venous) sinus has _surrounding dense tissue_ replacing the tunica media and tunica externa. So sinuses have the *(structure, but not function? function, but not structure?)* of veins. List two places where such sinuses are found. _spleen, bone marrow, anterior pituitary parathyroid glands_

B. Hemodynamics: physiology of circulation (cardiac output, resistance, and blood pressure) **(pages 630–645)**

■ **B1.** Do this exercise about blood flow.

a. Figure 21.8 (page 630 of your text) shows that velocity of blood flow is greatest in *(arteries? capillaries? veins?)*. Flow is slowest in _capillaries_. As a result, ample time is available for exchange between capillary blood and tissues.

b. As blood moves from capillaries into venules and veins, its velocity *(increases? decreases?)*, enhancing venous return.

c. Although each individual capillary has a cross-sectional area which is microscopic, the body contains so many capillaries that their total cross-sectional area is enormous. (For comparison, look at the eraser end on one pencil; you are seeing the cross-sectional area of that one pencil. If you bind together 100 pencils with a strong rubber band and observe their eraser ends as a group, you see that the individual pencils contribute to a relatively large total cross-sectional area.) All of the capillaries in the body have a total cross-sectional area of about _4500–6000_ cm^2.

d. *(A direct? An indirect?)* relationship exists between cross-sectional area and velocity of blood flow. This is evident in the aorta also. This vessel has a cross section of only about _3–5_ cm^2 and has a very *(high? low?)* velocity.

e. *A clinical challenge.* Mr. Fey has an aortic aneurysm. The diameter of his aorta has increased from normal size to about 20 cm. The region of the aorta with the aneurysm therefore has a *(larger? smaller?)* cross-sectional area. One would expect the velocity of Mr. Fey's aortic blood flow to *(in, de)* ___-crease. This change may put Mr. Fey at risk for _clot formation_.

f. On the average *(1? 5? 10?)* minute(s) is normally required for a blood cell to travel one complete circulatory circuit, for example, heart → lung → heart → foot → heart.

■ **B2.** Do this exercise about important hemodynamic relationships.

a. In Chapter 20 we discussed two factors that control cardiac output. Show these by completing the equation:

Cardiac output (CO) = _____*HR*_____ × _____*SV*_____

$$= 70 \text{ ml/beat} \times 75 \text{ beats/min}$$

$$= \underline{\quad 5.25 \quad} \text{ l/min}$$

b. Show how two other factors regulate cardiac output by completing the equation:

Cardiac output (CO) = _____*MABP*_____ ÷ _____*R*_____

In other words, as resistance to flow increases, cardiac output __*in*__-creases.

c. Keep in mind that the equation above can be rewritten as:

Cardiac output × systemic vascular resistance = mean arterial BP

or

$$CO \times SVR = MABP$$

Note: MABP is sometimes called MAP (mean arterial pressure).
Draw an arrow on the blank line to show the correct relationship between cardiac output (CO) and blood pressure (BP).

$$\uparrow CO \text{ (when SVR is constant)} \rightarrow BP$$

■ **B3.** Refer to Figure 21.9 (page 631 of the text), and complete this exercise about blood pressure shown in that figure.

a. In the aorta and brachial artery blood pressure (BP) is about 120 mm Hg immediately following ventricular contraction. This is called *(systolic? diastolic?)* BP. As

ventricles relax (or go into _____), blood is no longer ejected into these arteries. However, the normally *(elastic? rigid?)* walls of these vessels

recoil against blood, pressing it onward with a diastolic BP of _____ mm Hg.

b. Write the normal ranges of BP values for the following types of vessels: arterioles,

_____ mm Hg; capillaries, _____ mm Hg; venules and veins, _____ mm Hg;

venae cavae, _____ mm Hg.

c. On Figure 21.9 the steepest decline in the pressure curve occurs as blood passes through the *(aorta? arterioles? capillaries?)*. This indicates that the greatest resistance to flow is present in the *(aorta? arterioles? capillaries?)*.

d. The pressure in the right atrium is normally about _____ mm Hg. If the right side of the heart is weakened and cannot adequately pump blood into the lungs, or if pulmonary vessels are sclerosed, offering resistance to blood flow, blood will tend to backlog in the right atrium, causing a(n) *(increase? decrease?)* in right atrial pressure.

■ **B4.** *A clinical challenge.* Do this exercise about Mr. Tyler, who has been depressed since his wife died two years ago and has gained 50 pounds during that time.

a. Which one of the following factors that increase systemic vascular resistance is most likely to be higher in Mr. Tyler?
   A. Blood viscosity
   B. Blood vessel length
   C. Blood vessel radius
   This factor *(does? does not?)* put Mr. Tyler at higher risk for hypertension.

b. Since Mr. Tyler's stress level is often high, his *(sympathetic? parasympathetic?)* nerves are likely to be more active. Which of the three factors (A–C) above is most likely to be affected? *(A? B? C?)* If his arterioles decrease in diameter by 50 percent, the resistance to flow increases *16* times. This factor is likely to *in*crease his blood pressure.

■ **B5.** Identify the types of transport involved in capillary exchange by doing this exercise. Use these answers:

> B. Bulk flow   V. Vesicular transport
> D. Diffusion

_____ a. Passage of lipid-soluble chemicals such as oxygen, carbon dioxide, and steroid hormones through the phospholipid bilayer of capillary membrane endothelial cells

_____ b. Movement of large plasma proteins (such as fibrinogen or albumin) made in liver cells through intercellular clefts in sinusoids to reach plasma

_____ c. Movement of antibody proteins from maternal blood to fetal blood

_____ d. Passive movements of large numbers of ions, molecules, and particles in the same direction, largely due to opposing forces of blood pressure and osmotic pressure

■ **B6.** Recall that pressure is the principal factor causing blood to move through the vessels of the body. Pressure is also the major factor determining movement of substances across capillary membranes. Refer to Figure LG 21.2 and do this activity.

a. Color arrows A–D on Figure LG 21.2 to identify the four pressures involved in capillary exchange.

b. These pressures interact in a kind of tug-of-war that results in a net filtration pressure (NFP) of _____ mm Hg. In other words, at the arterial end of the capillary, substances tend to move *(into? out of?)* the vessel.

c. Which value is substantially different at the venous end of capillaries? _____
   Calculate the NFP there: _____ mm Hg.

d. The near-equilibrium at the two ends of capillaries is based on
   _____'s law of the capillaries. Which system "mops up" the slight amount of fluid that is not drawn back into veins? _____

e. Excessive amounts of fluid accumulating in interstitial spaces is the condition known as _____. This may result from high blood pressure in which *(BCOP? BHP?)* is increased, loss of plasma protein in which *(BCOP? BHP?)* is decreased, or *(increase? decrease?)* in capillary permeability.

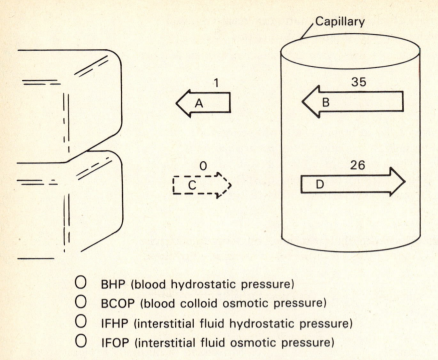

Capillary

○ BHP (blood hydrostatic pressure)
○ BCOP (blood colloid osmotic pressure)
○ IFHP (interstitial fluid hydrostatic pressure)
○ IFOP (interstitial fluid osmotic pressure)

Figure LG 21.2 Practice diagram showing forces controlling capillary exchange and for calculation of NFP at arterial end of capillary. Numbers are pressure values in mm Hg. Color arrows as directed in Checkpoint B6.

■ **B7.** *A clinical challenge.* On a random review of patient charts from a hospital, five clients were noted as having edema. Identify the physiological cause that was most likely the cause of each case of edema. Use these answers:

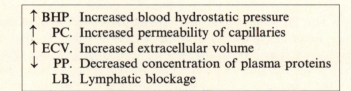

↑ BHP. Increased blood hydrostatic pressure
↑   PC. Increased permeability of capillaries
↑ ECV. Increased extracellular volume
↓   PP. Decreased concentration of plasma proteins
    LB. Lymphatic blockage

_____ a. Maureen S, age 40, had axillary lymph nodes removed for biopsy following right-sided mastectomy; she stated that her "right arm seems more swollen" than her left.

_____ b. Paul P, age 72, has a diagnosis of congestive heart failure and experiences what he describes as "spells when I just can't get my breath."

_____ c. Tommy R, age 7, received second-degree burns on his left leg, which is swollen and seeping fluids.

_____ d. Elaine B, age 41, has kidney failure; she has gained 11 pounds since her last hemodialysis treatment 3 days ago. She reported that she "went overboard eating and drinking at a couple of parties over the weekend." (Two answers)

_____ e. Tracey D, who reports that she is allergic to iodine, accidentally ate an entree containing some seafood. She is experiencing pharyngeal and esophageal edema.

B8. Explain how these two factors increase venous return during exercise.

a. Skeletal muscle pump (For help, refer to Chapter 20, Checkpoint F4c, page 412.)

b. Respiratory pump

■ **B9.** Imagine that your blood pressure (BP) is slightly lower than your body needs right now. Do this exercise to see how your body is likely to respond to increase your blood pressure.

a. Refer to Figure LG 21.3 and fill in box 1 to describe types of receptors that trigger changes in blood pressure. Baroreceptors are sensitive to changes in

_____, and chemoceptors are activated by chemicals such as

_____. These receptors are located in three locations:

_____, _____, and

_____.

b. Nerve messages from the carotid sinuses travel to the brain via cranial nerves ____, whereas messages from the aorta and right atrium travel by cranial nerves ____. Write in these cranial nerves (C.N.) on the figure.

c. This *input* reaches the cardiovascular center located within the

_____ of the brain. Fill in box 3 to indicate this region of the brain. This center receives input from other parts of the brain also. Note three of these brain parts in box 2 in the figure.

d. Consider the cardiovascular center as comparable to the control center of an automobile. The center has an accelerator portion (like a gas pedal) and an inhibitor portion

(like the _____ of a car). When baroreceptors report that BP is too low (as if the car is running too slowly), the *(cardiostimulatory? cardioinhibitory?)* center must be activated; in the car analogy, you would *(step on the gas pedal? step on the brakes?)*. At the same time, the cardioinhibitory center must be deactivated; continuing with the car analogy, what response do you make?

_____ What is the outcome for the car/heart?

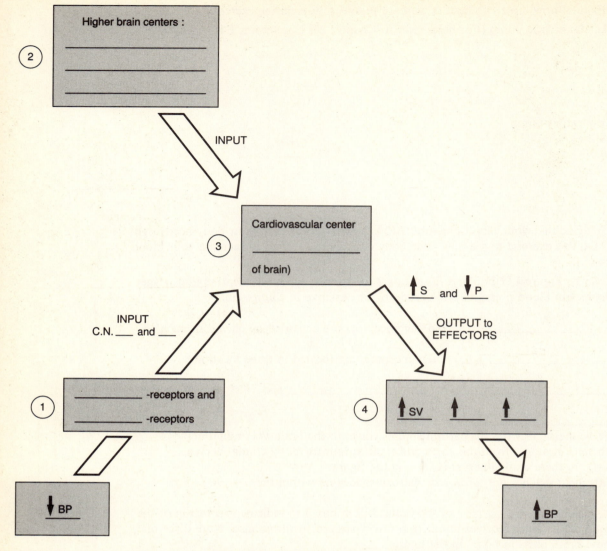

Figure LG 21.3 Overview of input and output regulating blood pressure. Complete as directed in Checkpoint B9.

e. In addition a vasomotor portion of the cardiovascular center responds to lower blood pressure by *(stimulating? inhibiting?)* smooth muscle of blood vessels, resulting in *(vasoconstriction? vasodilatation?)*. This helps to increase BP by ____-creasing systemic vascular resistance (SVR).

f. Summarize nervous *output* to *effectors* from the cardiovascular center by writing in box 4 in Figure LG 21.3 the three factors that are altered. The result of this output is an attempt to restore homeostasis by ____-creasing your blood pressure.

g. Overall, Figure LG 21.3 demonstrates *(positive? negative?)* feedback mechanisms for regulating BP, since a decrease in BP initiates factors to increase BP.

■ **B10.** On Figure LG 21.4. fill in blanks and complete arrows to show details of all of the nervous output, hormonal action, and other factors that increase blood pressure.

■ **B11.** *A clinical challenge.* Complete this exercise about blood pressure disorders.

a. Hypertension means *(high? low?)* blood pressure. It may be caused by an excess of any of the factors shown in Figure LG 21.4. If directions of arrows in Figure LG 21.4 are reversed, blood pressure can be *(increased? decreased?)*.

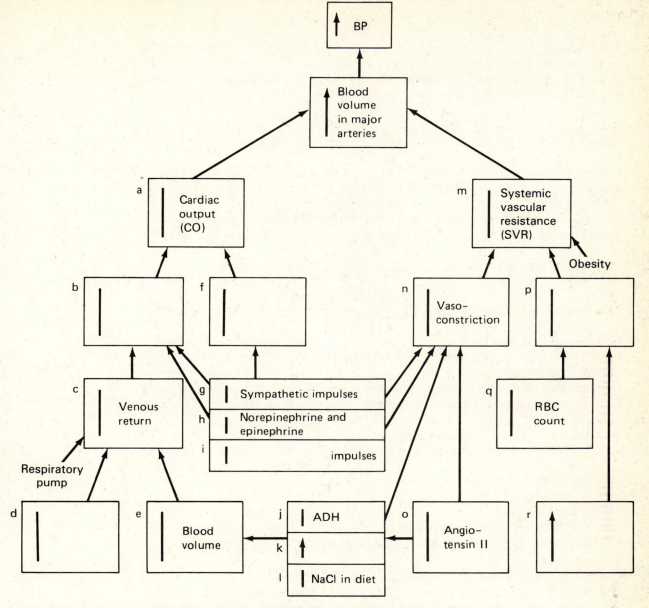

Figure LG 21.4 Summary of factors controlling blood pressure (BP). Arrows indicate increase (↑) or decrease (↓). Complete as directed in Checkpoint B10.

b. Circle all the factors listed below that tend to decrease blood pressure.
 A. Increase in cardiac output, as by increased heart rate or stroke volume
 B. Increase in (vagal) impulses from cardioinhibitory center
 C. Decrease in blood volume, as following hemorrhage
 D. Increase in blood volume, by excess salt intake and water retention
 E. Increased systemic vascular resistance due to vasoconstriction of arterioles (blood prevented from leaving large arteries and entering small vessels)
 F. Decrease in sympathetic impulses to the smooth muscle of arterioles
 G. Decreased viscosity of blood via loss of blood protein or red blood cells, as by hemorrhage
 H. Use of medications called dilators since they dilate arterioles, especially in areas such as abdomen and skin

■ **B12.** *For extra review.* Do the following Checkpoint on additional details on blood pressure regulation.

a. Vasomotor _____ refers to a moderate level of vasoconstriction normally maintained by the medulla and sympathetic nerves. A sudden loss of sympathetic nerve output, for example, by head trauma or emotional trauma, can reduce this tone and cause BP to *(skyrocket? drop precipitously?)*. This may lead to

_____ (fainting) or even to shock.

b. Smooth muscle of most small arteries and arterioles contains *(alpha? beta?)* receptors; in response to the neurotransmitter _____, these vessels vaso-*(constrict? dilate?)*. Sympathetic stimulation of most veins causes them to vaso-*(constrict? dilate?)*, thereby ____-creasing venous return.

c. Vasodilator medications therefore tend to ____-crease BP as blood pools in nonessential areas (such as GI tract or skin). Name two vasodilator chemicals that may be

produced by the body: _____ from cells of the atria, and

_____ from mast cells during inflammation.

d. *(The Bainbridge? Marey's?)* reflex helps to increase heart rate and force of contraction when receptors in the _____ detect increased return of blood to the heart, for example, during exercise. This is therefore a *(positive? negative?)* feedback mechanism.

e. Baroreceptors involved with the aortic reflex primarily maintain *(cerebral? general systemic?)* BP, whereas receptors that trigger the carotid sinus reflex primarily

control _____ BP.

f. Carotid bodies and aortic bodies are sites of *(baro? chemo?)*-receptors that provide

input for control of BP. Low oxygen (known as _____) or high

carbon dioxide (called _____) will tend to cause *(vasoconstriction? vasodilatation?)* and consequently ____-crease BP. As a result, blood can move more rapidly to the lungs to pick up oxygen and give up carbon dioxide.

■ **B13.** Describe the process of autoregulation in this activity.

a. Autoregulation is a mechanism for regulating blood flow through specific tissue areas. Unlike cardiac and vasomotor mechanisms, autoregulation occurs by *(local? autonomic nervous system?)* control.

b. In other words, when a tissue (such as an active muscle) is hypoxic, that is, has a *(high? low?)* oxygen level, the cells of that tissue release *(vasoconstrictor? vasodilator?)* substance. These may include _____ acid, built up by active muscle. Other products of metabolism that serve as dilator substances include:

c. The fact that warming of an area causes *(vasoconstriction? vasodilation?)* is the principle behind application of heat to ____-crease local blood flow for healing.

d. Recall from Chapter 19 that the initial step in hemostasis is *(vasodilation? vasospasm?)*. Name two chemicals that produce this effect.

■ **B14.** *A clinical challenge.* Write arrows to indicate signs and symptoms of shock.
(↑ = increase; ↓ = decrease)

a. Sweat ____

b. Heart rate (pulse) ____

c. BP ____

d. Blood pH ____

e. Mental status ____

f. Urine volume ____

■ **B15.** Describe causes of shock and compensation for shock in this activity.

a. Shock is said to occur when tissues receive inadequate _____
supply to meet their demands for oxygen, nutrients, and waste removal. Shock resulting from decrease in circulating blood volume is known as

_____ shock. Write two or more causes of this type of shock.

b. Compensatory mechanisms activate *(sympathetic? parasympathetic?)* nerves that ____-
crease both heart rate and systemic vascular resistance. These attempts to increase
BP normally work for blood losses up to ____ percent of total blood volume. One
common indicator of this compensatory mechanism is *(warm, pink, and dry? cool,*

pale, and clammy?) skin related to vaso-_____, and ____-
creased sweating.

c. Chemical vasoconstrictors include _____, released from the

adrenal medulla; angiotensin II, formed when _____ is

secreted by kidneys; _____ made by the adrenal cortex; and the

posterior pituitary hormone, _____.
d. Which hormones cause water retention which helps to increase BP in shock?

e. How does hypoxia help to compensate for shock?

B16. Explain what is likely to happen in severe, prolonged shock:

a. Stage II

b. Stage III

C. Checking circulation   (pages 645–646)

C1. Define *pulse*.

C2. In general, where may pulse best be felt? List six places where pulse can readily be palpated. Try to locate pulses at these points on yourself or on a friend.

■ **C3.** Normal pulse rate is about _____ beats per minute. *Tachycardia* means *(rapid?*

slow?) pulse rate; _____ means slow pulse rate.

■ **C4.** Describe the method commonly used for taking blood pressure by answering these questions about the procedure.

a. Name the instrument used to check BP. _____

b. The cuff is usually placed over the _____ artery. How can you tell that this artery is compressed after the bulb of the sphygmomanometer is squeezed?

c. As the cuff is deflated, the first sound heard indicates _____ pressure, since reduction of pressure from the cuff below systolic pressure permits blood to begin spurting through the artery during diastole.

d. Diastolic pressure is indicated by the point at which the sounds

_____. At this time blood can once again flow freely through the artery.

e. Sounds heard while taking blood pressure are known as _____ sounds.

■ **C5.** *A clinical challenge.* Mrs. S has a systolic pressure of 128 mm Hg and diastolic of 83 mm Hg. Write her blood pressure in the form in which BP is usually expressed:

$\frac{128}{83}$ _____. Determine her pulse pressure. *45*

Would this be considered a normal pulse pressure? _*yes*_

D. Circulatory routes: systemic arteries (pages 646–660)

D1. On Figure LG 21.1, label the *pulmonary artery, pulmonary capillaries,* and a *pulmonary vein.* All blood vessels in the body other than pulmonary vessels are known as

*systemic* vessels since they supply a variety of body systems.

■ **D2.** Use Figure LG 21.5 to do this learning activity about systemic arteries.

a. The major artery from which all systemic arteries branch is the

*aorta*. It exits from the chamber of the heart known as the *(right? (left?))* ventricle. The aorta can be divided into three portions. They are the _*ascending*_ (N1 on the figure), the _*arch of aorta*_ (N2), and the _*descending*_ (N3).

b. The first arteries to branch off of the aorta are the _*coronary*_ arteries. Label these (at O) on the figure. These vessels supply blood to the _*heart*_.

c. Three major arteries branch from the aortic arch. Write the names of these (at K, L, and M) on the figure, beginning with the vessel closest to the heart.

d. Locate the two branches of the brachiocephalic artery. The right subclavian artery

(letter _*K*_) supplies blood to _*head + arm*_ _____.

Vessel E is named the *(right? left?)* _*common carotid*_ artery. It supplies the right side of the head and neck.

e. As the subclavian artery continues into the axilla and arm, it bears different names (much as a road does when it passes into new towns). Label G and H. Note that H is the vessel you studied as a common site for measurement of

_*BP*_ _____.

f. Label vessels I and J, and then continue the drawing of these vessels into the forearm and hand. Notice that they anastomose in the hand. Vessel *((I?) J?)* is often used for checking pulse in the wrist.

g. Besides supplying the arms, subclavian arteries each send a branch that ascends the neck through foramina in cervical vertebrae. This is the _*vertebral*_ artery. The right and left vertebral arteries join at the base of the brain to form the _*basilar*_ artery. (Refer to D and C on Figure LG 21.6.)

brachiocep
left com.
carotid
lft sub
clavia

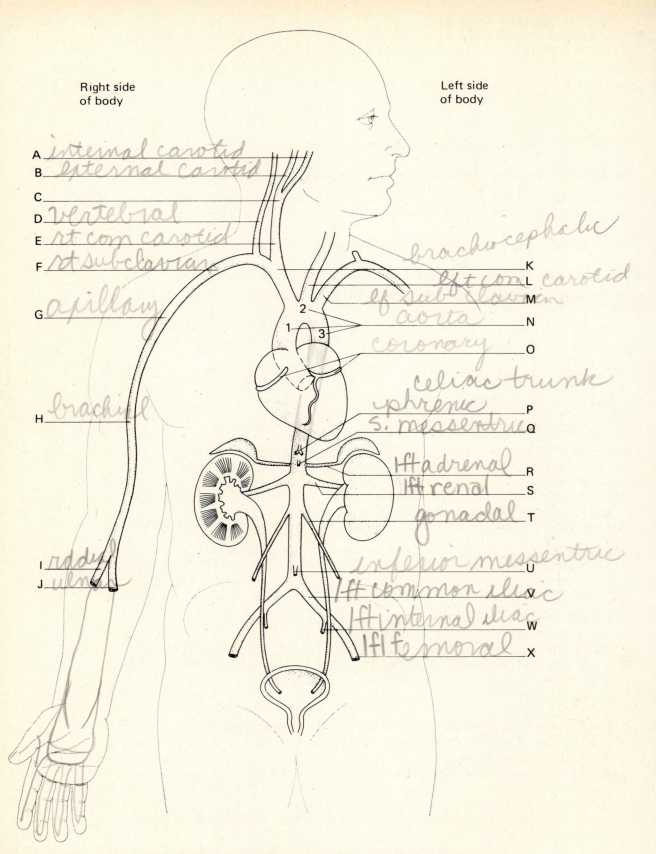

Right side
of body

Left side
of body

A internal carotid
B external carotid
C
D vertebral
E rt com carotid
F rt subclavian
G axillary
H brachial
I radial
J ulnar

K brachiocephalic
L lft com carotid
M lf subclavian
N aorta
O coronary
P celiac trunk
phrenic
Q s. mesenteric
R lft adrenal
S lft renal
T gonadal
U inferior mesenteric
V lft common iliac
W lft internal iliac
X lft femoral

1 2 3

Figure LG 21.5 Major systemic arteries. Identify letter labels as directed in Checkpoints D2, D3, and D6–D8.

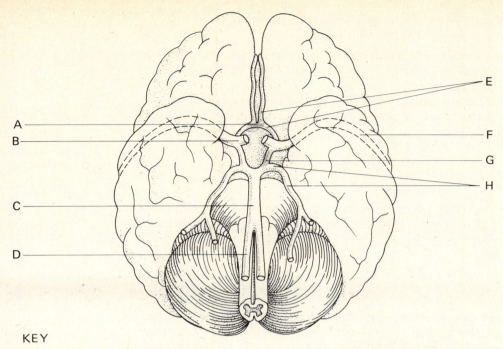

KEY

A. Anterior communicating cerebral
    artery (1)
B. Internal carotid arteries (cut) (2)
C. Basilar artery (1)
D. Vertebral arteries (2)

E. Anterior cerebral arteries (2)
F. Middle cerebral arteries (2)
G. Posterior communicating cerebral
    arteries (2)
H. Posterior cerebral arteries (2)

Figure LG 21.6   Arterial blood supply to the brain including vessels of the cerebral arterial circle (of Willis). View is from the undersurface of the brain. Numbers in parentheses indicate whether the arteries are single (1) or paired (2). Refer to Checkpoints D2, D4, and D5.

■ **D3.** The right and left common carotid arteries ascend the neck in positions considerably *(anterior? posterior?)* to the vertebral arteries. Find a pulse in one of your own carotids. These vessels each bifurcate (divide into two vessels) at about the level of your mandible. Identify the two branches (A and B) on Figure LG 21.5. Which supplies the brain and eye? *(Internal? External?)* What structures are supplied by the external carotids? (See Exhibit 21.6, page 652 in the text, for help.)

■ **D4.** Refer to Figure LG 21.6 and answer these questions about the vital blood supply to the brain.

a. An anastomosis, called the cerebral arterial circle of __*Willis*__ is located just inferior to the brain. This circle closely surrounds the pituitary gland.

b. The circle is supplied with blood by two pairs of vessels: anteriorly, the *(internal?*

   external?) carotid arteries, and, posteriorly, the _____ arteries (joined to form the basilar artery).

c. The circle supplies blood to the brain by three pairs of cerebral vessels, identified on

   Figure LG 21.6 by letter labels _____, _____, and _____. Vessels A and G complete the circle; they are known as anterior and posterior

   _____ arteries. Now cover the key and correctly label each of the lettered vessels.

■ **D5.** *A clinical challenge.* Relate symptoms to the interruption of normal blood supply to the brain as described in each case. Explain your reasons.

a. A patient's basilar artery is sclerosed. Which part of the brain is most likely to be deprived of normal blood supply? *(Frontal lobes of cerebrum? Occipital lobes of cerebrum, as well as cerebellum and pons?)*

b. Paralysis and loss of some sensation are noted on a patient's *right* side. Angiography points to sclerosis in a common carotid artery. Which one is more likely to be sclerosed? *(Right? Left?)* Explain.

c. The middle cerebral artery which passes between temporal and frontal lobe (see broken lines on Figure LG 21.6) is blocked. Which sense is more likely to be affected? *(Hearing? Vision?)* Explain.

d. Anterior cerebral arteries pass anteriorly from the circle of Willis and then arch backward along the medial aspect of each cerebral hemisphere. Disruption of blood flow in the right anterior cerebral artery will be most likely to cause effects in the *(right? left?)* *(hand? foot?)*. *(Hint:* The affected brain area would include region Z in Figure LG 15.1 and/or region R in Figure LG 15.2, pages LG 290 and 294. Note that these regions are on the medial aspect of the right hemisphere.)

■ **D6.** Complete the table describing visceral branches of the abdominal aorta. Identify these on Figure LG 21.5 (vessels *P* to *U*). Refer also to Figure LG 24.4, page LG 517.

Artery	Letter	Structures Supplied

a. Celiac (three major branches)
   1.
 	
   2.
   3.
		1. Liver
		
		2.
		3.

b.	Q	Small intestine and part of large intestine and pancreas
c.		Adrenal (suprarenal) glands
d. Renal	S	
e.		Testes (or ovaries)
f. Inferior mesenteric		

■ **D7.** The aorta ends at about level _____ vertebra by dividing into right and left

_____ arteries. Label these at V on Figure LG 21.5.

D8. Label the two branches (W and X on Figure LG 21.5) of each common iliac artery. Name structures supplied by each.

a. Internal iliac artery

b. External iliac artery

■ **D9.** *For extra review.* Suppose that the right femoral artery were completely occluded. Describe an alternate (or collateral) route that would enable blood to pass from the right common iliac artery to the little toe of the right foot.

Right common iliac artery →

E. Circulatory routes: systemic veins   (pages 660–670)

■ **E1.** Name the three main vessels that empty venous blood into the right atrium of the

heart. _____ _____ _____

■ **E2.** Do this exercise about venous return from the head.

a. Venous blood from the brain drains into vessels known as

_____. These are *(structurally? functionally?)* like veins.

b. Name the sinus that lies directly beneath the sagittal suture:

_____.

c. Blood from all of the cranial vascular sinuses eventually drains into the

_____ veins which descend in the neck. These veins are

positioned close to the _____ arteries.

d. Where are the jugular arteries and carotid veins located?

■ **E3.** Imagine yourself as a red blood cell currently located in the *left brachial vein.* Answer these questions.

a. In order for you to flow along in the bloodstream to the *left brachial artery,* which of these must you pass through? *(A brachial capillary? The heart and a lung?)*

b. In order for you to flow from the left brachial vein to the *right arm,* must you pass through the heart? *(Yes? No?)* Both sides of the heart, that is, right and left? *(Yes? No?)* One lung? *(Yes? No?)*

■ **E4.** Once you are familiar with the pathways of systemic arteries, you know much about the veins that accompany these arteries. Demonstrate this by tracing routes of a drop of blood along the following deep vein pathways.

a. From the thumb side of the left forearm to the right atrium

b. From the medial aspect of the left knee to the right atrium

c. From the right kidney to the right atrium

■ **E5.** Contrast veins with arteries in this exercise.

a. Veins have *(higher? lower?)* blood pressure than arteries and a *(faster? slower?)* rate of blood flow.

b. Veins are therefore *(more? less?)* numerous, in order to compensate for slower blood flow in each vein. The "extra" veins consist of vessels located just under the skin (in

subcutaneous fascia); these are known as _____ veins.

■ **E6.** On text Figures 21.27 to 21.29 (pages 665–669), locate the following veins. Identify the location at which each empties into a deep vein.

a. Basilic _____

c. Great saphenous _____

b. Median cubital _____

d. Small saphenous _____

E7. Veins connecting the superior and inferior venae cavae are named

_____ veins. Identify the azygos system of veins on Figure 21.28 (page 667) in the text. What parts of the thorax do they drain? Explain how these veins may help to drain blood from the lower part of the body if the inferior vena cava is obstructed.

■ **E8.** Name the visceral veins that empty into the abdominal portion of the inferior vena cava. Then color them blue on Figure LG 26.1 (page LG 559). Contrast accompanying systemic arteries by coloring them red on that figure. Note on Figure LG 26.1 that the inferior vena cava normally lies on the *(right? left?)* side of the aorta. One way to remember this is that the inferior vena cava is returning blood to the *(right? left?)* side of the heart.

F. Circulatory routes: pulmonary, hepatic portal, and fetal circulation (pages 670–676)

■ **F1.** Refer to Figure LG 21.1 and do the following activity.

a. Label the following vessels: *gastric, splenic,* and *mesenteric* vessels. Note that all of these vessels drain directly into the vessel named *(7, inferior vena cava? H, portal vein to the liver?)*. Label that vessel on Figure LG 21.1.

b. Label the vessel that carries oxygenated blood into the liver.

c. In the liver, blood from both the portal vein and hepatic artery mixes as it passes

through tortuous capillary-like vessels known as _____. On Figure LG 21.1 label a vessel that collects this blood and transports it to the inferior vena cava.

d. What functions are served by the special hepatic portal circulation?

■ **F2.** Contrast pulmonary circulation with systemic circulation in this activity. Write P (pulmonary) or S (systemic).

_____ a. Which is considered a lower resistance system? List structural features that explain this lower resistance.

_____ b. Which arteries are likely to have a higher blood pressure (BP)? Write typi-

cal BP values for these vessels: aorta _____ mm Hg; pulmonary artery

_____ mm Hg.

_____ c. Which capillaries are likely to have a hydrostatic pressure (BP) of about 10 mm Hg? Write one consequence of increased pulmonary capillary BP.

_____ d. Which system has its blood vessels constrict in response to hypoxia? Write one consequence of this fact.

F3. What aspects of fetal life require the fetus to possess special cardiovascular structures not needed after birth?

■ **F4.** Complete the following table about fetal structures. (Note that the structures are arranged in order of blood flow from fetal aorta back to fetal aorta.)

Fetal Structure	Structures Connected	Function	Fate of Structure After Birth
a.		Carries fetal blood low in oxygen and nutrients and high in wastes.	
b. Placenta	(Omit)		
c.	Placenta to liver and ductus venosus		
d.		Branch of umbilical vein, bypasses liver	
e. Foramen ovale			
f.			Becomes ligamentum arteriosum

■ **F5.** For each of the following pairs of fetal vessels, circle the vessel with higher oxygen content.

a. Umbilical artery/umbilical vein
b. Femoral artery/femoral vein
c. Aorta/thoracic portion of inferior vena cava

G. Aging and developmental anatomy (page 676)

■ **G1.** Draw arrows next to each factor listed below to indicate whether it increases (↑) or decreases (↓) with normal aging.

_____ a. Size and strength of cardiac muscle cells

_____ b. Cardiac output (CO)

_____ c. Total cholesterol and LDLs in blood

_____ d. Systolic blood pressure

G2. Write a paragraph describing embryonic development of blood vessels. Include the following key terms: *15 to 16 days, mesoderm, mesenchyme, blood islands, spaces, endothelial, tunica.*

G3. List four sites of blood formation in the embryo and fetus.

H. Disorders, medical terminology (pages 677–678)

H1. Contrast primary hypertension and secondary hypertension.

■ **H2.** Match each organ listed in the box with the related disorder that may cause secondary hypertension.

AC. Adrenal cortex	K. Kidney
AM. Adrenal medulla	

_____ a. Excessive production of renin which catalyzes formation of angiotensin II, a powerful vasoconstrictor

_____ b. Pheochromocytoma, a tumor that releases large amounts of norepinephrine and epinephrine

_____ c. Release of excessive amounts of aldosterone which promotes salt and water retention

H3. Describe possible effects of hypertension upon the following body structures:

a. Heart

b. Cerebral blood vessels

c. Kidneys

H4. Explain how each of the following therapeutic measures can help control hypertension.

a. Sodium restriction

b. Cessation of smoking

c. Exercise

d. Diuretics

e. Vasodilators

f. A.C.E. inhibitors

g. Stress reduction

H5. Define *aneurysm* and list four possible causes of aneurysms.

■ **H6.** *A clinical challenge.* Explain how a deep vein thrombosis (DVT) of the left femoral vein can lead to a pulmonary embolism.

ANSWERS TO SELECTED CHECKPOINTS

A1. (a) 1, aorta; 2, small artery; 3, arteriole; 4, capillary; 5, venule; 6, small vein; 7, inferior vena cava. (b) See Figure 21.19 (page 647 of the text. (c) Closed; blood is carried from the heart to tissues and back to the heart within a closed system of tubes which contributes to hemodynamics, for example, by large arteries converting stored energy into kinetic energy of blood so that blood continues to move on through smaller vessels.

A2. (a) L. (b) TI. (c) A. (d) TM. (e) TE. (f) VV.

A3. (a) Capillary. (b) Arteriole. (c) Capillary. (d) Venule, small vein, inferior vena cava (actually all veins and venules). (e) Aorta and other large arteries.

A5. (a) Thinner; less; an even flow. (b) Valves. (c) Varicose veins; superficial; skeletal muscles around deep veins limit overstretching. (d) Surrounding dense connective tissue (such as the dura mater); function, but not structure; intracranial (such as superior sagittal venous sinus) and coronary sinus.

B1. (a) Arteries; capillaries. (b) Increases. (c) 4500–6000. (d) An indirect; 3 to 5, high. (e) Larger; de; clot formation (also shock since pooled blood there does not circulate to tissues). (f) 1.

B2. CO = stroke volume (SV) × heart rate (HR) = 5.2 l/min. (b) CO = mean arterial blood pressure (MABP) ÷ resistance; de. (c) ↑ (CO) (when SVR is constant) → ↑ (BP).

B3. (a) Systolic; diastole; elastic, 70 to 80. (b) 85 to 35; 35–16; 16–0; close to 0. (c) Arterioles; arterioles. (d) 0; increase.

B4. (a) B; does. Recall that the body grows an average of about 300 km (200 miles) of new blood vessels for each extra pound of weight. (b) Sympathetic; C; 16; in.

B5. (a–b) D. (c) V. (d) B.

B6. (a) A, IFOP; B, BHP; C, IFHP; D, BCOP. (b) +10 out of. (c) BHP is about 16 mm Hg; −9. (d) Starling; lymphatic. (e) Edema; BHP, BCOP, increase.

B7. (a) LB. (b) ↑ BHP (pulmonary hypertension due to backlog of blood from a failing left ventricle. (c) ↓ PP (as plasma proteins leaked out of damaged vessels); also ↑ PC. (d) ↑ ECV and ↑ BHP. (e) ↑ PC (related to release of histamine in allergic reaction).

B9. (a) Box 1: baro, chemo. Blood pressure; H^+, CO_2, and O_2; carotid arteries, aorta, and right atrium. (b) IX, X. (c) Box 3: medulla. Box 2: cerebral cortex, limbic system, and hypothalamus. (d) Brakes; cardiostimulatory; step on the gas pedal; take your foot off the gas; car goes faster/heart beats faster and with greater force of contraction. (e) Stimulating, vasoconstriction; in. (f) ↑ stroke volume (SV), ↑ heart rate (HR), and ↑ systemic vascular resistance (SVR); in. (g) Negative.

B10. (a) ↑ CO. (b) ↑ Stroke volume. (c) ↑ Venous return. (d) ↑ Exercise. (e) ↑ Blood volume. (f) ↑ Heart rate. (g) ↑ Sympathetic impulses. (h) ↑ Norepinephrine and epinephrine. (i) ↓ Vagal impulses. (j) ↑ ADH. (k) ↑ Aldosterone. (l) ↑ NaCl in diet. (m) ↑ Systemic vascular resistance (SVR). (n) ↑ Vasoconstriction. (o) ↑ Angiotensin II. (p) ↑ Viscosity of blood. (q) ↑ RBC count. (r) ↑ Plasma proteins.

B11. (a) High; decreased. (b) B C F G H.

B12. (a) Tone; drop precipitously; syncope. (b) Alpha; norepinephrine, constrict; constrict, in. (c) De; atrial natriuretic peptide (ANP), histamine. (d) The Bainbridge, right atrium; positive. (e) General systemic, cerebral. (f) Chemo; hypoxia, hypercapnia, vasoconstriction, in.

B13. (a) Local. (b) Low, vasodilator; lactic; EDRF (nitric oxide), K^+, H^+, and adenosine. (c) Vasodilation, in. (d) Vasospasm; eicosanoids such as thromboxane A_2 and prostaglandin F_{2a}.

B14. (a–b) ↑. (c–f) ↓.

B15. (a) Blood; hypovolemic; loss of blood or other body fluids, or shifting of fluids into interstitial spaces (edema) as with burns. (b) Sympathetic, in; 10; cool, pale, and clammy, constriction, in. (c) Epinephrine, renin, aldosterone, ADH (vasopressin). (d) Aldosterone and ADH. (e) It causes local vasodilation.

C3. 60–100, rapid, bradycardia.

C4. (a) Sphygmomanometer. (b) Brachial; no pulse is felt or no sound is heard via a stethoscope. (c) Systolic. (d) Change or cease. (e) Korotkoff sounds.

C5. 128/83, 45 (128–83); yes, it is in the normal range.

D2. (a) Aorta; left; ascending aorta, arch of the aorta, descending aorta. (b) Coronary; heart. (c) K, brachiocephalic; L, left common carotid; M, left subclavian. (d) F; in general, the right extremity and right side of thorax, neck, and head; right common carotid. (e) G, Axillary; H, brachial; blood pressure. (f) I, radial; J, ulnar; I. (g) Vertebral; basilar.

D3. Anterior; internal; external portions of head (face, tongue, ear, scalp) and neck (throat, thyroid).

D4. (a) Willis. (b) Internal, vertebral. (c) E, F, H; communicating.

D5. (a) Occipital lobes of cerebrum, as well as cerebellum and pons. (b) Left, since this artery supplies the left side of the brain. (Remember that neurons in tracts cross to the opposite side of the body, but arteries do not.) (c) Hearing, since it is perceived in the temporal lobe. (Visual areas are in occipital lobes). (d) Left foot.

D6.

Artery	Letter	Structures Supplied
a. Celiac (three major branches) 1. **Hepatic** 2. **Gastric** 3. **Splenic**	P	1. Liver, **gallbladder and parts of stomach, duodenum, pancreas** 2. **Stomach and esophagus** 3. **Spleen, pancreas, and stomach**
b. **Superior mesenteric**	Q	Small intestine and part of large intestine and pancreas
c. **Adrenal (suprarenal)**	R	Adrenal (suprarenal) glands
d. Renal	S	**Kidney**
e. **Gonadal (testicular or ovarian)**	T	Testes (or ovaries)
f. Inferior mesenteric	U	**Parts of colon and rectum**

D7. L-4, common iliac.

D9. Right external iliac artery → right lateral circumflex artery (right descending branch) → right anterior tibial artery → right dorsalis pedis artery → to arch and digital arteries in foot.

E1. Superior vena cava, inferior vena cava, coronary sinus.

E2. (a) Intracranial vascular (venous) sinuses; functionally. (b) Superior sagittal. (c) Internal jugular; common carotid. (d) There are no such vessels.

E3. (a) The heart and a lung. (b) Yes; yes; yes.

E4. (a) Left radial vein → left brachial vein → left axillary vein → left subclavian vein → left brachiocephalic vein → superior vena cava. (Note that there are left and right brachiocephalic veins, but only one artery of that name.) (b) Small veins in medial aspect of left knee → left popliteal vein → left femoral vein → left external iliac vein → left common iliac vein → inferior vena cava. (c) Right renal vein → inferior vena cava.

E5. (a) Lower, slower. (b) More; superficial.

E6. (a) Axillary. (b) Axillary. (c) Femoral. (d) Popliteal.

E8. See key, page 559 of the Guide. Right, right.

F1. (a) E, gastric; F, splenic; I, mesenteric; H, portal vein to the liver. (b) D, hepatic artery. (c) Sinusoids; G, hepatic vein (there are two but only one is shown on Figure LG 21.1). (d) While in liver sinusoids, blood is cleaned, modified, detoxified. Ingested nutrients and other chemicals are metabolized and stored.

F2. (a) P; pulmonary arteries have larger diameters and shorter length (a shorter circuit), thinner walls, less elastic tissue. (b) S; 120/80; 25/8 (systolic BP is about one-fifth of systemic BP). (c) P; pulmonary edema with shortness of breath and impaired gas exchange. (d) P; an advantage is the diversion of blood to well-ventilated regions. One disadvantage is that chronic vasoconstriction in persons with chronic hypoxia (as in emphysema or chronic bronchitis) may lead to cor pulmonale (see page 620 of your text).

F4.

Fetal Structure	Structures Connected	Function	Fate of Structure After Birth
a. Umbilical arteries (2)	Fetal internal iliac arteries to placenta	Carries fetal blood low in oxygen and nutrients and high in wastes.	Medial umbilical ligaments
b. Placenta	(Omit)	Site where maternal and fetal blood exchange gases, nutrients, and wastes	Delivered as "afterbirth"
c. Umbilical vein (1)	Placenta to liver and ductus venosus	Carries blood high in oxygen and nutrients, low in wastes	Round ligament of the liver (ligamentum teres)
d. Ductus venosus	Umbilical vein to inferior vena cava	Branch of umbilical vein, bypasses liver	Ligamentum venosum
e. Foramen ovale	Right and left atria	Bypasses lungs	Fossa ovalis
f. Ductus arteriosus	Pulmonary artery to aorta	Bypasses lungs	Becomes ligamentum arteriosum

F5. (a) Umbilical vein. (b) Femoral artery. (c) Thoracic portion of inferior vena cava (since umbilical vein blood enters it via ductus venosus).

G1. (a) ↓. (b) ↓. (c) ↑. (d) ↑.

H2. (a) K. (b) AM. (c) AC.

H6. Embolus (a "clot-on-the-run") travels through femoral vein to external and common iliac veins to inferior vena cava to right side of heart. It lodges in pulmonary arterial branches, blocking further blood flow into the pulmonary vessels.

MASTERY TEST: Chapter 21

Questions 1–2: Circle the letters preceding all correct answers to each question.

1. In the most direct route from the left leg to the left arm of an adult, blood must pass through all of these structures:
   A. Inferior vena cava
   B. Brachiocephalic artery
   C. Capillaries in lung
   D. Hepatic portal vein
   E. Left subclavian artery
   F. Right ventricle of heart
   G. Left external iliac vein

2. In the most direct route from the fetal right ventricle to the fetal left leg, blood must pass through all of these structures:
   A. Aorta
   B. Umbilical artery
   C. Lung
   D. Ductus arteriosus
   E. Ductus venosus
   F. Left ventricle
   G. Left common iliac artery

Questions 3–9: Circle the letter preceding the one best answer to each question. Note whether the underlined word or phrase makes the statement true or false.

3. Choose the *false* statement.
   A. In order for blood to pass from a vein to an artery, it must pass through chambers of the <u>heart.</u>
   B. In its passage from an artery to a vein a red blood cell must ordinarily travel through a <u>capillary.</u>
   C. The wall of the femoral artery is <u>thicker</u> than the wall of the femoral vein.
   D. Most of the smooth muscle in arteries is in the <u>tunica interna.</u>
4. Choose the *false* statement.
   A. Arteries <u>contain valves, but veins do not.</u>
   B. Decrease in the size of the lumen of a blood vessel by contraction of smooth muscle is called <u>vasoconstriction.</u>
   C. Most capillary exchange occurs by <u>simple diffusion.</u>
   D. Sinusoids are <u>wider and more tortuous (winding)</u> than capillaries.
5. Choose the *false* statement.
   A. Cool, clammy skin is a sign of shock that results from <u>sympathetic stimulation of blood vessels and sweat glands.</u>
   B. Nicotine is a <u>vasodilator that helps control</u> hypertension.
   C. <u>Orthostatic hypotension</u> refers to a sudden, dramatic drop in blood pressure upon standing or sitting up straight.
   D. Phlebitis is an <u>inflammation of a vein.</u>
6. Choose the *false* statement.
   A. The vessels that act as the major regulators of blood pressure are <u>arterioles.</u>
   B. The only blood vessels that carry out exchange of nutrients, oxygen, and wastes are <u>capillaries.</u>
   C. Blood in the umbilical artery is normally <u>more highly</u> oxygenated than blood in the umbilical vein.
   D. At any given moment more than 50 perent of the blood in the body is in the <u>veins.</u>

7. Choose the *true* statement.
   A. The basilic vein is located <u>lateral</u> to the cephalic vein.
   B. The great saphenous vein runs along the <u>lateral</u> aspect of the leg and thigh.
   C. The vein in the body most likely to become varicosed is the <u>femoral.</u>
   D. The hemiazygos and accessory hemiazygos veins lie to the <u>left</u> of the azygos.
8. Choose the *true* statement.
   A. A normal blood pressure for the average adult is <u>160/100.</u>
   B. During exercise, blood pressure will tend to <u>increase.</u>
   C. Sympathetic impulses to the heart and to arterioles tend to <u>decrease blood pressure.</u>
   D. Decreased cardiac output causes <u>increased</u> blood pressure.
9. Choose the *true* statement.
   A. Cranial venous sinuses are composed of the <u>typical three layers</u> found in walls of all veins.
   B. Cranial venous sinuses all eventually empty into the <u>external jugular veins.</u>
   C. Internal and external jugular veins <u>do</u> unite to form common jugular veins.
   D. Most parts of the body supplied by the internal carotid artery are ultimately drained by the <u>internal jugular vein.</u>

Questions 10–14: Circle the letter preceding the one best answer to each question.

10. Hepatic portal circulation carries blood from:
    A. Kidneys to liver
    B. Liver to heart
    C. Stomach, intestine, and spleen to liver
    D. Stomach, intestine, and spleen to kidneys
    E. Heart to lungs
    F. Pulmonary artery to aorta

11. Which of the following is a parietal (rather than a visceral) branch of the aorta?
    A. Superior mesenteric artery
    B. Bronchial artery
    C. Celiac artery
    D. Lumbar artery
    E. Esophageal artery

12. All of these vessels are in the leg or foot *except:*
    A. Saphenous vein
    B. Azygos vein
    C. Peroneal artery
    D. Dorsalis pedis artery
    E. Popliteal artery
13. All of these are superficial veins *except:*
    A. Median cubital   D. Great saphenous
    B. Basilic   E. Brachial
    C. Cephalic

14. Vasomotion refers to:
    A. Local control of blood flow to active tissues
    B. The Doppler method of checking blood flow
    C. Intermittent contraction of small vessels resulting in discontinuous blood flow through capillary networks
    D. A new aerobic dance sensation

Questions 15–20: Arrange the answers in correct sequence.

_____ _____ _____ _____ 15. Route of a drop of blood from the right side of the heart to the left side of the heart:
    A. Pulmonary artery
    B. Arterioles in lungs
    C. Capillaries in lungs
    D. Venules and veins in lungs

Questions 16–18, use the following answers:

A. Aorta and other arteries	C. Capillaries
B. Arterioles	D. Venules and veins

_____ _____ _____ _____ 16. Blood pressure in vessels, from highest to lowest:

_____ _____ _____ _____ 17. Total cross-sectional areas of vessels, from highest to lowest:

_____ _____ _____ _____ 18. Velocity of blood in vessels, from highest to lowest:

_____ _____ _____ _____ _____ 19. Route of a drop of blood from small intestine to heart:
    A. Superior mesenteric vein
    B. Hepatic portal vein
    C. Small vessels within the liver
    D. Hepatic vein
    E. Inferior vena cava

_____ _____ _____ 20. Layers of blood vessels, from most superficial to deepest:
    A. Tunica interna
    B. Tunica externa
    C. Tunica media

Questions 21–25: fill-ins. Complete each sentence or answer the question with the word(s) or phrase that best fits.

_____ 21. A weakened section of a blood vessel forming a balloonlike sac is known as a(n) ____.

_____ 22. Name the major factor that creates blood osmotic pressure (BCOP) that prevents excessive flow of fluid from blood to interstitial areas.

_____ 23. Name the two blood vessels that are locations of baroreceptors and chemoreceptors for regulation of blood pressure.

_____ 24. In taking a blood pressure, the cuff is first inflated over an artery. As the cuff is then slowly deflated, the first sounds heard indicate the level of ____ blood pressure.

_____ 25. Most arteries are paired (one on the right side of the body, one on the left). Name five or more vessels that are unpaired.

ANSWERS TO MASTERY TEST: ■ Chapter 21

Multiple Answers

1. A C E F G
2. A D G

Multiple Choice

3. D	9. D
4. A	10. C
5. B	11. D
6. C	12. B
7. D	13. E
8. B	14. C

Arrange

15. A B C D
16. A B C D
17. C D B A
18. A B D C
19. A B C D E
20. B C A

Fill-ins

21. Aneurysm
22. Plasma protein such as albumin
23. Aorta and carotids
24. Systolic
25. Aorta, anterior communicating cerebral, basilar, brachio-cephalic, celiac, gastric, hepatic, splenic, superior and inferior mesenteric, middle sacral

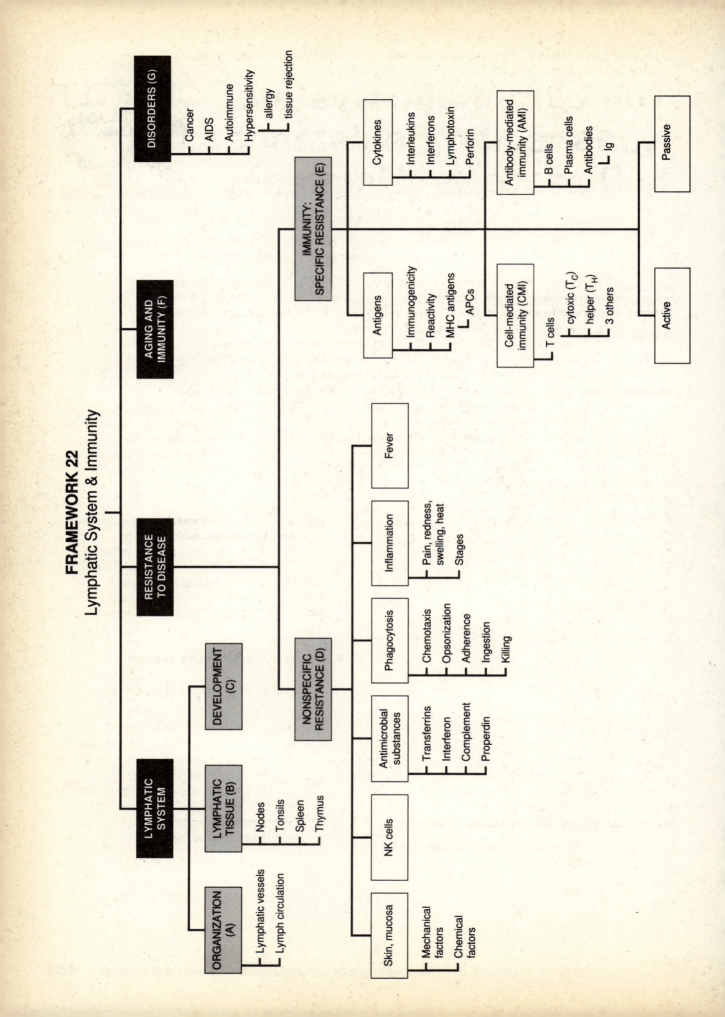

FRAMEWORK 22
Lymphatic System & Immunity

LYMPHATIC SYSTEM
— ORGANIZATION (A)
— Lymphatic vessels
— Lymph circulation

— LYMPHATIC TISSUE (B)
— Nodes
— Tonsils
— Spleen
— Thymus

— DEVELOPMENT (C)

RESISTANCE TO DISEASE
— NONSPECIFIC RESISTANCE (D)
— Skin, mucosa
— Mechanical factors
— Chemical factors
— NK cells
— Antimicrobial substances
— Transferrins
— Interferon
— Complement
— Properdin
— Phagocytosis
— Chemotaxis
— Opsonization
— Adherence
— Ingestion
— Killing
— Inflammation
— Pain, redness, swelling, heat
— Stages
— Fever

— IMMUNITY: SPECIFIC RESISTANCE (E)
— Antigens
— Immunogenicity
— Reactivity
— MHC antigens
— APCs
— Cytokines
— Interleukins
— Interferons
— Lymphotoxin
— Perforin
— Cell-mediated immunity (CMI)
— T cells
— cytotoxic (T_C)
— helper (T_H)
— 3 others
— Antibody-mediated immunity (AMI)
— B cells
— Plasma cells
— Antibodies
— Ig
— Active
— Passive

AGING AND IMMUNITY (F)

DISORDERS (G)
— Cancer
— AIDS
— Autoimmune
— Hypersensitivity
— allergy
— tissue rejection

The Lymphatic System, Nonspecific Resistance to Disease, and Immunity

The human body is continually exposed to foreign (nonself) entities: bacteria or viruses, pollens, chemicals in a mosquito bite, or foods or medications that provoke allergic responses. Resistance to specific invaders is normally provided by a healthy immune system made up of antibodies and battalions of T-lymphocytes.

Nonspecific resistance is offered by a variety of structures and mechanisms, such as skin, mucus, white blood cells assisting in inflammation, and even by a rise in temperature (fever) that limits the survival of invading microbes. The lymphatic system provides sentinels at points of entry (as in tonsils) and at specific sites (lymph nodes) near entry to body cavities (inguinal and axillary regions). The lymphatic system also wards off the detrimental effects of edema by returning proteins and fluids to blood vessels.

Start your study of the physical defense mechanisms of the body by previewing the Chapter 22 Framework and the key terms it contains.

TOPIC OUTLINE AND OBJECTIVES

A. Lymphatic vessels and lymph circulation

1. Describe the components of the lymphatic system and list their functions.
2. Discuss how edema develops.

B. Lymphatic tissue

C. Developmental anatomy of the lymphatic system

3. Describe the development of the lymphatic system.

D. Nonspecific resistance to disease

4. Discuss the roles of the skin and mucous membranes, antimicrobial substances, phagocytosis, inflammation, and fever in nonspecific resistance to disease.

E. Immunity (specific resistance to disease)

5. Define immunity and describe how T cells and B cells arise.

6. Explain the relationship between an antigen (Ag) and an antibody (Ab).
7. Describe the roles of antigen presenting cells, T cells, and B cells in cell-mediated and antibody-mediated immunity.
8. Explain how self-tolerance occurs.
9. Discuss the relationship of immunology to cancer.

F. Aging and the immune system

G. Disorders and medical terminology

10. Describe the clinical symptoms of the following disorders: acquired immune deficiency syndrome (AIDS), autoimmune diseases, systemic lupus erythematosus (SLE), severe combined immunodeficiency (SCID), hypersensitivity (allergy), tissue rejection, and Hodgkin's disease (HD).
11. Define medical terminology associated with the lymphatic system.

WORDBYTES

Now study the following parts of words that may help you better understand terminology in this chapter.

Wordbyte	Meaning	Example	Wordbyte	Meaning	Example
anti-	against	*anti*body	hetero-	other	*hetero*graft
auto-	self	*auto*immune	iso-	same	*iso*graft
axilla-	armpit	*axilla*ry nodes	lact-	milk	*lact*eals
chyl-	juice	cisterna *chyl*i	meta-	beyond	*meta*stasis
gen-	to produce	anti*gen*; *gen*etic	xeno-	foreign	*xeno*graft

CHECKPOINTS

A. Lymphatic vessels and lymph circulation (pages 683–686)

■ **A1.** List the components of the lymphatic system. *lymph, lymphatic vessels spleen bone marrow tonsils lymph nodes*

■ **A2.** Describe the functions of the lymphatic system in this exercise.

a. Lymphatic vessels drain from tissue spaces *plasma proteins* that are too large to readily pass into blood capillaries. They also transport *lipid & lipid soluble vitamins* from the gastrointestinal (GI) tract to the bloodstream.

b. The defensive functions of the lymphatic system are carried out largely by white blood cells named *lympho* -cytes. (B? (T?)) lymphocytes destroy foreign substances directly, whereas B lymphocytes lead to the production of *antibodies* .

■ **A3.** Write S next to specific immune responses and NS next to nonspecific responses.

NS a. Barriers to infection, such as skin and mucous membranes

NS b. Inflammation and phagocytosis

NS c. Chemicals such as gastric or vaginal secretions that are both acidic

S d. Antigen-antibody responses

S e. Destruction of an intruder by a cytotoxic T lymphocyte

NS f. Antimicrobial substances such as interferons (IFNs) or complement

A4. Contrast in terms of structure and location:

a. Blood capillaries/lymph capillaries

lymph v. have thinner walls + more valves

b. Veins/lymphatics

c. Lymph/interstitial fluid

A5. Describe the general pathway of lymph circulation. (*For extra review*. Refer to Checkpoint D2 and Figure LG 1.1, page 8 in the Learning Guide.)

arteries (blood) → blood capillaries → interstitial spaces (If) → lymphatic cap. → lymphatic ducts → subclavian vein → (blood)

■ **A6.** The lymphatic system does not have a separate heart for pumping lymph. Describe two principal factors that are responsible for return of lymph from the entire body to major blood vessels in the neck. (Note that these same factors also facilitate venous return.)

milking action of skeletal muscle

respiratory movement creates a pressure gradient between two ends of lymphatic system

inhale → lymph flows from abdominal region where pressure is higher toward thoracic region where pressure is ↓

■ **A7.** Answer the following questions about the largest lymphatics.

a. In general, the thoracic duct drains (*three-fourths?* one-half? one-fourth?) of the body's lymph. This vessel starts in the lumbar region as a dilation known as the
cisterna chyli. This cisterna chyli receives lymph from which areas of the body? *lft side of head, neck, chest lft upper extremity, + entire body below the ribs*

b. In the neck the thoracic duct receives lymph from three trunks. Name these.
lft jugular, lft subclavian, lft broncho mediast

c. One-fourth of the lymph of the body drains into the _right lymphatic_ duct. The regions of the body from which this vessel collects lymph are: *rt side of head + neck, rt upper extremity, rt side of thorax, rt lung + rt h li*

d. The thoracic duct empties into the junction of the blood vessels named left
_____ and left _jugular_ veins. The right lymphatic duct empties into veins of the same name on the right side.

e. In summary, lymph fluid starts out originally in plasma or cells and circulates as
interstitial fluid. Then as lymph it undergoes extensive cleaning as it passes through nodes. Finally fluid and other substances in lymph are returned to
_____ in veins of the neck.

A8. *A clinical challenge.* Explain why a person with a broken left clavicle might exhibit these two symptoms:

a. Edema, especially of the left arm and both legs

b. Excess fat in feces
lymph transports lipids from GI to bloodstream

B. Lymphatic tissue (pages 686–690)

B1. Contrast the following terms:
diffuse lymphatic tissue/lymphatic nodule/lymphatic organ
not enclosed in capsule
oval shaped concentrations of tissue
bone marrow + thymus gland

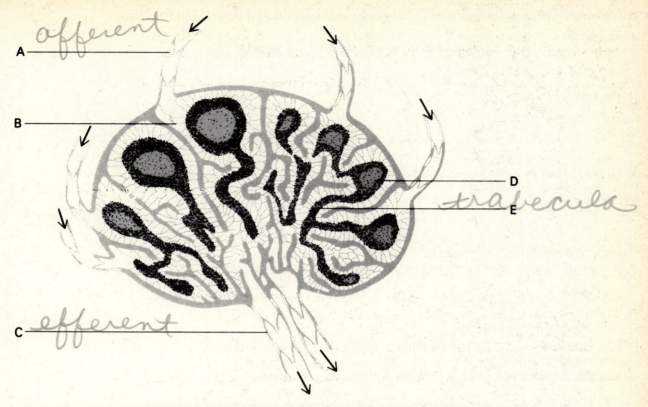

A _____ *afferent*

B _____

D _____

E _____ *trabecula*

C _____ *efferent*

Figure LG 22.1   Structure of a lymph node. Color and label as directed in Checkpoint B2.

■ **B2.** Label structures A–E on Figure LG 22.1. Then color each of those structures according to descriptions below.

(a) ○ Vessels for entrance of lymph into node

(b) ○ Vessel for exit of lymph out of node

(c) ○ Site of B lymphocyte proliferation into antibody-secreting plasma cells

(d) ○ Channel for flow of lymph in the cortex of the node

(e) ○ Dense strand of lymphocytes, macrophages, and plasma cells

B3. Describe how lymph is "processed" as it circulates through lymph nodes. How does this function of lymph nodes explain the fact that nodes ("glands") become enlarged and tender during infection?

filter foreign substances from lymph trapped by reticular fibers macrophages destroy

■ **B4.** Spread, or _____ *metastasis* _____, of cancer may occur via the lymphatic system. Lymph nodes that have become cancerous tend to be enlarged, firm, and *(tender? nontender?)*.

■ **B5.** *A clinical challenge.* Which tonsils are most commonly removed in a tonsillectomy?

__Palatine_____ State one disadvantage of loss of tonsillar tissue.

tonsils are situated to protect against invasion of foreign substance that are inhaled or ingested

■ **B6.** The spleen is located in the *(upper right? upper left? lower right? lower left?)*

quadrant of the abdomen, immediately inferior to the _____. List three functions of the spleen.

B cell proliferation *phagocytosis of bacteria* *Stores & releases blood*

■ **B7.** Describe the thymus gland in this exercise.

a. It is located in the *(neck? mediastinum?)*. Its position is just posterior to the

___sternum_____ bone. Its size is relatively *(large? small?)* in childhood,

de-creasing in size after age 10 to 12 years.

b. Like lymph nodes, the thymus contains a medulla surrounded by a

___cortex_____. Within the cortex are densely packed

___lympho_____-cytes. The thymus is the site of maturation of *(B? T?)*
cells *(only in childhood? throughout life?)*.

C. Developmental anatomy of the lymphatic system   (pages 690–691)

■ **C1.** The lymphatic system derives from ___meso_____-derm, beginning about the *(fifth? seventh? tenth?)* week of gestation.

■ **C2.** Lymphatic vessels arise from lymph sacs that form from early *(arteries? veins?)*.
Name the embryonic lymph sac that develops vessels to each of the following regions of the body.

a. Pelvic area and lower extremities: _____ lymph sac

b. Upper extremities, thorax, head, and neck: _____ lymph sac.
The left jugular lymph sac eventually develops into the upper portion of the main

collecting lymphatic vessel, the _____ duct.

D. Nonspecific resistance to disease   (pages 692–697)

D1. Briefly contrast two types of resistance to disease: *specific* vs. *nonspecific*. (For help, refer back to Checkpoint A3.)

nonspecific comprise general response to wide range of pathogens.
Specific production of specific antibodies or activation of T cells against a particular pathogen or foreign substance

■ **D2.** *A clinical challenge.* Match the following conditions or factors with the nonspecific resistance that is lost in each case.

Nonspecific Resistance Lost

_C_ a. Cleansing of oral mucosa

_D_ b. Cleansing of vaginal mucosa

_F_ c. Flushing of microbes from urinary tract

_E_ d. Lacrimal fluid with lysozyme

_A_ e. Respiratory mucosa with cilia

_G_ f. Closely packed layers of keratinized cells

_H_ g. HCl production; pH 1 or 2

_B_ h. Sebum production

Condition/Factor

A. Long-term smoking
B. Excessive dryness of skin
C. Use of anticholinergic medications that decrease salivation
D. Vaginal atrophy related to normal aging
E. Dry eyes related to decreased production of tears, as in aging
F. Decreased urinary output, as in prostatic hypertrophy
G. Skin wound
H. Partial gastrectomy (removal of part of stomach)

■ **D3.** *For extra review.* Describe nonspecific defenses more fully in this exercise.

a. Interferons (IFNs) are proteins produced by cells infected with

viruses. Alpha interferon (alpha-IFN) is classified as a *(type I?* ~~type II?)~~ interferon. Against which virus-associated disorders is this substance used?

AIDS genital warts, hepatitis C

The gene that codes for alpha-IFN is missing in persons who have one type of

leukemia.

b. NK cells, or _natural_ _killer_ cells, are a type of *(lymphocyte?* ~~neutrophil?)~~. Type II (gamma) IFNs stimulate the

cytolitic activity of these cells. NK cells are less functional in

patients with _AIDS or cancer_.

c. Two types of white blood cells that are involved in phagocytosis are

neutrophils, and to a lesser extent, _eosinophils_.

Other leukocytes, known as _monocytes_, travel to infection sites and

develop into highly phagocytic _macrophage_ cells. Macrophages

compose the reticuloendothelial, or _mononuclear_

phagocytic, system.

D4. Write a paragraph describing roles of the complement system. Include these key terms: *C3, histamine, opsonization,* and *membrane attack complex (MAC).*

A group of 20 normally inactive proteins in plasma & on cell surfaces make up the compliment system. When activated these proteins enhance certa immune, allergic & inflammatory response. A plasma pro called C3 can be activated in two ways Classicly & alternatively Once activated C3 turns on other compliments Histamine increases the permeability of blood capillaries allowing to penetrate to combat infection or allergy. Compliment C3b bind the surface of a microbe and enhances phagocytosis this is opsonization Several compliment proteins come together to form a membran attack complex that punch holes in microbe that rupture

■ **D5.** Match each of the steps in phagocytosis with the related description. Note that descriptions are listed in chronological sequence.

A. Adherence	I. Ingestion
C. Chemotaxis	K. Killing

__C__ a. Phagocytic cells "sniff" the delectable fragrance of invading microbes and especially savor bacteria with the tasty *complement* coating. "Ah, dinner soon!"

__A__ b. Phagocytes trap the invaders, possibly wedging them against a blood vessel or a clot, and then attach to the surface of the invader.

__I__ c. Phagocytic pseudopods like tentacles surround the tasty morsels forming a culinary treat, the *phagocytic vesicle.* "Slurp!"

__K__ d. Lysosomal enzymes, oxidants, and *defensins* pounce on the microbial meal, a has-been within a half hour. *Residual bodies,* like the pits of prunes, are discarded. And the phagocytic face smiles in contentment.

I. aids disposal of microbes + toxins at the site of the injury, prevents their spread to other organs & prepares the site for tissue repair

D6. Do this exercise on inflammation.

a. Defend or dispute this statement: "Inflammation is a process that can be both helpful and harmful."

b. List the four cardinal signs or symptoms of inflammation.

redness   *pain*

heat   *swelling*

c. Briefly describe the three stages of inflammation.

1. *vasodilation + increased permeability of blood vessels vasodilation allows more blood to flow to the area + increase permeability permits defensive materials through in blood to enter the injured area.*

2. *Phagocytic migration*

3. *repair*

■ **D7.** *For extra review.* Select the number that best fits each description of a phase of inflammation. Use these answers:

> 1. Vasodilation and increased capillary permeability
> 2. Phagocytic migration
> 3. Repair

__*1*__ a. Brings defensive substances to the injured site and helps remove toxic wastes.

__*1*__ b. Histamine, prostaglandins, kinins, leukotreines, and complement enhance this process.

__*1*__ c. Causes redness, warmth, and swelling of inflammation.

__*2*__ d. Localizes and traps invading organisms.

__*2*__ e. Occurs within an hour after initiation of inflammation; involves margination, diapedesis, chemotaxis, and leukocytosis.

__*3*__ f. White blood cell and debris formation that may lead to abscess or ulcer.

■ **D8.** Identify the chemicals that fit descriptions below.

C. Complement	LPF. Leukocytosis-promoting factor
H. Histamine	LT. Leukotrines
I. Interferon	PG. Prostaglandins
K. Kinins	T. Transferrins

PG a. Produced by damaged or inflamed tissues; aspirin and ibuprofen neutralize these pain-inducing chemicals.

LT, H b. Produced by mast cells and basophils.

LPF c. Stimulates increase in production of neutrophils.

K d. Induces vasodilation, permeability, chemotaxis, and irritation of nerve endings (pain).

K C e. Neutrophils are attracted by these chemicals present in the inflamed area (two answers).

T f. Proteins that inhibit bacterial growth by depriving microbes of iron.

C g. Group of 20 serum proteins that stimulate release of histamine and also make bacteria "tastier" to phagocytes by opsonization.

I h. Made by some leukocytes and fibroblasts; stimulates cells near an invading virus to produce antiviral proteins that interfere with survival of the virus.

E. Immunity (specific resistance to disease) (pages 697–712)

■ **E1.** Complete this exercise about cells that carry out immune responses.

a. Name the two categories of cells that carry out immune responses.

_____*T cells*_____ and _____*B cells*_____ Where do lymphocytes originate? *Stem cells*

b. Some immature lymphocytes migrate to the thymus and become (B? *T?*) cells. Here T cells develop immunocompetence, meaning that these cells have the ability to

_____*perform*_____. Some T cells become CD4+ cells and others become

_____*CD8+*_____ cells, based on the type of _____ in their plasma membranes.

c. Where do B cells mature into immune cells? _____*bone marrow*_____.

■ **E2.** Contrast two types of immunity by writing AMI before descriptions of *antibody-mediated immunity* and CMI before those describing *cell-mediated immunity*.

_____ a. Especially effective against microbes that enter cells, such as viruses and parasites

_____ b. Especially effective against bacteria present in extracellular fluids

_____ c. Involves plasma cells (derived from B cells) that produce antibodies

_____ d. Utilizes killer T cells (derived from CD8+ T cells) that directly attack the antigen.

_____ e. Facilitated by helper T (CD4+) cells

■ **E3.** Do this exercise on antigens.

  a. An antigen is defined as "any chemical substance which, when introduced into the

   body, _____." In general, antigens are *(parts of the body? for-*
   eign substances?).

  b. Complete the definitions of the two properties of antigens:

   1. Immunogenicity: ability to _____ specific antibodies or
      specific T cells

   2. Reactivity: ability to _____ specific antibodies or specific
      T cells

  c. An antigen with both of these characteristics is called a(n)

   _____.

  d. A partial antigen is known as a _____. It displays *(immuno-*

   genicity? reactivity?) but not _____. For example, for the hapten penicillin to evoke an immune response (immunogenicity), it must form a com-

   plete antigen by combining with a _____. Persons who have

   this particular protein are said to be _____ to penicillin.

  e. Describe the chemical nature of antigens.

  f. Explain why plastics used for valves or joints are not likely to initiate an allergic re-
     sponse and be rejected.

  g. Can an entire microbe serve as an antigen? *(Yes? No?)* List the parts of microbes
     which may be antigenic.

  h. If you are allergic to pollen in the spring or fall or to certain foods, the pollen or
     foods serve as *(antigens? antibodies?)* to you.

  i. Antibodies or specific T cells form against *(the entire? only a specific region of the?)*

   antigen. This region is known as the _____. Most antigens
   have *(only one? only two? a number of?)* antigenic determinant sites.

E4. Explain how humans have the ability to recognize, bind to, and then evoke an immune response against over a billion different antigenic determinants. Describe two aspects:

a. Genetic recombination

b. Somatic mutations

■ **E5.** Describe roles of major histocompatibility complex (MHC) antigens in this exercise.

a. Describe the good news (how they help you) and the bad news (how they may be harmful) about MHC antigens.

b. Contrast two classes of MHC antigens according to types of cells incorporating the MHC into plasma membrane and the lengths of amino acid fragments picked up by each MHC:
MHC-I

MHC-II

c. Which of the following do T cells normally ignore?
A. MHC with peptide fragment from foreign protein
B. MHC with peptide fragment from self-protein

■ **E6.** Describe the processing and presenting of antigens in this activity.

a. Most cells of the body can process and present *(endogenous? exogenous?)* antigens, such as proteins made by body cells infected by viruses.

b. Define *exogenous antigens.*

Cells that present these antigenic proteins are called APCs, or

a_____ p_____

c_____. These cells are generally found *(in lymph nodes? at sites where antigens are likely to enter the body, such as skin or mucous membranes?).* Name three types of APCs.

c. Place in correct sequence the six boxed codes that describe the steps APCs or other body cells take to initiate an immune response to intruder antigens. Use lines provided.

Bind to MHC-II.	Bind peptide fragment to MHC-II
Exo/insert.	Exocytosis and insertion of antigen fragment-MHC-II into APC plasma membrane
Fusion.	Fusion of vesicles of peptide fragments with MHC-II
Partial dig.	Partial digestion of antigen into peptide fragments
Phago/endo.	Phagocytosis or endocytosis of antigen
Present to T.	Present antigen to T cell in lymphatic tissue

1. _____ → 2. _____ → 3. _____ →

4. _____ → 5. _____ → 6. _____

■ **E7.** Match the name of each cytokine with the description that best fits.

Gamma-IFN.	Gamma interferon
IL-2.	Interleukin-2
IL-4.	Interleukin-4
M.	Monokine
MMIF.	Macrophage migration inhibiting factor

_____ a. General name for a cytokine made by a monocyte

_____ b. Made by activated helper T cells; stimulates B cells, leading to IgE production

_____ c. Known as T cell growth factor, it is made by helper T cells; needed for almost all immune responses; causes proliferation of cytotoxic T cells (as well as B cells) and activates NK cells

_____ d. Formerly called *macrophage activating factor* (MAF), stimulates phagocytosis in macrophages and neutrophils; enhances AMI and CMI responses

_____ e. Prevents macrophages from leaving the site of infection

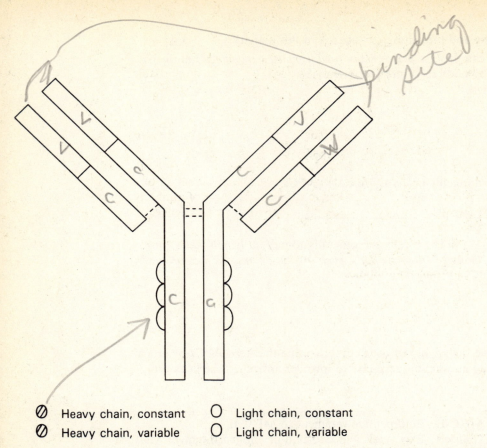

Symbol	Label	Symbol	Label
⊘	Heavy chain, constant	○	Light chain, constant
⊘	Heavy chain, variable	○	Light chain, variable

Figure LG 22.2 Diagram of an antibody molecule. Color and label as directed in Checkpoint E8.

■ **E8.** Refer to Figure LG 22.2 and check your understanding of antibody structure in this Checkpoint.

a. Chemically all antibodies are composed of glycoproteins named

Globulins. They consist of *(2? 4? 8?)* polypeptide chains. The two

heavy chains contain about _450_ amino acids; light chains contain more than _200_ amino acids. Crosshatch the *heavy chains* on the figure.

b. *(Constant? Variable?)* portions are diverse for different antibodies since these serve as binding sites for different antigens. *(Constant? Variable?)* portions *(differ? are identical?)* for all antibodies within the same class. Color constant and variable portions of both chains as indicated by color code ovals on the figure. Also label antigen *hinge region, binding sites, disulfide bonds,* and *carbohydrates.*

■ **E9.** Since glycoproteins forming antibodies are involved in immunity, they are called

immunoglobulins, abbreviated ___Ig___. The different classes of Ig's are distinguishable by the (constant? variable?) portions of the antibody structure. Write the name of the related class of Ig next to each description.

___M___ a. The first antibodies to be secreted after initial exposure to antigen, they are short-lived; destroy invading microbe by agglutination and lysis.

___G___ b. The only type of Ig to cross the placenta, it provides specific resistance of newborns. Also significantly enhances phagocytosis and neutralization of toxins in persons of all ages since it is the most abundant type of antibody.

___A___ c. Found in secretions such as mucus, saliva, and tears, so protects against oral, vaginal, and respiratory infections.

___E___ d. Located on mast and basophil cells and involved in allergic reactions, for example, to certain foods, pollen, or bee venom.

___D___ e. Activate B cells to produce antibodies

E10. Describe the stages of cell-mediated immunity (CMI) by completing the following outline that provides key terms. (*Suggestion:* Refer to Figure 22.16, page 706 in the text.)

a. Antigen recognition (car-starting analogy)
  1. T cell receptors

  2. Costimulators

  (What is the meaning of *anergy?*)

b. Proliferation and differentiation of effector cells
  1. Sensitized T cells

  2. Clones of five different T cells (See Checkpoint E11 also.)

c. Elimination of the intruder by chemicals from T_C cells
   1. Perforin

   2. Lymphotoxin

   3. Gamma-IFN

■ **E11.** Match T cells to descriptions of their roles. (*Suggestion:* Refer to Checkpoint E7 above.)

DTH. Delayed-type-hypersensitivity cells	T_H. Helper T cells
M. Memory T cells	T_S. Suppressor T cells
T_C. Cytotoxic T cells	

_____ a. Cells that may produce chemicals that inhibit production of B cells and T cells.

_____ b. Programmed to recognize the original invader; can initiate dramatic responses to reappearance of the intruder

_____ c. Called T8 cells since developed from cells with CD8 protein; to become cytolytic, must be costimulated by IL-2 from T_H cells

_____ d. Called T4 cells since developed from cells with CD4 protein; activated by APGs and costimulated by IL-1 and IL-2; produce IL-2

_____ e. Mediate allergic responses by secreting cytokines (such as gamma-IFN) that activate macrophages

E12. Describe antibody-mediated immunity (AMI) in this activity.

a. Where do B cells perform their functions? *(At sites where invaders enter the body? In lymphatic tissue?)*

b. Now describe the steps in AMI, using these key terms or phrases as a guide.
   1. Activation of B cells

   2. Follicular dendritic cells

3. Antigen-MHC-II

4. Costimulation

5. Plasma cells

6. Four or five days

7. Memory B cells

8. Antibodies with identical structure to original antigen receptors

9. Complement enzymes activated

E13. *A clinical challenge.* Define monoclonal antibodies (MAb's). These antibodies are

produced by _____ cells. Explain how a *hybridoma* cell is produced.

Describe three or more clinical uses of monoclonal antibodies.

■ **E14.** Do this exercise on immunizations.

a. Once a specific antigen has initiated an immune response, either by infection or by a(n) *(initial? booster dose?)* immunization, the person produces some long-lived B

and T cells called _____.

b. The level of antibodies rises *(slowly? rapidly?)* after the initial immunization. This

level measured in serum is known as the antibody _____.

Which antibody titer rises first? *(IgA? IgG? IgM?)* Which rises next? _____

c. Upon subsequent exposure to the same antigen, such as during another infection or

by a _____ dose of vaccine, memory cells provide a *(more?*
less?) intense response. Antibodies produced by this secondary response, also known

as _____ _____, have a *(higher? lower?)*
affinity for the antigen compared to antibodies produced in the primary response.

■ **E15.** *A clinical challenge.* Identify types of immunity described below by selecting from
these answers:

AAAI. Artificially acquired active immunity	**NAAI.** Naturally acquired active immunity
AAPI. Artificially acquired passive immunity	**NAPI.** Naturally acquired passive immunity

_____ a. As 28-year-old Bud tore apart an old shed, a rusty nail entered his left hand. Bud, who reported that he had not had a tetanus shot since he was "about 14," received a shot of tetanus immunoglobulins at the emergency room.

_____ b. Kelly has provided her 3-month-old baby Crystal with temporary immunity by the antibodies that crossed over the placenta during Kelly's pregnancy and also by antibodies in milk from breast-feeding.

_____ c. Kim took her baby Jamie to the clinic for Jamie's regularly scheduled MMR (measles-mumps-rubella) immunization.

■ **E16.** The ability of your T cells to avoid reacting with your own body proteins is

known as immunological _____. B cells *(also? do not?)* display

this tolerance. Loss of such tolerance results in _____ disorders.

E17. Describe each of the processes listed below which are steps in development of immunological tolerance involving T cells.

a. Positive selection

b. Deletion

c. Anergy

Which of the above steps *(a? b? c?)* appears to be the main mechanism that B cells use to prevent responses to self-proteins?

E18. "The ability to recognize tumor antigens as nonself" is a definition of the characteristic of immunological *(surveillance? tolerance?)*. Describe how each of the following forms of therapy may help cancer patients whose own immunological surveillance is failing.

a. Adoptive cellular immunotherapy

b. Cytokine therapy

E19. Define the term *psychoneuroimmunology (PNI)*.

Describe several examples of physiological bases for body-mind relationships.

F. Aging and the immune system   (page 712)

■ **F1.** Complete arrows to indicate immune changes that usually accompany aging.
 | a. Number of helper T cells
 | b. Risk of autoimmune conditions by production of antibodies against self
 | c. Response to vaccines

F2. Write several health practices that can help you to maximize your resistance against disease.

G. Disorders and medical terminology   (pages 712–717)

■ **G1.** Answer these questions about AIDS.

a. AIDS refers to a _____ i_____

 d_____ s_____. This condition was first
 recognized in the United States by Centers for Disease Control (CDC) in the year
 19___ following increase of two relatively rare conditions in AIDS patients. Name

 those conditions: p_____ c_____

 p_____ (PCP), a respiratory infection, and

 K_____ s_____ (KS), a skin cancer.

b. The causative agent for AIDS has been identified as the h_____

 i_____ v_____. What is the major effect
 of this virus?

c. List the groups of persons that have been most infected in the United States.

 In the United States, 9 out of 10 AIDS patients are *(female? male?)*. Worldwide, 75 percent of persons with AIDS are thought to have gotten AIDS through *(heterosexual? homosexual?)* contacts.

d. The incubation period (from HIV infection to full-blown AIDS) is about 7 to 10 *(days? months? years?)*.

e. Like other viruses, the HIV virus *(does? does not?)* depend on host cells for replica-

tion. The HIV virus is a _____-virus, since its genetic code is
carried in *(DNA? RNA?)*. What role does reverse transcriptase play in a retrovirus?

Name AIDS drugs that work by inhibiting this enzyme.

f. HIV primarily affects *(B? cytotoxic T? helper T?)* lymphocytes, since these cells

display the receptor or "_____ protein," known as the ____

molecule. A normal CD4 (helper-T) lymphocyte count is _____/mm^3; a count of

under _____/mm^3 was designated in 1992 as diagnostic of AIDS.
g. Name several other cells that may be attacked by the HIV virus.

h. Formation of antibodies in the infected person usually occurs within ____ weeks after viral exposure. The presence of HIV *(antigens? antibodies?)* is commonly used to
diagnose AIDS. If a person was infected with AIDS 18 days ago, the person is likely
to test *(positive? negative?)* for AIDS but *(can? cannot?)* transmit the virus.
i. Circle the four fluids through which the HIV virus has been found to be transmitted
in sufficient quantities to be infective.
Blood   Breast milk   Mosquito venom   Saliva
Semen   Tears   Vaginal fluids
Health care personnel who use barrier precautions *(are? are not?)* at high risk for
contracting AIDS.

G2. Discuss the following aspects of AIDS.
a. Describe stages of AIDS, including major infections, signs, and symptoms that commonly occur.

b. Discuss development of an AIDS vaccine.

c. List several ways of avoiding contracting AIDS.

G3. Do this exercise on autoimmune disease.

a. Normally the body *(does? does not?)* produce antibodies and T cells against its own

tissues. Such self-recognition is called _____.

b. Describe possible mechanisms by which autoimmune disease may develop.

c. List five or more autoimmune diseases.

■ **G4.** Match the four types of allergic responses in the box with the correct descriptions below.

> I. Type I
> II. Type II
> III. Type III
> IV. Type IV

_____ a. Known as *cell-mediated* reactions, or delayed-type-hypersensitivity reactions, these lead to responses such as those that occur after exposure to poison ivy or after tuberculin testing (Mantoux test)

_____ b. Antigen-antibody *immune complexes* trapped in tissues activate complement and lead to inflammation, as in lupus (SLE) or rheumatoid arthritis (RA)

_____ c. Caused by IgG or IgM antibodies directed against RBCs or other cells, as in, for example, response to an incompatible transfusion; called a *cytotoxic* reaction

_____ d. Involves IgEs produced by mast cells and basophils; effects may be localized (such as swelling of the lips) or systemic, such as acute *anaphylaxis*.

■ **G5.** Transplantation is likely to lead to tissue rejection because the transplanted organ or tissue serves as an *(antigen? antibody?)*. The body tries to reject this foreign tissue

by producing _____.

■ **G6.** Match types of transplants in the box with descriptions below.

> A. Allograft   I. Isograft   X. Xenograft

_____ a. Between animals of different species

_____ b. The most successful type of transplant

_____ c. Between individuals of same species, but of different genetic backgrounds, such as a mother's kidney donated to her daughter or a blood transfusion

■ **G7.** Name one drug often used for transplant patients since it is selectively

immunosuppressant. _____

■ **G8.** Match the condition in the box with the description below.

A. Autoimmune disease	SCID. Severe combined immune deficiency
H. Hodgkin's disease	SLE. Systemic lupus erythematosus
L. Lymphangioma	Sp. Splenomegaly

_H_ a. A curable malignancy usually arising in lymph nodes

_A_ b. Multiple sclerosis (MS), rheumatoid arthritis (RA), and systemic lupus erythematosus (SLE) are examples

_L_ c. Benign tumor of lymph vessels

SCID d. Characterized by lack of both B and T cells

SLE e. Autoimmune, inflammatory disease with skin and joint changes, and possibly serious effects on organs such as kidneys; skin lesions may resemble wolf bites, and part of name is the Latin word for "wolf"

_Sp_ f. Enlarged spleen

ANSWERS TO SELECTED CHECKPOINTS

A1. Lymph, lymph vessels, and lymph organs (such as tonsils, spleen, and lymph nodes).

A2. (a) Proteins; lipids. (b) Lympho; T, antibodies.

A3. (a–c) NS. (d–e) S. (f) NS.

A6. Skeletal muscle contraction squeezing lymphatics which contain valves that direct flow of lymph. Respiratory movements.

A7. (a) Three-fourths; cisterna chyli; digestive and other abdominal organs and both lower extremities. (b) Left jugular, left subclavian, and left bronchomediastinal. (c) Right lymphatic; right upper extremity and right side of thorax, neck, and head. (d) Internal jugular, subclavian. (e) Interstitial; plasma.

A8. Fracture of this bone might block the thoracic duct which enters veins close to the left clavicle. Lymph therefore backs up with these results: (a) Extra fluid remains in tissue spaces normally drained by vessels leading to the thoracic duct. (b) Lymph capillaries in the intestine normally absorb fat from foods. Slow lymph flow can decrease such absorption leaving fat in digestive wastes.

B2. (a) A, afferent lymphatic vessel. (b) C, efferent lymphatic vessel. (c) D, germinal center. (d) B, cortical sinus. (e) E, medullary cord.

B4. Metastasis; nontender.

B5. Palatine; as lymphatic tissue does become inflamed at times, it does protect against foreign substances.

B6. Upper left, diaphragm. B lymphocyte and antibody production; blood reservoir; phagocytosis of microbes, red blood cells, and platelets.

B7. (a) Mediastinum; sternum; large, de. (c) Cortex; lympho; T, throughout life.

C1. Meso, fifth.

C2. Veins. (a) Posterior. (b) Jugular; thoracic.

D2. (a) C. (b) D. (c) F. (d) E. (e) A. (f) G. (g) H. (h) B.

D3. Viruses; type I; Kaposi's sarcoma, occurring in persons with AIDS (caused by the HIV virus), genital warts, and hepatatis C; leukemia. (b) Natural killer, lymphocyte; cytolytic; AIDS and some forms of cancer. (c) Neutrophils, eosinophils; monocytes; macrophages; mononuclear phagocytic.

D5. (a) C. (b) A. (c) I. (d) K.

D7. (a–d) 1. (e) 2. (f) 3.

D8. (a) PG. (b) H, LT. (c) LPF. (d) K. (e) K, C. (f) T. (g) C. (h) I.

E1. (a) Lymphocytes and antigen presenting cells; from stem cells in bone marrow. (b) T; perform immune functions against specific antigens if properly stimulated; $CD8^+$, antigen-receptor protein. (c) Bone marrow.

E2. (a) CMI. (b) AMI. (c) AMI. (d) CMI. (e) AMI and CMI

E3. (a) Is recognized as foreign; foreign substances. (b1) Stimulate production of; (b2) React with (and potentially be destroyed by). (c) Complete antigen, or immunogen. (d) Hapten; reactivity, immunogenicity; body protein; allergic. (e) Large, complex molecules, such as proteins, nucleoproteins, lipoproteins, glycoproteins, or complex polysaccharides. (f) They are made of simple, repeating subunits that are not likely to be antigenic. (g) Yes; flagella, capsules, cell walls, as well as toxins made by bacteria. (h) Antigens. (i) Only a specific region of the; antigenic determinant (epitope); a number of.

E5. (a) Good news: they help T cells to recognize foreign invaders since antigenic proteins must be processed and presented in association with MHC antigens before T cells can recognize the antigenic proteins; bad news: MHC antigens present transplanted tissue as "foreign" to the body, causing it to be rejected. (b) MHC-I molecules are in plasma membranes of all body cells except RBCs; MHC-I antigens pick up short (8–9 amino acid) peptide fragments to present to T cells; MHC-II antigens pick up longer (13–17 amino acid) peptides and are found on antigen-presenting cells (APCs), thymic cells, and T cells activated by previous exposure to the antigen. (c) B (basis of the principle of self-tolerance).

E6. (a) Endogenous. (b) Antigens from intruders outside of body cells, such as bacteria, pollen, foods, cat hair; antigen presenting cells; at sites where antigens are likely to enter the body, such as skin or mucous membranes; macrophages, B cells, and dendritic cells (including Langerhans cells in skin). (c) 1. Phago/endo. 2. Partial dig. 3. Fusion. 4. Bind to MHC-II. 5. Exo/insert. 6. Present to T.

E7. (a) M. (b) IL-4. (c) IL-2. (d) Gamma-IFN. (e) MMIF.

E8. (a) Globulins; 4; 450; 200. See Figure LG 22.2A. (b) Variable; constant, are identical. See Figure LG 22.2A.

E9. Ig's; constant. (a) M. (b) G. (c) A. (d) E. (e) D.

E11. (a) T_S. (b) M. (c) T_C. (d) T_H. (e) DTH.

E14. (a) Initial, memory cells. (b) Slowly; titer; IgM; IgG. (c) Booster, more; immunological memory, higher.

E15. (a) AAPA. (b) NAPI. (c) AAAI.

E16. Tolerance; do; autoimmune.

F1. (a) ↓. (b) ↑. (c) ↓.

G1. (a) Acquired immune deficiency disease; 81; pneumocystis carinii pneumonia, Kaposi's sarcoma. (b) Human immunodeficiency virus; by destroying the immune system, the disease places the patient at high risk for opportunistic infections (normally harmless), such as PCP, and for other conditions such as KS. (c) Homosexual men who have not used safe sexual practices, IV drug users, hemophiliacs who received contaminated blood products before 1985, and heterosexual partners of HIV-infected persons; male; heterosexual. (d) Years. (e) Does; retro, RNA; directs DNA synthesis in host cells modeled on the viral RNA code; AZT, ddl, and ddC. (f) Helper T, docking CD4; 1200, 200. (g) Macrophages and brain cells. (h) 3–20; antibodies; negative, can. (i) Blood, semen, vaginal fluids, and breast milk; are not.

G4. (a) IV. (b) III. (c) II. (d) I.

G5. Antigen; antibodies or T lymphocytes against the tissue.

G6. (a) X. (b) I. (c) A.

G7. Cyclosporine.

G8. (a) H. (b) A. (c) L. (d) SCID. (e) SLE. (f) Sp.

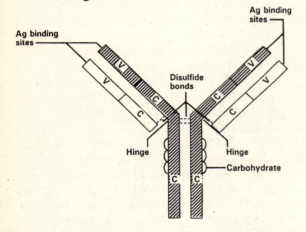

Figure LG 22.2A Diagram of an antibody molecule.
C. Constant portion   V. Variable portion

Questions 1–2: Arrange the answers in correct sequence.

_____ _____ _____ _____   1. Flow of lymph through a lymph node:
   A. Afferent lymphatic vessel
   B. Medullary sinus
   C. Cortical sinus
   D. Efferent lymphatic vessel

_____ _____ _____ _____ _____   2. Activities in humoral immunity, in chronological order:
   A. B cells develop in bone marrow or other part of the body.
   B. B cells differentiate and divide (clone), forming plasma cells.
   C. B cells migrate to lymphoid tissue.
   D. B cells are activated by specific antigen that is presented.
   E. Antibodies are released and are specific against the antigen that activated the B cell.

Questions 3–16: Circle the letter preceding the one best answer to each question.
Take particular note of underlined terms.

3. Both the thoracic duct and the right lymphatic duct empty directly into:
   A. Axillary lymph nodes
   B. Superior vena cava
   C. Cisterna chyli
   D. Subclavian arteries
   E. Junction of internal jugular and subclavian veins

4. Which cells are the most dominant microphages?
   A. Histiocytes   D. Neutrophils
   B. Basophils   E. Eosinophils
   C. Lymphocytes

5. All of these are examples of nonspecific defenses *except:*
   A. Antigens and antibodies
   B. Saliva
   C. Complement
   D. Interferon
   E. Skin
   F. Phagocytes

6. All of the following correctly match lymphocytes with their functions *except:*
   A. Helper T cells: stimulate B cells to divide and differentiate
   B. Cytotoxic (killer) T cells: produce lymphokines that attract and activate macrophages and lymphotoxins that directly destroy antigens
   C. Natural killer (NK) cells: suppress action of cytotoxic T cells and B cells
   D. B cells: become plasma cells that secrete antibodies

7. All of the following match types of allergic reactions with correct descriptions *except:*
   A. Type I: anaphylactic shock, for example, from exposure to iodine or bee venom
   B. Type II: transfusion reaction in which recipient's IgG or IgM antibodies attack donor's red blood cells
   C. Type III: Involves Ag-Ab-complement complexes that cause inflammations, as in lupus (SLE) or rheumatoid arthritis (RA)
   D. Type IV: cell-mediated reactions that involve immediate responses by B cells, as in the TB skin test.

8. Choose the one *false* statement about T cells.
   A. T_C cells are best activated by antigens associated with both MHC-I and MHC-II molecules.
   B. T_H cells are activated by antigens associated with MHC-II molecules.
   C. For T_C cells to become cytolytic, they need costimulation by IL-2.
   D. Perforin is a chemical released by T_H cells.

9. Choose the *false* statement about lymphatic vessels.
   A. Lymph capillaries are more permeable than blood capillaries.
   B. Lymphatics have thinner walls than veins.
   C. Like arteries, lymphatics contain no valves.
   D. Lymph vessels are blind-ended.

10. Choose the *false* statement about nonspecific defenses.
    A. Complement functions protectively by being converted into histamine.
    B. Histamine increases permeability of capillaries so leukocytes can more readily reach the infection site.
    C. Complement and properdin are both enzymes found in serum.
    D. Opsonization enhances phagocytosis.

11. Choose the *false* statement about T cells.
    A. Some are called memory cells.
    B. They are called T cells because they are processed in the thymus.
    C. They are involved primarily in antibody-mediated immunity (AMI).
    D. Like B cells, they originate from stem cells in bone marrow.

12. Choose the *false* statement about lymph nodes.
    A. Lymphocytes are produced here.
    B. Lymph nodes are distributed evenly throughout the body, with equal numbers in all tissue.
    C. Lymph may pass through several lymph nodes in a number of regions before returning to blood.
    D. Lymph nodes are shaped roughly like kidney (or lima) beans.

13. Choose the *false* statement about lymphatic organs.
    A. The palatine tonsils are the ones most often removed in a tonsillectomy.
    B. The spleen is the largest lymphatic organ in the body.
    C. The thymus reaches its maximum size at age 40.
    D. The spleen is located in the upper left quadrant of the abdomen.

14. Choose the *false* statement.
    A. Skeletal muscle contraction aids lymph flow.
    B. Skin is normally more effective than mucous membranes in preventing entrance of microbes into the body.
    C. Interferon is produced by viruses.
    D. An allergen is an antigen, not an antibody.

15. Choose the one *true* statement about AIDS.
    A. It appears to be caused by a virus that carries its genetic code in DNA.
    B. Persons at highest risk are homosexual men and women.
    C. Persons with AIDS experience a decline in helper T cells.
    D. AIDS is transmitted primarily through semen and saliva.

16. All of the following are cytokines known to be secreted by helper T cells *except:*
    A. Gamma interferon
    B. Interleukin-1
    C. Interleukin-2
    D. Interleukin-4

Questions 17–20: Circle T (true) or F (false). If the statement is false, change the underlined word or phrase so that the statement is correct.

T F 17. Antibodies are usually composed of one light and one heavy polypeptide chain.

T F 18. T_C cells are especially active against slowly growing bacterial diseases, some viruses, cancer cells associated with viral infections, and transplanted cells.

T F 19. Immunogenicity means the ability of an antigen to react with a specific antibody or T cell.

T F 20. A person with autoimmune disease produces fewer than normal antibodies.

Questions 21–25: fill-ins. Complete each sentence with the word or phrase that best fits.

_____ 21. ____ lymphocytes provide antibody-mediated immunity, and ____ lymphocytes and macrophages offer cellular (cell-mediated) immunity.

_____ 22. Helper T cells are also known as ____ cells.

_____ 23. List the four fundamental signs or symptoms of inflammation.

_____ 24. The class of immunoglobins most associated with allergy are Ig ____.

_____ 25. Three examples of mechanical factors that provide nonspecific resistance to disease are ____.

ANSWERS TO MASTERY TEST: ■ Chapter 22

Arrange

1. A C B D   2. A C D B E

Multiple Choice

3. E	10. A
4. D	11. C
5. A	12. B
6. C	13. C
7. D	14. C
8. D	15. C
9. C	16. B

True-False

17. F. Two light and two heavy
18. T
19. F. Stimulate formation of
20. F. Antibodies against the individual's own tissues

Fill-ins

21. B, T
22. T_H cells, T4 cells, and CD4 cells.
23. Redness, warmth, pain, and swelling
24. E
25. Skin, mucosa, cilia, epiglottis; also flushing by tears, saliva, and urine

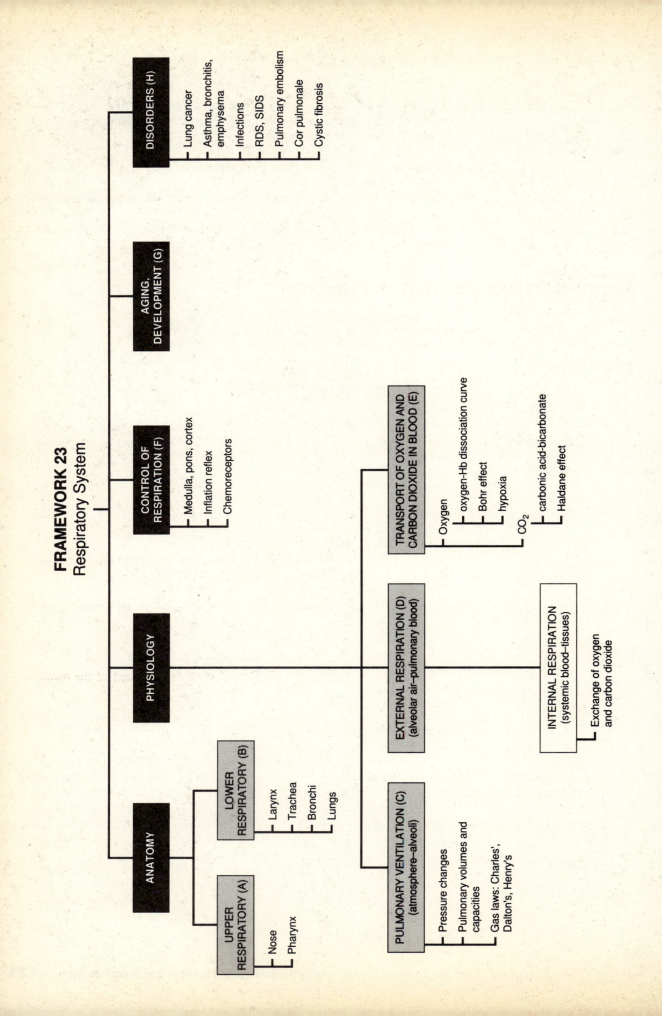

FRAMEWORK 23
Respiratory System

PHYSIOLOGY

ANATOMY

UPPER RESPIRATORY (A)
- Nose
- Pharynx

LOWER RESPIRATORY (B)
- Larynx
- Trachea
- Bronchi
- Lungs

CONTROL OF RESPIRATION (F)
- Medulla, pons, cortex
- Inflation reflex
- Chemoreceptors

AGING, DEVELOPMENT (G)

DISORDERS (H)
- Lung cancer
- Asthma, bronchitis, emphysema
- Infections
- RDS, SIDS
- Pulmonary embolism
- Cor pulmonale
- Cystic fibrosis

PULMONARY VENTILATION (C)
(atmosphere–alveoli)
- Pressure changes
- Pulmonary volumes and capacities
- Gas laws: Charles', Dalton's, Henry's

EXTERNAL RESPIRATION (D)
(alveolar air–pulmonary blood)

TRANSPORT OF OXYGEN AND CARBON DIOXIDE IN BLOOD (E)
- Oxygen
  - oxygen–Hb dissociation curve
  - Bohr effect
  - hypoxia
- CO_2
  - carbonic acid–bicarbonate
  - Haldane effect

INTERNAL RESPIRATION (systemic blood–tissues)
- Exchange of oxygen and carbon dioxide

The Respiratory System

Oxygen is available all around us. But to get an oxygen molecule to a muscle cell in the stomach or a neuron of the brain requires coordinated efforts of the respiratory system with its transport adjunct, the cardiovascular system. Oxygen traverses a system of ever-narrowing and diverging airways to reach millions of air sacs (alveoli). Each alveolus is surrounded by a meshwork of pulmonary capillaries, much like a balloon encased in nylon stocking. Here oxygen changes places with the carbon dioxide wastes in blood, and oxygen travels through the blood to reach the distant stomach or brain cells. Interference via weakened respiratory muscles (as in normal aging) or extremely narrowed airways (as in asthma or bronchitis) can limit oxygen delivery to lungs. Decreased diffusion of gases from alveoli to blood (as in emphysema, pneumonia, or pulmonary edema) can profoundly affect total body function and potentially lead to death.

Take a couple of deep breaths, and look at the Chapter 23 Framework and the key terms for each section.

TOPIC OUTLINE AND OBJECTIVES

A. Upper respiratory passageways

1. Identify the organs of the respiratory system and describe their functions.

B. Lower respiratory passageways, lungs

2. Explain the structure of the alveolar-capillary (respiratory) membrane and its function in the diffusion of respiratory gases.

C. Physiology of respiration: pulmonary ventilation

3. Describe the events involved in inspiration and expiration.

D. Physiology of respiration: exchange of oxygen and carbon dioxide

E. Transport of oxygen and carbon dioxide in blood

4. Explain how respiratory gases are carried by blood.

F. Control of respiration

5. Describe the various factors that control the rate of respiration.

G. Aging and development of the respiratory system

6. Describe the effects of aging on the respiratory system.
7. Describe the development of the respiratory system.

H. Disorders, medical terminology

8. Define asthma, bronchitis, emphysema, bronchogenic carcinoma (lung cancer), pneumonia, tuberculosis (TB), respiratory distress syndrome (RDS) of the newborn, respiratory failure, sudden infant death syndrome (SIDS), coryza (common cold), influenza (flu), pulmonary embolism, pulmonary edema, cystic fibrosis (CF), and smoke inhalation injury as disorders of the respiratory system.
9. Define medical terminology associated with the respiratory system.

Now study the following parts of words that may help you better understand terminology in this chapter.

Wordbyte	Meaning	Example	Wordbyte	Meaning	Example
atele-	incomplete	*atele*ctasis	intra-	within	*intra*alveolar
dia-	through	*dia*phragm	-pnea	breathe	a*pnea*
dys-	bad, difficult	*dys*pnea	pneumo-	air	*pneumo*thorax
-ectasis	dilation	atel*ectasis*	pulmo-	lung	*pulmo*nary
ex-	out	*ex*piration	rhin-	nose	*rhin*orrhea
in-	in	*in*spiration	spir-	breathe	in*spir*ation

CHECKPOINTS

A. Upper respiratory passageways (pages 721–725)

A1. Explain how the respiratory and cardiovascular systems work together to accomplish gaseous exchange among the atmosphere, blood, and cells.

■ **A2.** Name the structures of the nose that are designed to carry out each of the following functions.

a. Warm, moisten, and filter air

Conchae

b. Sense smell

olfactory epithelium

c. Assist in speech

~~tongue~~ sinus

■ **A3.** Write LP (laryngopharynx), NP (nasopharynx), or OP (oropharynx) to indicate locations of each of the following structures.

NP a. Adenoids

OP b. Palatine tonsils

OP c. Lingual tonsils

OP d. Opening (fauces) from oral cavity

LP e. Opening into larynx and esophagus

A4. On Figure LG 23.1, cover the key and identify all structures associated with the nose, palate, pharynx, and larynx. Color structures with color code ovals.

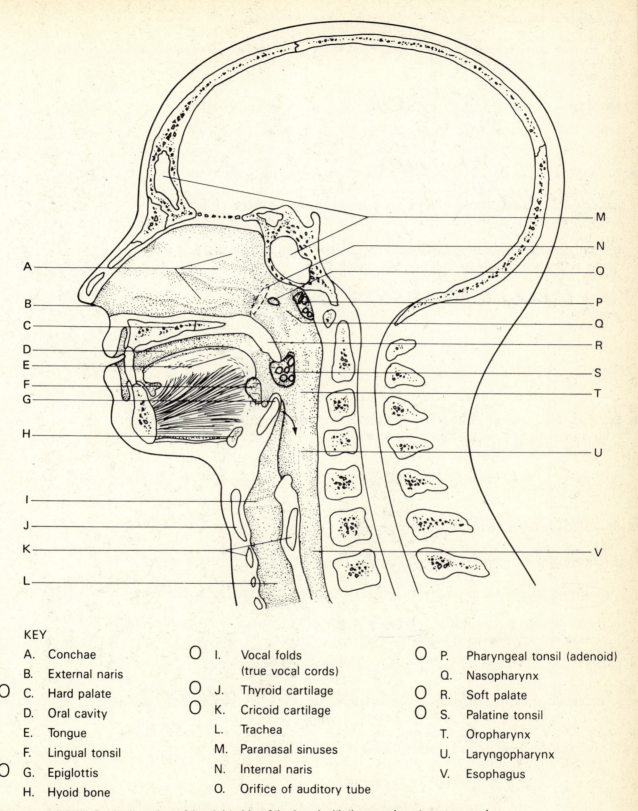

KEY

A. Conchae
B. External naris
○ C. Hard palate
D. Oral cavity
E. Tongue
F. Lingual tonsil
○ G. Epiglottis
H. Hyoid bone

○ I. Vocal folds
       (true vocal cords)
○ J. Thyroid cartilage
○ K. Cricoid cartilage
L. Trachea
M. Paranasal sinuses
N. Internal naris
O. Orifice of auditory tube

○ P. Pharyngeal tonsil (adenoid)
Q. Nasopharynx
○ R. Soft palate
○ S. Palatine tonsil
T. Oropharynx
U. Laryngopharynx
V. Esophagus

Figure LG 23.1 Sagittal section of the right side of the head with the nasal septum removed.
Color and identify labeled structures as directed in Checkpoint A4. Arrow refers to Checkpoint B1.

B. Lower respiratory passageways, lungs (pages 725–736)

B1. Explain how the larynx prevents food from entering the trachea. (*Hint:* notice the arrow on Figure LG 23.1.)

Epiglottis

■ **B2.** The *rima glottidis* is a space between the true vocal cords. The *arytenoid* cartilages are pyramid-shaped cartilages of the larynx attached to the vocal cords.

B3. Tell how the larynx produces sound. Explain how pitch is controlled and what causes male pitch usually to be lower than female pitch.

true vocal cords are thicker

■ **B4.** *A clinical challenge.* After a larynx is removed (for example, due to cancer), what other structures help the laryngectomee to speak?

■ **B5.** Complete this Checkpoint about the trachea.

a. The trachea is commonly known as the *windpipe*. It is located (*anterior?* posterior?) to the esophagus. About *12* cm (*4.5* inches) long, the trachea terminates at the *bronchus carina*, which is a Y-shaped intersection with the primary bronchi.

b. The tracheal wall is lined with a *mucous* membrane, and strengthened by 16–18 C-shaped rings composed of *cartilage*.

■ **B6.** *A clinical challenge.* Anna has aspirated a small piece of candy. Dr. Lennon expects to find it in the right bronchus rather than in the left. Why?

B7. Identify tracheobronchial tree structures *J–N* on Figure LG 23.2, and color each of those structures indicated with color code ovals.

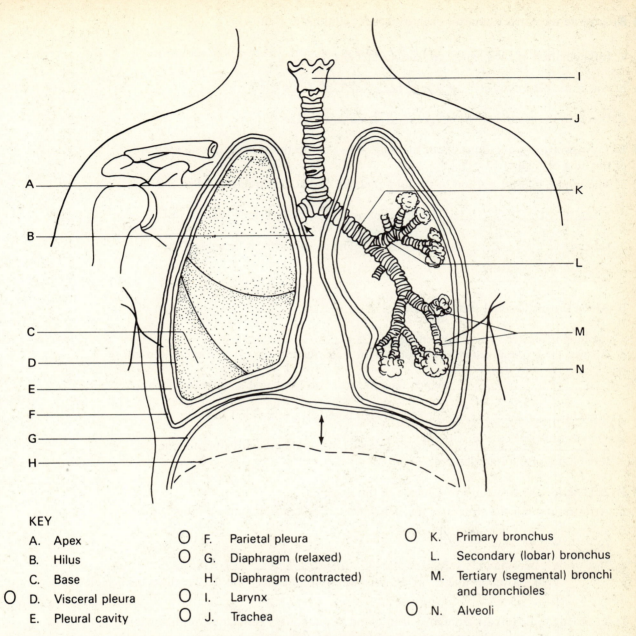

KEY

	A.	Apex	O	F.	Parietal pleura	
	B.	Hilus	O	G.	Diaphragm (relaxed)	
	C.	Base		H.	Diaphragm (contracted)	
O	D.	Visceral pleura	O	I.	Larynx	
	E.	Pleural cavity	O	J.	Trachea	

O	K.	Primary bronchus
	L.	Secondary (lobar) bronchus
	M.	Tertiary (segmental) bronchi and bronchioles
O	N.	Alveoli

Figure LG 23.2 Diagram of lungs with pleural coverings and bronchial tree. Color and identify labeled structures as directed in Checkpoints B7, B9, and B11.

■ **B8.** Bronchioles *(do?* (do not?)) have cartilage rings. How is this fact significant during an asthma attack?

muscle spasms can collapse airway

B9. On Figure LG 23.2, color the two layers of the pleura and the diaphragm according to color code ovals. Also cover the key and identify all other parts of the diagram from A to I.

■ **B10.** Write the correct term for each of these conditions.

a. Inflammation of the pleura: *pleurisy*

b. Air in the pleural cavity: *pneumothorax*

c. Blood in the pleural cavity: *hemothorax*

■ **B11.** Answer these questions about the lungs. (As you do the exercise, locate the parts of the lung on Figure LG 23.2.)

a. The broad, inferior portion of the lung that sits on the diaphragm is called the

_____. The upper narrow apex of each lung extends just

superior to the _____. The costal surfaces lie against the

_____.

b. Along the mediastinal surface is the _____ where the root of

the lung is located. This root consists of _____.

c. Answer these questions with *right* or *left*. In which lung is the cardiac notch located?

_____ Which lung has just two lobes and so only two lobar

bronchi? _____ Which lung has a horizontal fissure?

_____ Write S (superior), M (middle), and I (inferior) on the
three lobes of the right lung.

■ **B12.** Describe the structures of a lobule in this exercise.

a. Arrange in order the structures through which air passes as it enters a lobule en

route to alveoli. *T B A*

A. Alveolar ducts	B. Respiratory bronchiole	T. Terminal bronchiole

b. In order for air to pass from alveoli to blood in pulmonary capillaries, it must pass

through the *alveolar - capillary* (respiratory)
membrane. Identify structures in this pathway: A to E on Figure LG 23.3.

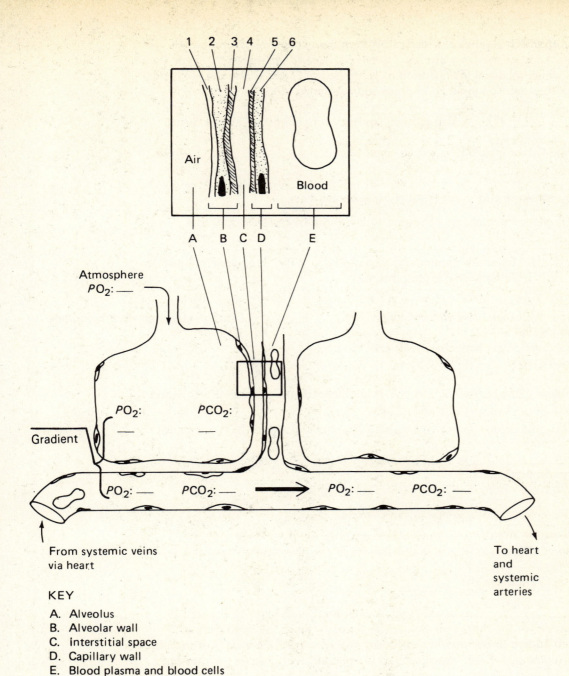

Figure LG 23.3 Diagram of alveoli and pulmonary capillary. Insert enlarges alveolar-capillary membrane. Numbers and letters refer to Checkpoint B12. Complete partial pressures (in mm Hg) as directed in Checkpoint D3.

KEY

A. Alveolus
B. Alveolar wall
C. Interstitial space
D. Capillary wall
E. Blood plasma and blood cells

c. *For extra review.* Now label layers 1–6 of the alveolar-capillary membrane in the insert in Figure LG 23.3.

d. *A clinical challenge.* Normally the respiratory membrane is extremely *(thick? thin?).* Suppose pulmonary capillary blood pressure rises dramatically, pushing extra fluid out of blood. The presence of excess fluid in interstitial areas and ultimately in

alveoli is known as pulmonary _____. Diffusion occurs *(more? less?)* readily through the fluid-filled membrane.

e. The alveolar epithelial layer of the respiratory membrane contains three types of cells. Forming most of the layer are flat *(squamous? septal? macrophage?)* (Type I alveolar) cells which are ideal for diffusion of gases. Septal (Type II alveolar) cells produce an important phospholipid and lipoprotein substance called

_____, which reduces tendency for collapse of alveoli due to

surface tension. Macrophage cells help to remove _____ from lungs.

C. Physiology of respiration: ventilation (pages 736–742)

■ **C1.** Match each of the three phases of respiration with the correct description.

> E. External (pulmonary) respiration   PV. Pulmonary ventilation
> I. Internal (tissue) respiration

_____ a. Exchange of air between atmosphere and alveoli

_____ b. Exchange of gas between alveoli and blood

_____ c. Exchange of gas between blood and cells

■ **C2.** Refer to Figure 23.14 (page 739 in the text) and describe the process of ventilation in this exercise.

a. In diagram (a), just before the start of inspiration, pressure within the lungs (called

_____) is _____ mm Hg. This is *(more than? less than? the same as?)* atmospheric pressure.

b. At the same time, pressure within the pleural cavity (called

_____ pressure) is _____ mm Hg. This is *(more than? less than? the same as?)* alveolar and atmospheric pressure.

c. The first step in inspiration occurs as the muscles in the floor and walls of the

thorax contract. These are the _____ and

_____ muscles. Note in diagram (b) (and also in Figure LG 23.2) that the size of the thorax *(increases? decreases?)*. Since the two layers of pleura tend to adhere to one another, the lungs will *(increase? decrease?)* in size also.

d. Increase in volume of a closed space such as the pleural cavity causes the pressure

there to *(increase? decrease?)* to _____ mm Hg. Since the lungs also increase in size (due to pleural cohesion), alveolar pressure also *(increases? decreases?)* to

_____ mm Hg. This inverse relationship between volume and pressure is a

statement of _____'s law.

e. A pressure gradient is now established. Air flows from high pressure area *(alveoli? atmosphere?)* to low pressure area *(alveoli? atmosphere?)*. So air flows *(into? out of?)* lungs. Thus *(inspiration? expiration?)* occurs. By the end of inspiration, sufficient air will have moved into the lungs to make pressure there equal to atmospheric pressure, that is, _____ mmHg.

f. Inspiration is a(n) *(active? passive?)* process, whereas expiration is normally a(n) *(active? passive?)* process. In certain respiratory disorders, such as COPD, accessory muscles help to force air out. Name two of these: _____

_____

■ **C3.** Answer these questions about compliance.

a. Imagine trying to blow up a new balloon. Initially, the balloon resists your efforts. Compliance is the ability of a substance to yield elastically to a force; in this case it is the ease with which a balloon can be inflated. So a new balloon has a *(high? low?)* level of compliance, whereas a balloon that has been inflated many times has *(high? low?)* compliance.

b. Similarly, alveoli that inflate easily have *(high? low?)* compliance. The presence of a coating called _____ lining the inside of alveoli prevents alveolar walls from sticking together during ventilation, and so *(increases? decreases?)* compliance.

c. Surfactant production is especially developed during the final weeks before birth. A premature infant may lack adequate surfactant; this disorder is known as

_____.

d. Collapse of all or part of a lung may occur as a result of lack of surfactant or other factors. This lung collapse is known as _____.

C4. Refer to Exhibit 23.1 (page 741 in your text). Read carefully descriptions of modified respiratory movements such as sobbing, yawning, sighing, or laughing while demonstrating those actions yourself.

C5. Each minute the average adult takes _____ breaths (respirations). Check your own respiratory rate and write it here: _____ breaths per minute.

■ **C6.** Maureen is breathing at the rate of 15 breaths per minute. She has a tidal volume of 480 ml per breath. Her *minute volume of respiration (MVR)* is

_____.

■ **C7.** Of the total amount of air that enters the lungs with each breath, about *(99 percent? 70 percent? 30 percent? 5 percent?)* actually enters the alveoli. The remaining amount of air is much like the last portion of a crowd trying to rush into a store: it does not succeed in entering the alveoli during an inspiration, but just reaches airways and then is quickly ushered out during the next expiration. Such air is known as anatomic _____ and constitutes about _____ ml of a typical breath.

■ **C8.** Match the lung volumes and capacities with the descriptions given. You may find it helpful to refer to Figure 23.16 (page 742 in the text).

ERV. Expiratory reserve volume	TLC. Total lung capacity
FRC. Functional residual capacity	TV. Tidal volume
IRV. Inspiratory reserve volume	VC. Vital capacity
RV. Residual volume	

_____ a. The amount of air taken in with each inspiration during normal breathing is called ____.

_____ b. At the end of a normal expiration the volume of air left in the lungs is called ____. Emphysemics who have lost elastic recoil of their lungs cannot exhale adequately, so this volume will be large.

_____ c. Forced exhalation can remove some of the air in FRC. The maximum volume of air that can be expired beyond normal expiration is called ____. This volume will be small in emphysema patients.

_____ d. Even after the most strenuous expiratory effort, some air still remains in the lungs; this amount, which cannot be removed vountarily, is called ____.

_____ e. The volume of air that represents a person's maximum breathing ability is called ____. This is the sum of ERV, TV, and IRV.

_____ f. Adding RV to VC gives ____.

_____ g. The excess air a person can take in after a normal inhalation is called ____.

■ **C9.** Indicate normal volumes for each of the following.

a. TV = _____ ml

b. TLC = _____ ml (_____ liters)

c. VC = _____ ml

D. Physiology of respiration: exchange of oxygen and carbon dioxide (pages 743–746)

■ **D1.** Check your understanding of gas laws by matching the correct law with the condition that it explains.

B. Boyle's law	D. Dalton's law	H. Henry's law

_____ a. If a patient breathes air highly concentrated in oxygen (as in a hyperbaric chamber), a higher percentage of oxygen will dissolve in the blood and tissues.

_____ b. The total atmospheric pressure (760 mm Hg) is due mostly to pressure caused by nitrogen, partly to PO_2, and slightly to PCO_2.

_____ c. Under high PN_2 more nitrogen dissolves in blood. The *bends* occurs when pressure decreases and nitrogen forms bubbles in tissue as it comes out of solution.

_____ d. As the size of the thorax increases, the pressure within it decreases.

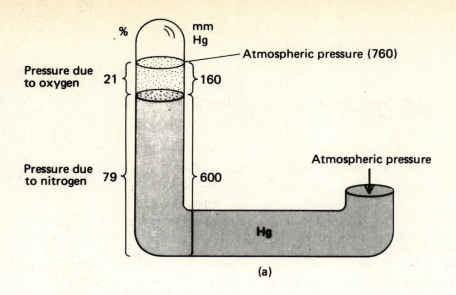

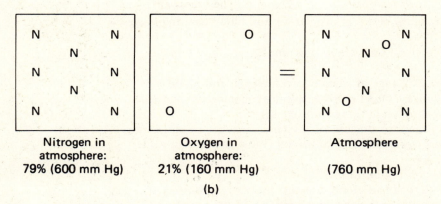

Figure LG 23.4 Atmospheric pressure. (a) Effect of gas molecules on column of mercury (Hg). (b) Partial pressures contributing to total atmospheric pressure. Color according to directions in Checkpoint D2.

■ **D2.** Refer to Figure LG 23.4 and do this exercise about Dalton's Law.

a. The atmosphere contains enough gaseous molecules to exert pressure upon a column

of mercury to make it rise about ____ mm. The atmosphere is said to have a pressure of 760 mm Hg.

b. Air is about ____ percent nitrogen and ____ percent oxygen. To show this, circle the nitrogen molecules in *green* and oxygen molecules in *red* in Figure LG 23.4b. Note

that carbon dioxide is not represented in the figure since only about ____ percent of air is CO_2.

c. Dalton's law explains that of the total 760 mm of atmospheric pressure, a certain amount is due to each type of gas. Determine the portion of the total pressure due to nitrogen molecules:

$$79\% \times 760 \text{ mm Hg} = \text{_____ mm Hg}$$

This is the partial pressure of nitrogen, or _____.

d. Calculate the partial pressure of oxygen.

e. In Figure LG 23.4a, color the part of the Hg column due to N_2 pressure (up to the 600 mm Hg mark) *green*. Then color the portion of the Hg column due to O_2 pressure *red*.

■ **D3.** Answer these questions about external and internal respiration.

a. A primary factor in the diffusion of gas across a membrane is the difference in

   concentration of the gas (reflected by _____ pressures) on the two sides of the membrane. On the left side of Figure LG 23.3 write values for pO_2 (in mm Hg) in each of the following areas. (Refer to text Figure 23.17, page 745, for help.)
      Atmospheric air (Recall this value from Figure LG 23.4.)
      Alveolar air (Note that this value is lower than that for atmospheric pO_2 since
         some alveolar O_2 enters blood.)
      Blood entering lungs

b. Calculate the pO_2 difference (gradient) between alveolar air and blood entering lungs.

$$\underline{\quad} \text{ mm Hg} - \underline{\quad} \text{ mm Hg} = \underline{\quad} \text{ mm Hg}$$

c. Three other factors that increase exchange of gases between alveoli and blood are: *(large? small?)* surface area of lungs; *(thick? thin?)* respiratory membrane; and *(increased? decreased?)* blood flow through lungs (as in exercise).

d. By the time blood leaves lungs to return to heart and systemic arteries, its pO_2 is normally *(greater than? the same as? less than?)* pO_2 of alveoli. Write the correct value on Figure LG 23.3.

e. Now fill in all three pCO_2 values on Figure LG 23.3.

f. *A clinical challenge.* Jenny, age 42, has had blood drawn from her radial artery to determine her arterial blood gases. Her pO_2 is 56 and her pCO_2 is 48. Are these

   typical values for a healthy adult? _____

■ **D4.** With increasing altitude the air is "thinner," that is, gas molecules are farther apart, so atmospheric pressure is lower. Atop a 25,000-foot mountain, this pressure is only 282 mm Hg. Oxygen still accounts for 21 percent of the pressure. What is pO_2 at that level?

From this calculation you can see limitations of life (or modifications that must be made) at high altitudes. If atmospheric pO_2 is 59.2, neither alveolar nor blood pO_2 could surpass that level.

E. Transport of oxygen and carbon dioxide in blood (pages 746–750)

■ **E1.** Answer these questions about oxygen transport. Refer to Figure 23.18 (page 746 in your text).

a. One hundred ml of blood contains about _20_ ml of oxygen. Of this, about 19.7 ml is carried as _hemoglobin_. Only a small amount of oxygen is carried in the dissolved state since oxygen has a *(high?* (low?)) solubility in blood or water.

b. Oxygen is attached to the _heme_ atoms in hemoglobin. The chemical formula for oxyhemoglobin is _HbO_. When hemoglobin carries all of the oxygen it can hold, it is said to be fully _saturated_. High pO_2 in alveoli will tend to *(increase?* decrease?) oxygen saturation of hemoglobin.

c. Refer to Figure 23.19 (page 747 in your text). Note that arterial blood, with a pO_2 of about 105 mm Hg, has its hemoglobin _98_ % saturated with oxygen. (This may be expressed as $SO_2 = $ _98_ %.)

d. List four factors that will enhance the dissociation of oxygen from hemoglobin so that oxygen can enter tissues. (*Hint:* Think of conditions within active muscle tissue.)

acidity – Bohr effect   _temperature_
pCO_2 .   _BPG_

e. *For extra review.* Demonstrate the effect of temperature on the oxygen-hemoglobin dissociation curve. Draw a vertical line on Figure 23.21 (page 748 of the text) at pO_2 = 40 (the value for venous blood). Compare the percent saturation of hemoglobin (SO_2) at these two body temperatures:

38°C (100.4°F) _67_ % SO_2 43°C (109.4°F) _40_ % SO_2

In other words, at a higher body temperature, for example in high fever, *(more?* (less?)) oxygen will be attached to hemoglobin, while ((more?) *less?*) oxygen will enter tissues to fuel metabolism.

f. By the time blood enters veins to return to the heart, its oxygen saturation of hemoglobin (SO_2) is about _75_ %. Note on Figure 23.19 (page 747 of your text) that although pO_2 drops from 105 in arterial blood to 40 in venous blood, oxygen saturation drops *(more?* (less?)) dramatically, that is, from 98% to 75%. Of what significance is this?

■ **E2.** Carbon monoxide has about _200_ times the affinity that oxygen has for hemoglobin. State the significance of this fact.

~~heme~~ Hb won't have room for O_2

E3. Contrast anemic hypoxia with stagnant hypoxia.

anemic hypoxia has too little functioning Hb

in stagnant hypoxia blood can't travel fast enough to supply O_2 to tissue

■ **E4.** Complete this Checkpoint about transport of carbon dioxide.

a. Write the percentage of CO_2 normally carried in each of these forms: _70_ percent is present in bicarbonate ion (HCO_3^-); _23_ percent is bound to the globin portion of hemoglobin; _7_ percent is dissolved in plasma.

b. Carbon dioxide (CO_2) produced by cells of your body diffuses into red blood cells (RBCs) and combines with water to form _carbonic acid_.

c. Carbonic acid tends to dissociate into two products. One is H+ which binds to

_Hb_. The other product is _HCO_3^-_ (bicarbonate), which is carried in *(RBCs?* ~~plasma?~~*)* in exchange for a *(K+?* ⟨Cl-?⟩*)* ion that shifts into the RBC.

d. Now write the entire sequence of reactions described in (b) and (c). Be sure to include the enzyme that catalyzes the first reaction. Notice that the reactions show that increase in CO_2 in the body (as when respiratory rate is slow) will tend to cause a buildup of acid (H+) in the body.

$$CO_2 \ + \ \underbrace{H_2O}_{\text{(enzyme) } \textit{carbonic anhydrase}} \quad \underbrace{H_2CO_3}_{\textit{(carbonic acid)}} \rightarrow$$

$$\underset{\substack{\downarrow \\ \text{Binds} \\ \text{to Hb}}}{H^+} \quad + \quad \underset{\substack{\downarrow \\ \text{Shifts to plasma in} \\ \text{exchange for Cl}^-}}{HCO_3^-} \textit{(bicarbonate)}$$

E5. Note that as the red blood cells reach lung capillaries, the same reactions you just studied occur, but in reverse. Study Figure 23.23b (page 750 in the text) carefully. Then list the major steps that occur in the lungs so that CO_2 can be exhaled.

F. Control of respiration (pages 750–755)

■ **F1.** Complete the table about respiratory control areas. Indicate whether the area is located in the medulla *(M)* or pons *(P)*.

Name	M/P	Function
a.	M	Controls rhythm; consists of inspiratory and expiratory areas
b. Pneumotaxic		
c.		Prolongs inspiration and inhibits expiration

■ **F2.** Answer these questions about respiratory control.

a. The main chemical change that stimulates respiration is increase in blood level of

___, which is directly related to *(decrease in pO_2? increase in pCO_2?)* of blood.

b. Cells most sensitive to changes in blood CO_2 are located in the *(medulla? pons? aorta and carotid arteries?)*.

c. An increase in arterial blood pCO_2 is called _____. Write an ar-

terial PCO_2 value that is hypercapnic. ___ mm Hg *(Even slight? Only severe?)* hypercapnia will stimulate the respiratory system, leading to *(hyper? hypo?)*-ventilation.

d. State two locations of chemoreceptors sensitive to changes in pO_2.

(Even slight? Only large?) decreases in pO_2 level of blood will stimulate these chemoreceptors and lead to hyperventilation. Give an example of a pO_2 low enough

to evoke such a response. ___ mm Hg

e. Increase in body temperature (as in fever), as well as stretching of the anal sphincter, will cause ___-crease in the respiratory rate.

f. Take a deep breath. Imagine the _____ receptors in your airways being stimulated. These will cause *(excitation? inhibition?)* of the inspiratory and apneustic areas, resulting in expiration. This reflex, known as the

_____ reflex, prevents overinflation of lungs.

G. Aging and development of the respiratory system (pages 755–756)

■ **G1.** Describe possible effects of these age-related changes.

a. Pulmonary blood vessels become sclerosed, so are more rigid and resistant to blood flow.

This could lead to the pathological condition called _____.

b. Chest wall becomes more rigid as bones and cartilage lose flexibility.

c. Decreased macrophage and ciliary action of lining of respiratory tract.

■ **G2.** Describe development of the respiratory system in this exercise.

a. The laryngotracheal bud is derived from _____-derm. List structures formed from this bud.

b. Identify portions of the respiratory system derived from mesoderm.

H. Disorders, medical terminology (pages 756–761)

H1. A host of effects upon the respiratory system are associated with smoking. Describe these two.

a. Bronchogenic carcinoma (Include the roles of basal cells and excessive mucus production.)

b. Emphysema (Note that as walls of alveoli break down, surface area of the respiratory membrane ____-creases, so amount of gas diffused ____-creases also.)

■ **H2.** Match the condition with the correct description.

> A. Asthma   P. Pneumonia
> CB. Chronic bronchitis   RF. Respiratory failure
> D. Dyspnea   TB. Tuberculosis
> E. Emphysema

_____ a. Permanent inflation of lungs due to loss of elasticity; rupture and merging of alveoli, followed by their replacement by fibrous tissue

_____ b. Inflammation of bronchi with excessive mucus production for at least three months per year for at least two consecutive years

_____ c. Acute infection or inflammation of alveoli which fill with fluid

_____ d. Spasms of small passageways with wheezing and dyspnea

_____ e. Caused by a species of Mycobacterium; lung tissue is destroyed and replaced with inelastic connective tissue

_____ f. Difficult, painful breathing; shortness of breath

_____ g. Occurs when pO_2 drops below 50 mm Hg and pCO_2 rises above 50 mm Hg

_____ h. A group of conditions known as chronic obstructive pulmonary disease (COPD) (three answers)

■ **H3.** Discuss respiratory distress syndrome (RDS) in this exercise.

a. Before birth fetal lungs are filled with *(air? fluid?)*. Inflation of lungs after birth depends largely on the presence of a chemical known as _____. This chemical is produced by *(basal cells of bronchi? septal cells of alveoli?)*.

b. Surfactant *(raises? lowers?)* surface tension so that alveolar walls are less likely to stick together. Thus surfactant *(facilitates? inhibits?)* lung inflation.

c. RDS especially targets *(premature? full-term?)* infants. Explain why.

H4. To what does SIDS refer? _____
What is the mortality rate of SIDS?

Discuss possible causes of SIDS.

H5. Explain how the abdominal thrust (Heimlich maneuver) helps to remove food that might otherwise cause death by choking.

■ **H6.** Write the ABCs of CPR.

A2. (a) Mucosa lining nose, septum, conchae, meati and sinuses, lacrimal drainage, and coarse hairs in vestibule. (b) Olfactory region lies superior to superior nasal conchae. (c) Sounds resonate in nose and paranasal sinuses.

A3. (a) NP. (b-d) OP. (e) LP.

B2. Glottis; arytenoid.

B4. Oral and nasal cavities, pharynx and paranasal sinuses continue to act as resonating chambers. Muscles of face, tongue, lips, pharynx, and esophagus also help in forming words.

B5. (a) Windpipe; anterior; 12 (4.5), carina. (b) Mucous, cartilage.

B6. The right bronchus is more vertical and slightly wider than the left.

B8. Do not; muscle spasms can collapse airways.

B10. (a) Pleurisy. (b) Pneumothorax. (c) Hemothorax.

B11. (a) Base; clavicle; ribs. (b) Hilus; bronchi, pulmonary vessels and nerves. (c) Left; left; right; refer to Figure 23.9, page 733 of the text.

B12. (a) T R A. (b) Alveolar-capillary. See KEY to Figure LG 23.3. (c) 1, Surfactant; 2, alveolar epithelium; 3, epithelial basement membrane; 4, interstitial space; 5, capillary basement membrane; 6, capillary endothelium. (d) Thin; edema; less. (e) Squamous; surfactant; debris.

C1. (a) PV. (b) E. (c) I.

C2. (a) Alveolar or intrapulmonic, 760; the same as. (b) Intrapleural, 756; less than. (c) Diaphragm, external intercostal; increases; increase. (d) Decrease, 754; decreases, 758; Boyle. (e) Atmosphere, alveoli; into; inspiration; 760. (f) Active, passive; abdominal and internal intercostals.

C3. (a) Low, high. (b) High; surfactant, increases. (c) Respiratory distress syndrome (RDS) or hyaline membrane disease (HMD). (d) Atelectasis.

C6. 7200 ml/min (= 15 breaths/min × 480 ml/breath).

C7. 70% (350 ml/500 ml); dead space, 150.

C8. (a) TV. (b) FRC. (c) ERV. (d) RV. (e) VC. (f) TLC. (g) IRV.

C9. (a) 500 (0.5). (b) 6,000, 6. (c) 4,800.

D1. (a) H. (b) D. (c) H. (d) B.

D2. (a) 760. (b) 79, 21; 0.04. (c) 600.4; pN_2. (d) 21% × 760 mm Hg = about 160 mm Hg = pO_2.

D3.

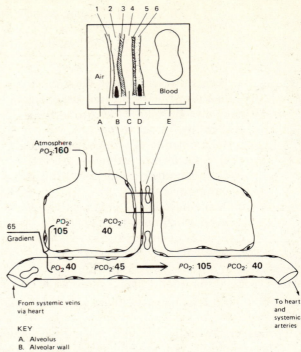

Figure LG 23.3A Diagram of alveoli and pulmonary capillary. Insert enlarges alveolar-capillary membrane.

(a) Partial; see Figure LG 23.3A. (b) 105 − 40 = 65. (c) Large; thin; increased. (d) The same as; 105. (e) See Figure LG 23.3A. (f) No. Typical radial arterial values are same as for alveoli or blood leaving lungs: pO_2 = 105, pCO_2 = 40. These values indicate inadequate gas exchange.

D4. 282 × 21% = 59.2 mm Hg.

E1. (a) 20; oxyhemoglobin; low. (b) Iron; HbO_2, saturated; increase. (c) 98; 98. (d) Increase in temperature, pCO_2, acidity, and BPG. (e) 62; 38; less, more. (f) 75; less. In the event that respiration is temporarily halted, even venous blood has much oxygen attached to hemoglobin and available to tissues.

E2. 200; oxygen-carrying capacity is drastically reduced in carbon monoxide poisoning.

E4. (a) 70, 23, 7. (b) H_2CO_3 (carbonic acid). (c) Hemoglobin (as H•Hb); HCO_3^-, plasma, Cl^-. (d) $CO_2 + H_2O \xrightarrow{\text{carbonic anhydrase}} H_2CO_3$ (carbonic acid) → $H^+ + HCO_3^-$ (bicarbonate).

F1.

Name	M/P	Function
a. **Medullary rhythmicity**	M	Controls rhythm; consists of inspiratory and expiratory areas
b. Pneumotaxic	P	**Limits inspiration and facilitates expiration**
c. **Apneustic**	P	Prolongs inspiration and inhibits expiration

F2. (a) H$^+$, increase in pCO$_2$. (b) Medulla. (c) Hypercapnia; any value higher than 40, even slight, hyper. (d) Aortic and carotid bodies, only large; usually below 60. (e) In. (f) Stretch; inhibition; inflation (Hering-Breuer).

G1. (a) Blood backs up into right ventricle, causing it to pump extra hard and hypertrophy; cor pulmonale. (b) Decreased vital capacity and so decreased arterial pO$_2$. (c) Risk of pneumonia.

G2. (a) Endo; lining of larynx, trachea, bronchial tree, and alveoli. (b) Smooth muscle, cartilage, and other connective tissues of airways.

H2. (a) E. (b) CB. (c) P. (d) A. (e) TB. (f) D. (g) RF. (h) A, CB, E.

H3. (a) Fluid; surfactant; septal cells of alveoli. (b) Lowers; facilitates. (c) Premature. Septal cells do not produce adequate amounts of surfactant until between weeks 28 and 32 of the 39-week human gestational period. A baby born at 7 months (30 weeks), for example, would be at high risk for RDS.

H6. Establish Airway, ventilate (Breathing), and reestablish Circulation.

MASTERY TEST: Chapter 23

Questions 1–4: Arrange the answers in correct sequence.

_____ _____ _____ 1. From first to last, the steps involved in inspiration:
   A. Diaphragm and intercostal muscles contract.
   B. Thoracic cavity and lungs increase in size.
   C. Alveolar pressure decreases to 758 mm Hg.

_____ _____ _____ 2. From most superficial to deepest:
   A. Parietal pleura
   B. Visceral pleura
   C. Pleural cavity

_____ _____ _____ 3. From superior to inferior:
   A. Bronchioles
   B. Bronchi
   C. Larynx
   D. Pharynx
   E. Trachea

_____ _____ _____ _____ _____ _____ 4. Pathway of inspired air:
   A. Alveolar ducts
   B. Bronchioles
   C. Lobar bronchi
   D. Primary bronchi
   E. Segmental bronchi
   F. Alveoli

Questions 5–9: Circle the letter preceding the one best answer to each question.

5. All of the following terms are matched correctly with descriptions *except:*
   A. Internal nares: choanae
   B. External nares: nostrils
   C. Posterior of nose: vestibule
   D. Pharyngeal tonsils: adenoids

6. Which of these values (in mm Hg) would be normal for pO_2 of blood in the femoral artery?
   A. 40   D. 160
   B. 45   E. 760
   C. 100   F. 0

7. Pressure and volume in a closed space are inversely related, as described by ____ law.
   A. Boyle's   C. Dalton's
   B. Starling's   D. Henry's

8. Choose the correct formula for carbonic acid:
   A. HCO_3^- D. HO_3C_2
   B. H_3CO_2 E. H_2C_3O
   C. H_2CO_3

9. A procedure in which an incision is made in the trachea and tube inserted into the trachea is known as a(n):
   A. Tracheostomy   C. Intubation
   B. Bronchogram   D. Pneumothorax

Questions 10–20: Circle T (true) or F (false). If the statement is false, change the underlined word or phrase so that the statement is correct.

T   F   10. In the chloride shift Cl^- moves into red blood cells in exchange for H+.

T   F   11. When chemoreceptors sense increase in pCO_2 or increase in acidity of blood (H+), respiratory rate will normally be stimulated.

T   F   12. Both increased temperature and increased acid content tend to cause oxygen to bind more tightly to hemoglobin.

T   F   13. Under normal circumstances intrapleural pressure is always less than atmospheric.

T   F   14. The pneumotaxic and apneustic areas controlling respiration are located in the pons.

T   F   15. Fetal hemoglobin has a lower affinity for oxygen than maternal hemoglobin does.

T   F   16. Most CO_2 is carried in the blood in the form of bicarbonate.

T   F   17. The alveolar wall does contain macrophages that remove debris from the area.

T   F   18. Intrapulmonic pressure means the same thing as intrapleural pressure.

T   F   19. Inspiratory reserve volume is normally larger than expiratory reserve volume.

T   F   20. The pO_2 and pCO_2 of blood leaving the lungs are normally about the same as pO_2 and pCO_2 of alveolar air.

Questions 21–25: fill-ins. Complete each sentence with the word or phrase that best fits.

_____ 21. The process of exchange of gases between alveolar air and blood in pulmonary capillaries is known as ____.

_____ 22. Take a normal breath and then let it out. The amount of air left in your lungs is the capacity called ____ and it usually measures about ____ ml.

_____ 23. The Bohr effect states that when more H^+ ions are bound to hemoglobin, less ____ can be carried by hemoglobin.

_____ 24. The epiglottis, thyroid, and cricoid cartilages are all parts of the ____.

_____ 25. ____ is a chemical that lowers surface tension and therefore increases inflatability (compliance) of lungs.

ANSWERS TO MASTERY TEST: ■ Chapter 23

Arrange

1. A B C
2. A C B
3. D C E B A
4. D C E B A F

Multiple Choice

5. C   8. C
6. C   9. A
7. A

True-False

10. F. HCO_3^-
11. T
12. F. Dissociate from
13. T
14. T
15. F. Higher
16. T
17. T
18. F. Alveolar
19. T
20. T

Fill-ins

21. External respiration (or pulmonary respiration, or diffusion)
22. Functional residual capacity (FRC), 2400
23. Oxygen
24. Larynx
25. Surfactant

FRAMEWORK 24
Digestive System

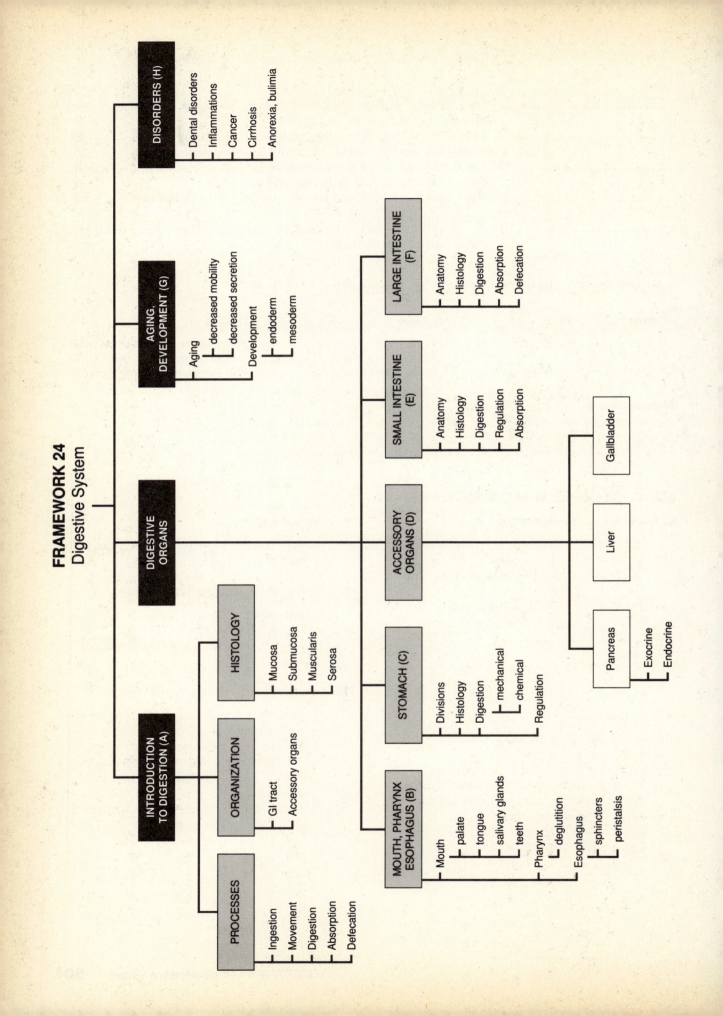

INTRODUCTION TO DIGESTION (A)

PROCESSES
- Ingestion
- Movement
- Digestion
- Absorption
- Defecation

ORGANIZATION
- GI tract
- Accessory organs

HISTOLOGY
- Mucosa
- Submucosa
- Muscularis
- Serosa

DIGESTIVE ORGANS

MOUTH, PHARYNX ESOPHAGUS (B)
- Mouth
  - palate
  - tongue
  - salivary glands
  - teeth
- Pharynx
  - deglutition
- Esophagus
  - sphincters
  - peristalsis

STOMACH (C)
- Divisions
- Histology
- Digestion
  - mechanical
  - chemical
- Regulation

ACCESSORY ORGANS (D)
- Pancreas
  - Exocrine
  - Endocrine
- Liver
- Gallbladder

SMALL INTESTINE (E)
- Anatomy
- Histology
- Digestion
- Regulation
- Absorption

LARGE INTESTINE (F)
- Anatomy
- Histology
- Digestion
- Absorption
- Defecation

AGING, DEVELOPMENT (G)
- Aging
  - decreased mobility
  - decreased secretion
- Development
  - endoderm
  - mesoderm

DISORDERS (H)
- Dental disorders
- Inflammations
- Cancer
- Cirrhosis
- Anorexia, bulimia

The Digestive System

Food comes in very large pieces (like whole oranges, stalks of broccoli, and slices of bread) that must fit into very small spaces in the human body (like liver cells or brain cells). The digestive system makes such change possible. Foods are minced and enzymatically degraded so that absorption into blood en route to cells becomes a reality. The gastrointestinal (GI) tract provides a passageway complete with mucous glands to help food slide along, muscles that propel food forward and muscles (sphincters and valves) that regulate flow, and secretions that break apart foods or modify pH to suit local enzymes. Accessory structures such as teeth, tongue, pancreas, and liver lie outside the GI tract, yet each contributes to the mechanical and chemical dismantling of food from bite-sized to right-sized pieces. The ultimate fate of the absorbed products of digestion is the story line of metabolism (in Chapter 25).

You will find food for thought in the Chapter 24 Framework and its terminology.

TOPIC OUTLINE AND OBJECTIVES

A. Introduction to digestion

1. Identify the organs of the gastrointestinal (GI) tract and the accessory organs of digestion and their functions in digestion.

B. Mouth, pharynx, and esophagus

C. Stomach

D. Accessory organs: pancreas, liver, gallbladder

E. Small intestine

2. Describe the mechanical movements of the gastrointestinal tract.
3. Explain how salivary secretion, gastric secretion, gastric emptying, pancreatic secretion, bile secretion, and small intestinal secretion are regulated.
4. Define absorption and explain how the end products of digestion are absorbed.

F. Large intestine

5. Define the processes involved in the formation of feces and defecation.

G. Aging and developmental anatomy

6. Describe the effects of aging on the digestive system.
7. Describe the development of the digestive system.

H. Disorders and medical terminology

8. Describe the clinical symptoms of the following disorders: dental caries, periodontal disease, peritonitis, peptic ulcer disease (PUD), appendicitis, gastrointestinal tumors, diverticulitis, cirrhosis, hepatitis, gallstones, anorexia nervosa, and bulimia.
9. Define medical terminology associated with the digestive system.

WORDBYTES

Now study the following parts of words that may help you better understand terminology in this chapter.

Wordbyte	Meaning	Example	Wordbyte	Meaning	Example
amyl-	starch	*amyl*ase	gingiv-	gums	*gingiv*itis
-ase	enzyme	malt*ase*	ileo-	ileum	*ileo*cecal valve
caec-, cec-	blind	*cec*um	jejun-	jejunum	*jejun*oileostomy
chole-, cholecyst-	gallbladder	*chole*cystectomy	lith-	stone	chole*lith*iasis
			or-	mouth	*or*al
chym-	juice	*chym*otrypsin	-ose	sugar	lact*ose*
crypt-	hidden	*crypt*s of Lieberkuhn	retro-	behind	*retro*peritoneal
			-rhea	to flow	diar*rhea*
dent-	tooth	*dent*ures	stoma-	mouth, opening	colo*stomy*
-ectomy	removal of	append*ectomy*			
entero-	intestine	*entero*endocrine	taen-	ribbon	*taen*ia coli
gastro-	stomach	*gastr*in, *gastr*ectomy	vermi-	worm	*vermi*form appendix

CHECKPOINTS

A. Introduction to digestion (pages 766–770)

A1. Explain why food is vital to life. Give three specific examples of uses of foods in the body. *food is source of energy that drives chemical reactions, provide matter to form new tissue, muscle contractions, conduct of nerve impulses*

■ **A2.** List the five basic activities of the digestive system.

ingestion   absorption   defecation   movement of food   digestion

■ **A3.** *Chemical* digestion occurs by action of enzymes (such as those in saliva) and intestinal secretion, whereas *mechanical* digestion involves action of the teeth and muscles of the stomach and intestinal wall.

■ **A4.** Cover the key and identify all digestive organs on Figure LG 24.1. Then color the structures with color code ovals; these five structures are all *(parts of the gastrointestinal tract?* accessory structures?).

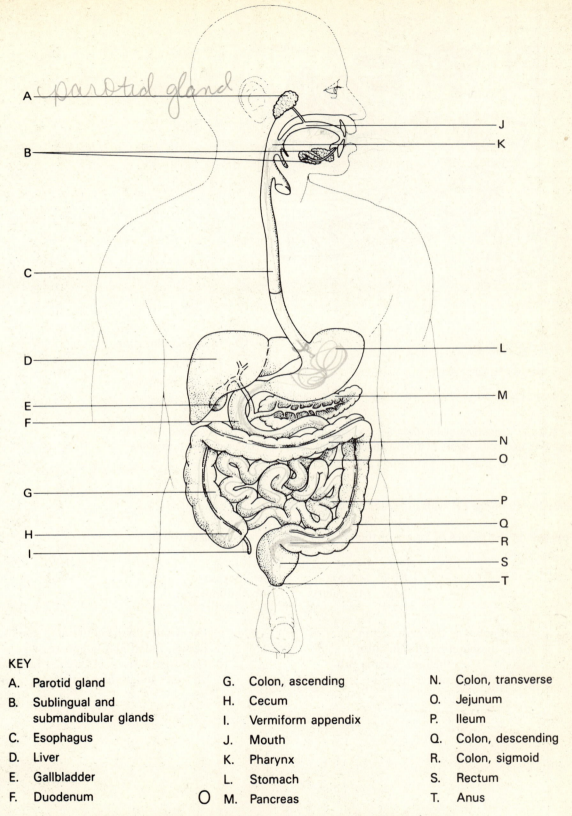

A ~~parotid gland~~

KEY

- A. Parotid gland
- B. Sublingual and submandibular glands
- C. Esophagus
- D. Liver
- E. Gallbladder
- F. Duodenum

- G. Colon, ascending
- H. Cecum
- I. Vermiform appendix
- J. Mouth
- K. Pharynx
- L. Stomach
- M. Pancreas

- N. Colon, transverse
- O. Jejunum
- P. Ileum
- Q. Colon, descending
- R. Colon, sigmoid
- S. Rectum
- T. Anus

Figure LG 24.1 Organs of the digestive system. Color, draw arrows, and identify labeled structures as directed in Checkpoints A4, B3, and F1.

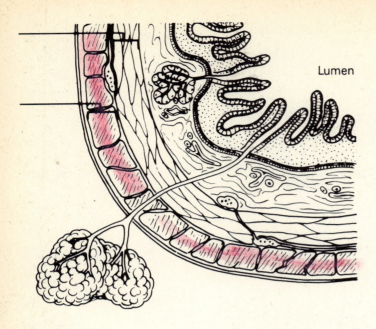

Lumen

KEY
○ Mucous membrane
○ Muscularis
○ Serous membrane
○ Submucosa layer

Figure LG 24.2 Gastrointestinal (GI) tract and related gland seen in cross section. Refer to Checkpoint A5 and color layers of the wall according to color code ovals. Label muscularis layers.

■ **A5.** Color layers of the GI wall indicated by color code ovals on Figure LG 24.2.

■ **A6.** Now match the names of layers of the GI wall with the correct description.

Muc. Mucosa	Ser. Serosa
Mus. Muscularis	Sub. Submucosa

Ser a. Also known as the peritoneum, it forms mesentery and omentum

Muc b. Consists of epithelium, lamina propria, and muscularis mucosa

Sub c. Connective tissue containing glands, nerves, blood and lymph vessels

Mus d. Consists of an inner circular layer and an outer longitudinal layer

■ **A7.** Match the names of these peritoneal extensions with the correct descriptions.

> F. Falciform ligament   M. Mesentery
> G. Greater omentum   Meso. Mesocolon
> L. Lesser omentum

___F___ a. Attaches liver to anterior abdominal wall

M ___M___ b. Binds intestine to posterior abdominal wall; provides route for blood and lymph vessels and nerves to reach small intestine

___Meso___ c. Binds part of large intestine to posterior abdominal wall

___G___ d. "Fatty apron"; covers and helps prevent infection in small intestine

___L___ e. Suspends stomach and duodenum from liver

A8. *A clinical challenge.* Define the following terms:

a. Ascites   peritoneal cavity becomes destended by several liters of serous fluid

b. Peritonitis

inflammation of the peritoneum

B. Mouth, pharynx, and esophagus   (pages 770–780)

■ **B1.** Refer to Figure LG 24.3 and do this activity.
a. Label the following structures:
   Hard palate   Palatine tonsils   Uvula of soft palate
   Lingual frenulum   Palatoglossal arch   Vestibule
b. Color teeth in lower right of figure as indicated by color code ovals.
c. Number teeth (1–8) on the upper right of the diagram according to times of eruptions. One (the first one to erupt) is done for you.

B2. Contrast terms in each pair:
a. Intrinsic muscles of the tongue/extrinsic muscles of the tongue   alter shape + size
   in. originate + insert inside tongue
   Ex. originate outside tongue insert into it
   side to side in + out
b. Fungiform papillae/filiform papillae
taste buds red dots
       white dots

B3. Identify the three salivary glands on Figure LG 24.1 and visualize their locations on yourself.
    parotid
    sublingual
    submandibular

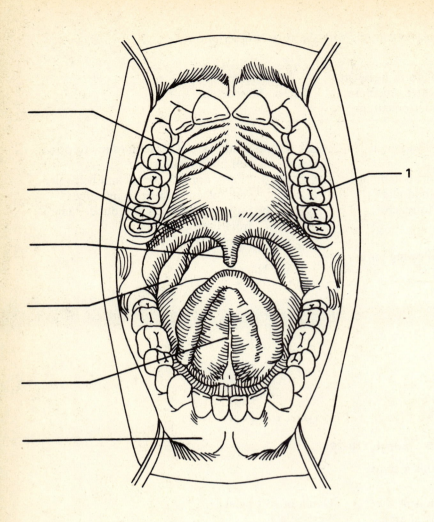

○ Cuspids	○ Molars
○ Incisors	○ Premolars (bicuspids)

Figure LG 24.3   Mouth (oral cavity). Complete as directed in Checkpoint B1.

■ **B4.** Complete this exercise about salivary glands.

a. Which glands are largest? *(Parotid? Sublingual? Submandibular?)*.

b. Which secrete the thickest secretion due to presence of much mucus?

____Sublingual____

c. About 1 to 1½ *(tablespoons? cups? liters?)* of saliva are secreted daily.

d. State three functions of saliva.

keeps mucus membranes moist
digestion

e. The pH of the mouth is appropriate for action of salivary amylase. This is about pH *(2? 6.5? 9?)*.

pH 6.35 – 6.85

■ **B5.** Arrange the following in correct sequence:

_____ _____ _____ a. From most superficial to deepest:

> A. Root   B. Neck   C. Crown

_____ _____ _____ b. From most superficial to deepest within a tooth:

> A. Enamel   B. Dentin   C. Pulp cavity

_____ _____ _____ c. From hardest to softest:

> A. Enamel   B. Dentin   C. Pulp cavity

■ **B6.** What is the function of the periodontal ligament?

■ **B7.** Salivary amylase is an enzyme that digests ___Starches___. Most of the starch ingested *(is? is not?)* broken down by the time food leaves the mouth. What inactivates amylase in the stomach?

Stomach acids

■ **B8.** The term *deglutition* means ___Swallowing___. Match each stage of deglutition with the correct description.

> E. Esophageal   P. Pharyngeal   V. Voluntary

___P___ a. Soft palate and epiglottis close off respiratory passageways.

___V___ b. Tongue pushes food back into oropharynx.

___E___ c. Peristaltic contractions push bolus from pharynx to stomach.

■ **B9.** Failure of the lower esophageal sphincter to close results in the sensation of ___heartburn___. Consequently, the esophageal lining may be irritated by *(acidic? basic?)* contents of the stomach that enter the esophagus. Failure of this sphincter to relax is a condition known as ___achalasia___. Resulting distension of the esophagus causes pain that may be confused with ___cardiac___ pain.

B10. Summarize digestion in the mouth, pharynx, and esophagus by completing parts a and b of Table 24.1.

HCl

Table LG 24.1 Summary of Design

Digestive Organs	Carbohydrate	Protein
a. Mouth, salivary glands	Salivary amylase: digests starch to maltose	
b. Pharynx, esophagus	*no enzyme*	*no enzyme*
c. Stomach		
d. Pancreas		
e. Intestinal juices		
f. Liver	No enzymes for digestion of carbohydrates	
g. Large intestine	No enzymes for digestion of carbohydrates	No enzymes for digestion of proteins

Lipid	Mechanical	Other Functions
no	Deglutition, peristalsis	
		1. Secretes intrinsic factor 2. Produces hormone stomach gastrin
Pancreatic lipase: digests about 80% of fats		
No enzymes for digestion of lipids		

C. Stomach (pages 780–787)

■ **C1.** Refer to Figure LG 24.1 and complete this Checkpoint.

a. Arrange these regions of the stomach according to the pathway of food from first to last:

body	fundus
cardia	pylorus

cardia → *fundus* → *body* → *pylorus*

b. Match the terms in the box with the correct description.

GC. Greater curvature	LC. Lesser curvature

LC More lateral and inferior in location   _____ Attached to lesser omentum

_____ Attached to greater omentum

■ **C2.** Complete this table about gastric secretions.

Name of Cell	Type of Secretion	Function of Secretion
a. Chief (zygomatic)	*Pepsinogen Gastric lipase*	
b. Mucous	*Mucus*	*protects stomach lining*
c. *Parietal*	HCl	
d. *G cells*	Gastrin	

■ **C3.** How does the structure of the stomach wall differ from that in other parts of the GI tract?

a. Mucosa

b. Muscularis

■ **C4.** Answer these questions about chemical digestion in the stomach.

a. Pepsin is most active at very (*acid?* *alkaline?*) pH.

b. State two factors that enable the stomach to digest protein without digesting its own cells (which are composed largely of protein).

layer of protective mucus 1-3mm
pepsinogen is secreted in an inactive
state, HCL or pepsin molecules activate

c. If mucus fails to protect the gastric lining, the condition known as

_____ *ulcer* _____ may result.

d. Another enzyme produced by the stomach is _____ *gastric lipase* _____, which

digests _____. In adults it is quite (*effective?* *ineffective?*). Why?

■ **C5.** Answer these questions about control of gastric secretion.

a. Name the three phases of gastric secretion: _____ *cephalic* _____,

_____ *gastric* _____, and _____ *intestinal* _____. Which of these causes

gastric secretion to begin when you smell or taste food? _____ *cephalic* _____.

b. Gastric glands are stimulated mainly by the _____ *vagus* _____ nerves.
Their fibers are (*sympathetic?* *parasympathetic?*). Two stimuli that cause vagal

impulses to stimulate gastric activity are _____ *sight* _____ and

_____ *smell* _____. Three emotions that inhibit production of gastric

secretions are _____ *fear* _____, _____ *anger* _____, and

_____ *anxiety* _____.

c. Distension of the stomach triggers the __gastric__ phase. Two effects of this stimulus are (increase? ~~decrease?~~) in parasympathetic impulses and

release of the hormone __gastrin__. What else triggers release of gastrin? __presence of partially digested proteins__

Gastrin travels through the blood to all parts of the body; its target cells are gastric glands and gastric smooth muscle, which are (stimulated? inhibited?), and the pyloric and ileocecal sphincters, which are (contracted? relaxed?)

d. *A clinical challenge.* Both vagal nerves and gastrin (stimulate? inhibit?) HCl production. Their effect is (greater? less?) in the presence of histamine, which is made by

__~~G cells~~ mast__ cells in the stomach wall. Describe the effect of H_2 blockers such as cimetidine (Tagamet) on the stomach. __decreased HCl production__

e. When food (chyme) reaches the intestine, nerves initiate the

__enterogastric__ reflex, which (stimulates? inhibits?) further gastric secretion. Three hormones released by the intestine also inhibit gastric secretion as well as gastric motility, so the effect of these hormones is to (stimulate? delay?) gastric

emptying. Name these three hormones: __GIP__,

__secretin__, and __CCK__.

- C6. Food stays in the stomach for about __4__ hours. Which food type leaves the

stomach most quickly? __Carbohydrate__ Which type stays in the stomach

longest? __triglycerides__

- C7. Answer these questions about vomiting (emesis).
  a. Identify the two strongest stimuli for vomiting __irritation & distention__

  b. Such stimuli are transmitted to the __medulla__, which is the site of the vomiting center. Nerve impulses then convey instructions to (contract? relax?) abdominal muscles and (contract? relax?) esophageal sphincters.
  c. Explain how prolonged vomiting can lead to serious disturbances in homeostasis.

  __loss of gastric juices & fluids can lead to disturbances in fluid + acid-base balance__

■ **C8.** The stomach is responsible for *(much? little?)* absorption of foods. What types of substances are absorbed by the stomach?

water, electrolytes, asprin + alcohol

C9. Complete part c of Table LG 24.1, describing the role of the stomach in digestion.

D. Accessory structures: pancreas, liver, gallbladder *(pages 787–797)*

D1. Refer to Figure LG 24.4 and color all of the structures with color code ovals according to colors indicated.

■ **D2.** Complete these statements about the pancreas.

a. The pancreas lies posterior to the *greater curvature of stomach*
b. The pancreas is shaped roughly like a fish, with its head in the curve of the *small intestine* and its tail nudging up next to the

_____.

c. The pancreas contains two kinds of glands. Ninety-nine percent of its cells produce *(endocrine? exocrine?)* secretions. One type of these secretions is *(acid? alkaline?)* fluid to neutralize the chyme entering from the stomach.
d. One enzyme in the pancreatic secretions is trypsin; it digests *(fats? carbohydrates? proteins?)* Trypsin is formed initially in the inactive form (trypsinogen) and is activated by *(HCl? NaHCO₃? enterokinase?)*. Trypsin itself serves as an activator in the

formation of the protease named *Carb* _____ and the peptidase

named _____.
e. Most of the amylase and lipase produced in the body are secreted by the pancreas. Describe the functions of these two enzymes.

f. All exocrine secretions of the pancreas empty into ducts (____ and ____ on Figure

LG 24.4). These empty into the _____.
g. *A clinical challenge.* In most persons the pancreatic duct also receives bile flowing

through the _____ (*F* on the figure). State one possible complication that may occur if a gallstone blocks the pancreatic duct (as at point *H* on the figure).

h. The endocrine portions of the pancreas are known as the _____

_____. List hormones they secrete. _____
Typical of all hormones, these pass into *(ducts? blood vessels?)*, specifically into

vessels that empty into the _____ vein.

D3. Fill in part d of Table LG 24.1, describing the role of the pancreas in digestion.

■ **D4.** Describe regulation of the pancreas in this exercise.

a. The pancreas is stimulated by *(sympathetic? parasympathetic?)* nerves, specifically

the _____ nerves, and also by hormones produced by cells

located in the wall of the _____.

b. The hormone cholecystokinin (CCK) activates the pancreas to secrete fluid rich in *(HCO$_3$⁻? digestive enzymes such as trypsin, amylase, and lipase?)*. What is the nature of the pancreatic fluid produced under the influence of secretin?

■ **D5.** Answer these questions about the liver.

a. This organ weighs about ~~1.4~~ kg (**3** pounds). It lies in the

__UPPER RT__ quadrant of the abdomen. Of its __2__ lobes, the

__Rt__ is the largest.

b. The __falciform__ ligament separates the right and left lobes. In the edge of this ligament is the ligamentum teres (round ligament of the liver), which is

the obliterated __umbilical__ vein.

c. Blood enters the liver via vessels named __hepatic portal vein__ and __hepatic artery__. In liver lobules, blood mixes in channels called __sinusoids__ before leaving the liver via vessels named __hepatic veins__ __central vein__

■ **D6.** Complete Figure LG 24.4 as directed.

a. Identify the pathway of bile from liver to intestine by following the structures you colored green in that figure: C D F H. The general direction of bile is from *(superior to inferior? inferior to superior?)*.

b. Now draw arrows to show the direction of blood through the liver. In order to reach the inferior vena cava, blood must flow *(superiorly? inferiorly?)* through the liver.

PORTAL TRIAD
H. ARTERY
H. PORTAL VE
BILE DUC

produces bile for digest
stores iron
breaks down proteins

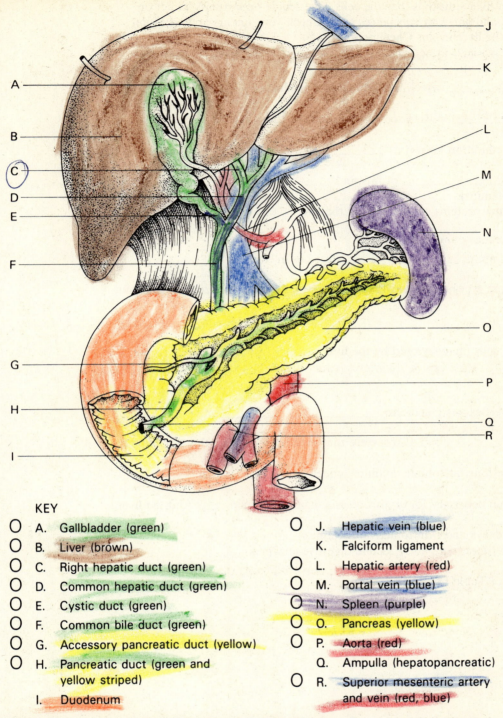

KEY

○ A. Gallbladder (green)
○ B. Liver (brown)
○ C. Right hepatic duct (green)
○ D. Common hepatic duct (green)
○ E. Cystic duct (green)
○ F. Common bile duct (green)
○ G. Accessory pancreatic duct (yellow)
○ H. Pancreatic duct (green and yellow striped)
  I. Duodenum

○ J. Hepatic vein (blue)
  K. Falciform ligament
○ L. Hepatic artery (red)
○ M. Portal vein (blue)
○ N. Spleen (purple)
○ O. Pancreas (yellow)
○ P. Aorta (red)
  Q. Ampulla (hepatopancreatic)
○ R. Superior mesenteric artery and vein (red, blue)

Figure LG 24.4 Liver, gallbladder, pancreas, and duodenum, with associated blood vessels and ducts. (Stomach has been removed). Refer to Checkpoints D1, D2, D6, and D9; color and draw arrows as directed.

■ **D7.** While blood is in liver sinusoids, hepatic cells have ample opportunity to act on this blood, modify it, and add new substances to it. Check your understanding of important functions of the liver in this Checkpoint.

a. Bile secreted from the liver is composed largely of the pigment named

_____, which is a breakdown product of

_____ cells. Excessive amounts of this pigment give skin a

yellowish color, a condition known as _____. Two functions of

bile are _____ of fats and _____ of fats
(and fat-soluble vitamins).

b. Name several plasma proteins synthesized by the liver.

c. Identify two or more types of chemicals that are detoxified (metabolized) by the liver.

d. List three types of cells phagocytosed by the liver.

e. The liver can convert excess glucose to _____ and

_____, and also reverse that process. (More on this in Chapter 25.)

f. The liver stores the four fat-soluble vitamins named ____, ____, ____, and ____. It

also stores a vitamin necessary for erythropoiesis, namely vitamin ____. And the liver

cooperates with kidneys and skin to activate vitamin ____.

g. Note that the liver *(does? does not?)* secrete digestive enzymes.

D8. Complete part f of Table LG 24.1

■ **D9.** Imagine a terrific one-act play that you might attend on an upcoming Saturday night. The cast of characters includes all chemicals and organs identified by CAPITAL letters. The stage is set with organs/characters in position much as in Figure LG 24.4.

a. CHYME strolls in from the stomach (located offstage) and moves into the duodenum. Note that the doorway through which CHYME entered must therefore be the

_____ sphincter. CHYME today is particularly rich in food

types such as _____ (which are plastered all over CHYME's costume).

b. The presence of CHYME bearing these foods serves as a stimulus to the walls of the

_____ to secrete the hormone CCK, whose full name is

_____. (CCK enters and waves.)

c. CCK then exits from the intestinal area by way of *(common bile duct? portal vein?).* CCK *(goes directly to the gallbladder without passing any other organ? meanders in the bloodstream through all parts of the body?).* (CCK now takes a scenic trip through the blood vessels of arms, legs, ears, heart, as desired.)

d. At last CCK arrives at the GALLBLADDER, "knowing" to attach there since GALLBLADDER has specific receptors for CCK. CCK directs GALLBLADDER to *(contract? relax?).*

e. GALLBLADDER harbors many molecules of BILE SALTS. Contraction of GALL-BLADDER expels BILE SALTS. These pass into the *(portal vein? cystic duct?)* and via the common bile duct to the duodenum.

f. A GALLSTONE (biliary calculus) becomes lodged in the bile pathway, such as the common bile duct. If scenes a–d of this play are repeated, how will scene e differ?

g. A cholecystectomy may need to be performed. If so, which player must terminate

his/her role in the cast? _____

h. In addition to the bile-related activities already performed by one of our stars, CCK, what other functions might CCK also portray in upcoming productions?

E. Small intestine (pages 797–806)

■ **E1.** Describe the structure of the intestine in this Checkpoint.

a. The average diameter of the small intestine is:
  A. 2.5 cm (1 inch)   B. 5.0 cm (2 inches)

b. Its average length is about:
  A. 3 m (10 feet)   B. 6 m (20 feet)

c. Name its three main parts (in sequence from first to last):

_____ → _____ → _____

■ **E2.** Match the correct term related to the intestine to the description that fits.

DBG. Duodenal (Brunner's) glands	**PP.** Peyer's patches
MV. Microvilli (brush border)	**V.** Villi
PC. Plicae circulares	

_____ a. Fingerlike projections up to 1 mm high that give the intestinal lining a velvety appearance and increase absorptive surface

_____ b. Fingerlike projections of plasma membrane

_____ c. Circular folds that further increase intestinal surface area

_____ d. Aggregated lymphatic follicles located primarily in the wall of the ileum

_____ e. Submucosal glands that secrete protective alkaline fluids

E3. Name four or more brush border enzymes.

Explain why this name is given to these enzymes.

Describe the action of one of these enzymes, α-*dextrinase*.

E4. Contrast segmentation and peristalsis, the two types of movement of the small intestine.

■ **E5.** Write the main steps in the digestion of each of the three major food types.
a. *Carbohydrates*

Polysaccharides → _disaccharides_ → monosaccharides
b. *Proteins*

Proteins → _poly peptides_ → _dipep_ → _amino_
c. *Lipids*

Neutral fats → emulsified fats → _fatty acid_ _glycerol_

■ **E6.** *A clinical challenge.* Persons with lactose intolerance lack the enzyme

_____, which is necessary for digestion of lactose found in foods

such as _____. Write a rationale for the usual symptoms of this
condition.

E7. Complete the description of functions of intestinal juices by filling in part e of Table LG 24.1. Then review digestion of each of the three major food groups by reading your columns vertically.

■ **E8.** *For extra review* of roles of GI organs in digestion, identify which chemicals in the list in the box are made by each of the following organs.

Alb. Albumin	HCl. Hydrochloric acid
Amy. Amylase	IF. Intrinsic factor
B. Bile	Ins. Insulin
CCK. Cholecystokinin	KB. Ketone bodies
Chol. Cholesterol	L. Lipase
E. Enterokinase	M. Mucus
FP. Fibrinogen and prothrombin	MLSD. Maltase, lipase, sucrase, α-dextrinase
Gas. Gastrin	N. Nucleosidases
GIP. Gastric inhibitory peptide	P. Pepsinogen
Glu. Glucagon	S. Secretin
Gly. Glycogen	TCP. Trypsinogen, chymotrypsinogen, procarboxypeptidase

m Amy lipase a. Salivary glands and tongue (three answers)

_____ m _____ b. Pharynx and esophagus (one answer)

Gas, HCl, IF c. Stomach (six answers)

_____ d. Pancreas (six answers)

_____ e. Small intestine (seven answers)

alb, _____ f. Liver (six answers)

■ **E9.** *For extra review* of secretions involved with digestion, match names of secretions listed for Checkpoint E8 with descriptions below. Blank lines follow some descriptions. On these indicate whether the secretion is classified as an enzyme (E), hormone (H), or neither of these (N). The first one is done for you.

___CCK___ a. Causes contraction of gallbladder and relaxation of the sphincter of the hepatopancreatic ampulla so that bile enters duodenum __H__

_____ b. Stimulates production of alkaline pancreatic fluids and bile ____

_____ c. Stimulates pancreas to produce secretions rich in enzymes such as lipase and amylase ____

_____ d. Promote growth and maintenance of pancreas; enhance effects of each other (two answers)

_____ e. Increases gastric activity (secretion, motility, and growth) ____

_____ f. Inhibit stomach motility and/or secretions (three answers)

_____ g. Precursors to protein-digesting enzymes (two answers)

_____ h. Activates pepsinogen ____

_____ i. The active form of this enzyme starts protein digestion in the GI tract

_____ j. Activates the inactive precursor to form trypsin ____

_____ k. Plasma proteins

_____ l. A storage form of carbohydrate

_____ m. Most effective of this type of enzyme produced by the pancreas; digests fats

_____ n. Emulsifies fats before they can be digested effectively ____

_____ o. Starch-digesting enzymes secreted by salivary glands and pancreas

_____ p. Intestinal enzymes that complete carbohydrate breakdown, resulting in simple sugars

_____ q. Digest DNA and RNA ____

_____ r. Brush border enzymes that carry out digestion on the surface of villi (two answers)

until in sm. intestine

inactive form

CHO ← amylase
fat ← lipase
protein ← chymo-trypsin / trypsin

■ **E10.** Describe the absorption of end products of digestion in this exercise.

a. Almost all absorption takes place in the *(large? small?)* intestine. Glucose and some amino acids move into epithelial cells lining the intestine by *(primary? secondary?)*

active transport coupled with active transport of ____. Glucose and other monosaccharides are transported into capillaries by the process of

_____.

b. Simple sugars, amino acids, and short-chain fatty acids are absorbed into *(blood?*

lymph?) capillaries located in _____ in the intestinal wall.

These vessels lead to the _____ vein and then to the

_____ for storage or metabolism.

c. Long-chain fatty acids and monoglycerides first combine with

_____ salts to form *(micelles? chylomicrons?)*. This enables fatty acids and monoglycerides to enter epithelial cells in the intestinal lining and soon enter lacteals leading to the *(portal vein? thoracic duct?)*. Most bile salts are ultimately *(eliminated in feces? recycled to the liver?)*.

d. Aggregates of fats coated with _protein_ are known as chylomicrons. After traveling through lymph and blood, they reach the

_____ where they undergo conversions to form

APO LIPO -proteins such as LDLs, VLDLs or

HDL. Name three types of fats found in lipoproteins.

e. *A clinical challenge.* Which vitamins are absorbed with the help of bile? *(Water-soluble? Fat-soluble?)* Circle the fat-soluble vitamins. A B₁₂ C D E K Obstruction of bile pathways may lead to symptoms related to deficiencies of fat-soluble vitamins. Name two. (For help, refer to Exhibit 25.6, pages 850–851 of the text.)

f. About _9_ liters (or quarts) of fluids are ingested or secreted into the GI tract each day. Of this, all but about 0.5 to 1.0 liter is reabsorbed into blood capillaries in the walls of the *(small? large?)* intestine. Several hundred milliliters of fluid are also reabsorbed each day into the *(small? large?)* intestine.

g. Normally about _•/_ ml of fluid exits each day in feces. When inadequate water reabsorption occurs, as in *(constipation? diarrhea?)*, then

_____ such as Na⁺ or Cl⁻ are also lost.

F. Large intestine   (pages 806–812)

F1. Identify the regions of the large intestine in Figure LG 24.1. Draw arrows to indicate direction of movement of intestinal contents.

■ **F2.** The total length of the large intestine is about ____ m (*5* feet). More than 90 percent of its length consists of the part known as the _____.

■ **F3.** Match each structure in the large intestine with the related description.

> AC. Anal canal   IS. Ileocecal sphincter (valve)
> HF. Hepatic flexure   VA. Vermiform appendix

_____ a. Valve between small and large intestine

_____ b. Blind-ended tube attached to cecum

_____ c. Portion of colon located between ascending and transverse colons; also called right colic flexure

_____ d. Terminal 3 cm (1 inch) of rectum

■ **F4.** Contrast different portions of the GI tract by identifying structures or functions associated with each. Use these answers:

> LI. Large intestine   Sto. Stomach
> SI. Small intestine

LI a. Has thickened bands of longitudinal muscle known as taeniae coli.

_____ b. These pouches give this structure its puckered appearance.

_____ c. Its fat-filled peritoneal attachments are known as epiploic appendages.

LI d. Bacteria here decompose bilirubin to urobilinogen, which gives feces its brown color.

Sto LI e. Involved in the gastrocolic reflex (two answers).

Sto f. Has rugae.

SI g. Has villi and microvilli.

■ **F5.** *A clinical challenge.* Certain groups of individuals lack the normal flora of bacteria in the intestine. Name two such groups.

Name one sign or symptom resulting from inadequate bacterial synthesis of vitamin K.

_____

F6. List the chemical components of feces.

F7. Describe the process of defecation. Include these terms: *stretch receptors, rectal muscles, sphincters, diaphragm,* and *abdominal muscles.*

■ **F8.** Describe helpful effects of dietary fiber in this activity.
a. (*Soluble? Insoluble?*) fibers tend to speed up passage of digestive wastes. Potential health benefits include lowering risk for a number of disorders, including

hemorrhoids (enlarged and inflamed rectal veins),

constipation, (hard, dry stool), and colon cancer. Circle the two best sources of insoluble fiber in the following list.

apples   broccoli   oats and oat bran   vegetable skins   wheat bran

b. Write one health benefit of soluble fiber.

Write three sources of soluble fiber from the list above.

F9. Complete part g of Table 24.1, describing the role of the large intestine in digestion.

G. Aging and developmental anatomy of the digestive system   (pages 812–813)

■ **G1.** *A clinical challenge.* List physical changes with aging that may lead to a decreased desire to eat among the elderly population.
a. Related to the upper GI tract (to stomach)

b. Related to the lower GI tract (stomach and beyond)

■ **G2.** Indicate which germ layer, endoderm (E) or mesoderm (M), gives rise to each of these structures.

_____ a. Epithelial lining and digestive glands of the GI tract

_____ b. Liver, gallbladder, and pancreas

_____ c. Muscularis layer and connective tissue of submucosa

G3. List GI structures derived from each portion of the primitive gut.
a. Foregut

b. Midgut

c. Hindgut

H. Disorders, medical terminology (pages 814–816)

H1. Explain how tooth decay occurs. Include the roles of bacteria, dextran, plaque, and acid. List the most effective known measures for preventing dental caries.

H2. Briefly describe these disorders, stating possible causes of each.
a. Periodontal disease

b. Appendicitis

c. Cirrhosis

■ **H3.** Explain why a patient may be given large doses of antibiotics prior to abdominal surgery.

■ **H4.** Which of the following is an inflammation? *(Diverticulosis? Diverticulitis?)*

■ **H5.** Match the terms with the descriptions.

A. Anorexia nervosa	Dys. Dysphagia
B. Bulimia	F. Flatus
Cho. Cholecystitis	Hem. Hemorrhoids
Colit. Colitis	Hep. Hepatitis
Colos. Colostomy	Htb. Heartburn
Con. Constipation	P. Peptic ulcer
Dia. Diarrhea	

Colos a. Incision of the colon, creating artificial anus

Hep b. Inflammation of the liver

Colitis c. Inflammation of the colon

Htb d. Burning sensation in region of esophagus and stomach; probably due to gastric contents in lower esophagus

Dia e. Frequent defecation of liquid feces

Cho f. Inflammation of the gallbladder

Con g. Infrequent or difficult defecation

P h. Craterlike lesion in the GI tract due to acidic gastric juices

F i. Excess air (gas) in stomach or intestine, usually expelled through anus

B j. Binge-purge syndrome

A k. Loss of appetite and self-imposed starvation

Dys l. Difficulty in swallowing

ANSWERS TO SELECTED CHECKPOINTS

A2. Ingestion, movement, digestion, absorption, and defecation.

A3. Chemical, mechanical.

A4. Accessory structures.

A5.

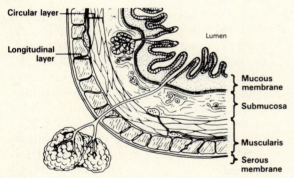

Figure LG 24.2A Gastrointestinal (GI) tract and related gland seen in cross section.

A6. (a) Ser. (b) Muc. (c) Sub. (d) Mus.

A7. (a) F. (b) M. (c) Meso. (d) G. (e) L.

B1.

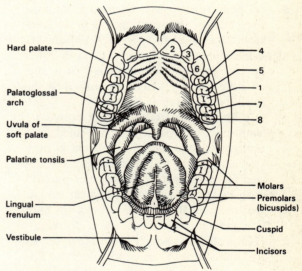

Figure LG 24.3A Mouth (oral cavity).

B4. (a) Parotid. (b) Sublingual. (c) Liters. (d) Dissolving medium for foods, lubrication, waste removal medium, and source of lysozyme, lingual lipase salivary amylase. (e) 6.5.

B5. (a) C B A. (b) A B C. (c) A B C.
B6. Anchors tooth to bone and acts as a shock absorber.
B7. Starch, is not; acidic pH of stomach.
B8. Swallowing. (a) P. (b) V. (c) E.
B9. Heartburn; acidic; achalasia; heart.
C1. (a) Cardia → fundus → body → pylorus. (b) GC; GC; LC.
C2.

Name of Cell	Type of Secretion	Function of Secretion
a. Chief (zygomatic)	Pepsinogen Gastric lipase	Precursor of pepsin
b. Mucous	Mucus	Protects gastric lining from acid and pepsin
c. Parietal (oxyntic)	HCl Intrinsic factor	Activates pepsinogen to pepsin. Facilitates absorption of vitamin B_{12}
d. G cells	Gastrin	Stimulates secretion of HCl and pepsinogen; contracts lower esophageal sphincter, increases gastric motility and relaxes pyloric sphincter

C3. (a) Arranged in rugae. (b) It has three, rather than two, layers of smooth muscle; the extra one is an oblique layer located inside the circular layer.
C4. (a) Acid. (b) Pepsin is released in the inactive state (pepsinogen); mucus protects the stomach lining from pepsin. (c) Ulcer. (d) Gastric lipase; emulsified fats, as in butter; ineffective; its optimum pH is 5 or 6, and most fats have yet to be emulsified (by bile from liver).
C5. (a) Cephalic, gastric, intestinal; cephalic. (b) Vagus; parasympathetic; sight, smell, or thought of food; presence of food in the stomach; anger, fear, and anxiety. (c) Gastric; increase; gastrin; presence of partially digested proteins, alcohol, and caffeine in the stomach; stimulated, relaxed. (d) Stimulate; greater, mast; H_2 blockers decrease HCl production which could otherwise lead to "acid stomach" or peptic ulcer. (e) Enterogastric, inhibits; delay; GIP, secretin, and CCK.
C6. Two to six; carbohydrate; fats (triglycerides).
C7. (a) Irritation and distension of the stomach. (b) Medulla; contract, relax. (c) Loss of electrolytes (such as HCl) as well as fluids can lead to fluid, electrolyte, and acid-base imbalances.

C8. Little; water, electrolytes, certain drugs (such as aspirin), and alcohol.
D2. (a) Stomach. (b) Duodenum, spleen. (c) Exocrine; alkaline. (d) Proteins; enterokinase; chymotrypsin, carboxypeptidase. (e) Amylase digest carbohydrates, including starch; lipase digests lipids. (f) G, H; duodenum. (g) Common bile duct; pancreatic proteases such as trypsin, chymotrypsin, and carboxypeptidase may digest tissue proteins of the pancreas itself. (h) Pancreatic islets (of Langerhans); insulin, glucagon, somatostatin, and pancreatic polypeptide; blood vessels, portal.
D4. (a) Parasympathetic, vagus, small intestine. (b) Digestive enzymes such as trypsin, amylase, and lipase; alkaline (HCO_3^-) fluid.
D5. (a) 1.4, 3.0; upper right; 4, right. (b) Falciform; umbilical. (c) Hepatic artery, portal vein; sinusoids, hepatic veins.
D6. (a) Superior to inferior. (b) Draw arrows from L and M toward B and then J; superiorly.
D7. (a) Bilirubin, red blood; jaundice; emulsification, absorption. (b) Clotting proteins (prothrombin and fibrinogen), albumin, and globulins. (c) Products of protein digestion, such as ammonia; medications such as sulfa drugs or penicillin; and steroid hormones. (d) Worn-out RBCs, WBCs, and some bacteria. (e) Glycogen and fat. (f) A, D, E, K; B_{12}, D. (g) Does not.
D9. (a) Pyloric; fats and partially digested proteins. (b) Small intestine, cholecystokinin. (c) Portal vein; meanders in the bloodstream through all parts of the body. (d) Contract. (e) Cystic duct. (f) Obstruction by the stone backs up bile, leading to spasms of smooth muscle (biliary colic) in bile pathways and entrance of new player: INTENSE PAIN. (g) GALLBLADDER. (h) Examples: stimulation of pancreatic secretions rich in digestive juices, movement of bile from gallbladder into duodenum, inducing a feeling of fullness (satiety) and inhibiting gastric emptying.
E1. (a) A. (b) B. (c) Duodenum → jejunum → ileum.
E2. (a) V. (b) MV. (c) PC. (d) PP. (e) DBG.
E5. (a) Shorter-chain polysaccharides and in some cases disaccharides. (b) Shorter-chain polypeptides → dipeptides → amino acids. (c) Fatty acids and glycerol.

E6. Lactase, milk, and other dairy products; gas (flatus) and bloating since bacteria ferment the undigested lactose present in the GI tract.

E8. (a) Amy, L, M. (b) M. (c) Gas, HCl, IF, L, M, P. (d) Amy, Glu, Ins, L, N, TCP. (e) CCK, E, GIP, M, MLSD, N, S. (f) Alb, B, Chol, FP, Gly, KB.

E9. (a) CCK, H. (b) S, H. (b) CCK, H. (d) CCK and S. (e) G, H. (f) CCK, GIP, and S. (g) P, TCP. (h) HCl, N. (i) P. (j) E. (k) Alb and FP. (l) Gly. (m) L. (n) B, N. (o) Amy. (p) MLSD. (q) N, E. (r) MLSD and N.

E10. (a) Small; secondary, Na⁺; facilitated diffusion. (b) Blood, villi; portal, liver. (c) Bile, micelles; thoracic duct; recycled to the liver. (d) Protein; liver, lipo, HDLs; triglycerides, phospholipids, and cholesterol. (e) Fat-soluble; A D E K; examples, A: night blindness and dry skin; D: rickets or osteomalacia due to decreased calcium absorption; K: excessive bleeding. (f) 9; small; large. (g) 100–200; diarrhea, electrolytes.

F2. 1.5 (5.0); colon.
F3. (a) IS. (b) VA. (c) HF. (d) AC.
F4. (a–d) LI. (e) Sto, LI. (f) Sto. (g) Sl.
F5. Newborns, persons who have been on long-term antibiotic therapy; excessive bleeding.
F8. (a) Insoluble; hemorrhoids (or "piles"), constipation; vegetable skins and wheat bran. (b) Lowers blood cholesterol level; apples, broccoli, oats, and oat bran.
G1. (a) Decreased taste sensations, gum inflammation (pyorrhea) leading to loss of teeth and loose-fitting dentures, and difficulty swallowing (dysphagia). (b) Decreased muscle tone and neuromuscular feedback may lead to constipation.
G2. (a) E. (b) E. (c) M.
H3. To kill intestinal bacteria and reduce the risk of peritonitis.
H4. Diverticulitis.
H5. (a) Colos. (b) Hep. (c) Colit. (d) Htb. (e) Dia. (f) Cho. (g) Con. (h) P. (i) F. (j) B. (k) A. (l) Dys.

MASTERY TEST: Chapter 24

Questions 1–10: Circle the letter preceding the one best answer to each question.

1. Which of these organs is not part of the GI tract, but is an accessory organ?
   A. Mouth   D. Small intestine
   B. Pancreas   E. Esophagus
   C. Stomach

2. The main function of salivary and pancreatic amylase is to:
   A. Lubricate foods
   B. Help absorb fats
   C. Digest polysaccharides to smaller carbohydrates
   D. Digest disaccharides to monosaccharides
   E. Digest polypeptides to amino acids

3. Which enzyme is most effective at pH 1 or 2?
   A. Gastric lipase   D. Salivary amylase
   B. Maltase   E. Pancreatic
   C. Pepsin   amylase

4. Choose the *true* statement about fats.
   A. They are digested mostly in the stomach.
   B. They are the type of food that stays in the stomach the shortest length of time.
   C. They stimulate release of gastrin.
   D. They are emulsified and absorbed with the help of bile.

5. Which of the following is a hormone, not an enzyme?
   A. Gastric lipase   C. Pepsin
   B. Gastrin   D. Trypsin

6. Which of the following is under only nervous (not hormonal) control?
   A. Salivation
   B. Gastric secretion
   C. Intestinal secretion
   D. Pancreatic secretion

7. All of the following are enzymes involved in protein digestion *except:*
   A. Amylase
   B. Trypsin
   C. Carboxypeptidase
   D. Pepsin
   E. Chymotrypsin

8. All of the following chemicals are produced by the walls of the small intestine *except:*
   A. Lactase   D. Trypsin
   B. Secretin   E. Brush border
   C. CCK   enzymes
9. Which type of movement is used primarily to propel chyme through the intestinal tract, rather than to mix chyme and enzymes.
   A. Peristalsis
   B. Rhythmic segmentation
   C. Haustral churning

10. Choose the *false* statement about layers of the wall of the GI tract.
    A. Most large blood and lymph vessels are located in the submucosa.
    B. The mysenteric plexus is part of the muscularis layer.
    C. Most glandular tissue is located in the layer known as the mucosa.
    D. The mucosa layer forms the peritoneum.

Questions 11–15: Arrange the answers in correct sequence.

_____ _____ _____ 11. From anterior to posterior:
   A. Palatoglossal arch
   B. Palatopharyngeal arch
   C. Palatine tonsils

_____ _____ _____ _____ 12. GI tract wall, from deepest to most superficial:
   A. Mucosa   C. Serosa
   B. Muscularis   D. Submucosa

_____ _____ _____ _____ _____ 13. Pathway of chyme:
   A. Ileum   D. Duodenum
   B. Jejunum   E. Pylorus
   C. Cecum

_____ _____ _____ _____ _____ 14. Pathway of bile:
   A. Bile canaliculi
   B. Common bile duct
   C. Common hepatic duct
   D. Right and left hepatic ducts
   E. Hepatopancreatic ampulla and duodenum

_____ _____ _____ _____ _____ 15. Pathway of wastes:
   A. Ascending colon   D. Descending colon
   B. Transverse colon   E. Rectum
   C. Sigmoid colon

Questions 16–20: Circle T (true) or F (false). If the statement is false, change the underlined word or phrase so that the statement is correct.

T   F   16. In general, the sympathetic nervous system <u>stimulates</u> salivation and secretions of the gastric and intestinal glands.

T   F   17. The principal chemical activity of the stomach is to begin digestion of <u>protein.</u>

T   F   18. The esophagus produces <u>no digestive enzymes or mucus.</u>

T   F   19. Cirrhosis and hepatitis are diseases of the <u>liver.</u>

T   F   20. The greater omentum <u>connects the stomach to the liver.</u>

Questions 21–25: fill-ins. Complete each sentence with the word or phrase that best fits.

_____ 21. Stomach motility is ____-creased by the enterogastric reflex, CCK, and GIP, and ____-creased by stomach gastrin and parasympathetic nerves.

_____ 22. Mumps involves inflammation of the ____ salivary glands.

_____ 23. Most absorption takes place in the ____, although some substances, such as ____, are absorbed in the stomach.

_____ 24. Mastication is a term that means ____.

_____ 25. Epithelial lining of the GI tract, as well as the liver and pancreas, are derived from ____-derm.

ANSWERS TO MASTERY TEST: ■ Chapter 24

Multiple Choice

1. B	6. A
2. C	7. A
3. C	8. D
4. D	9. A
5. B	10. D

Arrange

11. A C B
12. A D B C
13. E D B A C
14. A D C B E
15. A B D C E

True-False

16. F. Inhibits
17. T
18. F. No digestive enzymes, but it does produce mucus
19. T
20. F. Connects the stomach and duodenum to the transverse colon and drapes over the transverse colon and coils of the small intestine.

Fill-ins

21. De, in
22. Parotid
23. Small intestine; water, electrolytes, alcohol, and some drugs such as aspirin
24. Chewing
25. Endo

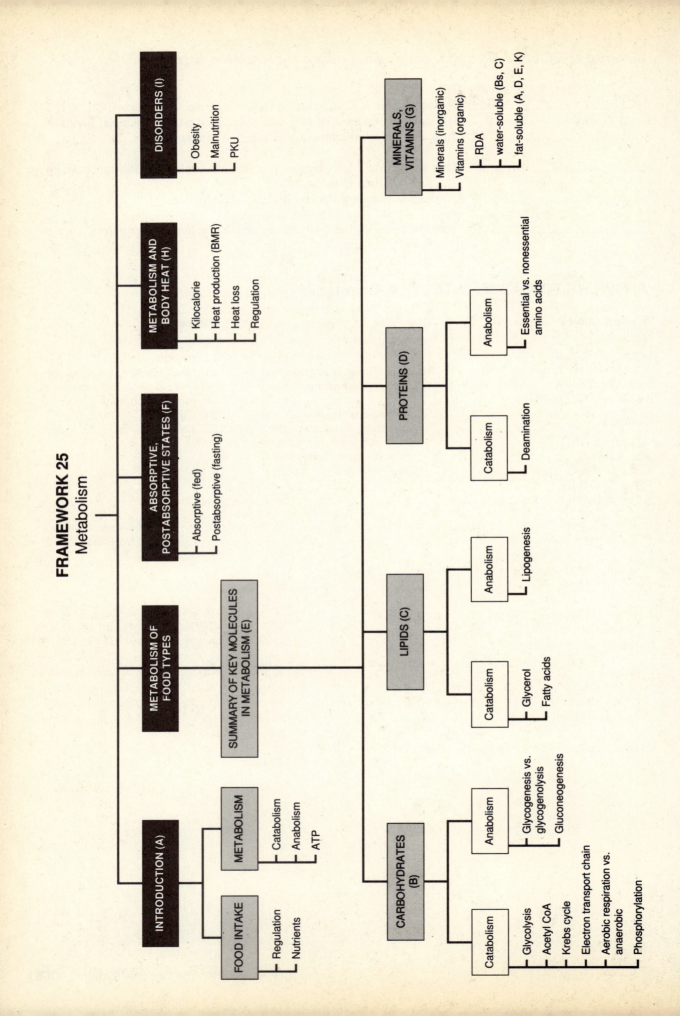

FRAMEWORK 25
Metabolism

Metabolism

Ingested foods, once digested, absorbed, and delivered, are used by body cells. This array of biochemical reactions within cells is known as metabolism. Products of digestion may be reassembled, as in formation of human protein from the amino acids of meat or beans, or in synthesis of fats from ingested oils. These reactions are examples of anabolism. The flip side of metabolic currency is catabolism, as in the breakdown of complex carbohydrates stored in liver or muscle to provide simple sugars for quick energy. Catabolism releases energy that maintains body heat and provides ATP to fuel activity. Vitamins and minerals play key roles in metabolism—for example, in enzyme synthesis and function.

The Chapter 25 Framework provides an organizational preview of the metabolism of the major food groups. Study the key terms and concepts it presents.

TOPIC OUTLINE AND OBJECTIVES

A. Introduction to metabolism

1. Explain how food intake is regulated.
2. Define a nutrient and list the functions of the six principal classes of nutrients.
3. Define metabolism and explain the role of ATP in anabolism and catabolism.

B. Carbohydrate metabolism

C. Lipid metabolism

D. Protein metabolism

4. Describe the metabolism of carbohydrates, lipids, and proteins.

E. Summary of key molecules of metabolism

F. Absorptive and postabsorptive states

5. Distinguish between the absorptive (fed) and postabsorptive (fasting) states.

G. Minerals, vitamins

6. Compare the sources, functions, and importance of minerals and vitamins in metabolism.

H. Metabolism and body heat

7. Describe the various mechanisms involved in heat production and heat loss.
8. Define basal metabolic rate (BMR) and explain several factors that affect it.
9. Explain how normal body temperature is maintained and describe fever, heat cramp, heatstroke, heat exhaustion, and hypothermia as abnormalities of temperature regulation.

I. Disorders

10. Define obesity, vitamin or mineral overdose, malnutrition, phenylketonuria (PKU), and celiac disease.

WORDBYTES

Now study the following parts of words that may help you better understand terminology in this chapter.

Wordbyte	Meaning	Example	Wordbyte	Meaning	Example
ana-	up	*ana*bolism	-gen	to form	gluconeo*gen*esis
calor-	heat	*Calor*ie	neo-	new	gluco*neo*genesis
cata-	down	*cata*bolism	-lysis	destruction	hydro*lysis*
de-	remove, from	*de*amination			

CHECKPOINTS

A. Introduction to metabolism (pages 822–825)

■ **A1.** Complete this Checkpoint about regulation of food intake.

a. The site within the brain that is the location of feeding and satiety centers is the *(medulla? hypothalamus?)*. When the satiety center is active, an individual will feel *(hungry? satiated or full?)*.

b. The feeding center is constantly active except when it is inhibited by the

_____ center. Circle all of the answers that will keep the feeding center active and enhance the desire to eat.

↓ Glucose in blood
↓ Amino acids in blood
↓ Fats entering intestine (with related ↓ in CCK release)
↓ Environmental temperature
↓ Stretching of stomach

c. Name the "satiety hormone." _____ Explain the reason for this name.

A2. List three primary functions of nutrients.

■ **A3.** List the six principal classes of nutrients.

A4. Think of the last time you had a drink of water today. List five functions of that water.

A5. Explain why metabolism might be thought of as an "energy-balancing act."

A6. Complete this table comparing catabolism with anabolism.

Process	Definition	Releases or Uses Energy	Examples
a. Catabolism			
b. Anabolism			

A7. ATP stands for a_____ t_____

p_____. Explain how it functions as the "energy currency" (or "money machine cash") of the cell. (For help, refer to Checkpoint D18, page 34 of the Learning Guide.)

■ **A8.** Complete this overview of catabolism (energy production).

a. Organic nutrients such as glucose are rich in hydrogen; energy is contained within the C-H bonds. In catabolism, much of the energy in glucose is ultimately released

and stored in the high-energy molecule named ____. Write the chemical formulas to show this overall catabolic conversion:

$$\underline{\hspace{3cm}} \rightarrow \underline{\hspace{4cm}} + \underline{\hspace{2cm}} + \text{energy stored in ATP}$$

   Glucose $\rightarrow$ carbon dioxide   +   water

Note that the inorganic compound carbon dioxide is energy-*(rich? poor?)* since it lacks C-H bonds.

b. The reaction shown is an oversimplification of the entire catabolic process, which actually entails many steps. Hydrogen atoms (H) removed from glucose consist of the

hydrogen ion (____) and an electron (e^-). Hydrogens removed from glucose must

first be trapped by H-carrying coenzymes, such as _____ or _____, and

ultimately the Hs combine with oxygen to form _____. In other words, glucose is dehydrogenated (loses H) while oxygen is reduced (gains H).

c. Catabolism involves a complex series of stepwise reactions in which hydrogens are transferred in the form of *(H? H^+ + e^-?)*. Since oxygen is the final molecule that serves as an oxidizer, catabolism is often described as biological *(reduction? oxidation?)*.

d. Along the way, energy from the original C-H bonds of glucose (and present in the

electrons of H) is trapped in _____. This process is known as

_____, since it involves the addition of a phosphate to ADP. A more complete expression of the total catabolism of glucose (shown briefly in (a) above) is included below. Write the correct labels under each chemical: *oxidized, oxidizer, reduced, reducer:*

Glucose   + O_2 +   ADP $\rightarrow$ CO_2 + H_2O +   ATP

(_____)   (_____)   (_____)   (_____)

■ **A9.** Check your understanding of oxidation-reduction reactions and generation of ATP involved in metabolism by circling the correct answer in each case.

a. Which is more reduced?
   A. Lactic acid ($C_3H_6O_3$)   B. Pyruvic acid ($C_3H_4O_3$)
b. Which molecule contains more chemical potential energy (can release more energy when catabolized)?
   A. Lactic acid ($C_3H_6O_3$)   B. Pyruvic acid ($C_3H_4O_3$)
c. Which is more likely to happen when a molecule within the human body is oxidized?
   A. The molecule loses H^+ and e^-.   B. The molecule gains H^+ and e^-.
d. Which is more oxidized?
   A. NADH + H^+ B. NAD^+
e. Which molecule contains more chemical potential energy (can release more energy when catabolized)?
   A. ATP   B. ADP
f. Which is an example of oxidative phosphorylation?
   A. Creatine-P + ADP $\rightarrow$ creatine + ATP
   B. Generation of ATP by energy released by the electron transfer chain
g. Which is the substrate in the following example of substrate-level phosphorylation?
   Creatine-P + ADP $\rightarrow$ creatine + ATP
   A. ATP   B. Creatine

B. Carbohydrate metabolism (pages 825–835)

■ **B1.** Answer these questions about carbohydrate metabolism.

a. The story of carbohydrate metabolism is really the story of

_____ metabolism, since this is the most common carbohydrate (and in fact the most common energy source) in the human diet.

b. What other carbohydrates besides glucose are commonly ingested? How are these converted to glucose?

c. Just after a meal, the level of glucose in the blood *(increases? decreases?)*. Cells use

some of this glucose; by _____ glucose, they release energy.

d. List three or more mechanisms by which excess glucose is used or discarded.

e. Increased glucose level of blood is known as _____. This can occur after a meal containing concentrated carbohydrate or in the absence of the

hormone _____, since this hormone *(facilitates? inhibits?)* entrance of glucose into cells. Lack of insulin occurs in the condition known as

_____.

f. As soon as glucose enters cells, glucose combines with _____ to

form glucose-6-phosphate. This process is known as _____, and

is catalyzed by enzymes called _____.

■ **B2.** Describe the process of glycolysis in this exercise.

a. Glucose is a ____-carbon molecule (which also has hydrogens and oxygens on it). During glycolysis each glucose molecule is converted to two ____-carbon molecules

named _____.

b. *(A lot? A little?)* energy is released from glucose during glycolysis. This process requires *(one? many?)* step(s) and *(does? does not?)* require oxygen.

c. The fate of pyruvic acid depends on whether sufficient _____ is available. If it is, pyruvic acid undergoes chemical change in phases 2 and 3 of glucose catabolism (see below). Those processes *(do? do not?)* require oxygen; that is, they are *(aerobic? anaerobic?)*.

d. If the respiratory rate cannot keep pace with glycolysis, insufficient oxygen is available to break down pyruvic acid. In the absence of sufficient oxygen, pyruvic acid is

temporarily converted to _____. This is likely to occur during

active _____.

e. Several mechanisms prevent the accumulation of an amount of lactic acid that might

be harmful. The liver changes some back to _____. Excess pCO_2 caused by exercise *(stimulates? inhibits?)* respiratory rate, so more oxygen is available for breakdown of the pyruvic acid.

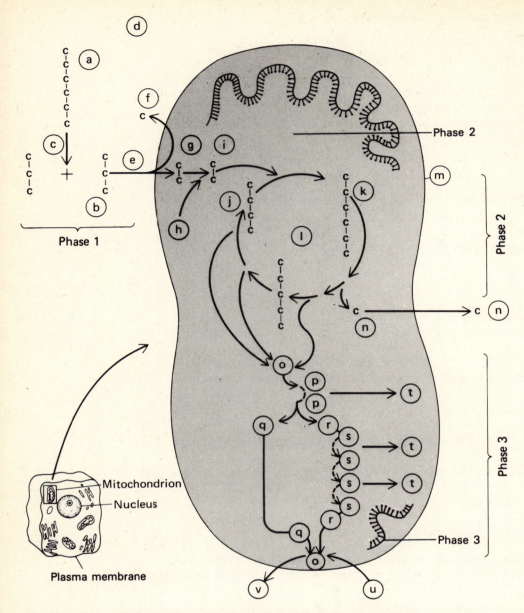

Figure LG 25.1 Summary of glucose catabolism, described in three phases. Phase 1 (glycolysis) occurs in the cytosol. Phase 2 (Krebs cycle) takes place in matrix of mitochondria. Phase 3 (electron transport) takes place on inner mitochondrial membrane. Letters refer to Checkpoint B3.

■ **B3.** Review glucose catabolism up to this point and summarize the remaining phases by doing this exercise. Refer to Figure LG 25.1. A six-carbon compound named

(a) _____ is converted to two molecules of

(b) _____ via a number of steps, together called

(c) _____ , occurring in the (d) _____ of the cell.

In order for pyruvic acid to be further broken down, it must undergo a

(e) _____ step. This involves removal of the one-carbon molecule,

(f) _____. The remaining two-carbon

(g) _____ group is attached to a carrier called

(h) _____, forming (i) _____.
This compound hooks on to a four-carbon compound called

(j) _____. The result is a six-carbon molecule of

(k) _____. The name of this compound is given to the cycle of

reactions called the (l) _____ cycle, occurring in the

(m) _____ .
Two main types of reactions occur in the Krebs cycle. One is the decarboxylation

reaction, in which (n) _____ molecules are removed. (Notice these locations on Figure 25.4, page 830 of your text.) (What happens to the CO_2 molecules that are removed?) As a result the six-carbon citric acid is eventually shortened to regenerate oxaloacetic, giving the cyclic nature to this process.
The other major type of reaction (oxidation-reduction) involves removal of

(o) _____ atoms during oxidation of compounds such as isocitric and α-ketoglutaric acids. (See Figure 25.4, page 830 in your text.) These hydrogen

atoms are carried off by two coenzymes named (p) _____ and "trapped" for use in phase 3.
In phase 3, hydrogen atoms on coenzymes NAD and FAD (and later coenzyme Q)

are ionized to (q) _____ and (r) _____. Electrons are shuttled along a chain of

(s) _____. (Refer to Checkpoint B4.) During electron transport, energy in these electrons (derived from hydrogen atoms in ingested foods) is "tapped"

and stored in (t) _____.
Finally, the electrons, depleted of some of their energy, are reunited with hydrogen

ions and with (u) _____ to form (v) _____.
Notice why your body requires oxygen—to keep drawing hydrogen atoms off of nutrients (in biological oxidation) so that energy from them can be stored in ATP for use in cellular activities.

■ **B4.** Now describe details of the electron transport chain in this activity.

a. The first step in this chain is transfer of high-energy electrons from NADH + H+ to the coenzyme *(FMN? Q?)*. As a result this carrier is *(oxidized? reduced?)* to

_____. From which vitamin is this carrier derived? ____

b. As electrons are passed on down the chain, $FMNH_2$ is *(oxidized? reduced?)* back to

_____.

c. Which mineral is the portion of cytochromes on which electrons are ferried? ____ Cytochromes are positioned toward the *(beginning? end?)* of the transport chain.

d. Circle two other minerals that are known to be involved in the electron transport chain and are hence necessary in the diet.
   calcium   copper   iodine   sulfur   zinc

e. The last carrier in the chain is *(coenzyme Q? cytochrome a₃?)*. It passes H+ and e⁻ to

_____, the final oxidizing molecule. This is a key step in the process of *(aerobic? anaerobic?)* cellular respiration.

f. Where are the electron carriers located? Circle the correct answer.
   MM. Mitochondrial matrix
   IMM. Inner mitochondrial membrane (cristae)
   OMM. Outer mitochondrial membrane

g. The carriers are grouped into *(two? three? four?)* complexes. Each complex acts to pump *(protons = H+? electrons = e⁻?)* from the *(matrix? space between inner and*

outer mitochondrial membranes?) into the _____.

h. As a result, a gradient of H+ is set up across the *(inner? outer?)* mitochondrial

membrane, resulting in potential energy known as the _____

_____ force. How do these protons (H+) travel back into the mitochondrial matrix?

i. The channels through which H+ diffuse contain the enzyme known as ATP

_____. Since the formation of the resulting ATP is driven by an energy source that consists of a gradient and subsequent diffusion of a *chemical*

(specifically _____), this mechanism is known as _____ generation of ATP.

B5. Write a chemical equation showing the overall reaction for aerobic respiration.

■ **B6.** Most of the 36 to 38 ATPs generated from the total oxidation of hydrogens derive from *(glycolysis? the Krebs cycle?)*. Of those generated by the Krebs cycle, most are formed from the oxidation of hydrogens carried by *(NAD? FAD?)*. About what percent of the energy originally in glucose is stored in ATP after glucose is completely catabolized? *(10 percent? 40 percent? 60 percent? 99 percent?)*

■ **B7.** Complete the exercise about the reaction shown below.

        (1)

Glucose ⇌ Glycogen

        (2)

a. You learned earlier that excess glucose may be stored as glycogen. In other words,

   glycogen consists of large branching chains of _____. Name the process of glycogen formation by labeling (1) in the chemical reaction above.

b. Between meals, when glucose is needed, glycogen can be broken down again to release glucose. Label (2) above with the name of this process. [Note that a number of steps are actually involved in both processes, (1) and (2).]

c. Which of these two reactions is anabolic? _____

d. Where is most (75 percent) of glycogen in the body stored?

   _____

e. Identify a hormone that stimulates reaction (1). Write its name on the upper arrow in the reaction above. As more glucose is stored in the form of glycogen, the blood

   level of glucose _____-creases. Now write below the lower arrow the names of two hormones that stimulate glycogenolysis.

f. In order for glycogenolysis to occur, a _____ group must be added to glucose as it breaks away from glycogen. The enzyme catalyzing this

   reaction is known as _____. Does the reverse reaction [shown in (2)] require phosphorylation also? *(Yes? No?)*

B8. Define *gluconeogenesis* and briefly discuss how it is related to other metabolic reactions.

Name five hormones that stimulate gluconeogenesis.

■ **B9.** Circle the processes at left and the hormones at right that lead to increased blood glucose level (hyperglycemia).

Processes
Glycogenesis
Glycogenolysis
Glycolysis
Gluconeogenesis

Hormones
Insulin
Glucagon
Epinephrine
Cortisol
Thyroid hormone
Growth hormone

For extra review, refer to Figures LG 18.3 and LG 18.7, pages 354 and 364 of the Learning Guide.

C. Lipid metabolism   (pages 836–840)

C1. List six examples of structural or functional roles of fats in the body.

_____   _____   _____

_____   _____   _____

C2. Is the fat stored in your subcutaneous tissue "the same fat" that was in that location two years ago? Explain.

■ **C3.** Do this exercise about fat metabolism.

a. The initial step in catabolism of triglycerides is their breakdown into

_____ and _____. Glycerol can then be

converted into _____; this molecule can enter glycolytic pathways. If the cell does not need to generate ATP, then

glyceraldehyde 3-phosphate can be converted into _____. This

anabolic step is an example of gluco- _____.

b. Recall that fatty acids are long chains of carbons, with attached hydrogens and a few oxygens. Two-carbon pieces are "snipped off" of fatty acids by a process called

_____, occurring in the _____.

c. These two-carbon pieces (_____) attach to coenzyme

_____. The resulting molecules, named _____, may enter the Krebs cycle.

d. When large numbers of acetyl CoA's form, they tend to pair chemically:

Acetic acid + acetic acid → _____
    (2C)   (2C)   (4C)

The presence of excessive amounts of these acids in blood *(raises? lowers?)* blood pH. Since these acids are called "keto" acids, this condition is known as

_____.

e. Slight alterations of aetoacetic acid leads to formation of β-hydroxybutyric acid and

_____. Collectively, these three chemicals are known as

_____ bodies. Formation of ketone bodies is known as

_____; it takes place in the _____.

f. Ketogenesis occurs when cells are forced to turn to fat catabolism. State two or more reasons why cells might carry out excessive fat catabolism leading to ketogenesis and ketosis.

g. *A clinical challenge.* Explain why a person whose diabetes is out of control might have sweet breath.

D. Protein metabolism   (pages 840–841)

■ **D1.** Which of the three major nutrients (carbohydrates, lipids, and proteins) fulfills each function?

a. Most direct source of energy; stored in body least: _____
b. Constitutes almost all of energy reserves, but also used for body-building:

_____

c. Usually used least for fuel; used most for body structure and for regulation:

_____

■ **D2.** Contrast the caloric values of the three major food types.

a. Carbohydrates and proteins each produce _____ Cal/g or (since a pound contains

454 g) _____ Cal/lb.

b. Fats produce _____ Cal/g or _____ Cal/lb.

D3. Throughout your study of systems of the body so far, you have learned about a variety of roles of proteins. List at least six functions of proteins in the body, using one or two words for each function. Include some structural and some regulatory roles.

_____   _____

_____   _____

_____   _____

■ **D4.** Name two hormones that enhance transport of amino acids into cells and also stimulate protein synthesis.

■ **D5.** Refer to Figure LG 25.2 and describe the uses of protein.

Two sources of proteins are shown: (a) _____ and

(b) _____. Amino acids from these sources may undergo a process

known as (c) _____ to form new body proteins. If other energy
sources are used up, amino acids may undergo catabolism. The first step is

(d) _____, in which an amino group (NH_2) is removed and

converted (in the liver) to (e) _____, a component of

(f) _____, which exits in urine. The remaining portion of the

amino acid may enter (g) _____ pathways at a number of points
(Figure 25.12, page 841 in your text). In this way, proteins can lead to formation of

(h) _____.

D6. Contrast essential amino acids with nonessential amino acids.

How many amino acids are considered essential to humans? ____

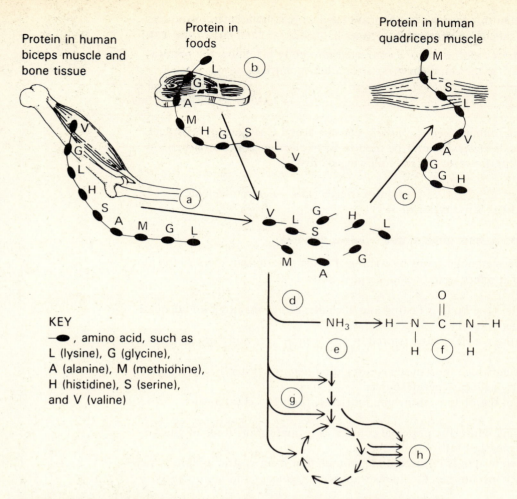

Protein in human biceps muscle and bone tissue

Protein in foods

(b)

Protein in human quadriceps muscle

M
L
S
L
V
A
G
G
H

(c)

V
G
L
H
S
A M G L

(a)

L
G
A
M
H G S
L V

V L G H L
S
M A G

(d)

KEY

—●, amino acid, such as
L (lysine), G (glycine),
A (alanine), M (methionine),
H (histidine), S (serine),
and V (valine)

$NH_3 \longrightarrow$

$$H-N-C-N-H$$

with O double-bonded to C above, and H below each N

(e) (f)

(g)

(h)

Figure LG 25.2   Metabolism of protein. Lowercase letters refer to Checkpoint D5.

E. Summary of key molecules in metabolism   (pages 841–843)

■ **E1.** Refer to Figure 25.13 (page 842 in your text). Identify major roles of the three molecules that play most key roles in metabolism. Use these answers:

Acetyl CoA. Acetyl coenzyme A   PA. Pyruvic acid
G-6-P. Glucose-6-phosphate

_____ a. Entrance molecule for fatty acids and ketone bodies into Krebs cycle

_____ b. Formed from pyruvic acid under aerobic conditions

_____ c. Converts to lactic acid under anaerobic conditions

_____ d. Similar to amino acid alanine except for one amino group; molecule by which alanine can enter catabolic pathways

_____ e. Molecule into which glucose is converted when glucose enters any body cell

_____ f. Molecule by which blood glucose is converted to oxaloacetic acid (OAA) to enter the Krebs cycle

_____ g. Molecule by which glycogen can be utilized as fuel by liver or muscle cells

■ **E2.** *For extra review.* Do this exercise on the main steps in metabolic interconversions of carbohydrates and fats. (For help, refer to pages 835, 842–843, and 845 of the text.)

  a. Excessive carbohydrate calories may be converted into body fat. Show the pathway by filling in blanks. Use the same answers as for Checkpoint E1.

  Glycogen → _____ → _____ → _____ → Fatty acid portion of fats

  b. Mammals, including humans, *(can? cannot?)* convert acetyl CoA into pyruvic acid. As a result, fatty acids *(can? cannot?)* be used to form glucose. Gluconeogenesis involving fats can occur, however, by the following pathway:

  The _____ portion of triglycerides → glyceraldehyde 3-phosphate (G 3-P) → pyruvic acid →

F. Absorptive and postabsorptive states (pages 843–846)

■ **F1.** Contrast these two metabolic states by writing *A* if the description refers to the absorptive state and *P* if it refers to the postabsorptive state.

  _____ a. Period when the body is fasting and is challenged to maintain an adequate blood glucose level
  _____ b. State during which glucose (stored in glycogen in liver and muscle) is released
  _____ c. State during which most systems (excluding nervous system) switch over to use of fatty acids as energy sources
  _____ d. Time when the body is absorbing nutrients from the GI tract

  _____ e. State during which the principal concern is formation of stores of glucose (as glycogen) and fat.
  _____ f. State when insulin is released under stimulation from gastric inhibitory peptide (GIP); insulin then facilitates transport of glucose and amino acids into cells
  _____ g. Period dominated by anti-insulin hormones, such as glucagon

■ **F2.** Do this exercise about maintenance of glucose level between meals.

  a. The normal blood glucose level in the postabsorptive state is about *(30–50? 70–110? 150–170? 600–700?)* mg glucose/100 ml blood.
  b. Now list four sources of glucose that may be called upon during the postabsorptive state.

  _____   _____

  _____   _____

  c. Can fatty acids serve as a source of glucose? *(Yes? No?)* Explain.

  d. Fats can be used to generate ATP by being broken down into

  _____, which can then enter the Krebs cycle. Therefore, fatty acids are said to provide a glucose- *(sparing? utilizing?)* mechanism between meals.

F3. *For extra review.* Complete the table summarizing how metabolism is hormonally regulated. (Refer to Exhibit 25.4, page 847 of your text.)

Hormone	Source	Action
a.		Stimulates glycogenesis in liver; stimulates glucose uptake in other cells
b.	Alpha cells of islets of Langerhans in pancreas	
c. Human growth hormone (hGH)		
d. Glucocorticoids		
e.	Thyroid gland	
f.	Adrenal medulla	

■ **F4.** Circle the hormones in Checkpoint F3 above that are hyperglycemic, that is, tend to *increase* blood glucose.

G. Minerals, vitamins   (pages 846–852)

■ **G1.** Define *minerals.*

Minerals make up about _____ percent of body weight and are concentrated in the

_____.

■ **G2.** Study Exhibit 25.5 (pages 847–849 in your text). Then check your understanding of minerals by doing this matching exercise.

Ca. Calcium	Fe. Iron	Na. Sodium
Cl. Chlorine	I. Iodine	P. Phosphorus
Co. Cobalt	K. Potassium	S. Sulfur
F. Fluorine	Mg. Magnesium	

_____ a. Main anion in extracellular fluid, part of HCl in stomach; component of table salt

_____ b. Involved in generation of nerve impulse, helps to regulate osmosis, acts in buffer systems

_____ c. Most abundant cation in the body, found mostly in bones and teeth; necessary for normal muscle contraction and for blood clotting

_____ d. Important component of hemoglobin and cytochromes

_____ e. Main cation inside of cells; used in nerve transmission

_____ f. Essential component of thyroxin

_____ g. Constituent of vitamin B_{12}, so necessary for red blood cell formation

_____ h. Important component of amino acids, vitamins, and hormones

_____ i. Improves tooth structure

_____ j. Found mostly in bones and teeth; important in buffer system and in ATP processes; component of DNA and RNA.

■ **G3.** Circle the correct answer. The primary purpose of vitamins is:
A. Synthesis of body structures
B. Regulation of body activities

■ **G4.** Vitamins are *(organic? inorganic?)*. Most vitamins *(can? cannot?)* be synthesized in the body. In general, what are the functions of vitamins?

■ **G5.** Contrast the two principal groups of vitamins, and list the main vitamins in each group.

a. Fat-soluble _____

b. Water-soluble _____

■ **G6.** The initials U.S. RDA on labels of foods and vitamin or mineral supplements refer

to: U _____   S_____

R_____   D_____

A_____.

G7. Defend or dispute this statement: "Most persons who eat a balanced diet do need to take vitamin or mineral supplements."

■ **G8.** Select the vitamin that fits each description.

A	B₁	B₂	B₁₂	C	D	E	K

_____ a. This serves as a coenzyme that is essential for blood clotting, so it is called the antihemorrhagic vitamin; synthesized by intestinal bacteria.

_____ b. Its formation depends upon sunlight on skin and also on kidney and liver activation; necessary for calcium absorption.

_____ c. Riboflavin is another name for it; a component of FAD; necessary for normal integrity of skin, mucosa, and eye.

_____ d. This vitamin acts as an important coenzyme in carbohydrate metabolism; deficiency leads to beriberi.

_____ e. Formed from carotene, it is necessary for normal bones and teeth; it prevents night blindness.

_____ f. This substance is also called ascorbic acid; deficiency causes anemia, poor wound healing, and scurvy.

_____ g. This coenzyme, the only B vitamin not found in vegetables, is necessary for normal erythropoiesis; absorption from GI tract depends on intrinsic factor.

_____ h. Also known as tocopherol, it is necessary for normal red blood cell membranes; deficiency is associated with sterility in some animals.

H. Metabolism and body heat (pages 852–857)

■ **H1.** Contrast calorie and kilocalorie.

If you ingest, digest, absorb, and metabolize a slice of bread, the energy released from the bread equals about 80 to 100 *(cal? kcal?)*.

■ **H2.** Answer these questions about catabolism.
Catabolism of foods *(uses? releases?)* energy. Most of the heat produced by the body comes from catabolism by *(oxidation? reduction?)* of nutrients.

H3. Discuss factors that affect how rapidly your body catabolizes, that is, your metabolic rate. Include: *exercise, nervous system, hormones,* and *body temperature, specific dynamic action (SDA), age,* and *pregnancy.*

H4. Define *basal metabolic rate (BMR)*.

■ **H5.** Discuss BMR in this exercise.

    a. BMR may be determined by measuring the amount of _____
    consumed within a period of time, since oxygen is necessary for the metabolism of
    foods. Normally for each liter of oxygen consumed, the body metabolizes enough

    food to release about _____ kcal.
    b. Suppose you consume 15 liters of oxygen in an hour. Your body would release

    _____ kcal of heat in that hour.
    c. A BMR of 20 percent below the standard value is likely to be due to *(hyper? hypo?)*-
    thyroidism.

■ **H6.** Your body temperature usually *(does? does not?)* rise much as a result of catabolism. Why?

H7. Write the percent of heat loss by each of the following routes (at room temperature). Then write an example of each of these types of heat loss. One is done for you.
    a. Radiation ____
    b. Conduction and convection **15** Cooling by draft while taking a shower
    c. Evaporation ____

■ **H8.** During an hour of active exercise Bill produces 1 liter of sweat. The amount of

heat loss (cooling) that accompanies this much evaporation is _____ Cal. On a humid day, *(more? less?)* evaporation of Bill's sweat occurs, so Bill would be cooled *(more? less?)* on such a day.

H9. Lil is outside on a snowy day without a warm coat. Her skin is pale and chilled and she begins to shiver. Explain how these responses are attempts of her body to maintain homeostasis of temperature. What other responses might raise her body temperature? Include these words in your answers: *sympathetic, metabolism, adrenal medulla, thermogenesis, thyroid,* and *coat.*

■ **H10.** Arrange in order the events believed to occur in the production of fever of

39.4°C (103°F) and recovery from this state. ____ ____ ____ ____ ____ ____

A. 39.4°C (103°F) temperature is reached and maintained.

B. Prostaglandins (PGs) cause the hypothalamic "thermostat" to be reset from normal 37.5°C (98.6°F) to a higher temperature such as 39.4°C (103°F).

C. Vasoconstriction and shivering occur as heat is conserved in order to raise body temperature to the 39.4°C (103°F) level directed by the hypothalamus.

D. Source of pyrogens is removed (for example, bacteria are killed by antibiotics), lowering hypothalamic "thermostat" to 37.5°C (98.6°F).

E. "Crisis" occurs. Obeying hypothalamic orders, the body shifts to heat loss mechanisms such as sweating and vasodilation, returning body temperature to normal.

F. Infectious organisms cause phagocytes to release interleukin-1, which then causes hypothalamic release of PGs.

H11. In what ways is a fever beneficial?

H12. Contrast these temperature disorders: heatstroke/heat exhaustion

H13. Explain why elderly persons are at greater risk for hypothermia.

I. Metabolic disorders   (pages 857–858)

I1. Contrast the terms in the following pairs.

a. Obesity/morbid obesity

b. Toxicity of water-soluble vitamins/toxicity of fat-soluble vitamins

■ **I2.** *A clinical challenge.* Which condition involves deficiency in both protein and calories? *(Kwashiorkor? Marasmus?)* Explain why an individual with kwashiorkor may have a large abdomen even though the person's diet is inadequate.

■ **I3.** Individuals with PKU are unable to convert _____ to

_____. What are results of toxic levels of phenylalanine?

■ **I4.** Persons with celiac disease require a diet that is lacking in most *(meats? vegetables? grains?).* The problem stems from a water-insoluble protein named

_____, which causes destruction of the intestinal lining. Name two grains that are acceptable in the diet of a patient with celiac disease.

■■■

ANSWERS TO SELECTED CHECKPOINTS

A1. (a) Hypothalamus; satiated or full. (b) Satiety; all answers. (c) CCK; this hormone, secreted when triglycerides enter the small intestine, causes a feeling of fullness (inhibits eating).

A3. Carbohydrates, proteins, lipids, minerals, vitamins, and water.

A8. (a) ATP; $C_6H_{12}O_6 \rightarrow CO_2 + H_2O + energy$ stored in ATP; poor. (b) H^+; NAD or FAD, H_2O. (c) $H^+ + e^-$; oxidation. (d) ATP; phosphorylation;
Glucose + O_2 + ADP $\rightarrow$
(reducer)   (oxidizer)
  CO_2 + H_2O + ATP
(oxidized)   (reduced)

A9. (a) A. (b) A. (c) A. (d) B. (c) A. (f) B. (g) B.

B1. (a) Glucose. (b) Starch, sucrose, lactose; by enzymes located mostly in liver cells. (c) Increases, oxidizing. (d) Stored as glycogen, converted to fat or protein and stored, excreted in urine. (e) Hyperglycemia; insulin, facilitates; diabetes mellitus. (f) Phosphate; phosphorylation, kinases.

B2. (a) Six; three, pyruvic acid. (b) A little; many, does not. (c) Oxygen; do, aerobic. (d) Lactic acid; exercise. (e) Pyruvic acid; stimulates.

B3. (a) Glucose. (b) Pyruvic acid. (c) Glycolysis. (d) Cytosol. (e) Transition. (f) CO_2. (g) Acetyl. (h) Coenzyme A. (i) Acetyl coenzyme A. (j) Oxaloacetic acid. (k) Citric acid. (l) Citric acid (or Krebs). (m) Matrix of mitochondria. (n) CO_2, which are exhaled. (o) Hydrogen. (p) NAD and FAD. (q) H^+. (r) Electrons (e^-). (s) Cytochromes or electron transfer system. (t) ATP. (u) Oxygen. (v) Water.

B4. (a) FMN; reduced, $FMNH_2$; B_2 (riboflavin). (b) Oxidized, FMN. (c) Iron (Fe); end. (d) Copper and sulfur. (e) Cytochrome a_3; oxygen; aerobic. (f) IMM. (g) Three; protons = H^+, matrix, space between inner and outer mitochondrial membranes. (h) Inner, proton motive; across special channels in the inner mitochondrial membrane. (i) Synthetase; H^+, chemiosmotic.

B6. The Krebs cycle; NAD; 40.

B7. (a) Glucose; 1: glycogenesis. (b) 2: Glycogenolysis. (c) Glycogenesis. (d) Muscles. (e) Insulin; de; glucagon and epinephrine. (f) Phosphate; phosphorylase; yes.

B9. Processes: glycogenolysis and gluconeogenesis; hormones: all except insulin.

C3. (a) Fatty acids and glycerol; glyceraldehyde 3-phosphate; glucose; neogenesis. (b) Beta oxidation, matrix of mitochondria. (c) Acetic acid, A; acetyl CoA (or acetyl coenzyme A). (d) Acetoacetic acid; lowers; ketoacidosis. (e) Acetone; ketone; ketogenesis; liver. (f) Examples of reasons: starvation, fasting diet, lack of glucose in cells due to lack of insulin (diabetes mellitus), excess growth hormone (GH) which stimulates fat catabolism. (g) As acetone (formed in ketogenesis) passes through blood in pulmonary vessels, some is exhaled and detected by its sweet aroma.

D1. (a) Carbohydrates. Exception: cardiac muscle and cortex of the kidneys utilize acetoacetic acid from fatty acids preferentially over glucose. (b) Lipids. (c) Proteins.

D2. (a) 4, about 1800. (b) 9, about 4000.

D4. Insulin and hGH. (Thyroid hormone also stimulates protein synthesis.)

D5. (a) Worn-out cells as in bone or muscle. (b) Ingested foods. (c) Anabolism. (d) Deamination. (e) Ammonia. (f) Urea. (g) Glycolytic or Krebs cycle. (h) $CO_2 + H_2O$ + ATP.

E1. (a–b) Acetyl CoA. (c–d) Pyruvic acid. (e–g) G-6-P.

E2. (a) Glycogen → G-6-P → pyruvic acid → acetyl CoA → fatty acid portion of fats. (b) Cannot; cannot; the glycerol portion of triglycerides → G 3-P → pyruvic acid → glucose.

F1. (a–c) P. (d–f) A. (g) P.

F2. (a) 70–110. (b) Liver glycogen (4-hour supply); muscle glycogen and lactic acid during exercise; glycerol from fats; as a last resource, amino acids from tissue proteins. (c) No. See Checkpoint E2b in this chapter. (d) Acetic acids (and then acetyl CoA); sparing.

F4. (b) Glucagon. (c) hGH. (d) Glucocorticoids. (e) Thyroxine. (f) Epinephrine.

G1. Inorganic substances; 4, skeleton.

G2. (a) Cl. (b) Na. (c) Ca. (d) Fe. (e) K. (f) I. (g) Co. (h) S. (i) F. (j) P.

G3. B.

G4. Organic; cannot; most serve as coenzymes, maintaining growth and metabolism.

G5. (a) A D E K. (b) B complex and C.

G6. United States Recommended Daily Allowance.

G8. (a) K. (b) D. (c) B_2. (d) B_1. (e) A. (f) C. (g) B_{12}. (h) E.

H1. A kilocalorie (kcal) is the amount of heat required to raise 1000 ml (1 liter) of water 1°C; a kilocalorie (kcal) = 1000 calories (Cal); kcal.

H2. Releases; oxidation.

H5. (a) Oxygen; 4.9. (b) 73.5 (= 4.9 kcal/liter oxygen × 15 liters/hour). (c) Hypo.

H6. Does not. About half of the energy released by catabolism is stored in ATP and used for body activities. The remaining energy is released as heat and regulated by heat loss mechanisms.

H8. 580 (= 0.58 Cal/ml water × 1000 ml/liter); less; less.

H10. F B C A D E.

I2. Marasmus; protein deficiency (due to lack of essential amino acids) decreases plasma proteins. Blood has less osmotic pressure, so fluids exit from blood. Movement of these into the abdomen (ascites) increases the size of the abdomen. Fatty infiltration of the liver also adds to the abdominal girth.

I3. Phenylalanine, tyrosine; toxicity to the brain with possible mental retardation.

I4. Grains; gluten; rice and corn.

MASTERY TEST: Chapter 25

Questions 1–12: Circle the letter preceding the one best answer to each question.

1. Which of these processes is anabolic?
   A. Pyruvic acid $\rightarrow CO_2 + H_2O$ + ATP
   B. Glucose $\rightarrow$ pyruvic acid + ATP
   C. Protein synthesis
   D. Digestion of starch to maltose
   E. Glycogenolysis

2. Which of the following can be represented by the equation: Glucose $\rightarrow$ pyruvic acids + small amount ATP?
   A. Glycolysis
   B. Transition step between glycolysis and Krebs cycle
   C. Krebs cycle
   D. Electron transport system

3. Which hormone is said to be hypoglycemic since it tends to lower blood sugar?
   A. Glucagon   D. Insulin
   B. Glucocorticoids   E. Epinephrine
   C. Growth hormone

4. The complete oxidation of glucose yields all of these products *except:*
   A. ATP   C. Carbon dioxide
   B. Oxygen   D. Water

5. All of these processes occur exclusively or primarily in liver cells *except* one, which occurs in virtually all body cells. This one is:
   A. Gluconeogenesis
   B. Beta oxidation of fats
   C. Ketogenesis
   D. Deamination and conversion of ammonia to urea
   E. Krebs cycle

6. The process of forming glucose from amino acids or glycerol is called:
   A. Gluconeogenesis   D. Deamination
   B. Glycogenesis   E. Glycogenolysis
   C. Ketogenesis

7. All of the following are parts of fat absorption and fat metabolism *except:*
   A. Lipogenesis
   B. Beta oxidation
   C. Chylomicron formation
   D. Ketogenesis
   E. Glycogenesis

8. Which of the following vitamins is water-soluble? A, C, D, E, K?

9. At room temperature most body heat is lost by:
   A. Evaporation   C. Conduction
   B. Convection   D. Radiation

10. Choose the *false* statement about vitamins.
    A. They are organic compounds.
    B. They regulate physiological processes.
    C. Most are synthesized by the body.
    D. Many act as parts of enzymes or coenzymes.

11. Choose the *false* statement about temperature regulation.
    A. Some aspects of fever are beneficial.
    B. Fever is believed to be due to a "resetting of the body's thermostat."
    C. Vasoconstriction of blood vessels in skin will tend to conserve heat.
    D. Heat-producing mechanisms which occur when you are in a cold environment are primarily parasympathetic.

12. Which of the following processes involves a cytochrome chain?
    A. Glycolysis
    B. Transition step between glycolysis and Krebs cycle
    C. Krebs cycle
    D. Electron transport system

Question 13: Arrange the answers in correct sequence.

_____ _____ _____ _____ _____ _____

13. Steps in complete oxidation of glucose:
    A. Glycolysis takes place.
    B. Pyruvic acid is converted to acetyl coA.
    C. Hydrogens are picked up by NAD and FAD; hydrogens ionize.
    D. Krebs cycle releases CO_2 and hydrogens
    E. Oxygen combines with hydrogen ions and electrons to form water.
    F. Electrons are transported along cytochromes, and energy from electrons is stored in ATP.

Questions 14–20: Circle T (true) or F (false). If the statement is false, change the underlined word or phrase so that the statement is correct.

T F 14. Coenzyme A is derived from pantothenic acid, also known as <u>vitamin C</u>.

T F 15. Complexes of carrier molecules utilized in the electron transport system are located in the <u>mitochondrial matrix</u>.

T F 16. Catabolism of carbohydrates involves <u>oxidation</u>, which is a process of <u>addition of hydrogens</u>.

T F 17. Nonessential amino acids are those that <u>are not used in the synthesis of human protein</u>.

T F 18. The complete oxidation of glucose to CO_2 and H_2O yields <u>4</u> ATPs.

T F 19. Anabolic reactions are <u>synthetic reactions that release energy</u>.

T F 20. <u>Carbohydrates, proteins, and vitamins</u> provide energy and serve as building materials.

Questions 21–25: fill-ins. Complete each sentence with the word or phrase that best fits.

_____ 21. Catabolism of each gram of carbohydrate or protein results in release of about ____ kcal, whereas each gram of fat leads to about ____ kcal.

_____ 22. The body's "favorite" source of energy (since it is most easily catabolized) is ____, whereas the least favorite is ____.

_____ 23. ____ is the mineral that is most common in extracellular fluid (ECF); it is also important in osmosis, buffer systems, and in nerve impulse conduction.

_____ 24. Aspirin and acetaminophen (Tylenol) reduce fever by inhibiting synthesis of ____ so that the body's "thermostat" located in the ____ is reset to a lower temperature.

_____ 25. In uncontrolled diabetes mellitus, the person may go into a state of acidosis due to the production of ____ acids resulting from excessive breakdown of ____.

ANSWERS TO MASTERY TEST: ■ Chapter 25

Multiple Choice

1. C
2. A
3. D
4. B
5. E
6. A
7. E
8. C
9. D
10. C
11. D
12. D

Arrange

13. A B D C F E

True-False

14. F. One of the B vitamins (or B complex vitamins)
15. F. Mitochondrial inner (cristae) membrane
16. F. Oxidation, removal of hydrogens
17. F. Can be synthesized by the body
18. F. 36 to 38
19. F. Synthetic reactions that require energy
20. F. Carbohydrates and proteins but not vitamins

Fill-ins

21. 4, 9
22. Carbohydrate, protein
23. Sodium (Na^+)
24. Prostaglandins (PG), hypothalamus
25. Acetoacetic (keto-), fats

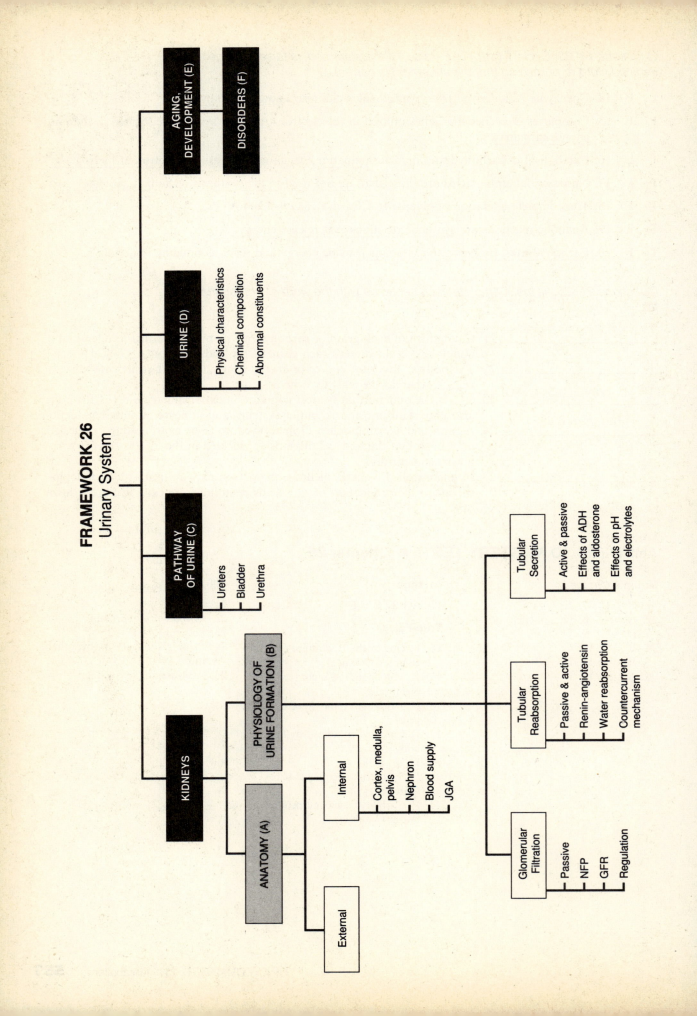

FRAMEWORK 26
Urinary System

AGING, DEVELOPMENT (E)

DISORDERS (F)

URINE (D)
- Physical characteristics
- Chemical composition
- Abnormal constituents

PATHWAY OF URINE (C)
- Ureters
- Bladder
- Urethra

KIDNEYS

ANATOMY (A)

External

Internal
- Cortex, medulla, pelvis
- Nephron
- Blood supply
- JGA

PHYSIOLOGY OF URINE FORMATION (B)

Glomerular Filtration
- Passive
- NFP
- GFR
- Regulation

Tubular Reabsorption
- Passive & active
- Renin-angiotensin
- Water reabsorption
- Countercurrent mechanism

Tubular Secretion
- Active & passive
- Effects of ADH and aldosterone
- Effects on pH and electrolytes

The Urinary System

CHAPTER 26

Wastes like urea or hydrogen ions (H^+) are products of metabolism that must be removed or they will lead to deleterious effects. The urinary system works with lungs, skin, and the GI tract to eliminate wastes. Kidneys are designed to filter blood and selectively determine which chemicals stay in the blood or exit in the urine. The process of urine formation in the kidney requires an intricate balance of factors such as blood pressure and hormones (ADH and aldosterone). Examination of urine (urinalysis) can provide information about the status of the blood and possible kidney malfunction. A healthy urinary system requires an exit route for urine through ureters, bladder, and urethra.

Start your study of the urinary system with a look at the Chapter 26 Framework and familiarize yourself with the key concepts and terms.

TOPIC OUTLINE AND OBJECTIVES

A. Kidneys: anatomy

1. Describe the functions of the kidneys.
2. Identify the external and internal gross anatomical features of the kidneys.
3. Define the structural adaptations of a nephron for urine formation.

B. Kidneys: physiology of urine formation

4. Discuss the process of urine formation through glomerular filtration, tubular reabsorption, and tubular secretion.
5. Describe how the kidneys produce dilute and concentrated urine.
6. Explain the principle of hemodialysis.

C. Ureters, urinary bladder, urethra

7. Discuss the structure and physiology of the ureters, urinary bladder, and urethra.

D. Urine

8. List and describe the physical characteristics, normal chemical constituents, and abnormal constituents of urine.

E. Aging and development of the urinary system

9. Describe the effects of aging on the urinary system.
10. Describe the development of the urinary system.

F. Disorders, medical terminology

11. Discuss the causes of glomerulonephritis, pyelitis, pyelonephritis, cystitis, nephrosis, polycystic disease, renal failure, urinary tract infections (UTIs), and diabetes insipidus.
12. Define medical terminology associated with the urinary system.

WORDBYTES

Now study the following parts of words that may help you better understand terminology in this chapter.

Wordbyte	Meaning	Example	Wordbyte	Meaning	Example
anti-	against	*anti*diuretic	nephr-	kidney	*nephr*ectomy
calyx	cup	*calyc*es	pyelo-	pelvis (or)	intravenous
-cele	herniation	vesico*cele*		kidney)	*pyelo*gram (IVP)
cyst-	bladder	*cyst*oscopy	ren-	kidney	*ren*al failure
insipid	without	diabetes	urin-	urinary	*urin*alysis
	taste	*insipid*us	vesic-	bladder	*vesic*ocele
juxta-	near	*juxta*glomerular			

CHECKPOINTS

A. Kidneys: anatomy   (pages 864–874)

■ **A1.** Describe functions of the urinary system in this exercise.

a. List waste products eliminated through urine.

b. Which ion do kidneys excrete when blood pH is too low? _____

c. Kidneys produce _____, which regulates blood pressure, and

_____, which is vital to normal red blood cell formation. Kidneys also activate vitamin ____.

A2. Besides the urinary organs, what other body structures perform excretory functions? List three and name the substances they eliminate.

Structure	Substances Eliminated
1. _____	_____
2. _____	_____
3. _____	_____

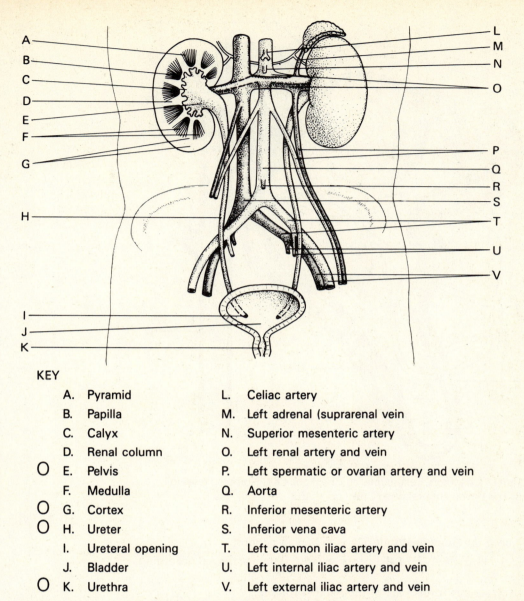

A ————
B ————
C ————
D ————
E ————
F ————
G ————
H ————
I ————
J ————
K ————

L
M
N
O
P
Q
R
S
T
U
V

KEY

	A.	Pyramid	L.	Celiac artery
	B.	Papilla	M.	Left adrenal (suprarenal vein
	C.	Calyx	N.	Superior mesenteric artery
	D.	Renal column	O.	Left renal artery and vein
O	E.	Pelvis	P.	Left spermatic or ovarian artery and vein
	F.	Medulla	Q.	Aorta
O	G.	Cortex	R.	Inferior mesenteric artery
O	H.	Ureter	S.	Inferior vena cava
	I.	Ureteral opening	T.	Left common iliac artery and vein
	J.	Bladder	U.	Left internal iliac artery and vein
O	K.	Urethra	V.	Left external iliac artery and vein

Figure LG 26.1 (*Left side of diagram*) Organs of the urinary system. Structures A to G are parts of the kidney. Identify and color as directed in Checkpoints A3 and C. (*Right side of diagram*) Blood vessels of the abdomen. Color as directed in Chapter 21, Checkpoint E8, page 443 of the Learning Guide.

■ **A3.** Identify the organs that make up the urinary system on Figure LG 26.1 and answer the following questions about them.

a. The kidneys are located at about *(waist? hip?)* level, between _____ and _____

vertebrae. Each kidney is about _____ cm (_____ inches) long. Visualize kidney location and size on yourself.

b. The kidneys are in an extreme *(anterior? posterior?)* position in the abdomen. They

are described as _____ since they are posterior to the peritoneum.

c. Identify the parts of the internal structure of the kidney on Figure LG 26.1. Then color all structures indicated by color code ovals.

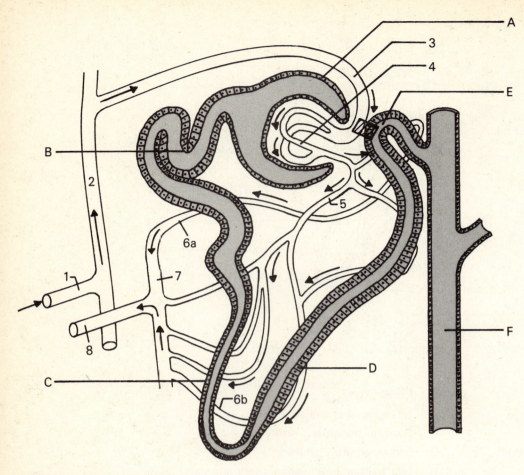

Figure LG 26.2 Diagram of a nephron. Letters refer to Checkpoint A4. Numbers refer to Checkpoints A4 and A8. The boxed area refers to Checkpoint A6.

■ **A4.** Do this exercise about the functional unit of the kidney, a nephron.

a. On Figure LG 26.2, label parts A–F of the nephron, with letters arranged according to the flow of urine.

b. Which one of those structures forms part of the renal corpuscle?

The other part of the renal corpuscle is a cluster of capillaries known as a

_____ and numbered *(4? 6B?)* in the figure.

c. Arrange in order the layers that fluid or solutes pass through as they move from blood into the forming urine (filtrate).

  A. Filtration slits between podocytes of visceral layer of glomerular capsule

  B. Basement membrane of the glomerulus

  C. Endothelial pores of capillary membrane

d. List the functional advantages of these structures: *endothelial pores* and *filtration slits*.

e. Identify the part of the renal tubule on the figure that consists of cuboidal epithelium with a brush border of many microvilli that increase absorptive surface.

_____ Which part consists of simple squamous epithelium? _____
f. In which regions are *principal cells* located? *(B and C? E and F?)* These cells are

sensitive to two hormones. Name them. _____ and

_____ The role of intercalated cells is to secrete _____.

■ **A5.** Contrast two types of nephrons by writing *C* for cortical nephrons and *JM* for jux-
tamedullary nephrons.

_____ a. Also known as short-looped nephrons

_____ b. Constitute about 15–20 percent of nephrons

_____ c. Have a thin (squamous epithelial) layer in ascending limb of loop of Henle

_____ d. Permit variations in concentration of urine

■ **A6.** Describe the juxtaglomerular apparatus (JGA) in this exercise.
  a. As shown in the boxed structure in Figure LG 26.2, the JGA consists of *(two?
     four?)* types of cells. One type is located in the final portion of the *(ascending? de-
     scending?)* limb of the loop of Henle or in the distal convoluted tubule. These cells,

known as _____

cells, monitor concentration of _____ in the tubular lumen.

  b. The second type of cell is a _____ cell located in the wall of
     the *(afferent arteriole? glomerulus?)*. These cells are known as

_____ cells; they secrete the chemical known as

_____, wnich regulates _____.

■ **A7.** About what percent of cardiac output passes through the kidneys each minute?

____ percent. This amounts to about _____ ml/min.

■ **A8.** Follow the pathway that blood takes through kidneys by naming vessels 1–8 on
  Figure LG 26.2. Some arrows are shown; add more to reinforce your understanding of
  the pathway of blood.

1. _____   3. _____   5. _____   7. _____

2. _____   4. _____   6. _____   8. _____

■ **A9.** Explain what is unique about the blood supply of kidneys in this activity.

   a. In virtually all tissues of the body, blood in capillaries flows into vessels named

   _____. However, in kidneys, blood in glomerular capillaries

   flows directly into _____.

   b. Which vessel has a larger diameter? *(Afferent? Efferent?)* arteriole. State one effect
   of this difference.

B. Physiology of urine formation   (pages 874–891)

■ **B1.** In what main ways is blood modified as it passes through the kidneys?

■ **B2.** List the three steps in urine production:

   _____   _____   _____

■ **B3.** Now describe the first step in urine production in this Checkpoint.

   a. Glomerular filtration is a process of *(pushing? pulling?)* fluids and solutes out of

   _____ and into the fluid known as _____.

   b. *(All types? All small substances? Only selected small substances?)* are forced out of
   blood. Refer to Figure LG 26.3, area A (plasma in glomeruli) and area B (filtrate
   just beyond the capsule). Compare the composition of these two fluids (described in
   the first two columns of Exhibit 26.1, page 880 in your textbook). Note that during

   filtration all solutes are freely filtered from blood except _____.
   Why is so little protein filtered?

   c. Blood pressure (or hydrostatic pressure) in glomerular capillaries is about _____
   mm Hg. This value is *(higher? lower?)* than that in other capillaries of the body.
   This extra pressure is accounted for by the fact that the diameter of the efferent arteriole is *(larger? smaller?)* than that of the afferent arteriole. Picture three garden
   hoses connected to each other, representing the afferent arteriole, glomerular capillaries, and efferent arteriole. The third one (efferent arteriole) is extremely narrow; it
   creates such resistance that pressure builds up in the first two. Fluids will be forced
   out of pores in the middle (glomerular) hose.

   d. List three structural features of renal corpuscles that enhance their filtration capacity.

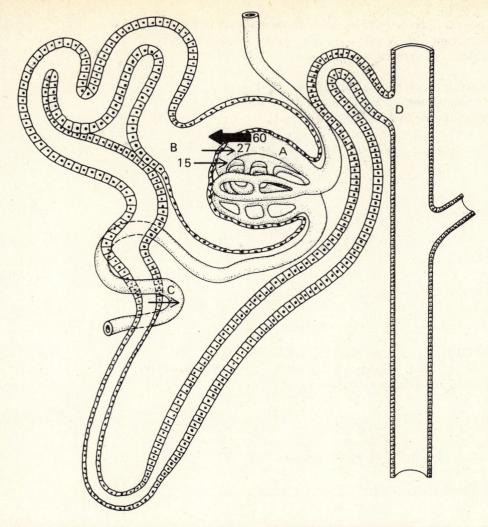

Figure LG 26.3 Diagram of a nephron with renal tubule abbreviated. Numbers refer to Checkpoint B4; letters refer to Checkpoints B3, B11, and B12.

■ **B4.** Blood hydrostatic pressure is not the only force determining the amount of filtration occurring in glomeruli.

a. Name two other forces (shown on Figure 26.10, page 876 in your textbook).

_____   _____

b. Normal values for these three forces are given on Figure LG 26.3. Write the formula for calculating effective filtration pressure NFP; draw an arrow beneath each term to show the direction of the force in Figure LG 26.3. Then calculate a normal NFP using the values given.

NFP = (glomerular blood hydrostatic pressure) – (_____ + _____),

Normal NFP = (_____ mm Hg) – (_____ mm Hg + _____ mm Hg)

= _____ mm Hg

Notice that blood (hydrostatic) pressure pushes *(out of? into?)* glomerular blood and is largely counteracted by the other two forces.

■ **B5.** *A clinical challenge.* Show the effects of alterations of these pressures in a pathological situation. Determine NFP of the patient whose pressure values are shown below. Note which values are abnormal and suggest causes.

> Glomerular blood hydrostatic pressure = 42
> Blood colloid osmotic pressure = 27
> Capsular filtrate hydrostatic pressure = 15

> NFP = _____

■ **B6.** Define GFR in this exercise.

a. To what do the letters GFR refer? g_____

f_____ r_____

b. Write a normal value for GFR. _____ ml/min (_____ liters/day)
c. If NFP decreases (as in the patient in Checkpoint B5), GFR____-creases. What consequences may be expected to accompany a decrease in GFR?

■ **B7.** List three mechanisms that regulate GFR: _____

_____ _____

■ **B8.** *A clinical challenge.* Mrs. K's blood pressure has dropped slightly. Explain how her renal autoregulation mechanism functions to ensure adequate renal blood flow.

a. When Mrs. K's systemic blood pressure decreases, her NFP and GFR are likely to *(increase? decrease?)*, causing filtrate to move more *(rapidly? slowly?)* through the renal tubules. As a result, *(more? less?)* fluid and salts are resorbed in the PCT and loop of Henle.
b. Consequently, the amount of fluid slowly passing the macula densa is *(high? low?)* in volume and concentration of salts. This condition causes JGA cells to produce *(more? fewer?)* of certain vasoconstrictor substances. As afferent arterioles then dilate, *(more? less?)* blood flows into glomerular capillaries, causing NFP and GFR to *(increase? decrease?)*.
c. This restoration of homeostasis by renal autoregulation occurs by a *(positive? negative?)* feedback mechanism. The name *autoregulation* indicates that neither the autonomic nor any chemical made outside the kidneys is involved. The nervous system control *(is? is not?)* intrinsic to kidneys.

■ **B9.** Now describe chemicals involved in other mechanisms that regulate GFR by completing this checkpoint. Select the chemical that matches each description. Use the following list of answers. (It may help to refer to Figure LG 18.6, page 360 in the Learning Guide.)

ACE. Angiotensin converting enzyme	Ang II. Angiotensin II
ADH. Antidiuretic hormone	ANP. Atrial natriuretic peptide (ANP)
Ald. Aldosterone	Epi. Epinephrine
Ang. I. Angiotensin I	R. Renin

_____ a. An enzyme secreted by JG cells of the kidney

_____ b. Produced by action of renin on a plasma protein made by liver

_____ c. Made by lungs, this enzyme converts angiotensin I to angiotensin II

_____ d. A vasoconstrictor that also stimulates thirst and release of aldosterone and ADH

_____ e. Produced by adrenal cortex, it causes collecting ducts to reabsorb more Na+ and water

_____ f. Released from the posterior pituitary, this hormone also increases blood volume

_____ g. The "third factor," it is produced by cells of the heart

_____ h. This hormone increases GFR; it opposes actions of renin, aldosterone, and ADH and suppresses secretion of those chemicals

_____ i. May have clinical applications in reducing edema by causing diuresis and natriuresis

_____ j. Released under sympathetic stimulation; lead to arteriolar vasoconstriction and decreased GFR (two answers)

B10. _A clinical challenge._ Joyce, an accident victim, was admitted to the hospital 3 hours ago. Her chart indicates that she had been hemorrhaging at the scene of the accident. Joyce's nurse monitors her urinary output and notes that it is currently 12 ml/hour. (Normal output is 30–50 ml/hour.) Explain why a person in such severe stress is likely to have decreased urinary output (oliguria).

■ **B11.** Do this exercise about the second step of urine formation.

a. Glomerular filtration is a(n) _(passive, nonselective? active, selective?)_ process. If this were the only step in urine formation, all of the substances in the filtrate would leave the body in urine. Note from Exhibit 26.1 (page 880 in your text) that the body

would produce _____ liters of urine each day! And it would contain many valuable substances. Obviously some of these "good" substances must be drawn back into blood and saved.

b. Recall that blood in most capillaries flows into vessels named

_____, but blood in glomerular capillaries flows into

_____ and _____. This unique arrangement permits blood to recapture some of the substances indiscriminately pushed out during filtration. This occurs during the second step of urine formation, called

_____. This process moves substances in the _(same? opposite?)_ direction as glomerular filtration, as shown by the direction of the arrow at area C of Figure LG 26.3.

■ **B12.** Refer to areas C and D on Figure LG 26.3 and to the two columns on the far right of Exhibit 26.1, page 880 of your text. Discuss the effects of tubular reabsorption in this learning activity.

a. Which solutes are 100 percent reabsorbed into area C, so that virtually none remains in urine (area D)?

b. About what percent of water that is filtered out of blood is reabsorbed back into

blood? _____ percent

c. Which chemicals are about 90 percent reabsorbed? _____

d. Which solute is about 50 percent absorbed? _____
e. Identify the solute that is filtered from blood and not reabsorbed.

_____ State the clinical significance of this fact.

f. Tubular reabsorption is a *(nonselective? discriminating?)* process which enables the body to save valuable nutrients, ions, and water.

B13. *For extra review.* Describe reabsorption mechanisms in this exercise.

a. About _____ percent of total resting metabolic energy is used for reabsorption of Na^+ in renal tubules by *(primary? secondary?)* active transport mechanisms. Explain how active transport of Na^+ also promotes osmosis of water and diffusion of substances such as K^+, Cl^-, HCO_3^-, and urea.

b. Reabsorption of glucose, amino acids, and lactic acid occurs almost entirely in the *(PCT? DCT?)*, and the process utilizes Na^+ *(symporters? antiporters?)*. This mechanism is an example of *(primary? secondary?)* active transport since it is the sodium pump, not the symporter, that uses ATP. If blood concentration of a substance such

as glucose is abnormally high, the renal _____ is surpassed. In other words, the blood level of glucose is greater than the T_m (or

_____ _____).

■ **B14.** *A clinical challenge.* Eileen and Marlene are both "spilling sugar" in urine; that is,

they have _____-uria. Discuss possible causes of glycosuria in each case, considering their blood sugar levels.

a. Eileen: glucose of 340 mg/100 ml plasma

b. Marlene: glucose of 80 mg/100 ml plasma

■ **B15.** Identify the part of the renal tubule that best fits the description. Use these answers.

```
    A.  Ascending limb of the loop of Henle       DCT.  Distal convoluted tubule
   CD.  Collecting duct                           PCT.  Proximal convoluted tubule
    D.  Descending limb of the loop of Henle
```

_____ a. This portion of the renal tubule with a prominent brush border permits tubular reabsorption of most filtered water and almost 100 percent of filtered nutrients.

_____ b. This is the first site where osmosis of water occurs without coupled reabsorption of other filtrates.

_____ c. Little or no water is reabsorbed here since these cells are impermeable to water.

_____ d. Variable amounts of Na⁺ and water may be reabsorbed here under the influence of aldosterone and ADH. (two answers)

■ **B16.** The third step in urine production is _____. It involves movement of substances from *(blood to urine? urine to blood?)*. In other words, tubular secretion is movement of substances in *(the same? opposite?)* direction as movement occurring in filtration. List four substances that are secreted by the process of tubular secretion.

■ **B17.** Explain how tubular secretion helps to control pH. Refer to Figure 26.17 (pages 885–886 in the text) and do this exercise.

a. Jenny's respiratory rate is slow. Her blood therefore contains high levels of

_____. It is also somewhat *(acidic? alkaline?)*. Explain why.

b. Jenny's kidneys can assist in restoring normal pH (Figure 26.17a and b). The presence of acid (H^+) stimulates the PCT cells to eliminate H^+. In order for this to happen, another cation present in urine must exchange places with H^+. What ion enters

kidney tubule cells? _____ It joins HCO_3^- to form _____.
This process requires *(antiporters? symporters?)*.

c. Figure 26.17c shows another possible mechanism for reducing acidity of blood. Intercalated cells located in *(PCT? collecting ducts?)* pump H^+ out into the lumen, causing urine to be up to *(10? 1000?)* times more acidic than blood. This pump *(does? does not?)* require ATP; the pump is an example of *(primary? secondary?)* active

transport. The leftover HCO_3^- (from H_2CO_3) may now enter peritubular blood, ____-creasing blood pH even more.

d. In addition, when blood is acidic, more ammonia (_____) is produced as a byproduct of *(lipid? protein?)* metabolism. H^+ in the collecting duct may then be buffered

by combining with ammonia to form ammonium (_____), which is excreted into

urine. Alternatively, H^+ may combine with _____ to form

_____, which is also excreted in urine.

■ **B18.** *A clinical challenge.* Of the following factors circle the ones that are likely to increase above normal in a patient with kidney failure. State the rationale for your answer.

      BUN   Plasma creatinine   GFR   Renal plasma clearance of glucose

■ **B19.** For a chemical to work effectively in renal clearance tests for evaluation of renal function, that chemical *(must? must not?)* be readily filtered and *(must? must not?)* be readily reabsorbed. Two chemicals that are often used for testing renal clearance are

_____ and _____.

■ **B20.** State an example of a situation in which your body needs to produce dilute urine in order to maintain homeostasis.

Now describe how your body can accomplish this by doing the following learning activity.

a. To form dilute urine, kidney tubules must reabsorb *(more? fewer?)* solutes than usual. Two factors facilitate this. One is permeability of the ascending limb to Na^+, K^+, and Cl^- *(and also to? but not to?)* water. Thus solutes enter interstitium and then capillaries, but water stays in urine.

b. The second requirement for dilute urine is a *(high? low?)* level of ADH. The function of ADH is to ____-crease permeability of distal and collecting tubules to water. Less ADH forces more water to stay in urine, leading to a *(hyper? hypo?)*-osmotic urine.

■ **B21.** Summarize factors that result in concentrated urine in this exercise. The counter-current mechanism provides a means for producing *(hypotonic? hypertonic?)* urine

during times when the body is _____. This occurs because the *(ascending? descending?)* limb of the loop of Henle is completely impermeable to H_2O, and it is also related to vasa recta permeability. As a result, the interstitial fluid of the renal

_____ is extremely hypertonic, that is, has a *(high? low?)*

osmolality, so pulls H_2O from urine in the _____ tubules. These tubules are made more permeable at this time under the influence of the hormone

_____. As a result, a *(large? small?)* volume of concentrated urine is produced, and the body retains more fluid.

B22. Explain why the *countercurrent mechanism* is so named.

B23. Compare and contrast hemodialysis and continuous ambulatory peritoneal dialysis (CAPD) as methods of cleansing blood.

C. Ureters, urinary bladder, and urethra (pages 891–895)

■ **C1.** Refer to Figure LG 26.1 as you complete the following exercise.

a. Ureters connect _____ to _____. Ureters

are about _____ cm (_____ inches) long.

b. These tubes enter the urinary bladder at two of the three corners of the

_____. The third corner marks the opening into the

_____.

c. The bladder is lined with _____ epithelium. Explain how the
presence of this type of epithelium serves the primary function of the urinary bladder.

d. The urinary bladder is located in the *(abdomen? true pelvis?)*. Two sphincters lie just
inferior to it. The *(internal? external?)* sphincter is under voluntary control.

e. Urine leaves the bladder through the _____. In females the

length of this tube is ____ cm (____ inches); in males it is about ____ cm (____
inches). Write in correct sequence (according to pathway of urine) the three portions

of the male urethra: _____ → _____ →

_____ →

■ **C2.** Ureters, urinary bladder, and urethra are lined with _____
membrane. What is the clinical significance of that fact?

■ **C3.** In the micturition reflex *(sympathetic? parasympathetic?)* nerves stimulate the

_____ muscle of the urinary bladder and cause relaxation of the
internal sphincter. What stimulus initiates the micturition reflex?

C4. Contrast urinary incontinence and retention. List possible causes of each.

D. Urine (pages 895–898)

■ **D1.** Review factors that influence urine volume in this Checkpoint.

a. *Blood pressure.* When blood pressure drops, _____ cells of the

kidney release _____, which leads to formation of

_____. This substance *(increases? decreases?)* blood pressure in
two ways: directly, since it serves as a *(vasoconstrictor? vasodilator?)*, and indirectly,

since it stimulates release of _____, which causes retention of

both _____ and _____. Increase in fluid
volume will *(increase? decrease?)* blood pressure.

b. *Blood concentration.* The hypothalamic hormone _____ helps
to *(conserve? eliminate?)* fluid. When body fluids are hypertonic, ADH is released,
causing *(increased? decreased?)* tubular reabsorption of water back into blood.

c. *Medications.* A chemical that mimics secretion of ADH is called a(n)

_____. One that inhibits (or is antagonistic to) ADH is called

a(n) _____. Diuretics cause diuresis which means a *(large?
small?)* volume of urine. Removal of extra ECF in this manner is likely to *(increase?
decrease?)* blood pressure.

d. *Temperature.* On hot days skin gives up *(large? small?)* volumes of fluid via sweat
glands. As blood becomes more concentrated, the posterior pituitary responds by releasing *(more? less?)* ADH, so urine volume *(increases? decreases?)*.

D2. *A clinical challenge.* Mrs. P is taking furosemide (Lasix) to lower her blood pressure. Answer the following questions.

a. Why is this diuretic called a *loop diuretic?*

b. How does action of this diuretic lower blood pressure?

c. Mrs. P's nurse reminds her to eat a banana or orange each day. Why?

d. Some time later Mrs. P is given a prescription for spironolactone (Aldactone)? Speculate as to why Mrs. P's doctor changed her diuretic.

D3. Describe the following characteristics of urine. Suggest causes of variations from the normal in each case.

a. Volume: _____ ml/day

b. Color: _____

c. Turbidity: _____

d. Odor: _____

e. pH: _____

f. Specific gravity: _____

D4. The following substances are not normally found in urine. Explain what the presence of each might indicate.

a. Albumin

b. Bilirubin

c. Casts

d. High levels of hippuric acid

D5. Contrast the following pairs of conditions. Describe what is present in urine in each case.

a. Gluosuria/acetonuria

b. Hematuria/pyuria

c. Vaginitis due to *Candida albicans*/vaginitis due to *Trichomonas vaginalis*

E. Aging and development of the urinary system (pages 898–900)

■ **E1.** GFR ____-creases with age. Two other common problems associated with aging

are inability to control voiding (_____) and increased frequency of

_____. Explain why UTIs may occur more often in elderly males.

Nocturia may also occur more among the aged population. Define *nocturia*.

■ **E2.** Do this exercise about urinary system development.

a. Kidneys form from _____-derm, beginning at about the

_____ week of gestation.

b. Which develops first? *(Pro? Meso? Meta?)*-nephros. Which extends most superiorly

in location? _____-nephros. Which one ultimately forms the

kidney? _____

c. The urinary bladder develops from the original _____, which is

derived from _____-derm.

F. Disorders, medical terminology (pages 900–901)

■ **F1.** Match each of the terms with the correct description.

> C. Cystitis
> D. Dysuria
> E. Enuresis
> Gl. Glomerulonephritis
>
> Po. Polyuria
> Py. Pyelitis
> U. Uremia

_____ a. Painful urination

_____ b. Inflammation of the urinary bladder

_____ c. Inflammation of the pelvis and caly-
            ces of kidney

_____ d. Bed-wetting

_____ e. Excessive urine

_____ f. Inflammation of the kidney involving
            glomeruli; may follow strep infection

F2. Discuss and contrast two types of renal failure in this learning activity.

a. *(Acute? Chronic?)* kidney failure is an abrupt cessation (or almost) of kidney function. List several causes of acute renal failure (ARF):

b. *(Acute? Chronic?)* renal failure is progressive and irreversible. Describe changes that
   occur during the three stages of CRF.

■ **F3.** Urinary tract infections (UTIs) are most often caused by Gram-*(positive?*

negative?) bacteria named _____. List three or more signs or
symptoms of UTIs.

ANSWERS TO SELECTED CHECKPOINTS

A1. (a) Nitrogen-containing products of protein catabolism, such as ammonia and urea; also certain ions and excessive water. (b) H^+. (c) Renin, erythropoietin, D.

A3. (a) Waist; T12, L3; 10–12 cm (4–5 inches). (b) Posterior; retroperitoneal. (c) See key to Figure LG 26.1.

A4. (a) A, glomerular (Bowman's) capsule; B, proximal convoluted tubule (PCT); C, descending limb of the loop of the nephron (loop of Henle); D, ascending limb of the loop of the nephron (loop of Henle); E, distal convoluted tubule (DCT); F, collecting duct. (b) A, glomerular (Bowman's) capsule; glomerulus, 4. (c) C B A. (d) Their size restricts the passage of red blood cells and proteins from blood to urine. (e) B; C and the first part of D. (f) E and F; ADH and aldosterone; H^+.

A5. (a) C. (b–d) JM.

A6. (a) Two; ascending; macula densa, NaCl. (b) Modified smooth muscle, afferent arteriole; juxtaglomerular (JG); renin, blood pressure.

A7. 20–25; 1200 ml/min.

A8. 1, Arcuate artery; 2, interlobular artery; 3, afferent arteriole; 4, glomerulus; 5, efferent arteriole; 6a, peritubular capillaries; 6b, vasa recta; 7, interlobular vein; 8, arcuate vein.

A9. (a) Venules; efferent arterioles. (b) Afferent; increases glomerular blood pressure. (See Checkpoint B3.)

B1. Its volume, electrolyte content, and pH are adjusted; toxic wastes are removed.

B2. Glomerular filtration, tubular reabsorption, tubular secretion.

B3. (a) Pushing, blood, filtrate. (b) All small substances; protein molecules. They are too large to pass through filtration slits of healthy endothelial-capsular membranes. (c) 60; higher; smaller. (Refer to Checkpoint A9b.) (d) Glomerular capillaries are long, thin, and porous, and have high blood pressure.

B4. (a) Blood colloid osmotic pressure (BCOP) and capsular filtrate hydrostatic pressure (CHP) (b) NFP = (GBHP) – (CHP + BCOP); NFP = $(60) - (\overset{\leftarrow}{15} + \overset{\rightarrow}{27})$; $\overset{\rightarrow}{18}$; out of.

B5. (a) $\overset{\rightarrow}{NFP}$ = 0 mm Hg = (42) – (27 + 15). Anuria due to low glomerular blood pressure could be related to hemorrhage or stress.

B6. (a) Glomerular filtration rate. (b) 125; 180. (c) De; additional water and waste products (such as H^+ and urea) normally eliminated in urine are retained in blood, possibly resulting in hypertension and acidosis.

B7. Renal autoregulation, hormonal regulation, and neural regulation

B8. (a) Decrease, slowly; more. (b) Low; fewer; more, increase. (c) Negative; is.

B9. (a) R. (b) Ang. I. (c) ACE. (d) Ang. II. (e) Ald. (f) ADH. (g–i) ANP. (j) R, epi.

B11. (a) Passive, nonselective; 180. (b) Venules, peritubular capillaries and vasa recta; tubular reabsorption; opposite.

B12. (a) Glucose and bicarbonate (HCO_3^-). (b) 99. (c) Protein, Na^+, Cl^-, K^+, and uric acid. (d) Urea. (e) Creatinine; creatinine clearance is often used as a measure of glomerular filtration rate (GFR). 100 percent of the creatinine (a breakdown product of muscle protein) filtered or "cleared" out of blood will show up in urine since 0 percent is reabsorbed. (e) Discriminating.

B14. Glycos. (a) Eileen's hyperglycemia could be caused by ingestion of excessive amounts of carbohydrates or by hormonal imbalance, such as deficiency in effective insulin (diabetes mellitus) or excess growth hormone. Eileen's glucose level of 340 mg/100 ml plasma exceeds the renal threshold of about 200 mg/100 ml plasma since kidneys have a limited transport maximum (T_m) for glucose. (b) Although Marlene's blood glucose is in the normal (70–110 mg/per 100 ml) range, she may have a rare kidney disorder with a reduced transport maximum (T_m) for glucose.

B15. (a) PCT. (b) D. (c) A. (d) DCT and CD.

B16. Tubular secretion; blood to urine; the same. K^+, H^+, ammonium (NH_4^+), creatinine, penicillin, and para-aminohippuric acid.

B17. (a) CO_2; acidic; increasing levels of CO_2 tend to cause increase of H^+: $CO_2 + H_2O \rightarrow H_2CO_3 \rightarrow H^+ + HCO_3^-$. (Review content on transport of CO_2 in Chapter 23.) (b) Na^+; $NaHCO_3^-$ (sodium bicarbonate); antiporters. (c) Collecting ducts, 1000; does; primary; in. (d) NH_3, protein; NH_4; HPO_4^{2-}, $H_2PO_4^-$.

B18. BUN and plasma creatinine both increase in blood since less urea and creatinine are filtered from blood. Since GFR decreases, kidneys "clear out" less glucose from blood (renal plasma clearance of glucose decreases), resulting in hyperglycemia.

B19. Must, must not; creatinine and insulin.

B20. During periods of high water intake or decreased sweating. (a) More; but not to. (b) Low; in; hypo.

B21. Hypertonic, dehydrated; ascending; inner medulla, high, collecting; ADH; small.

C1. (a) Kidneys to urinary bladder; 25–30 (10–12) (b) Trigone; urethra. (c) Transitional; this type of epithelium stretches and becomes thinner as the bladder fills. (d) True pelvis; external. (e) Urethra; 3.8 (1.5); 20 (8); prostatic → membranous → spongy.

C2. Mucous; microbes can spread infection from the exterior of the body (at urethral orifice) along mucosa to kidneys.

C3. Parasympathetic; detrusor; stretching the walls of the bladder by more than 200–400 ml (about 0.8–1.6 cups) of urine.

D1. (a) Juxtaglomerular, renin, angiotensin II, increases, vasoconstrictor, aldosterone, Na^+ and H_2O; increase. (b) ADH, conserve; increased. (c) Antidiuretic; diuretic; large; decrease. (d) Large; more, decreases.

E1. De; incontinence, urinary tract infections (UTIs); increased incidence of prostatic hypertrophy leads to urinary retention; excessive urination at night.

E2. (a) Meso, third. (b) Pro; pro; meta. (c) Cloaca (urogenital sinus portion); endo.

F1. (a) D. (b) C. (c) Py. (d) E. (e) Po. (f) Gl.

F3. Negative; *E. coli;* burning or painful urination, urinary urgency and frequency, back pain, cloudy or blood-tinged urine, urethral discharge, fever, chills, nausea.

MASTERY TEST: Chapter 26

Questions 1–6: Arrange the answers in correct sequence.

_____ _____ _____ 1. From superior to inferior:
    A. Ureter
    B. Bladder
    C. Urethra

_____ _____ _____ _____ 2. From most superficial to deepest:
    A. Renal capsule
    B. Renal medulla
    C. Renal cortex
    D. Renal pelvis

_____ _____ _____ _____ _____ 3. Pathway of glomerular filtrate:
    A. Ascending limb of loop
    B. Descending limb of loop
    C. Collecting tubule
    D. Distal convoluted tubule
    E. Proximal convoluted tubule

_____ _____ _____ _____ _____ 4. Pathway of blood:
    A. Arcuate arteries
    B. Interlobular arteries
    C. Renal arteries
    D. Afferent arteriole
    E. Interlobar arteries

_____ _____ _____ _____ _____ 5. Pathway of blood:
    A. Afferent arteriole
    B. Peritubular capillaries and vasa recta
    C. Glomerular capillaries
    D. Venules and veins
    E. Efferent arteriole

_____ _____ _____ _____ _____ _____

6. Order of events to restore low blood pressure to normal:
   A. This substance acts as a vasoconstrictor and simulates aldosterone.
   B. Renin is released and converts angiotensinogen to angiotensin I.
   C. Angiotensin I is converted into angiotensin II.
   D. Decrease in blood pressure or sympathetic nerves stimulate juxtaglomerular cells.
   E. Under influence of this hormone Na^+ and H_2O reabsorption occurs.
   F. Increased blood volume raises blood pressure.

Questions 7–12: Circle the letter preceding the one best answer to each question.

7. Which of these is a normal constituent of urine?
   A. Albumin   D. Casts
   B. Urea   E. Acetone
   C. Glucose
8. Which is a normal function of the bladder?
   A. Oliguria   C. Calculi
   B. Nephrosis   D. Micturition
9. Which parts of the nephron are composed of simple squamous epithelium?
   A. Ascending and descending limbs of loop of the nephron
   B. Glomerular capsule (parietal layer) and descending limb of loop of the nephron
   C. Proximal and distal convoluted tubules
   D. Distal convoluted and collecting tubules
10. Choose the one *false* statement.
    A. An increase in glomerular blood hydrostatic pressure (GBHP) causes a decrease in net filtration pressure (NFP).
    B. Glomerular capillaries have higher blood pressure than other capillaries of the body.
    C. Blood in glomerular capillaries flows into arterioles, not into venules.
    D. Vasa recta pass blood from the efferent arteriole toward veins.

11. Choose the one *true* statement about renal corpuscle structure and function.
    A. Afferent arterioles offer more resistance to flow than do efferent arterioles.
    B. An efferent arteriole normally has a larger diameter than an afferent arteriole.
    C. The juxtaglomerular apparatus (JGA) consists of cells of the proximal convoluted tubule (PCT) and the afferent arteriole.
    D. Most water reabsorption normally takes place across the proximal convoluted tubule (PCT).
12. Which one of the hormones listed here causes increased urinary output and natriuresis?
    A. ANP   C. Aldosterone
    B. ADH   D. Angiotensin II

T F   13. Antidiuretic hormone (ADH) <u>decreases</u> permeability of distal and collecting tubules to water, so <u>increases</u> urine volume.

T F   14. A ureter is a tube that carries urine from <u>the bladder to the outside of the body.</u>

T F   15. Diabetes insipidus is a condition caused by a defect in production of <u>insulin</u> or lack of sensitivity to this hormone by principal cells in collecting ducts.

T F   16. A person taking diuretics is likely to urinate <u>less</u> than a person taking no medications.

T F   17. Blood colloid osmotic pressure (BCOP) is a force that tends to <u>push substances from blood into filtrate.</u>

T F   18. The ascending limb of the loop of the nephron is permeable to <u>Na^+ and Cl^-, but not to water.</u>

T F   19. The countercurrent multiplier mechanism permits production of <u>small volumes of concentrated urine.</u>

T F   20. In infants 2 years old and under, <u>retention</u> is normal.

Questions 21–25: fill-ins. Complete each sentence with the word or phrase that best fits.

_____ 21. The glomerular capsule and its enclosed glomerular capillaries constitute a ____.

_____ 22. A ____ consists of a renal corpuscle and a renal tubule.

_____ 23. A toxic level of urea in blood is a condition known as ____.

_____ 24. Sympathetic impulses cause greater constriction of ____ arterioles with resultant ____-crease in GFR and ____-crease in urinary output.

_____ 25. A person with chronic renal failure is likely to be in acidosis since kidneys fail to excrete ____, may be anemic since kidneys do not produce ____, and may have symptoms of hypocalcemia since kidneys do not activate ____.

ANSWERS TO MASTERY TEST: ■ Chapter 26

Arrange

1. A B C
2. A C B D
3. E B A D C
4. C E A B D
5. A C E B D
6. D B C A E F

Multiple Choice

7. B   10. A
8. D   11. D
9. B   12. A

True-False

13. F. Increases; decreases
14. F. A kidney to the bladder
15. F. ADH
16. F. More
17. F. Pull substances from filtrate into blood
18. T
19. T
20. F. Incontinence

Fill-ins

21. Renal corpuscle
22. Nephron
23. Uremia
24. Afferent, de, de
25. H^+, erythropoietic factor, vitamin D

FRAMEWORK 27
Fluid, Electrolyte, Acid-Base Balance

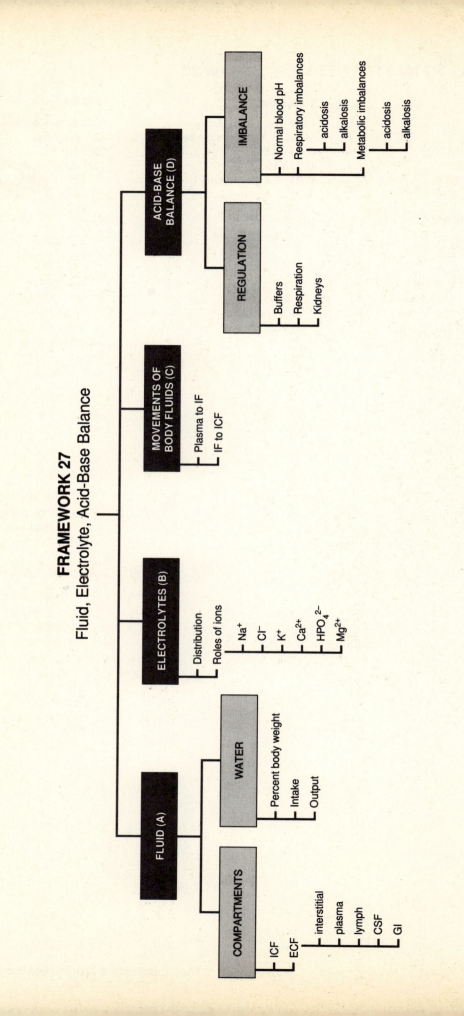

Fluid, Electrolyte, and Acid-Base Homeostasis

For the maintenance of homeostasis, levels of fluid, ions or electrolytes, and acids and bases must be kept within acceptable limits. Without careful regulation of pH, for example, nerves will become overactive (in alkalosis) or underfunction, leading to coma (in acidosis). Alterations in blood levels of electrolytes can be lethal: excess K^+ can cause cardiac arrest, and decrease in blood Ca^{2+} can result in tetany of the diaphragm and respiratory failure. Systems you studied in Chapter 23 (respiratory) and Chapter 26 (urinary), along with buffers in the blood, normally achieve fluid, electrolyte, and acid-base balance.

Look first at the Chapter 27 Framework and the essential terms.

TOPIC OUTLINE AND OBJECTIVES

A. Fluid compartments, water

1. Describe the various fluid compartments of the body.
2. Define the processes available for fluid intake and fluid output and how they are regulated.

B. Electrolytes

3. Contrast the electrolytic concentration of the three major fluid compartments.
4. Explain the functions of sodium, chloride, potassium, calcium, phosphate, and magnesium and regulation of their concentrations.

C. Movement of body fluids

5. Describe the factors involved in the movement of fluid between plasma and interstitial fluid, and between interstitial fluid and intracellular fluid.

D. Acid-base balance and imbalance

6. Compare the role of buffers, exhalation of carbon dioxide, and kidney excretion of H^+ in maintaining pH of body fluids.
7. Define acid-base imbalances and describe their effects on the body and how they are treated.

WORDBYTES

Now study the following parts of words that may help you better understand the terminology in this chapter.

Wordbyte	Meaning	Example	Wordbyte	Meaning	Example
hyper-	above	*hyper*ventilation	-osis	condition of	alkal*osis*
hypo-	below	*hypo*tonic			

CHECKPOINTS

A. Fluid compartments, water (pages 905–907)

■ **A1.** About 66 percent of body fluids are *(inside of? outside of?)* cells. Check your understanding of fluid compartments by circling all fluids that are considered *extracellular fluids (ECF):*

Blood plasma	Lymph
Aqueous humor and vitreous humor of the eyes	Pericardial fluid
Endolymph and perilymph of ears	Synovial fluid
Fluid within liver cells	Cerebrospinal fluid (CSF)
Fluid immediately surrounding liver cells	

For extra review, refer to Figure LG 1.1, page 8 of the Learning Guide.

■ **A2.** *A clinical challenge.* Is a 45-year-old woman who has a water content of 55 percent likely to be in fluid balance?

■ **A3.** About what percent of an adult's body weight consists of water? _____ Circle the person in each pair who is more likely to have a higher proportion of water content.
 a. 37-year-old woman/67-year-old woman
 b. Female/male
 c. Lean person/fat person

■ **A4.** Daily intake and output (I and O) of fluid usually both equal _____ ml

(_____ liters). Of the intake, about _____ ml are ingested in food and drink, and

_____ ml are a product of catabolism of foods (discussed in Chapter 25).

■ **A5.** Now list four systems of the body that eliminate water. Indicate the amounts lost by each system on an average day. One is done for you.

a. Integument (skin): 500 ml/day

b. _____ : _____ ml/day

c. _____ : _____ ml/day

d. _____ : _____ ml/day

On a day when you exercise vigorously, how would you expect these values to change?

■ **A6.** Think of the last time you felt very thirsty. What led you to drink fluids? List three mechanisms.

All three of these mechanisms ultimately signal the thirst center located in your *(hypothalamus? medulla? pharynx?)*. Your thirst dissipated due to presence of fluids in your mouth and pharynx. In addition, quenching of your thirst may have resulted from

inhibition of your thirst center by _____.

■ **A7.** Circle the factors that increase fluid output:

A. ADH
B. Aldosterone
C. Diarrhea
D. Increase in blood pressure (BP)

E. Fever
F. Hyperventilation
G. Atrial natriuretic peptide (ANP)

B. Electrolytes   (pages 907–911)

B1. Contrast terms in each pair

a. mEq/liter   mOsm/liter

b. osmotic effect of electrolyte   osmotic effect of nonelectrolyte

c. deciliter(dl)   liter

■ **B2.** Circle all of the answers that are classified as electrolytes:

A. Proteins in solutions   D. K^+ and Na^+

B. Anions   E. Compounds containing at least one ionic bond

C. Glucose

B3. List four general functions of electrolytes.

_____   _____

_____   _____

■ **B4.** On Figure LG 27.1a, write the chemical symbols for electrolytes that are found in greatest concentrations in each of the three compartments. Write one symbol on each line provided. Use the following list of symbols. (Hint: Refer to Figure 27.4, page 909 of the text.)

Cl^-. Chloride	Mg^{2+}. Magnesium
HCO_3^-. Bicarbonate	Na^+. Sodium
HPO_4^{2-}. Phosphate	Protein$^-$. Protein
K^+. Potassium	

■ **B5.** Analyze the information in Figure LG 27.1 by answering these questions.

a. Circle the two compartments that are most similar in electrolyte concentration.
cell   plasma   interstitial fluid (IF)

b. List two major differences in electrolyte content of these two similar compartments?

c. Most body protein is located in which compartment? _____

d. Protein anions in intracellular fluid are more likely to be bound to the cation *(Na$^+$?* *K$^+$?)*.

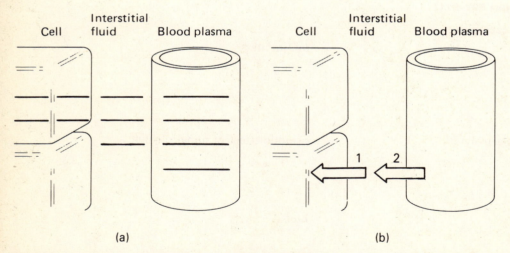

(a)   (b)

Figure LG 27.1   Diagrams of fluid compartments. (a) Write in electrolytes according to Checkpoint B4. (b) Direction of fluid shift resulting from Na$^+$ deficit. Color according to Checkpoint C3.

B6. Check your understanding of the functions and disorders related to major electrolytes by completing Table LG 27.1.

Table LG 27.1 Summary of Major Electrolytes

Electrolyte (Name and Symbol)	Normal Range (Serum Level), in mEq/liter	Principal Functions	Signs and Symptoms of Imbalances:	
			Hyper-	Hypo-
a.	136–142			
b.				Hypochloremia: muscle spasms, alkalosis, ¹respirations, coma
c.		Most abundant cation in ICF; helps maintain fluid volume of cells; functions in nerve transmission		
d.			Hypercalcemia: weakness, lethargy, confusion, stupor, and coma	
e.	1.7–2.6	Helps form bones, teeth; important components of DNA, RNA, ATP, and buffer system		
f. Mag-nesium (Mg2+)				

■ **B7.** *A clinical challenge.* Norine is having her serum electrolytes analyzed. Complete this exercise about her results.

a. Her Na⁺ level is 128. Norine's value indicates that she is more likely to be in a state of *(hyper? hypo?)*-natremia. Write two or more signs or symptoms of this electrolyte imbalance.

b. A diagnosis of hypochloremia would indicate that Norine's blood level of

_____ ion is lower than normal. A normal range for serum chloride (Cl⁻) level is

_____ mEq/liter. One cause of low chloride is excessive _____.

c. A normal range for K⁺ in serum is 3.8–5.0 mEq/liter. This range is considerably *(higher? lower?)* than that for Na⁺. This is reasonable since K⁺ is the main cation in *(ECF? ICF?)*, yet lab analysis of serum electrolytes is examining K⁺ in *(ECF? ICF?)*.

d. Norine's potassium (K^+) level is 5.6 mEq/liter, indicating that she is in a state of

hyper- _____. Hyperkalemia and hyponatremia (as described in a) may be caused by *(high? low?)* levels of aldosterone. Explain why.

If Norine's serum K^+ continues to increase, for example to 8.0 mEq/l, her most

serious risk, possibly leading to death, is likely to be _____.

e. A normal range for Mg^{2+} level of blood is 1.3–2.1 mEq/liter. Norine's electrolyte report indicates a value of 1.1 mEq/liter for Mg^{2+}. This electrolyte imbalance is known as _____. List three or more factors related to nutritional state that could lead to this condition.

f. Norine's Ca^{2+} level is 3.9 mEq/liter. A normal range is 4.6–5.5 mEq/liter. This

electrolyte imbalance, known as _____, is most closely linked

to _____-creased levels of parathyroid hormone (PTH) and _____-creased secre-

tion of the thyroid hormone, _____. Low blood levels of both Ca^{2+} and Mg^{2+} cause overstimulation of the central nervous system and muscles. List two symptoms of these electrolyte imbalances.

C. Movement of body fluids (pages 911–912)

■ **C1.** Four forces are involved in the movement of water between plasma and interstitial compartments and between interstitial and intracellular compartments. (They include, incidentally, the same types of forces involved in glomerular filtration.) Movement across capillary membranes due to these four forces is according to

_____ law of capillaries.

C2. Review Checkpoint B6 in Chapter 21 (page 429 in the Learning Guide) on calculation of net filtration pressure (NFP) that determines movement of fluids between blood and interstitial fluid.

■ **C3.** Show on Figure LG 27.1b the shift of fluid resulting from loss of Na^+, for example by excessive loss of sweat. Color arrows (1 and 2) according to these color code ovals:

O Shift of fluid possibly leading to hypovolemic shock
O Shift of fluid leading to overhydration or water intoxication of brain cells, possibly leading to convulsions and coma

D. Acid-base balance and imbalance (pages 912–917)

■ **D1.** Complete this Checkpoint about acid-base regulation.

a. The pH of blood and other ECF should be maintained between _____ and

_____.

b. Name the three major mechanisms that work together to maintain acid-base balance.

_____   _____   _____

c. List four important buffer systems in the body.

_____   _____

_____   _____

■ **D2.** Refer to Figure LG 27.2 and complete this exercise about buffers.

a. In Figure LG 27.2, you can see why buffers are sometimes called "chemical sponges." They consist of an anion (in this case bicarbonate, or HCO_3^-) that is combined with a cation, which in this buffer system may be either H^+ or Na^+. When H^+

is drawn to the HCO_3^- "sponge," the weak acid _____ is

formed. When Na^+ attaches to HCO_3^-, the weak base _____ is present. As you color H^+ and Na^+, imagine these cations continually trading places as they are "drawn in or squeezed out of the sponge."

b. Buffers are chemicals that help the body to cope with *(strong? weak?)* acids or bases that are easily ionized and cause harm to the body. When a strong acid, such as hydrochloric acid (HCl), is added to body fluids, buffering occurs, as shown in Figure LG 27.2b. The easily ionized hydrogen ion (H^+) from HCl is "absorbed by the sponge" as Na^+ is released from the buffer/sponge to combine with Cl^-. The two

products, _____ and _____, are weak (less easily ionized).

c. Now complete Figure LG 27.2c by writing the names of products resulting from buffering of a strong base, H^+ (and its arrow) and Na^+ (and its arrow) to show this exchange reaction.

d. When the body continues to take in or produce excess strong acid, the concentration

of the _____ member of this buffer pair will decrease as it is used up in the attempt to maintain homeostasis of pH.

e. Though buffer systems provide rapid response to acid-base imbalance, they are limited since one member of the buffer pair can be used up. And they can convert only strong acids or bases to weak ones; they cannot eliminate them. Two systems of the body that can actually eliminate acid or basic substances are

_____ and _____.

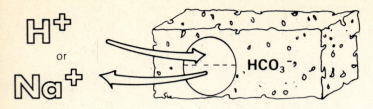

a. H⁺ and Na⁺: only one of you can join me at any one time.

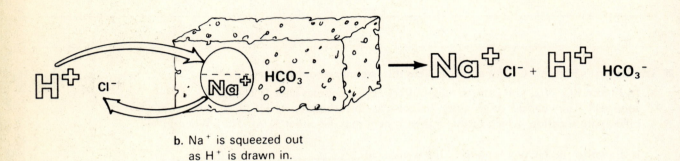

b. Na⁺ is squeezed out as H⁺ is drawn in.

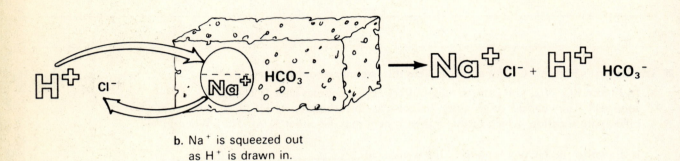

c. H⁺ is squeezed out as Na⁺ is drawn in.

○ H⁺ and arrows showing movement of H⁺

○ Na⁺ and arrows showing movement of Na⁺

Figure LG 27.2 Buffering action of carbonic acid–sodium bicarbonate system. Complete as directed in Checkpoint D2.

■ **D3.** *For extra review.* Check your understanding of the carbonic acid–bicarbonate buffer in this Checkpoint. When a person actively exercises, *(much? little?)* CO_2 forms, leading to H_2CO_3 production. Consequently, *(much? little?)* H^+ will be added to body tissues. In order to buffer this excess H^+ (to avoid drop in pH and damage to tissues), the *(H_2CO_3? $NaHCO_3$?)* component of the buffer system is called upon. The excess H^+

then replaces the _____ ion of $NaHCO_3$, in a sense "tying up" the potentially harmful H^+.

■ **D4.** Phosphates are more concentrated in *(ECF? ICF?)*. (For help, see Figure 27.4, page 909 in the text.) Name one type of cell in which the phosphate buffer system is

most important. _____

D5. Write the reaction showing how phosphate buffers strong acid.

■ **D6.** Which is the most abundant buffer system in the body? _____
Circle the part of the amino acid shown below that buffers acid. Draw a square around the part of the amino acid that buffers base.

$$NH_2—\overset{\displaystyle R}{\underset{\displaystyle H}{\overset{\displaystyle |}{\underset{\displaystyle |}{C}}}}—COOH$$

D7. Explain how the hemoglobin buffer system buffers carbonic acid, so that an acid even weaker than carbonic acid (that is, hemoglobin) is formed.

For extra review. See Chapter 23, Checkpoint E4, page 494 in the Learning Guide, and the diagram on page 914 in the text.

■ **D8.** Complete this Checkpoint about respiration and kidneys related to pH.

a. Hyperventilation will tend to *(raise? lower?)* blood pH, since as the person exhales

CO_2, less CO_2 is available for formation of _____ acid and free hydrogen ion.

b. A slight decrease in blood pH will tend to *(stimulate? inhibit?)* the respiratory center and so will *(increase? decrease?)* respirations.

c. The kidneys regulate acid-base balance by altering their tubular secretion of

_____ or elimination of _____ ion.

■ **D9.** An acid-base imbalance caused by abnormal alteration of the respiratory system is classified as *(metabolic? respiratory?)* acidosis or alkalosis. Any other cause of acid-base imbalance, such as urinary or digestive tract disorder, is identified as *(metabolic? respiratory?)* acidosis or alkalosis.

■ **D10.** *A clinical challenge.* Indicate which of the four categories of acid-base imbalances listed in the box is most likely to occur as a result of each condition described below.

MAcid. Metabolic acidosis	RAcid. Respiratory acidosis
MAlk. Metabolic alkalosis	RAlk. Respiratory alkalosis

_____ a. Decreased blood level of CO_2 as a result of hyperventilation

_____ b. Decreased respiratory rate in a patient taking overdose of morphine

_____ c. Excessive intake of antacids

_____ d. Prolonged vomiting of stomach contents

_____ e. Ketosis in uncontrolled diabetes mellitus

_____ f. Decreased respiratory minute volume in a patient with emphysema or fractured rib

_____ g. Excessive loss of bicarbonate from the body, as in prolonged diarrhea or renal dysfunction

D11. Briefly describe treatments that may be used for:

a. Respiratory acidosis

b. Metabolic acidosis

c. Respiratory alkalosis

d. Metabolic alkalosis

ANSWERS TO SELECTED CHECKPOINTS

A1. Inside of; all answers except fluid within liver cells.

A2. The percentage is within the normal range; however, more information is needed since the fluid must be distributed correctly among the three major compartments: cells (ICF) and ECF in plasma and interstitial fluid. For example, excessive distribution of fluid to interstitial space occurs in the fluid imbalance known as edema.

A3. About 60%. (a) 37-year-old woman. (b) Male. (c) Lean person.

A4. 2500 (2.5); 2300, 200.

A5. (Any order) (b) Kidney, 1500. (c) Lungs, 300. (d) GI, 200. Output of fluid increases greatly via sweat glands and slightly via respirations. To compensate, urine output decreases and intake of fluids increases.

A6. Thirst, increase of osmotic pressure, and release of renin that leads to production of angiotensin; hypothalamus; fluids stretching your stomach or intestine.

A7. C–G.

B2. All except C (glucose).

B4.

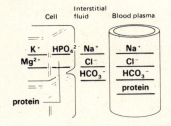

Figure LG 27.1A Diagram of fluid compartments.

B5. (a) Plasma and interstitial fluid. (b) Plasma contains more plasma protein, and slightly more sodium and less chloride than IF. (c) Intracellular (ICF). (d) K^+.

B7. (a) Normal serum Na^+ ranges from 136–142; hypo; examples: headache, low blood pressure, tachycardia, and weakness, possibly leading to confusion, stupor, and coma if her serum Na^+ continues to drop. (b) Cl^-; 95–103; emesis, dehydration or use of certain diuretics. (c) Lower; ICF, ECF. (d) Kalemia; low; aldosterone promotes reabsorption of Na^+ and H_2O and also Mg^{2+} into blood in kidneys, and secretion of K^+ into urine; ventricular fibrillation resulting in cardiac arrest. (e) Hypomagnesemia; malnutrition, malabsorption, diarrhea, alcoholism, or excessive lactation. (f) Hypocalcemia, de, in, calcitonin; tetany, convulsions.

C1. Starling's.

C3. 1, Shift of fluid leading to overhydration or water intoxication of brain cells, possibly leading to convulsions and coma; 2, shift of fluid from plasma to replace IF that entered cells. This may lead to hypovolemic shock.

D1. (a) 7.35–7.45. (b) Buffers, respiration, and kidney excretion. (c) Carbonic acid–bicarbonate, phosphate, hemoglobin-oxyhemoglobin, protein.

D2. (a) Carbonic acid, or H_2CO_3; sodium bicarbonate, or $NaHCO_3$. (b) Strong; the salt sodium chloride (NaCl) and the weak acid H_2CO_3 ($H \cdot HCO_3$). (c) Water (H_2O or $H \cdot OH$) and sodium bicarbonate ($NaHCO_3$). (d) $NaHCO_3$ or weak base. (e) Respiratory (lungs), urinary (kidneys).

D3. Much; much; $NaHCO_3^-$; Na^+.

D4. ICF; kidney cells.

D6. Protein;

$$\text{Buffers acid} - \boxed{NH_2} - \overset{\overset{\displaystyle R}{|}}{\underset{\underset{\displaystyle H}{|}}{C}} - \boxed{COOH} - \text{Buffers base}$$

D8. (a) Raise, carbonic. (b) Stimulate, increase. (c) H^+, HCO_3^-.

D9. Respiratory; metabolic.

D10. (a) RAlk. (b) RAcid. (c) MAlk. (d) MAlk. (e) MAcid. (f) RAcid. (g) MAcid.

MASTERY TEST: Chapter 27

Questions 1–16: Circle T (true) or F (false). If the statement is false, change the underlined word or phrase so that the statement is correct.

T F 1. Hyperventilation tends to <u>raise</u> pH.

T F 2. Carbonic acid is a <u>weaker</u> acid than reduced hemoglobin (H·Hb).

T F 3. Under normal circumstances fluid intake each day <u>is greater than</u> fluid output.

T F 4. During edema there is increased movement of fluid <u>out of plasma and into interstitial fluid.</u>

T F 5. Parathyroid hormone causes an increase in the blood level of <u>calcium</u>.

T F 6. When aldosterone is in <u>high</u> concentrations, sodium is conserved (in blood) and potassium is excreted (in urine).

T F 7. In general, electrolytes cause <u>greater</u> osmotic effects than nonelectrolytes.

T F 8. The body of an infant contains a <u>higher</u> percentage of water than the body of an adult.

T F 9. Starch, glucose, HCO_3^-, Na^+, and K^+ are all electrolytes.

T F 10. Loss of blood via hemorrhage is likely to decrease blood hydrostatic pressure <u>and also to decrease NFP.</u>

T F 11. <u>Protein</u> is the most abundant buffer system in the body.

T F 12. Sodium plays a much <u>greater</u> role than magnesium in osmotic balance of ECF since <u>more mEq/liter of sodium are in ECF.</u>

T F 13. The <u>carbonic acid</u> portion of the bicarbonate–carbonic acid system buffers base.

T F 14. The pH of arterial blood should be <u>higher</u> than that of venous blood and should be about pH <u>7.40 to 7.45.</u>

T F 15. Hydrostatic pressure of blood and hydrostatic pressure of interstitial fluid tend to <u>move substances in the same direction: out of blood and into interstitial fluid.</u>

T F 16. Interstitial fluid, blood plasma, and lymph are all <u>extracellular</u> fluids.

Questions 17–20: Circle the letter preceding the one best answer to each question.

17. On an average day the greatest volume of fluid output is via:
    A. Feces
    B. Sweat
    C. Lungs
    D. Urine

18. Which term refers to a lower than normal blood level of sodium?
    A. Hyperkalemia
    B. Hypokalemia
    C. Hypernatremia
    D. Hyponatremia
    E. Hypercalcemia

19. Which two electrolytes are in greatest concentration in ICF?
    A. Na^+ and Cl^-
    B. Na^+ and K^+
    C. K^+ and Cl^-
    D. K^+ and HPO_4^{2-}
    E. Na^+ and HPO_4^{2-}
    F. Ca^{2+} and Mg^{2+}

20. All of the following factors will tend to increase fluid output *except:*
    A. Fever
    B. Vomiting and diarrhea
    C. Hyperventilation
    D. Increased glomerular filtration rate
    E. Decreased blood pressure

Questions 21–25: fill-ins. Complete each sentence with the word or phrase that best fits.

_____ 21. List four factors that increase production of aldosterone.

_____ 22. About ____ percent of body fluids form part of ____-cellular fluid, and ____ percent are in ECF.

_____ 23. ____ are chemicals that have at least one ionic bond and dissociate into positive and negative ions.

_____ 24. To compensate for acidosis, kidneys will increase secretion of ____, and in alkalosis kidneys will eliminate more ____.

_____ 25. Normally blood serum contains about ____ mEq/liter of Na^+ and ____ mEq/liter of K^+.

ANSWERS TO MASTERY TEST: ■ Chapter 27

True-False

1. T
2. F. Stronger
3. F. Equals
4. T
5. T
6. T
7. T
8. T
9. F. HCO_3^-, Na^+, K^+ (Starch and glucose are nonelectrolytes.)
10. T
11. T
12. T
13. T
14. T
15. F. Move substances in opposite directions: BHP pushes substances out of blood and IFHP pulls substances from IF into blood.
16. T

Multiple Choice

17. D 19. D
18. D 20. E

Fill-ins

21. Decreased blood volume or cardiac output, decrease of Na^+ or increase of K^+ in ECF, or physical stress.
22. 67, intra-, 33
23. Electrolytes
24. H^+, HCO_3^-
25. 136 to 142, 3.8 to 5.0.

Continuity

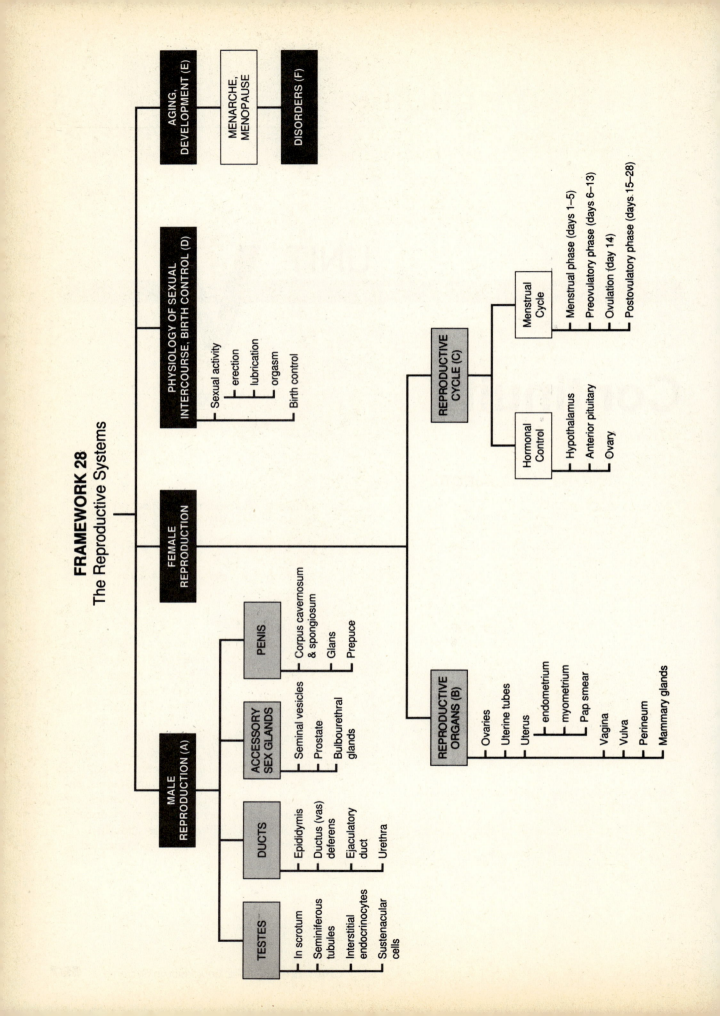

FRAMEWORK 28
The Reproductive Systems

AGING, DEVELOPMENT (E)
- MENARCHE, MENOPAUSE
- **DISORDERS (F)**

PHYSIOLOGY OF SEXUAL INTERCOURSE, BIRTH CONTROL (D)
- Sexual activity
  - erection
  - lubrication
  - orgasm
- Birth control

FEMALE REPRODUCTION

REPRODUCTIVE CYCLE (C)
- Menstrual Cycle
  - Menstrual phase (days 1–5)
  - Preovulatory phase (days 6–13)
  - Ovulation (day 14)
  - Postovulatory phase (days 15–28)
- Hormonal Control
  - Hypothalamus
  - Anterior pituitary
  - Ovary

REPRODUCTIVE ORGANS (B)
- Ovaries
- Uterine tubes
- Uterus
  - endometrium
  - myometrium
  - Pap smear
- Vagina
- Vulva
- Perineum
- Mammary glands

MALE REPRODUCTION (A)

TESTES
- In scrotum
- Seminiferous tubules
- Interstitial endocrinocytes
- Sustenacular cells

DUCTS
- Epididymis
- Ductus (vas) deferens
- Ejaculatory duct
- Urethra

ACCESSORY SEX GLANDS
- Seminal vesicles
- Prostate
- Bulbourethral glands

PENIS
- Corpus cavernosum & spongiosum
- Glans
- Prepuce

The Reproductive Systems

CHAPTER 28

Unit V introduces the systems in females and males that provide for continuity of the human species, not simply homeostasis of individuals. The reproductive system produces sperm and ova that may unite and develop to form offspring. In order for sperm and ova to reach a site for potential union, they must each pass through duct systems and be nourished and protected by secretions. External genitalia include organs responsive to stimuli and capable of sexual activity. Hormones play significant roles in the development and maintenance of the reproductive systems. The variations of hormones in ovarian and uterine (menstrual) cycles are considered in detail, along with methods of birth control. As we have done for all other systems, we discuss development of the reproductive system and normal aging changes. Start your study with the Chapter 28 Framework and become familiar with the terms presented there.

TOPIC OUTLINE AND OBJECTIVES

A. Male reproduction

1. Define reproduction and classify the organs of reproduction by function.
2. Explain the structure, histology, and functions of the testes, ducts, accessory sex glands, and penis.

B. Female reproductive organs

3. Describe the location, histology, and functions of the ovaries, uterine (fallopian) tubes, uterus, vagina, vulva, perineum, and mammary glands.

C. Female reproductive cycle

4. Compare the principal events of the menstrual and ovarian cycles.

D. Physiology of sexual intercourse; birth control

5. Explain the roles of male and female in sexual intercourse.
6. Contrast the various kinds of birth control (BC) and their effectiveness.

E. Aging and development of the reproductive systems

7. Describe the effects of aging on the reproductive systems.
8. Describe the development of the reproductive systems.

F. Disorders and medical terminology

9. Explain the symptoms and causes of sexually transmitted diseases (STDs) such as gonorrhea, syphilis, genital herpes, chlamydia, trichomoniasis, and genital warts.
10. Describe the symptoms and causes of male disorders (testicular cancer, prostate dysfunctions, impotence, and infertility) and female disorders [amenorrhea, dysmenorrhea, premenstrual syndrome (PMS), toxic shock syndrome (TSS), ovarian cysts, endometriosis, infertility, breast tumors, cervical cancer, pelvic inflammatory disease (PID), and vulvovaginal candidiasis].
11. Define medical terminology associated with the reproductive systems.

WORDBYTES

Now study the following parts of words that may help you better understand terminology in this chapter.

Wordbyte	Meaning	Example	Wordbyte	Meaning	Example
andro-	man	*andro*gen	mast-	breast	*mast*ectomy
-arche	beginning	men*arche*	meio-	smaller	*meio*sis
centesis	puncture	amnio*centesis*	men-	month	*men*opause,
cervix	neck	*cervix* of uterus			*men*ses
corp-	body	*corp*us cavernosum	-metrium	uterus	endo*metrium*
crypt-	hidden	*crypt*orchidism	mito	thread	*mito*sis
-genesis	formation	spermato*genesis*	oo-, ov-	egg	*oo*genesis, *ov*um
gyn-	woman	*gyn*ecologist	orchi-	testis	*orchi*dectomy
hyster(o)-	uterus	*hyster*ectomy	rete	network	*rete* testis
labia	lips	*labia* majora	salping-	tube, trumpet	*salping*itis
mamm-	breast	*mamm*ogram	troph-	nutrition	*troph*oblast

CHECKPOINTS

A. Male reproduction   (pages 923–936)

A1. Define each of these groups of reproductive organs.

a. Gonad

b. Ducts

c. Accessory glands

d. Supporting structures

■ **A2.** As you go through the chapter, note which structures of the male reproductive system are included in each of the above categories.

■ **A3.** Describe the scrotum and testes in this exercise.

a. Arrange coverings around testes from most superficial to deepest:

_____ _____ _____   D.   Dartos: smooth muscle that wrinkles skin
                                      TA. Tunica albuginea: dense fibrous tissue
                                      TV. Tunica vaginalis: derived from peritoneum

b. The temperature in the scrotum is about 3°C *(higher? lower?)* than the temperature in the abdominal cavity. Of what significance is this?

c. During fetal life, testes develop in the *(scrotum? posterior of the abdomen?)*.

Undescended testes is a condition known as _____. This

condition places the male at higher risk for _____.

d. Muscles that elevate testes closer to the warmth of the pelvic cavity are known as *(cremaster? dartos?)* muscles.

■ **A4.** Discuss spermatogenesis in this exercise.

a. Spermatozoa are one type of gamete; name the other type:

_____. Gametes are *(haploid? diploid?)*. Human gametes con-

tain ____ chromosomes.

b. Fusion of gametes produces a cell called a _____. This cell, and all cells of the organism derived from it, contain the *(haploid? diploid?)* number of chromosomes, written as *(n? 2n?)*. These chromosomes, received from both sperm

and ovum, are said to exist in _____ pairs.

c. Gametogenesis assures production of haploid gametes from otherwise diploid

humans. In males, this process, called _____-genesis, occurs in

the _____.

■ **A5.** Refer to Figure LG 28.1 and do the following Checkpoint. Use the following list of cells for answers:

Early spermatid   Secondary spermatocyte
Interstitial endocrinocyte (cell of Leydig)   Spermatogonium
Late spermatid   Spermatozoon (sperm)
Primary spermatocyte   Sustentacular (Sertoli) cells

a. Label cells A and B and 1–6.
b. Mark cells 1–6 according to whether they are diploid (2n) or haploid (n). Use lines on figure.

c. Meiosis is occurring in cells numbered _____ and _____. Color these meiotic cells according to color ovals on the figure. Which cell undergoes reduction division

(meiosis I)? _____ Which undergoes equatorial division *(meiosis II)?* _____ *For extra review* of the process of meiosis, review Chapter 3, Checkpoint F6, pages 54–55 in the Learning Guide.

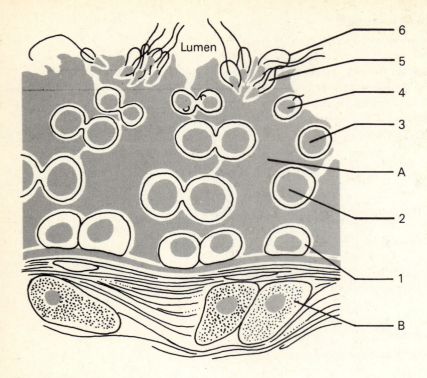

Lumen

6
5
4
3
A
2
1
B

○ Primary spermatocyte
○ Secondary spermatocyte

Figure LG 28.1 Diagram of a portion of a cross section of a seminiferous tubule showing the stages of spermatogenesis. Color and label as directed in Checkpoint A5.

d. From this figure, you can conclude that as cells develop and mature, they move *(toward? away from?)* the lumen of the tubule.

e. Which cells are in the process of spermiogenesis? _____ and _____.

f. Which cells secrete testosterone? _____

g. Which cells produce nutrients for developing sperm, phagocytose excess spermatid cytoplasm, form the blood-testis barrier, and produce the hormone inhibin? _____

h. Which cells lie dormant near the basement membrane until they divide mitotically beginning at puberty? They also serve as a reservoir of stem cells available for spermatogenesis throughout life. _____

A6. Explain how cytoplasmic bridges between cells formed during spermatogenesis may facilitate survival of Y-bearing sperm, and therefore permit generation of male offspring.

■ **A7.** Describe spermatozoa (sperm) in this Checkpoint.

a. Number produced in a normal male: _____/day

b. Life expectancy of sperm once inside female reproductive tract: _____ hours

c. Portion of sperm containing nucleus: _____

d. Part of sperm containing mitochondria: _____
e. Function of acrosome:

f. Function of tail: _____

■ **A8.** Match the hormones listed in the box with descriptions below. Answers may be used more than once.

DHT	Inhibin
FSH	LH
GnRH	Testosterone

a. Released by the hypothalamus; stimulates release of two anterior pituitary hormones:

_____

b. Inhibits release of GnRF:

_____

c. Stimulates seminiferous tubules to start sperm production; also stimulates sustantacular cells to nourish sperm:

_____

d. Stimulates release of testosterone:

_____

e. Principal male hormone and precursor of dihydrotestosterone (DHT):

_____

f. Testicular hormone that inhibits FSH production: _____

g. Before birth, stimulates development of external genitals: _____

h. At puberty, facilitate development of male sexual organs and secondary sex characteristics; contribute to libido (two hormones):

_____

A9. List five functions of androgens.

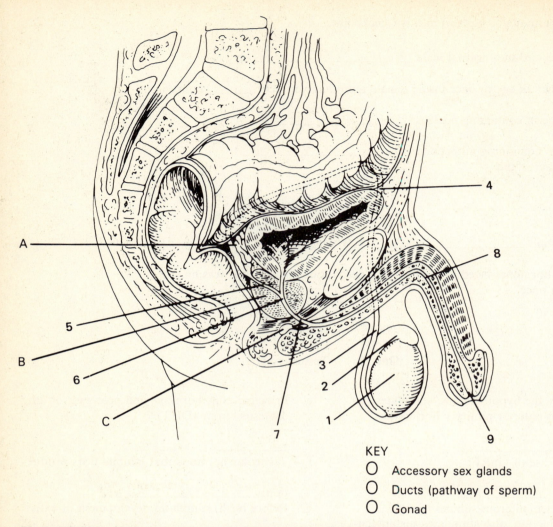

KEY
O Accessory sex glands
O Ducts (pathway of sperm)
O Gonad

Figure LG 28.2   Male organs of reproduction seen in sagittal section. Structures are numbered in order along the pathway taken by sperm. Color and label as directed in Checkpoints A10, A11, A15, and A20.

■ **A10.** Refer to Figure LG 28.2. Structures in the male reproductive system are numbered in order along the pathway that sperm take from their site of origin to the point where they exit from the body. Match the structures in the figure (using numbers 1 to 9) with descriptions below.

_____ a. Ejaculatory duct

_____ b. Epididymis

_____ c. Prostatic urethra

_____ d. Spongy (cavernous) urethra

_____ e. Membranous urethra

_____ f. Ductus (vas) deferens (portion within the abdomen)

_____ g. Ductus (vas) deferens (portion within the scrotum)

_____ h. Urethral orifice

_____ i. Testis

■ **A11.** Which structures numbered 1–9 in Figure LG 28.2 are paired? Circle their numbers:

1   2   3   4   5   6   7   8   9

Which structures are single? Write their numbers: _____

■ **A12.** Answer these questions about the ductus (vas) deferens.

a. The ductus deferens is about _____ cm (_____ inches) long.

b. It is located *(entirely? partially?)* in the abdomen. It enters the abdomen via the

_____. Protrusion of abdominal contents through this

weakened area is known as _____.

c. What structures besides the vas compose the seminal cord?

d. A vasectomy is performed at a point along the section of the vas that is within the

_____. Must the peritoneal cavity be entered during this procedure? *(Yes? No?)* Will the procedure affect testosterone level and sexual desire of the male? *(Yes? Not normally?)* Why?

■ **A13.** Which portion of the male urethra is longest? *(Prostatic? Membranous? Spongy or cavernous?)*

■ **A14.** Match the accessory sex glands with their descriptions.

| B. Bulbourethral glands | P. Prostate gland | S. Seminal vesicles |

_____ a. Paired pouches posterior to the urinary bladder

_____ b. Structures that empty secretions (along with contents of vas deferens) into ejaculatory duct

_____ c. Single doughnut-shaped gland located inferior to the bladder; surrounds and empties secretions into urethra

_____ d. Contribute about 60 percent of seminal fluid

_____ e. Pair of pea-sized glands located in the urogenital diaphragm

_____ f. Produces clotting enzymes and fibrinolysin

■ **A15.** Color parts of the male reproductive system on Figure LG 28.2 according to color code ovals. Label the *seminal vesicle, prostate gland,* and *bulbourethral gland* on the figure.

■ **A16.** *A clinical challenge.* Explain the clinical significance of the shape and location of the prostate gland.

■ **A17.** Complete the following exercise about semen.

a. The average amount per ejaculation is _____ ml.

b. The average range of number of sperm is _____/ml. When the

count falls below _____/ml, the male is likely to be sterile. Only one sperm fertilizes the ovum. Explain why such a high sperm count is required for fertility.

c. The pH of semen is slightly *(acid? alkaline?).* State the advantage of this fact.

d. What is the function of *seminalplasmin?*

e. What is the function of *fibrinolysin* in semen?

A18. List six or more characteristics of semen studied in a semen analysis.

■ **A19.** Refer to Figure 28.10 (page 935 in the text) and do this exercise.

a. The _____, or foreskin, is a covering over the

_____ of the penis. Surgical removal of the foreskin is known

as _____.

b. The names *corpus* _____ and *corpus* _____
   indicate that these bodies of tissue in the penis contain spaces that distend in the

presence of excess _____. As a result, blood is prohibited from
leaving the penis, since *(arteries? veins?)* are compressed. This temporary state is
known as the *(erect? flaccid?)* state.

c. The urethra passes through the corpus _____. The urethra

functions in the transport of _____ and

_____. During ejaculation, what prevents sperm from entering
the urinary bladder and urine from entering the urethra?

A20. Label these parts of the penis on Figure LG 28.2: *prepuce, glans, corpus caverno-sum, corpus spongiosum,* and *bulb of the penis.*

B. Female reproductive organs   (pages 936–949)

■ **B1.** Refer to Figure LG 28.3 and match the structures numbered 1–5 with their descriptions. Note that structures are numbered in order along the pathway taken by ova from the site of formation to the point of exit from the body.

_____ a. Uterus (body)   _____ d. Uterine (fallopian) tube

_____ b. Uterus (cervix)   _____ e. Vagina

_____ c. Ovary

■ **B2.** Indicate which of the five structures in Figure LG 28.3 are paired.
Circle the numbers: 1   2   3   4   5

Which structures are singular? Write their numbers: _____

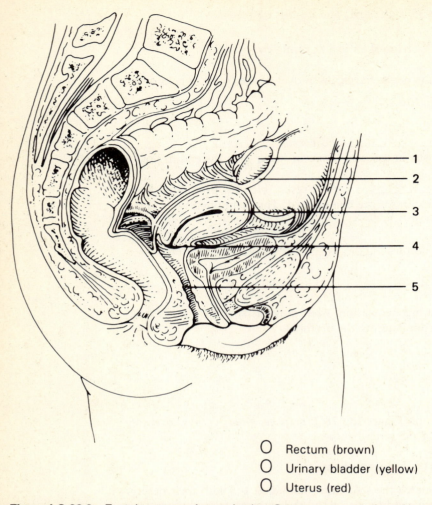

○ Rectum (brown)
○ Urinary bladder (yellow)
○ Uterus (red)

Figure LG 28.3 Female organs of reproduction. Structures are numbered in order in the pathway taken by ova. Color and label as directed in Checkpoints B1, B2, B7, B9, and B11.

■ **B3.** Describe the ovaries in this Checkpoint.

a. The two ovaries most resemble:
  A. Almonds (unshelled)   C. Pears
  B. Peas   D. Doughnuts

b. Ovaries descend to the brim of the pelvis during the _____ month of fetal life.
  A. Third   C. Ninth
  B. Sixth

c. Position of the ovaries is maintained by *(ligaments? muscles?)*. Name two or more of these structures.

d. Arrange in sequence according to chronological appearance, from first to develop to

  last: _____ _____ _____   CL. Corpus luteum
                                       OF. Ovarian follicle
                                       VOF. Vesicular ovarian (Graafian) follicle

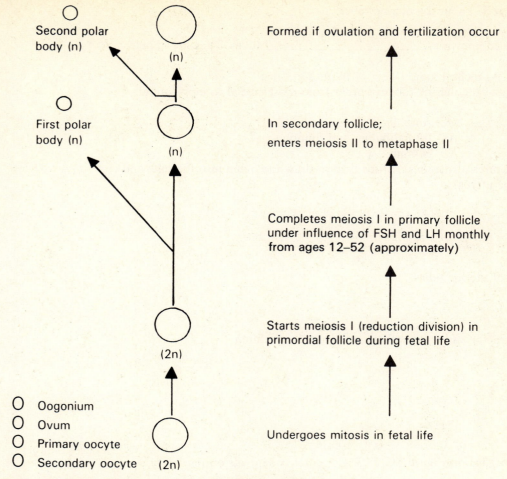

Figure LG 28.4 Oogenesis. Color as directed in Checkpoint B4.

■ **B4.** Color the four cells on Figure LG 28.4 according to the color code ovals.

■ **B5.** Describe exactly what is expelled during ovulation.

What determines whether the second meiotic (equatorial) division will occur?

■ **B6.** Complete this exercise about the uterine tubes.

a. Uterine tubes are also known as _____ tubes. They are about

_____ cm (_____ inches) long.

b. Arrange these parts of the uterine tubes in order from most medial to most lateral:

_____ _____ _____   A. Ampulla
                            B. Infundibulum and fimbriae
                            C. Isthmus

c. What structural features of the uterine tube help to draw the ovum into the tube and propel it toward the uterus?

d. List two functions of uterine tubes.

e. Define *ectopic pregnancy* and list possible causes of this condition.

B7. Color the pelvic organs on Figure LG 28.3. Use color code ovals on the figure.

■ **B8.** Answer these questions about uterine position.

a. The organ that lies anterior and inferior to the uterus is the

_____. The _____ lies posterior to it.

b. The rectouterine pouch lies *(anterior? posterior?)* to the uterus, whereas the

_____-uterine pouch lies between the urinary bladder and the uterus.

c. The fundus of the uterus is normally tipped *(anteriorly? posteriorly?)*. If it is malpositioned posteriorly, it would be *(anteflexed? retroflexed?)*.

d. The _____ ligaments attach the uterus to the sacrum. The

_____ ligaments pass through the inguinal canal and anchor

into external genitalia (labia majora). The _____ ligaments are broad, thin sheets of peritoneum extending laterally from the uterus.

e. Important ligaments in prevention of drooping (prolapse) of the uterus are the

_____ ligaments which attach the base of the uterus laterally to the pelvic wall.

■ **B9.** Refer to Figure LG 28.3 and do this exercise about layers of the wall of the uterus.

a. Most of the uterus consists of _____-metrium. This layer is *(smooth muscle? epithelium?)*. It should appear red in the figure (as directed above).

b. Now color the peritoneal covering over the uterus green. This is a *(mucous? serous?)* membrane.

c. The innermost layer of the uterus is the _____-metrium. Which portion of it is shed during menstruation? Stratum *(basalis? functionalis?)*. Which arteries supply the stratum functionalis? *(Spiral? Straight?)*

■ **B10.** *A clinical challenge.* Name the procedures described.

a. Dilation of the cervix of the uterus with scraping of the uterine lining:

_____.

b. Relatively painless procedure in which a small number of cells are removed from the cervical area and examined microscopically for possible changes indicating

malignancy: _____.

c. Surgical removal of the uterus: _____.

■ **B11.** Refer to Figure LG 28.3. In the normal position, at what angle does the uterus

join the vagina? _____

B12. Describe these vaginal structures.

a. Fornix

b. Hymen

c. Rugae

■ **B13.** The pH of vaginal mucosa is *(acid? alkaline?)*. What is the clinical significance of this fact?

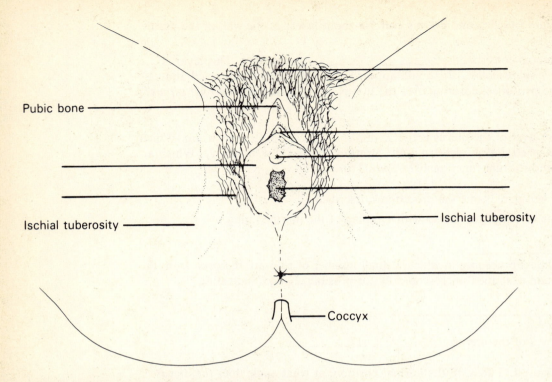

Pubic bone ──────

Ischial tuberosity ──────   ────── Ischial tuberosity

────── Coccyx

Figure LG 28.5 Perineum. Label as directed in Checkpoint B14.

■ **B14.** Refer to Figure LG 28.5 and do the following exercise.

a. Label these perineal structures: *anus, clitoris, labia major, labia minor, mons pubis, urethra,* and *vagina.*

b. Draw the borders of the perineum, using landmarks labeled on the diagram. Draw a line between the right and left ischial tuberosities. Label the triangles that are formed. In which triangle are all of the female external genitalia located?

_____

c. Draw a dotted line where an *episiotomy* incision would be performed for childbirth.

■ **B15.** Describe mammary glands in this Checkpoint.

Each breast consists of 15–20 _____ composed of lobules.

Milk-secreting cells, known as _____, empty into a duct system that terminates at the *(areola? nipple?).*

■ **B16.** *A clinical challenge.* Mrs. Rodriguez, a 55-year-old woman, tells the nurse that she is concerned about having breast cancer. Do this exercise about the nurse's conversation with Mrs. Rodriguez.

a. The nurse asks some questions to evaluate Mrs. Rodriguez's risk of breast cancer. Which of the following responses indicate an increased risk of breast cancer?
A. "My older sister had cancer when she was 40 and she had the breast removed."
B. "I've never had any kind of cancer myself."
C. "I have two daughters; they were born when I was 36 and 38."
D. "I don't smoke and I never did."
E. "I do need to cut back on fatty foods; I know I eat too many of them."

b. The nurse tells the client that the American Cancer Society recommends three methods of early detection of breast cancer. These are listed below. On the line next to each of these, write the frequency that the nurse is likely to suggest to Mrs. Rodriguez.

1. Self breast exam (SBE) _____

2. Mammogram _____

3. Breast exam by physician _____

C. Female reproductive cycle (pages 949–953)

C1. Contrast *menstrual cycle* and *ovarian cycle*.

■ **C2.** Select hormones produced by each of the endocrine glands listed below.

E. Estrogen	LH. Luteinizing hormone
FSH. Follicle-stimulating hormone	P. Progesterone
GnRH. Gonadotropin-releasing hormone	R. Relaxin
I. Inhibin	

Hypothalamus: _____   Ovary: _____

Anterior pituitary: _____

■ **C3.** Now select the hormones listed in C2 that fit descriptions below.

_____ a. Stimulates release of FSH.

_____ b. Stimulates release of LH.

_____ c. Inhibits release of FSH.

_____ d. Inhibits release of LH.

_____ e. Stimulates development and maintenance of uterus, breasts, and secondary sex characteristics; enhances protein anabolism.

_____ f. Prepares endometrium for implantation by fertilized ovum.

_____ g. Dilates cervix for labor and delivery.

_____ h. β-Estradiol, estrone, and estriol are forms of this hormone.

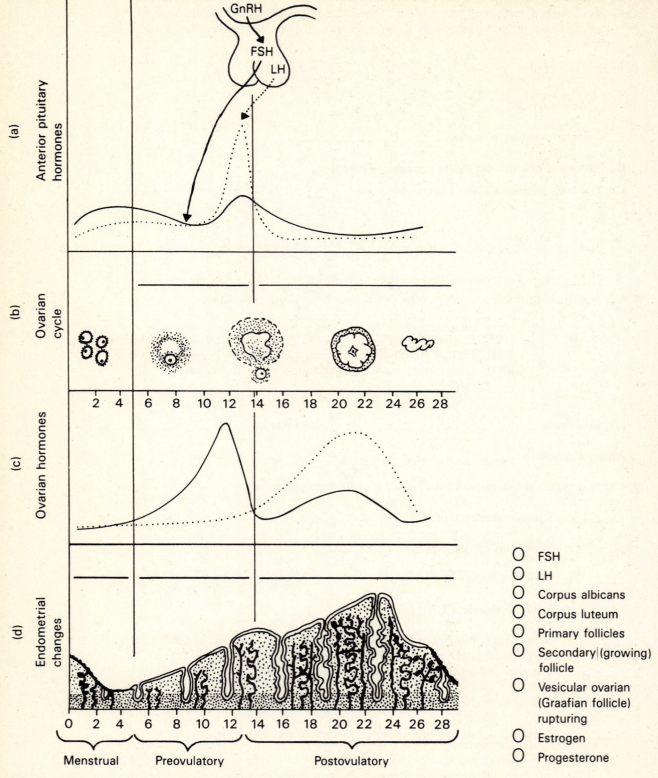

Figure LG 28.6 Correlations of female cycles. (a) Anterior pituitary hormones. (b) Ovarian changes: follicles and corpus luteum. (c) Ovarian hormomes. (d) Uterine (endometrial) changes. Color and label as directed in Checkpoint C4.

In the diagram:

(a) Anterior pituitary hormones

GnRH
FSH
LH

(b) Ovarian cycle

2 4 6 8 10 12 14 16 18 20 22 24 26 28

(c) Ovarian hormones

(d) Endometrial changes

0 2 4 6 8 10 12 14 16 18 20 22 24 26 28

Menstrual   Preovulatory   Postovulatory

O FSH
O LH
O Corpus albicans
O Corpus luteum
O Primary follicles
O Secondary (growing) follicle
O Vesicular ovarian (Graafian follicle) rupturing
O Estrogen
O Progesterone

■ **C4.** Refer to Figure LG 28.6, and check your understanding of events in female cycles in this checkpoint.

a. The average duration of the menstrual cycle is _____ days. The menstrual phase usually lasts about _____ days. On Figure LG 28.6d, color red the endometrium in the menstrual phase, and note the changes in its thickness. Also write *menses* on the line in that phase on Figure LG 28.6d.

b. Following the menstrual phase is the _____ phase. During both of these phases, _____ begin to develop in the ovary. At the birth of a female infant, her ovaries contain numerous *(primary? primordial?)* follicles. Each month about _____ of these follicles become secondary follicles, whereas only _____ usually matures each month.

c. Follicle development occurs under the influence of the hypothalamic hormone _____, which regulates the anterior pituitary hormones _____ and later _____.

d. What are the two main functions of follicles?

e. How does estrogen affect the endometrium?

For this reason, the preovulatory phase is also known as the _____ phase. Since follicles reach their peaks during this phase, it is also known as the _____ phase. Write those labels on Figure LG 28.6d and 28.6b.

f. A moderate increase in estrogen production by the growing follicles initiates a *(positive? negative?)* feedback mechanism that *(increases? decreases?)* FSH level. It is this change in FSH level that causes atresia of the remaining 19 or so partially developed follicles. *Atresia* means *(degeneration? regeneration?)*.

g. The principle of this negative feedback effect is utilized in oral contraceptives. By starting to take "pills" (consisting of target hormones estrogen and progesterone) early in the cycle (day 5), and continuing until day 25, a woman will maintain a very low level of the tropic hormone _____. Without FSH, follicles and ova will not develop, and so the woman will not _____.

h. The second anterior pituitary hormone released during the cycle is _____. The surge of this hormone results from the *(high? low?)* level of estrogens present just at the end of the preovulatory phase. This is an example of a *(positive? negative?)* feedback mechanism.

i. The LH surge causes the release of the ovum (actually secondary oocyte), the event

known as _____. The LH surge is the basis for a home test for *(ovulation? pregnancy?)*. Color all stages of follicle development through ovulation on Figure LG 28.6b according to color code ovals.

j. Following ovulation is the _____ phase. It lasts until day

_____ of a 28-day cycle. Under the influence of tropic hormone _____, follicle

cells are changed into the corpus _____. These cells secrete

two hormones: _____ and _____. Both

prepare the endometrium for _____. What preparatory changes occur?

k. What effect do rising levels of target hormones estrogen and progesterone have upon the GnRH and LH?

l. One function of LH is to form and maintain the corpus luteum. So as LH

____-creases, the corpus luteum disintegrates, forming a white scar on the ovary

known as the corpus _____. Color these two final stages of the original follicle on Figure LG 28.6b. Also write alternative names for the postovulatory phase on lines in Figure LG 28.6b and d.

m. The corpus luteum had been secreting _____ and

_____. With the demise of the corpus luteum, the levels of these hormones rapidly *(increase? decrease?)*. Since these hormones were maintaining the endometrium, endometrial tissue now deteriorates and will be shed during

the next _____ phase.

n. If fertilization should occur, estrogens and progesterone are needed to maintain the

endometrial lining. The _____ around the developing embryo

secretes a hormone named _____.
It functions much like LH in that it maintains the corpus

_____, even though LH has been inhibited (step k above). The corpus luteum continues to secrete estrogen and progesterone for several months until the placenta itself can secrete sufficient amounts. Incidentally, hCG is present in the urine of pregnant women only, and so is routinely used to detect

_____.

o. Review hormonal changes by coloring blood levels of hormones in Figure LG 28.6a and c, according to color code ovals. Then note uterine changes by coloring red the remainder of the endometrium in Figure LG 28.6d.

■ **C5.** To check your understanding of the events of these cycles, list letters of the major events in chronological order beginning on day 1 of the cycle. To do this exercise it may help to list on a separate piece of paper days 1 to 28 and write each event next to the approximate day on which it occurs.

___ ___ ___ ___ ___ ___ ___ ___ ___ ___ ___

A. Ovulation occurs.
B. Estrogen and progesterone levels drop.
C. FSH begins to stimulate follicles.
D. LH surge is stimulated by high estrogen levels.
E. High levels of estrogens and progesterone inhibit LH.
F. Estrogens are secreted for the first time during the cycle.
G. Corpus luteum degenerates and becomes corpus albicans.
H. Endometrium begins to deteriorate, leading to next month's menses.
I. Follicle is converted into corpus luteum.
J. Rising level of estrogen inhibits FSH secretion.
K. Corpus luteum secretes estrogens and progesterone.

■ **C6.** Describe three signs of ovulation in this Checkpoint.

a. Ovulation is likely to occur during the 24-hour period after the basal temperature ____-creases by about 0.4–0.6°F. This temperature change is related to increase in *(estrogens? progesterone?)*.

b. Under the influence of *(estrogen? progesterone?)* cervical mucus increases in quantity and becomes *(more stretchy? thicker?)* at midcycle.

c. Some women experience a pain around ovulation known as

_____.

D. Physiology of sexual intercourse; birth control   (pages 953–958)

D1. Define *sexual intercourse* or *coitus*.

■ **D2.** Answer these questions about the process.

a. *(Sympathetic? Parasympathetic?)* impulses from the *(brainstem? sacral cord?)* cause dilation of arteries of the penis, leading to the *(flaccid? erect?)* state.

b. *(Sympathetic? Parasympathetic?)* impulses cause sperm to be propelled into the urethra. Expulsion from the urethra to the exterior of the body is called

_____.

D3. Compare and contrast the phases of sexual activity in females and males by completing this table.

Phase	Female	Male
a. Erection		
b. Lubrication		
c. Orgasm		

■ **D4.** Match the methods of birth control listed in the box with descriptions below.

Ch. Chemical methods	OC. Oral contraceptive
Cl. Coitus interruptus	R. Rhythm
Con. Condom	RU 486. RU 486
D. Diaphragm	T. Tubal ligation
IUD. Intrauterine device	V. Vasectomy
N. Norplant	

_____ a. Spermicidal foams, jellies, and creams

_____ b. "The pill," combination of progester-one and estrogens, causing decrease in GnRH, FSH, and LH so ovulation does not occur

_____ c. Removal of a portion of the ductus (vas) deferens

_____ d. Tying off of the uterine tubes

_____ e. A natural method of birth control involving abstinence during the period when ovulation is most likely

_____ f. A mechanical method of birth control in which a dome-shaped structure is placed over the cervix; accompanied by use of a jelly or cream

_____ g. Small object, an intrauterine device, placed in the uterus by physician

_____ h. A rubber sheath over the penis

_____ i. Withdrawal

_____ j. Progestin-containing capsules placed under the skin for up to 5 years

_____ k. Competitive inhibitor of progesterone; a chemical that induces miscarriage

D5. Contrast benefits and risks associated with oral contraceptives.

E. Aging and development of the reproductive system (pages 958–961)

E1. Contrast changes occurring at puberty in females and males by completing this table.

	Female	Male
a. Age of onset		
b. Hormonal changes		
c. Changes in reproductive organs		
d. Secondary sex characteristics developed		

E2. Contrast these terms: menarche/menopause

E3. Describe normal aging changes that involve the reproductive system in:

a. Females

b. Males

■ **E4.** Which type of cancer is more common in women over 65? *(Cervical? Uterine?)*

■ **E5.** Name the reproductive structures (if any) that develop from the following embryonic ducts.

a. Mesonephric duct in females: _____

b. Mesonephric duct in males: _____

c. Paramesonephric (Müller's) duct in females: _____

d. Paramesonephric (Müller's) duct in males: _____

■ **E6.** Write E (endoderm) or M (mesoderm) next to each structure below to indicate its embryonic origin.

_____ a. Ovaries and testes

_____ b. Prostate and bulbourethral glands

_____ c. Greater (Bartholin's) and lesser vestibular glands

■ **E7.** Identify male homologues for each of the following.

a. Labia majora: _____

b. Clitoris: _____

c. Lesser vestibular (Skene's) glands: _____

d. Greater vestibular (Bartholin's) glands: _____

F. Disorders, medical terminology (pages 962–966)

F1. Define *sexually transmitted disease* and *venereal disease.*

F2. Describe the main characteristics of these sexually transmitted diseases by completing the table.

Disease	Causative Agent	Main Symptoms
a. Gonorrhea		
b.	*Treponema pallidum*	
c.	Type II herpes simplex virus	
d.	*Trichomonas vaginalis*	

F3. Discuss effects of syphilis on children born to mothers with untreated syphilis.

■ **F4.** Check your understanding of the disorders listed in the box by matching them with the correct descriptions.

A. Amenorrhea	G. Gonorrhea
D. Dysmenorrhea	S. Syphilis
E. Endometriosis	T. Trichomoniasis
F. Fibroadenoma	

_____ a. Painful menstruation, partly due to contractions of uterine muscle

_____ b. Possible cause of blindness in newborns if bacteria transmitted to eyes during birth

_____ c. Spreading of uterine lining into abdominopelvic cavity via uterine tubes

_____ d. Caused by bacterium *Treponema pallidum;* may involve many systems in tertiary stage

_____ e. Absence of menstrual periods

_____ f. Benign tumor with firm, rubbery consistency; easily moved within breast

_____ g. Caused by flagellated protozoan

■ **F5.** Fill in the term *(impotence* or *infertility)* that best fits the description. Then list several causes of each condition.

a. Inability to fertilize an ovum: _____.

b. Inability to attain and maintain an erection: _____.

F6. List six symptoms of *premenstrual syndrome (PMS).*

_____   _____

_____   _____

_____   _____

F7. Name the microorganism associated with *toxic shock syndrome (TSS).*

_____ Relate this condition to use of tampons.

F8. What is PID?

It is most commonly caused by the sexually transmitted disease (STD) known as

_____.

■ **F9.** Oophorectomy is removal of a(n) _____, whereas

_____ refers to removal of a uterine (fallopian) tube.

ANSWERS TO SELECTED CHECKPOINTS

A2. (a) Testes. (b) Seminiferous tubules, epididymis, vas deferens, ejaculatory duct, urethra. (c) Prostate, seminal vesicles, bulbourethral glands. (d) Scrotum, penis.

A3. (a) D TV TA. (b) Lower; lower temperature is required for normal sperm production. (c) Posterior of the abdomen; cryptorchidism; sterility or testicular cancer. (d) Cremaster.

A4. (a) Ova; haploid; 23. (b) Zygote; diploid, 2n; homologous. (c) Spermato, seminiferous tubules of testes.

A5. (a-b) See Figure LG 28.1A. (c) 2, 3; 2; 3. (d) Toward. (e) 4, 5. (f) B. (g) A. (h) 1.

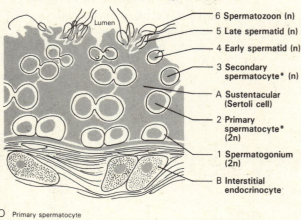

O Primary spermatocyte
O Secondary spermatocyte   *Cells in meiosis

Figure LG 28.1A Diagram of a portion of a cross section of a seminiferous tubule showing the stages of spermatogenesis.

A7. (a) 300 million. (b) 48. (c) Head. (d) Midpiece. (e) A specialized lysosome that releases enzymes that help the sperm to penetrate the secondary oocyte. (f) Propels sperm.

A8. (a) GnRF. (b) Testosterone. (c) FSH. (d) LH. (e) Testosterone. (f) Inhibin. (g) DHT. (h) Testosterone and DHT.

A10. (a) 5. (b) 2. (c) 6. (d) 8. (e) 7. (f) 4. (g) 3. (h) 9. (i) 1.

A11. Paired: 1–5; singular: 6–9.

A12. (a) 45, 18. (b) Partially; inguinal canal and rings; inguinal hernia. (c) Testicular artery, veins, lymphatics, nerves, and cremaster muscle. (d) Scrotum; no; not normally; hormones exit through testicular veins, which are not cut during vasectomy.

A13. Spongy or cavernous.

A14. (a) S. (b) S. (c) P. (d) S. (e) B. (f) P.

A15. Accessory sex glands: A (seminal vesicles), B (prostate), C (bulbourethral glands); ducts, 2–9; gonad, 1.

A16. Since it surrounds the urethra (like a doughnut), an enlarged prostate can cause painful and difficult urination (dysuria) and bladder infection (cystitis) due to incomplete voiding.

A17. 2.5–5 ml. (b) 50–150 million; 20 million; a high sperm count increases the likelihood of fertilization, since only a small percentage of sperm ever reaches the secondary oocyte; in addition, a large number of sperm are required to release sufficient enzyme to digest the barrier around the secondary oocyte. (c) Alkaline; protects sperm against acidity of the male urethra and the female vagina. (d) Serves as an antibiotic that destroys bacteria. (e) It liquefies the clotted semen within 20 minutes of the entrance of semen to the vagina, thereby permitting movement of sperm into the uterus.

A19. (a) Prepuce, glans; circumcision. (b) Cavernosum, spongiosum, blood; veins; erect. (c) Spongiosum; urine, sperm; sphincter action at base of bladder.

B1. (a) 3. (b) 4. (c) 1. (d) 2. (e) 5.

B2. 1, 2: paired; 3–5, singular.

B3. (a) A. (b) A. (c) Ligaments; mesovarian, ovarian, and suspensory ligament. (d) OF VOF CL.

B4.

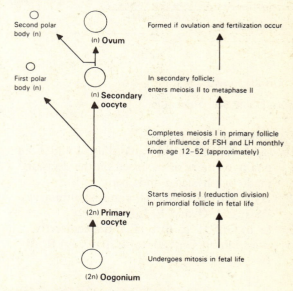

Figure LG 28.4A Oogenesis.

B5. The secondary oocyte along with the first polar body and some surrounding supporting cells are expelled. Meiosis II is completed only if the secondary oocyte is fertilized by a spermatazoon.

B6. (a) Fallopian; 10 (4). (b) C A B. (c) Currents produced by movements of fimbriae, ciliary action of columnar epithelial cells lining tubes, and peristalsis of the muscularis layer of the uterine tube wall. (d) Passageway for secondary oocyte, and sperm toward each other and site of fertilization. (e) Pregnancy following implantation at any location other than in the body of the uterus, for example, tubal pregnancy, or pregnancy on ovary, intestine, or in cervix of the uterus.

B8. (a) Bladder; rectum. (b) Posterior, vesico-. (c) Anteriorly; retroflexed. (d) Uterosacral; round; broad. (e) Cardinal.

B9. (a) Myo; smooth muscle. (b) Serous. (c) Endo; functionalis; spiral.

B10. (a) D&C (dilatation and curettage). (b) Pap (Papanicolaou) smear. (c) Hysterectomy.

B11. Close to 90° angle.

B13. Acid; retards microbial growth but may harm sperm cells.

B14. (a and c) See Figure LG 28.5A. (b) Urogenital triangle.

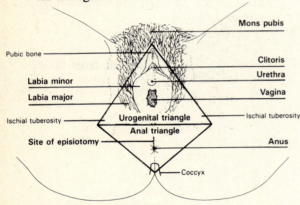

Figure LG 28.5A Perineum.

B15. Lobes; alveoli; nipple.

B16. (a) A, C, and E. (b) 1: monthly; 2 and 3: annually or more often based on the decision made by physician and client.

C2. Hypothalamus: GnRH; anterior pituitary: FSH, LH; Ovary: E, I, P, R.

C3. (a–b) GnRH. (c–d) I, (E and P indirectly since inhibit GnRH). (e) E. (f) P. (g) R. (h) E.

C4. (a) 28; 5; see Figure LG 28.6A. (b) Preovulatory; primary follicles; primordial; 20, 1. (c) GnRH, FSH, LH. (d) Secretion of estrogens and development of the ovum. (e) Estrogen stimulates replacement of the sloughed off stratum functionalis by growing (proliferating) new cells; proliferative; follicular. See Figure LG 28.6A. (f) Negative, decreases; degeneration. (g) FSH; ovulate. (h) LH; high; positive. (i) Ovulation; ovulation. See Figure LG 28.6A. (j) Postovulatory; 28; LH; luteum; progesterone, estrogens; implantation; thickening by growth and secretion of glands and increased retention of fluid. (k) Inhibition of these hormones. (l) De; albicans. See Figure LG 28.6A. (m) Progesterone, estrogens; decrease; menstrual. (n) Placenta, human chorionic gonadotropin (hCG); luteum; pregnancy. (o) See Figure LG 28.6A.

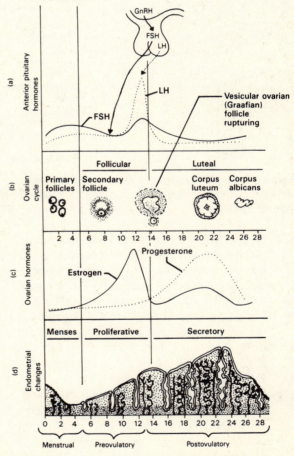

Figure LG 28.6A Correlations of female cycles. (a) Anterior pituitary hormones. (b) Ovarian changes: follicles and corpus luteum. (c) Ovarian hormones. (d) Uterine (endometrial) changes. Refer to Checkpoint C4.

C5. C F J D A I K E G B H.
C6. (a) In; progesterone. (b) Estrogen, more stretchy. (c) Mittelschmerz.
D2. (a) Parasympathetic, sacral cord, erect. (b) Sympathetic; ejaculation.
D4. (a) Ch. (b) OC. (c) V. (d) T. (e) R. (f) D. (g) IUD. (h) Con. (i) Cl. (j) N. (k) RU 486.
E4. Uterine.
E5. (a) None. (b) Ducts of testes and epididymides, ejaculatory ducts and seminal vesicles. (c) Uterus and vagina. (d) None.

E6. (a) M. (b) E. (c) E.
E7. (a) Scrotum. (b) Penis. (c) Prostate. (d) Bulbourethral glands.
F4. (a) D. (b) G. (c) E. (d) S. (e) A. (f) F. (g) T.
F5. (a) Infertility (or sterility). (b) Impotence.
F9. Ovary, salpingectomy.

MASTERY TEST: Chapter 28

Questions 1–10: Arrange the answers in correct sequence.

____ ____ ____ 1. From most external to deepest:
   A. Endometrium
   B. Serosa
   C. Myometrium

____ ____ ____ 2. Portions of the urethra, from proximal (closest to bladder) to distal (closest to outside of the body):
   A. Prostatic
   B. Membranous
   C. Spongy

____ ____ ____ 3. Thickness of the endometrium, from thinnest to thickest:
   A. On day 5 of cycle
   B. On day 13 of cycle
   C. On day 23 of cycle

____ ____ ____ 4. From anterior to posterior:
   A. Uterus and vagina
   B. Bladder and urethra
   C. Rectum and anus

____ ____ ____ ____ 5. Order in which hormones begin to increase level, from day 1 of menstrual cycle:
   A. FSH
   B. LH
   C. Progesterone
   D. Estrogen

____ ____ ____ ____ 6. Order of events in the female monthly cycle, beginning with day 1:
   A. Ovulation
   B. Formation of follicle
   C. Menstruation
   D. Formation of corpus luteum

____ ____ ____ ____ 7. From anterior to posterior:
   A. Anus
   B. Vaginal orifice
   C. Clitoris
   D. Urethral orifice

_____ _____ _____ _____ _____ 8. Pathway of sperm:
  A. Ejaculatory duct
  B. Testis
  C. Urethra
  D. Ductus (vas) deferens
  E. Epididymis

_____ _____ _____ _____ _____ 9. Pathway of milk in breasts:
  A. Ampullae
  B. Alveoli
  C. Secondary tubules
  D. Mammary ducts
  E. Lactiferous ducts

_____ _____ _____ _____ _____ _____ 10. Pathway of sperm entering female reproductive system:
  A. Uterine cavity
  B. Cervical canal
  C. External os
  D. Internal os
  E. Uterine tube
  F. Vagina

Questions 11–12: Circle the letter preceding the one best answer to each question.

11. Eighty to 90 percent of seminal fluid (semen) is secreted by the combined secretions of:
  A. Prostate and epididymis
  B. Seminal vesicles and prostate
  C. Seminal vesicles and seminiferous tubules
  D. Seminiferous tubules and epididymis
  E. Bulbourethral glands and prostate

12. In the normal male, there are two of each of the following structures *except:*
  A. Testis   D. Ductus deferens
  B. Seminal vesicle   E. Epididymis
  C. Prostate   F. Ejaculatory duct

Questions 13–15: Circle letters preceding all correct answers to each question.

13. Which of the following are produced by the testes?
  A. Spermatozoa   D. GnRH
  B. Testosterone   E. FSH
  C. Inhibin   F. LH

14. Which of the following are produced by the ovaries and then *leave* the ovaries?
  A. Follicle   D. Corpus albicans
  B. Secondary oocyte E. Estrogen
  C. Corpus luteum   F. Progesterone

15. Which of the following are functions of LH?
  A. Begin the development of the follicle.
  B. Stimulate change of follicle cells into corpus luteum cells.
  C. Stimulate release of secondary oocyte (ovulation).
  D. Stimulate corpus luteum cells to secrete estrogens and progesterone.
  E. Stimulate release of GnRH.

T F 16. <u>Spermatogonia and oogonia both</u> continue to divide throughout a person's lifetime to produce
new <u>primary spermatocytes or primary oocytes.</u>

T F 17. The menstrual cycle refers to a series of changes in the <u>uterus</u>, whereas the ovarian cycle is a series of changes in the <u>ovary.</u>

T F 18. <u>Meiosis is a part</u> of both oogenesis and spermatogenesis.

T F 19. Each oogonium beginning oogenesis produces <u>four mature ova.</u>

T F 20. An increased level of a target hormone (such as estrogen) will <u>inhibit</u> release of its related releasing factor and target hormone (GnRH and FSH).

Questions 21–25: fill-ins. Complete each sentence with the word or phrase that best
fits.

_____ 21. ____ is a term that means undescended testes.

_____ 22. Most oral contraceptives consist of the hormones ____ and ____.

_____ 23. ____ cells in testes phagocytose degenerating spermatogenic cells secrete inhibin, and also form the blood-testis barrier.

_____ 24. Menstrual flow occurs as a result of a sudden ____-crease in the hormones
____ and ____.

_____ 25. The three main functions of estrogens are promotion of growth and maintenance of ____, ____ balance, and stimulation of protein ____-bolism.

ANSWERS TO MASTERY TEST: ■ Chapter 28

Arrange

1. B C A
2. A B C
3. A B C
4. B A C
5. A D B C
6. C B A D
7. C D B A
8. B E D A C
9. B C D A E
10. F C B D A E

Multiple Choice

11. B
12. C

Multiple Answers

13. A, B, C
14. B, E, F
15. B, C, D

True-False

16. F. Spermatogonia; primary spermatocytes (Oogonia do not divide after birth of female baby.)
17. T
18. T
19. F. One mature ovum and three
polar bodies
20. T

Fill-ins

21. Cryptorchidism
22. Progesterone and estrogens
23. Sustentacular
24. De, progesterone, and estrogens
25. Reproductive organs, fluid and
electrolyte, ana

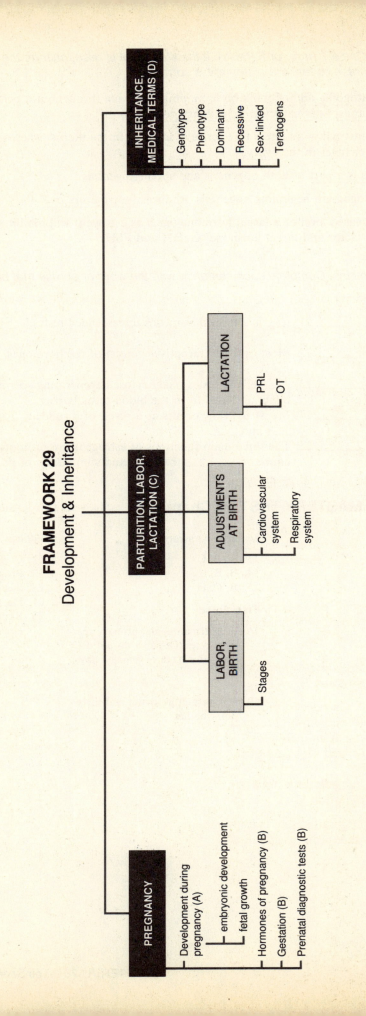

FRAMEWORK 29
Development & Inheritance

PREGNANCY
— Development during pregnancy (A)
— embryonic development
— fetal growth
— Hormones of pregnancy (B)
— Gestation (B)
— Prenatal diagnostic tests (B)

PARTURITION, LABOR, LACTATION (C)

LABOR, BIRTH
— Stages

ADJUSTMENTS AT BIRTH
— Cardiovascular system
— Respiratory system

LACTATION
— PRL
— OT

INHERITANCE, MEDICAL TERMS (D)
— Genotype
— Phenotype
— Dominant
— Recessive
— Sex-linked
— Teratogens

Development and Inheritance

Twenty-eight chapters ago you began a tour of the human body, starting with the structural framework and varieties of buildings (chemicals and cells) and moving on to the global communication systems (nervous and endocrine), transportation routes (cardiovascular), and refuse removal systems (respiratory, digestive, and urinary). This last chapter synthesizes all the parts of the tour: A new individual, with all those body systems, is conceived and develops during a nine-month pregnancy. Adjustments are made at birth for survival in an independent (outside of mother) world. But parental ties are still evident in inherited features, signs of intrauterine care (mother's avoidance of alcohol or cigarettes), and provision of nourishment, as in lactation.

Look at Framework 29 first. And congratulations on having completed your first tour through the wonders of the human body.

TOPIC OUTLINE AND OBJECTIVES

A. Development during pregnancy

1. Explain the processes associated with fertilization, morula formation, blastocyst development, and implantation.
2. Describe how in vitro fertilization (IVF) is performed.
3. Discuss the principal maternal body changes associated with embryonic and fetal growth.

B. Hormones of pregnancy; gestation; diagnostic techniques

4. Compare the sources and functions of the hormones secreted during pregnancy.
5. Describe the anatomical and physiological changes associated with gestation.

C. Parturition, labor, adjustments at birth, physiology of lactation

6. Explain the respiratory and cardiovascular adjustments that occur in an infant at birth.

7. Discuss the physiology and control of lactation.

D. Inheritance and medical technology

8. Define inheritance and describe the inheritance of several traits.
9. Discuss potential hazards to an embryo and fetus associated with chemicals and drugs, irradiation, alcohol, and cigarette smoking.
10. Define and describe Down syndrome, Turner's syndrome, metafemale syndrome, Klinefelter's syndrome, and fragile X syndrome.
11. Define medical terminology associated with development and inheritance.

WORDBYTES

Wordbyte	Meaning	Example		Wordbyte	Meaning	Example
lact-	milk	*lact*ation		natus	birth	post*natal*
morula	mulberry	*morula*				

CHECKPOINTS

A. Development and growth during pregnancy (pages 971–981)

■ **A1.** Do this exercise about events associated with fertilization.

a. How many sperm are ordinarily introduced into the vagina during sexual

   intercourse? _____ About how many reach the area where the

   developing ovum is located? _____

b. The name given to the developing ovum at the time of fertilization is the

   _____. List four or more mechanisms in the female and/or
   male that favor the meeting of sperm with oocyte so that fertilization can occur.

c. How long must sperm be in the female reproductive tract before they can fertilize
   the female gamete? *(10 minutes? 10 hours? 10 days?)* Describe the changes sperm undergo during this time?

   These changes in the sperm are known as _____.

d. Name the usual site of fertilization. _____ What do the corona
   radiata and zona pellucida have to do with fertilization?

628   UNIT V   Continuity

e. An _____ pregnancy is one that takes place in a site outside of the uterine cavity. Where do most of those pregnancies occur?

_____ List several other possible sites of such pregnancies.

f. How many sperm normally penetrate the developing egg? _____ What is the name

of this event? _____ What prevents further sperm from entering the female gamete?

■ **A2.** Describe roles of the following structures in fertilization.

   a. A male pronucleus present in a *(secondary spermatocyte? spermatozoon?)* combines with a female pronucleus from a *(secondary oocyte? ovum?)*. Each pronucleus contains the *(n? 2n?)* number of chromosomes.

   b. The product of fusion of the two pronuclei is known as a _____

   nucleus, containing _____ chromosomes. Completion of meiosis II by the secondary oocyte results in the final product of fertilization, consisting of segmentation nucleus, cytoplasm, and zona pellucida. This product of fertilization is known as the

   _____ .

■ **A3.** Contrast types of multiple births by choosing the answer that best fits each description.

> DTwin. Dizygotic twins   MTwin. Monozygotic twins

_____ a. Also known as fraternal twins

_____ b. Two infants derived from a single fertilized ovum that divides once

_____ c. Two infants derived from two secondary oocytes each fertilized by a different sperm

A4. Summarize the main events during the week following fertilization. Describe each of the following stages. Be sure to note how many days after fertilization each stage occurs. (For help, refer to Figures 29.2 to 29.4, pages 973–975 of the text.)

a. Cleavage (days _____)

b. Morula (days _____)

c. Blastocyst (days _____)

A5. Draw a diagram of a blastocyst. Label: *blastocoele, inner cell mass,* and *trophoblast.* Color red the cells that will become the embryo. Color blue the cells that will become part of the placenta. (Refer to Figure 29.3b, page 974 of the text, for help.)

■ **A6.** Describe the process of implantation in this exercise.

a. Implantation is likely to occur about _____ days after ovulation, or about day

_____ of a 28-day cycle. Note that the developing baby is implanted in the uterus even before the mother "misses" the first day of her next menstrual cycle.

b. During implantation, cells of the *(inner cell mass? trophoblast?)* secrete enzymes that enable the blastocyst to burrow into the uterine lining. Following implantation, the

blastocyst can absorb nutrients from _____.

c. Since the trophoblast is in contact with the mother's endometrial lining, it is reasonable that the developing embryo might be rejected as a "foreign graft" by the mother's immune system. State one mechanism that may explain why it is not rejected.

A7. Describe the following aspects of in vitro fertilization (IVF).

a. In what year was the first live birth as a result of this procedure? *(1962? 1970? 1978? 1986?)*

b. Briefly describe this procedure.

c. Contrast these two forms of IVF: *embryo transfer* and *GIFT*.

A8. Describe how the conditions of being underweight or overweight may affect reproductive hormones and fertility.

■ **A9.** The word *embryo* refers to the developing individual during the *(first two? last seven?)* months of development in utero. Define *fetus*.

■ **A10.** Complete this exercise about the embryonic period. Refer to Figure 29.5 (pages 976–977 in your text).

a. Identify the structures that ultimately form from these two parts of the implanted blastocyst.

Trophoblast   **Inner Cell Mass**

Between these two portions of the blastocyst is the _____ cavity.

b. The inner cell mass first forms the embryonic _____ consisting of two layers. The upper cells (closer to the implantation site toward the right on

Figure 29.5b, page 977 of the text) are called _____-derm cells. The lower cells (toward the left in the figure) are called

_____-derm cells. Eventually, a layer of

_____-derm will form between these. Collectively the three

layers are called _____ layers.

c. The process of formation of the primary germ layers is known as

_____, and it occurs by the end of the

_____ week following fertilization.

d. _____-derm cells line the amniotic cavity. The delicate fetal membrane formed from these cells that surrounds the fetus and is broken at birth is

called the _____. Another name for this membrane is the

"_____." (Ectoderm cells also form a number of fetal structures: see Checkpoint A12.)

e. The lower cells of the embryonic disc, called _____ cells, line

the primitive _____ and for a while extend out to form the

_____ sac. This sac *(is? is not?)* as significant in humans as it
is in birds.

f. Some mesoderm cells move around to line the original trophoblast cells (Figure 29.5,

page 976 of the text). Together these cells form the _____ layer
of the placenta. Which membrane is more superficial? *(Amnion? chorion?)* The space

between these layers is called the _____.

A11. Each of the primary germ layers is involved with formation of both fetal organs and fetal membranes. Name the major structures derived from each primary germ layer: endoderm, mesoderm, and ectoderm. Consult Exhibit 29.1 (page 978 in your text).

Endoderm	**Mesoderm**	**Ectoderm**

■ **A12.** *For extra review.* Do this matching exercise on primary germ layer derivatives.

Ecto. Ectoderm	Endo. Endoderm	Meso. Mesoderm

_____ a. Epithelial lining of all of digestive, respiratory, and genitourinary tracts except near openings to the exterior of the body

_____ b. Epidermis of skin, epithelial lining of entrances to the body (such as mouth, nose, and anus), hair, nails

_____ c. All of the skeletal system (bone, cartilage, joint cavities)

_____ d. All skeletal and cardiac muscle; most smooth muscle

_____ e. Blood and all blood and lymphatic vessels

_____ f. Entire nervous system, including posterior pituitary

_____ g. Thyroid, parathyroid, thymus, and pancreas

■ **A13.** Write the name of each fetal membrane next to its description.

a. Originally formed from ectoderm, this membrane encloses fluid that acts as a shock

   absorber for the developing baby: _____.

b. Derived from mesoderm and trophoblast, this membrane becomes the principal part

   of placenta: _____.

c. Endoderm-lined membrane serving as exclusive nutrient supply for embryos of some

   species: _____.

d. A small membrane that forms umbilical blood vessels: _____.

■ **A14.** Further describe the placenta in this exercise.

a. The placenta has the shape of a _____ embedded in the wall of

the uterus. Its fetal portion is the _____ fetal membrane.

Fingerlike projections, known as _____, grow into the uterine
lining. Consequently maternal and fetal blood are *(allowed to mix? brought into close
proximity but do not mix?)*. Fetal blood in umbilical vessels is bathed by maternal

blood in _____ spaces.
b. The maternal aspect of the placenta is derived from the *(chorion? decidua basalis?)*.

This is a portion of the _____-metrium of the uterus.
c. What is the ultimate fate of the placenta?

■ **A15.** *A clinical challenge.* Following a birth, vessels of the umbilical cord are carefully

examined. A total of _____ vessels should be found, surrounded by a mucous jellylike connective tissue. The vessels should include *(1? 2?)* umbilical artery(-ies) and *(1?
2?)* umbilical vein(s). Lack of a vessel may indicate a congenital malformation.

A16. List three uses of human placentas following birth.

A17. *A clinical challenge.* Describe the following conditions in this exercise. Define the condition, state the frequency of its occurrence, suggest possible causes, and list potential consequences for fetus and/or mother.

a. Placenta previa

b. Fetomaternal hemorrhage

c. Umbilical cord prolapse

■ **A18.** Study Exhibit 29.2 (pages 982–983 in your text). Write the number of the month when each of the following events occurs.

_____ a. Heart starts to beat.

_____ b. Heartbeat can be detected.

_____ c. Backbone and vertebral canal form.

_____ d. Eyes are almost fully developed; eyelids are still fused.

_____ e. Eyelids separate; eyelashes form.

_____ f. Limb buds develop.

_____ g. Ossification begins.

_____ h. Limb buds become distinct as arms and legs; digits are well formed.

_____ i. Nails develop.

_____ j. Fine (lanugo) hair covers body.

_____ k. Lanugo hair is shed.

_____ l. Fetus is capable of survival.

_____ m. Subcutaneous fat is deposited.

_____ n. Testes descend into scrotum.

B. Hormones of pregnancy; gestation; diagnostic techniques (pages 981-986)

■ **B1.** Complete this exercise about the hormones of pregnancy.

a. During pregnancy, the level of progesterone and estrogens must remain *(high? low?)* in order to support the endometrial lining. During the first three or four months or

so, these hormones are produced principally by the _____ lo-

cated in the _____, under the influence of the tropic hormone

_____. (It might be helpful to review the section on endocrine relations in Chapter 28.)

b. Human chorionic gonadotropin (hCG) reaches its peak between the

_____ and _____ month of pregnancy.

Since it is present in blood, it will be filtered into _____, where

it can readily be detected as an indication of pregnancy as early as _____ days after fertilization. Sometimes the urine of a woman who is not pregnant contains hCG. List several factors which could account for such a "false pregnancy test."

c. In addition, the chorion produces the hormone hCS; these initials stand for

_____ chorionic _____. List two functions of this hormone.

Peak levels of hCS occur at about week *(9? 20? 32?)* of gestation.

d. _____ is a hormone that relaxes pelvic joints and helps dilate the cervix near the time of birth. This hormone is produced by the

_____ and _____.

e. The hormone inhibin, produced by the placenta, functions to *(stimulate? inhibit?)*

secretion of FSH. This hormone is also made by the _____ in

nonpregnant women and in the _____ of males.

■ **B2.** The normal human gestation period is about _____ days, counting from *(fertili-*

zation? the beginning of the last menstrual period?) until birth. This is about _____
weeks.

■ **B3.** Complete this exercise about maternal adaptations during pregnancy. Describe normal changes that occur.

a. Pulse ____-creases.

b. Blood volume ____-creases by about _____ percent.

c. Tidal volume ____-creases about _____ percent.

d. Gastrointestinal (GI) motility ____-creases, which may cause *(diarrhea? constipation?)*.

e. Pressure upon the inferior vena cava ____-creases, which may lead to

_____.

f. Increase in weight of the uterus from _____ grams in nonpregnant state to about

_____ grams at term.

■ **B4.** About 10–15 percent of all pregnant women in the United States experience PIH,

which stands for p_____-i_____

h_____. This condition apparently results from impaired function

of _____. Write three or more signs or symptoms of PIH.

B5. Defend or dispute this statement: "All women who are pregnant or lactating
should avoid exercise because of its deleterious effects."

■ **B6.** Complete this exercise about prenatal diagnostic tests.

a. Amniocentesis involves withdrawal of _____ fluid. This fluid is
   constantly recycled through the fetus (by drinking and excretion); it *(does? does
   not?)* contain fetal cells and products of fetal metabolism. For what purposes is amniotic fluid collected and examined?

   For _____ (if performed at 14–16 weeks)

   For _____ (if performed at about 35 weeks)

b. *(Amniocentesis? Ultrasound?)* is by far the most common test for determining true fetal age if the date of conception is uncertain. The technique *(is also? is not?)* used
   for diagnostic purposes. Fetal ultrasonography involves *(use of a transducer on? a
   puncture into?)* the abdomen. For this procedure the mother's urinary bladder
   should be *(full? empty?)*.

c. CVS (meaning c_____ v_____

   s_____) is a diagnostic test that can be performed as early as
   (4? 8? 16? 22?) weeks of pregnancy. It *(does? does not?)* involve penetration of the

   uterine cavity. Cells of the _____ layer of the placenta embedded in the uterus are studied. These *(are? are not?)* derived from the same sperm
   and egg that united to form the fetus. Therefore this procedure, like amniocentesis,

   can be used to diagnose _____.

C. Parturition, labor, adjustments at birth, physiology of lactation (pages 986–990)

C1. Contrast these pairs of terms.

a. Labor/parturition

b. False labor/true labor

■ **C2.** Do the following hormones stimulate (S) or inhibit (I) uterine contractions during the birth process?

_____ a. Progesterone   _____ c. Oxytocin

_____ b. Estrogens

C3. What is the "show" produced at the time of birth?

■ **C4.** Identify descriptions of the three phases of labor (first, second, or third).

_____ a. Stage of expulsion: from complete cervical dilation through delivery of the baby

_____ b. Time after the delivery of the baby until the placenta ("afterbirth") is expelled; the placental stage

_____ c. Time from onset of labor to complete dilation of the cervix; the stage of dilation

C5. Define these terms related to birth and the postnatal period:

a. Dystocia

b. Involution

c. Lochia

C6. What adjustments must the newborn make at birth in attempting to cope with the new environment? Describe adjustments of the following:

a. Respiratory system (What serves as a stimulus for the respiratory center of the medulla?)

   The respiratory rate of a newborn is usually about *(one-third? three times?)* that of an adult.

b. Heart and blood vessels *(For extra review* go over changes in fetal circulation, Chapter 21, pages LG 443–444.)

c. Blood cell production

■ **C7.** A premature infant ("preemie") is considered to be one who weighs less than ____ grams (____ pounds ____ ounces) at birth. These infants are at high risk for the

RDS (or _____). *For extra review* of RDS, refer to Chapter 23, Checkpoint 43, page LG 497.

C8. Explain why as many as 50 percent of all normal newborns may experience a temporary state of jaundice soon after birth.

■ **C9.** Complete this exercise about lactation.

a. The major hormone promoting lactation is _____, which is

   secreted by the _____. During pregnancy the level of this hormone increases somewhat, but its effectiveness is limited while estrogens and progesterone are *(high? low?)*.

b. Following birth and the loss of the placenta, a major source of estrogens and progesterone is gone, so effectiveness of prolactin *(increases? decreases?)* dramatically.

c. The sucking action of the newborn facilitates lactation in two ways. What are they?

d. What hormone stimulates milk ejection (let-down)? _____

C10. Contrast *colostrum* with *maternal milk* with respect to time of appearance, nutritional content, and presence of maternal antibodies.

C11. Defend or dispute this statement: "Breast-feeding is an effective, reliable form of birth control."

C12. List five or more advantages offered by breast-feeding.

D. Inheritance, disorders, and medical terminology   (pages 990–997)

D1. Define these terms.

a. Inheritance

b. Genetic counseling

c. Homologues (homologous chromosomes)

D2. Compare and contrast:

a. Genotype/phenotype

b. Dominant allele/recessive allele

c. Homozygous/heterozygous

d. Angelman syndrome/Prader-Willi syndrome

■ **D3.** Answer these questions about genes controlling PKU. (See Figure 29.13, page 992 in the text.)

a. The dominant gene for PKU is represented by *(P? p?)*. The dominant gene is for the *(normal? PKU?)* condition.

b. The letters PP are an example of a genetic makeup or *(genotype? phenotype?)*. A person with such a genotype is said to be *(homozygous dominant? homozygous recessive? heterozygous?)*. The phenotype of that individual would *(be normal? be normal, but serve as a "carrier" of the PKU gene? have PKU?)*.

c. The genotype for a heterozygous individual is _____. The phenotype is

_____.

d. Use a Punnett square to determine possible genotypes of offspring when both parents are Pp.

■ **D4.** Do this exercise about genes and inheritance.

a. Define *genome*.

A human cell is estimated to have about _____ genes. *(Most? A small number?)* of these have been identified on the various human chromosomes.

b. In most cases, the gene for *(having a severe disorder? being normal?)* is dominant.

State one or more exceptions to this rule. _____

c. What is the meaning of the term *aneuploid?*

One example is *trisomy 21,* in which the individual has *(2? 3? 4?)* of chromosome 21. This alteration in the genome occurs in the condition known as

_____.

d. Sickle cell anemia (SCA) is a condition that is controlled by *(complete? incomplete?)* dominance. Match the correct genotype below with the related description.

1. Genotype present in persons who have SCA _____

2. Genotype present in persons who have the sickle cell trait _____
A. Homozygous recessive   B. Homozygous dominant   C. Heterozygous

e. List the three possible alleles for inheritance of the ABO group.

_____ Which of these is/are dominant? _____ List the possible genotypes of persons who produce A antigens (agglutinogens) on red blood cells.

_____ Describe the phenotype of persons with the OO genotype.

f. ABO blood group inheritance is known as *(multiple allele? polygenic?)* inheritance. List two or more examples of characteristics governed by polygenic inheritance.

■ **D5.** Complete the exercise about potential hazards to the embryo and fetus. Match the terms at right with the related descriptions.

C. Cigarette smoking	FAS. Fetal alcohol syndrome	T. Teratogen

_____ a. An agent or influence that causes defects in the developing embryo

_____ b. May cause defects such as small head, facial irregularities, defective heart, retardation

_____ c. Linked to a variety of abnormalities such as lower infant birth weight, higher mortality rate, GI disturbances, and SIDS

D6. Contrast *autosome* and *sex chromosome*.

■ **D7.** Answer these questions about sex inheritance.

a. All *(sperm? ova?)* contain the X chromosome. About half of the sperm produced by

a male contain the _____ chromosome, and half contain the _____.

b. Every cell (except those forming ova) in a normal female contains *(XX? XY?)* genes on sex chromosomes. All male cells (except those forming sperm) are said to be *(XX? XY?)*.

c. Who determines the sex of the child? *(Mother? Father?)* Show this by using a Punnett square.

d. Name the gene on the Y chromosome which determines that a developing individual will be male.

■ **D8.** Complete this exercise about X-linked inheritance.

a. Sex chromosomes contain other genes besides those determining sex of an individual.

Such traits are called _____ traits. Y chromosomes are shorter than X chromosomes and lack some genes. One of these is the gene controlling

ability to _____ red from green colors. Thus this ability is

controlled entirely by the _____ gene.

b. Write the genotype for females of each type: color-blind, _____; carrier, _____;

normal, _____.

c. Now write possible genotypes for males: color-blind, _____; normal, _____.

d. Determine the results of a cross between a color-blind male and a normal female.

e. Now determine the results of a cross between a normal male and a carrier female.

f. Color blindness and other X-linked, recessive traits are much more common in *(males? females?)*.

D9. Name another X-linked trait. _____
Discuss *fragile X syndrome*.

ANSWERS TO SELECTED CHECKPOINTS

A1. (a) Approximately 300 to 500 million; less than 1%. (b) Secondary oocyte; mechanisms in the female: peristaltic contractions and ciliary action of uterine tubes, uterine contractions promoted by prostaglandins in semen, production of a sperm-attracting chemical by the secondary oocyte; mechanisms in the male: presence of flagella on sperm and enhancement of motility by the acrosomal enzyme acrosin. (c) 10 hours; increased fragility of the acrosome which permits release of enzymes that can penetrate the coverings around the oocyte; capacitation. (d) Distal portion of the uterine tube; they are the coverings over oocyte which sperm must penetrate. (e) Ectopic; uterine tube; ovaries, abdomen, cervix portion of the uterus, and broad ligament. (f) One; syngamy; sperm penetration of the secondary oocyte which triggers release of calcium and then granules from the oocyte. These block entrance of other sperm.

A2. (a) Spermatozoon, secondary oocyte; n. (b) Segmentation, 46; zygote.

A3. (a) DTwin. (b) MTwin. (c) DTwin.

A6. (a) 6, 20. (b) Trophoblast; mother's uterine blood and glands. (c) Mother may produce antibodies that mask the antigens derived from paternal components of the developing baby.

A9. First two; developing individual during the last seven months of human gestation.

A10. (a) Trophoblast: part of chorion of placenta; inner cell mass: embryo and parts of placenta; amniotic. (b) Disc; ecto; endo; meso; primary germ. (c) Gastrulation, third. (d) Ecto; amnion; bag of waters. (e) Endoderm, gut, yolk; is not. (f) Chorion; chorion; extraembryonic coelom.

A12. (a) Endo. (b) Ecto. (c) Meso. (d) Meso. (e) Meso. (f) Ecto. (g) Endo.

A13. (a) Amnion. (b) Chorion. (c) Yolk sac. (d) Allantois.

A14. (a) Flattened cake (or thick pancake); chorion; chorionic villi; brought into close proximity but do not mix; intervillous. (b) Decidua basalis; endo. (c) It is discarded from mother at birth (as the "afterbirth"); separation of the decidua from the remainder of endometrium accounts for bleeding during and following birth.

A15. 3; 2; 1.
A18. (a) 1. (b) 3. (c) 1. (d) 3. (e) 6. (f) 1. (g) 2. (h) 2. (i) 3. (j) 5. (k) 9. (l) 7. (m) 8. (n) 8.
B1. (a) High; corpus luteum, ovary, hCG. (b) First, third; urine, 14; excess protein or blood in her urine, ectopic production of hCG by uterine cancer cells, or presence of certain diuretics or hormones in her urine. (c) Human (chorionic) somatomammotropin; stimulates development of breast tissue and alters metabolism during pregnancy (see page 983 in the text); 32. (d) Relaxin; placenta, ovaries. (e) Inhibit; corpus luteum (in ovary), sustentacular cells (in testis).
B2. 266, fertilization; 38.
B3. (a) In. (b) In, 30–50. (c) In, 30–40. (d) De, constipation. (e) In, varicose veins. (f) 60–80, 900–1200.
B4. Pregnancy-induced hypertension; kidneys; hypertension, fluid retention leading to edema with excessive weight gain, headaches and blurred vision, and proteinuria.
B6. (a) Amniotic; does; detecting suspected genetic abnormalities (14–16 weeks); assessing fetal maturity (35 weeks). (b) Ultrasound; is also; use of a transducer on; full. (c) Chorionic villus sampling, 8; does not; chorion; are; genetic and chromosomal defects.
C2. (a) I. (b) S. (c) S.
C4. (a) Second. (b) Third. (c) First.
C7. 2500 grams (5 pounds, 8 ounces); respiratory distress syndrome.
C9. (a) Prolactin (PRL), anterior pituitary; high. (b) Increases. (c) Maintains prolactin level by inhibiting PIH and stimulating release of PRH; stimulates release of oxytocin. (d) Oxytocin.

D3. (a) P; normal. (b) Genotype; homozygous dominant be normal. (c) Pp; normal but a carrier of PKU gene. (d) 25 percent PP, 50 percent Pp, and 25 percent pp.
D4. (a) The complete genetic makeup of a particular organism; 100,000; a small number. (b) Being normal; Huntington's chorea (or Huntington's disease, HD) or other disorders on the left side of Exhibit 29.3, page 992 in the text. (c) Having a chromosome number other than normal (which is 46 for humans); 3; Down syndrome. (d) Incomplete; 1, A; 2, C. (e) A, B, and O; A and B; AA or AO; produces neither A nor B antigens (agglutinogens) on red blood cells. (f) Multiple allele; skin color, eye color, height, or body build.
D5. (a) T. (b) FAS. (c) C.
D7. (a) Ova; X, Y. (b) XX; XY. (c) Father, as shown in the Punnett square (Figure 29.17, page 995 in the text), since only the male can contribute the Y chromosome. (d) SRY (sex-determining region of the Y chromosome).
D8. (a) X-linked; differentiate; X. (b) X^cX^c; X^cX^c; X^cX^C. (c) X^cY; X^CY. (d) All females are carriers and all males are normal. (e) Of females, 50 percent are normal and 50 percent are carriers; of males, 50 percent are normal and 50 percent are color-blind. (f) Males.

MASTERY TEST: Chapter 29

Questions 1–2: Arrange the answers in correct sequence.

———— ———— ———— ———— 1. From most superficial to deepest (closest to embryo):
    A. Amnion
    B. Amniotic cavity
    C. Chorion
    D. Decidua

———— ———— ———— ———— 2. Stages in development:
    A. Morula
    B. Blastocyst
    C. Zygote
    D. Fetus
    E. Embryo

3. During early months of pregnancy, hCG is produced by the ____ and it mimics the action of ____.
   A. Follicle; progesterone
   B. Corpus luteum; FSH
   C. Chorion of placenta; LH
   D. Amnion of placenta: estrogen

4. These fetal characteristics—weight of 8 ounces to 1 pound, length of 10 to 12 inches, lanugo covers skin, and brown fat is forming—are associated with which month of fetal life?
   A. Third month   C. Fifth month
   B. Fourth month   D. Sixth month

5. All of the following structures are developed from mesoderm *except:*
   A. Aorta   D. Humerus
   B. Biceps muscle   E. Sciatic nerve
   C. Heart   F. Neutrophil

6. Which part of the structures surrounding the fetus is developed from maternal cells (not from fetal)?
   A. Allantois   D. Decidua
   B. Yolk sac   E. Amnion
   C. Chorion

7. All of the following events occur during the first month of embryonic development *except:*
   A. Endoderm, mesoderm, and ectoderm are formed.
   B. Amnion and chorion are formed.
   C. Limb buds develop.
   D. Ossification begins.
   E. The heart begins to beat.

8. Implantation of a developing individual (blastocyst stage) usually occurs about ____ after fertilization.
   A. Three weeks   D. Seven hours
   B. One week   E. Seven minutes
   C. One day

9. The high level of estrogens present during pregnancy is responsible for all of the following *except:*
   A. Stimulates release of prolactin
   B. Prevents ovulation
   C. Maintains endometrium
   D. Prevents menstruation
   E. Stimulates uterine contractions

10. Which of the following is a term that refers to discharge from the birth canal during the days following birth?
    A. Autosome   D. Lochia
    B. Cautery   E. PKU
    C. Culdoscopy

11. Formation of the three primary germ layers is known as:
    A. Implantation
    B. Blastocyst formation
    C. Gastrulation
    D. Morula formation

T F   12. Oxytocin and progesterone both stimulate uterine contractions.

T F   13. Of the primary germ layers, only the ectoderm forms part of a fetal membrane.

T F   14. Maternal blood mixes with fetal blood in the placenta.

T F   15. Amniocentesis is a procedure in which amniotic fluid is withdrawn for examination at about 8–10 weeks during the gestation period.

T F   16. Stage 2 of labor refers to the period during which the "after-birth" is expelled.

T F   17. Maternal pulse rate, blood volume, and tidal volume all normally increase during pregnancy.

T F   18. The decidua is part of the endoderm of the fetus.

T F   19. Inner cell mass cells of the blastocyst form the chorion of the placenta.

T F   20. A phenotype is a chemical or other agent that causes physical defects in a developing embryo.

Questions 21–25: fill-ins. Complete each sentence with the word or phrase that best fits.

_____ 21. The newborn infant generally has a respiratory rate of ____, pulse of ____, and a WBC count that may reach ____, all values that are ____ than those of an adult.

_____ 22. As a prerequisite for fertilization, sperm must remain in the female reproductive tract for at least 10 hours so that ____ can occur.

_____ 23. hCG is a hormone made by the ____; its level peaks at about the end of the ____ month of pregnancy, when it can be used to detect ____.

_____ 24. Determine the probable genotypes of children of a couple in which the man has hemophilia and the woman is normal. ____

_____ 25. This is the ____ mastery test question in this book.

ANSWERS TO MASTERY TEST: ■ Chapter 29

Arrange

1. D C A B
2. C A B E D

Multiple Choice

3. C	8. B
4. C	9. A
5. E	10. D
6. D	11. C
7 D	

True-False

12. F. Oxytocin but not progesterone stimulates.
13. F. All three layers form parts of fetal membranes: ectoderm (trophoblasts) forms part of chorion; mesoderm forms parts of the chorion and the allantois; endoderm lines the yolk sac.
14. F. Does not mix but comes close to fetal blood
15. F. Amniocentesis, 14–16
16. F. Fetus
17. T
18. F. Endometrium of the uterus
19. F. Trophoblasts
20. F. Teratogen

Fill-ins

21. 45, 120–160, up to 45,000, higher
22. Capacitation (dissolving of the covering over the ovum by secretion of acrosomal enzymes of sperm)
23. Chorion of the placenta, second, pregnancy
24. All male children normal; all female children carriers
25. Last or final. Congratulations! Hope you learned a great deal.